T0395378

SOFT-COMPUTING
IN PHYSICAL AND
CHEMICAL SCIENCES
A SHIFT IN COMPUTING PARADIGM

SOFT-COMPUTING
IN **PHYSICAL** AND
CHEMICAL SCIENCES
A SHIFT IN COMPUTING PARADIGM

Kanchan Sarkar
Sankar Prasad Bhattacharyya

CRC Press
Taylor & Francis Group
Boca Raton London New York

CRC Press is an imprint of the
Taylor & Francis Group, an **informa** business

CRC Press
Taylor & Francis Group
6000 Broken Sound Parkway NW, Suite 300
Boca Raton, FL 33487-2742

International Standard Book Number-13: 978-1-4987-5593-1 (Hardback)

Library of Congress Cataloging-in-Publication Data

Names: Bhattacharyya, S. P. (Sankar Pradad), author. | Sarkar, Kanchan, author.
Title: Soft computing in chemical and physical sciences : a shift in computing paradigm / S.P. Bhattacharyya, Kanchan Sarkar.
Description: Boca Raton : CRC Press, Taylor & Francis, 2018. | Includes bibliographical references and index.
Identifiers: LCCN 2017048154| ISBN 9781498755931 (hardback : alk. paper) | ISBN 9781315152899 (ebook)
Subjects: LCSH: Chemical engineering--Data processing. | Physics--Data processing. | Soft computing.
Classification: LCC TP184 .B43 2018 | DDC 660--dc23
LC record available at https://lccn.loc.gov/2017048154

Visit the Taylor & Francis Web site at
http://www.taylorandfrancis.com

and the CRC Press Web site at
http://www.crcpress.com

Dedicated to the loving memory of Professor Mihir Chowdhury

Contents

Preface

Computers and computing have pervaded almost every sphere of research in chemistry, physics, and biology. Every year more and more practicing chemists of any hue are taking to computers to help them in their search for solutions of complex problems they are called upon to look for. It is no different with physicists either. New subjects such as computational materials science, computational chemistry and physics, and computational biology have emerged. The emphasis, until very recently, has been on the traditional ways of computing—hard computing based on analysis. During the years of growth of what can be loosely called in silico research in physics and chemistry, a host of methods collectively called soft computing have also appeared primarily due to the efforts of computer scientists for handling problems that are computationally hard—so hard that traditional methods fail to provide satisfactory answers using reasonable computing resources. These methods are often derived on the basis of the analogy that has roots in natural processes and phenomena and can be likened to computational models of nature's own way of solving complex problems. These methods can work even with incomplete or imprecise information or in the presence of uncertainty, and come up with solutions that are robust albeit approximate in a reasonable amount of computing time. There is certainly enough scope for crosstalk between traditional and new modes of computing in the context of problem-solving in chemistry and physics. Little seems to have taken place at the interface between the very different computing paradigms although interest in soft computing has been growing among computational chemists and material scientists of late, as a survey of contemporary literature would seem to indicate. The primary source of information on soft computing is scattered among journals and conference proceedings that are seldom consulted by chemists and physicists. A large gap therefore seems to exist between the scientific community and the primary sources of knowledge about soft computing.

This book is an attempt to provide an easy interface between the two computing paradigms, familiarize computing chemists and physicists with the art and science of soft computing techniques, and bring an awareness of the scope, versatility, and power of soft computing. Each method is described in detail and underlying mathematical principles clarified wherever possible. The basic aim has been to provide readers with enough material and insight to make them confident to experiment with these techniques in solving problems they are pursuing in their fields of research and come up with "designer methods" for specific types of problems.

After introducing the basic ideas relating to reduction of a problem at hand to an equivalent optimization problem, the construction of objective functions and fitness landscapes, and the way of incorporating constraints in the search for solutions (Chapter 1), several soft computing methods are presented in Chapters 2 to 5. The emphasis has been on explaining how and why a method works. Thus, starting with genetic algorithms in Chapter 2, evolutionary computing, differential evolution, and genetic computing are described in Chapter 3. Random mutation hill climbing and the method of simulated annealing are analyzed in Chapter 4, while Chapter 5 is devoted to the collective intelligence-based methods such as the particle swarm optimization, ant colony optimization, artificial bee colony optimization, and the fireflies algorithm. The next two chapters (Chapters 6 and 7) describe many applications of the techniques described in Chapters 2 to 5 to computational problems of physics and chemistry. Some of the applications are rudimentary and some are fairly advanced and include the problem of solving Schrodinger equation for atoms and molecules; finding global minima on the complex potential energy surfaces of clusters, direct determination of molecular structure, and electronic charge distribution; and designing molecules with targeted properties. Artificial neural networks are introduced in detail in Chapter 8 against the backdrop of pattern recognition and classification problems in chemistry. Starting with the perceptron model, several artificial neural networks such as the back propagation neural network, the Hopfield network, and the Kohonen network are presented along with the Hopfield–Tank neural model of optimizations (Chapter 8). The workings of the networks are illustrated with examples, some of them sophisticated and some of them simple, drawn from physics and chemistry.

Data clustering presents a difficult problem. The ideas of crisp and fuzzy clustering of data with the help of soft computing are examined in Chapter 9, which also presents a compact account of the second shift in computing paradigm that the theory of fuzzy sets and logic brings in. A few fuzzy-logic-based adaptive soft optimization techniques are explored here. Chapter 10 is devoted to presenting a brief account of a third possible shift in computing paradigm that quantum computing and automatic program development techniques could usher in. Quantum genetic algorithms and quantum genetic computing are explored in this context along with quantum adiabatic computing and quantum annealing. There is certainly enough scope for improvement in both breadth and depth.

We thank our coworkers 'Dr. P. Dutta, Dr. Pinaki Chaudhury, Dr. Subhajit Nandy, Dr. Rajendra Saha, and Dr. Rahul Sharma' whose sustained efforts kept the soft computing group in IACS going. We sincerely thank Professor D. Mukherjee, Professor D. S. Ray and Professor S. Adhikari of IACS, Professor B. L. Tembe (IIT Bombay), Dr. Mayank Dixit (IIT Bombay), and Moinak Choudhury (University of Minnesota) for help and encouragement, which enabled us to compile this book. We also thank our family members for continuous support. KS is supported by NSF grants NSF/EAR 1348066 and 1503084.

Kanchan Sarkar
University of Minnesota

Sankar Prasad Bhattacharyya
Indian Academy of Sciences

Authors

Kanchan Sarkar got his formal training in soft computing and machine intelligence methods from the Center for Soft Computing Research and Machine Intelligence Unit, Indian Statistical Institute, Kolkata. During doctoral training in Indian Association for the Cultivation of Science, Kolkata, he experimented with several pure and hybrid soft-computing techniques in the general context of computing minimum energy structures of atomic, molecular and ionic clusters, and undoped and bipolaron-doped oligomers. Currently, he is a postdoctoral associate in the Department of Chemical Engineering and Materials Science, University of Minnesota. His postdoctoral studies focus on the enormous possibilities that the techniques of evolutionary computing can offer a consistent procedure of generating projector augmented wave (PAW) data sets maintaining the same level of accuracy as all-electron full-potential linearized augmented plane-wave (AE-FLAPW) calculations up to high pressures. He has also introduced a new measure of atomic data set quality by considering performance uniformity over an extended pressure range. His other ongoing projects involve computing *ab initio* thermoelastic properties of materials under extremely high temperature and pressure conditions, and designing atomic data sets for lower mantle conditions to understand deep-earth processes. He is also an active contributor to the Virtual Laboratory of Earth and Planetary Materials (VLab) for high-temperature and high-pressure elasticity calculations.

Shankar Prasad Bhattacharyya, fellow of the Indian Academy of Sciences and Raja Ramanna Fellow (Department of Atomic Energy), retired as senior professor of physical chemistry, Indian Association for the Cultivation of Science, Kolkata. His main research interest is in the area of quantum chemistry, and his distinguished career in theoretical chemistry spans a period of over 35 years. His work on the development of new computational techniques, with particular expertise in global optimizations, for the treatment of complex problems in molecular chemistry and physics is well documented by an impressive list of highly influential publications in the field, devoted both to traditional mathematical techniques and, more recently, to soft computing-oriented approaches. He has supervised more than 20 PhD and 30 undergraduate project students. Professor Bhattacharyya is the author of more than 200 papers in peer-reviewed journals and a couple of book chapters.

1

A New Computing Paradigm

1.1 Introduction

Soft computing [1] is no longer just a scientific curiosity, but an emerging and refreshingly new computing paradigm that is under continuous development. It has left the confines of computer science and started spreading into the realm of physical, chemical, and other sciences as well. Scientists are taking a closer look at soft computing as an alternative paradigm to handle their difficult-to-solve problems. Soft computing has the advantage of being system independent and can be adapted to solve problems of any kind and complexity. The extreme generality of soft computing methods can be a source of its weakness in the sense that a method developed for handling problems of a specific type and refined over years may have an edge over methods of such generality. It is therefore necessary to understand clearly the nature of soft computing techniques, the domain of their applicability, their powers and pitfalls, and the way they can be interfaced with physical and chemical sciences and maximally exploited. We will briefly examine these issues in what follows.

1.2 What Is Soft Computing?

Computing has become an essential and a powerful tool at the disposal of physicists and chemists grappling with theoretical problems that are analytically intractable. Much of the computing done is, however, traditional, based as it is on analysis and backed up by a clear and valid mathematical formulation. The traditional methods have wide acceptability among scientists, although such methods often face insurmountable difficulties in handling problems that are plagued by nonlinearity, instability, and a lack of sufficient and precise information about the system. Soft computing represents a host of nontraditional means of effectively handling such problems [1,2]. Such methods may or may not succeed in predicting the correct (exact) solution, but can often come up with approximate solutions that are robust in the face of all the imprecision, incompleteness, and uncertainty in the information being processed or the system being examined, in a reasonable time. These methods are generally based on metaheuristics and are therefore system or problem independent. They have roots in some biological or at least natural processes and may be viewed as either bioinspired or more broadly speaking, as nature inspired. There are computing methods mimicking the processes of adaptation and biological [3] evolution, the foraging behavior of ants, bees, and birds, the flashing of fireflies, the way our immune system functions, or how neural networks [4] respond to incoming activation in the one hand (bioinspired) and for example, the way metals cool and anneal in the other (nature-inspired). These metaheuristics are exploited to develop practical problem-solving algorithms that often succeed in producing solutions of problems (e.g., traveling salesman problems) that appeared to be formidable or even intractable in the domain of traditional analysis-based computing. These methods are endowed with a built-in tolerance for imprecision and uncertainty, and in that sense they share some attributes of fuzzy systems. In fact, as an attribute fuzziness can be profitably exploited or has been exploited in soft computing algorithms.

Soft computing methods work by reducing a given problem to an equivalent problem of minimizing a carefully designed objective function or functional or maximizing a fitness function or functional without using any information about the derivatives of the objective or fitness function. They are thus function-only optimizers. In one or more steps of the relevant algorithm random numbers are used, introducing a stochastic element in the search for the optima. Soft computing methods are evidently nondeterministic or probabilistic in nature and are also called stochastic optimizers. Notwithstanding their stochastic nature, soft computing methods have some inbuilt strategy for biasing the search toward the most important region of the search space. No wonder that these methods have been regarded as being endowed with computational intelligence (CI) or artificial intelligence (AI). The hardcore practitioners of AI would be perhaps reluctant to accept soft computing techniques in the fold of AI-based computing. There are reasons for doing so. Intelligence has been broadly accepted as the ability of a human being to comprehend what is happening around, think, analyze a situation, reason, learn and respond, strategize, and even deceive. The lexicographers define AI as the ability of computers or computer programs to mimic in their operations, at least partly, the human thought process such as reasoning and learning. The definition of AI just given appears to be somewhat restrictive, specially when the possibility a computer and computer program performing a task intelligently without necessarily mimicking the thought process of human beings is considered. A broadened second definition that is inclusive of the first (more restrictive) definition has therefore emerged. According to the broader definition, AI is the ability of computer programs to do what humans require intelligence to accomplish. We propose to adopt the broader definition of AI. However, even under the expanded definition of AI, the nature of the task being performed or executed by the computer may often decide if it can be classified as intelligent computation. Thus, the most sophisticated numerical calculations are regarded as dumb or nonintelligent while the ability to perform symbolic integration and differentiation, binomial expansion, or spin algebra qualifies as intelligent computing. As it stands today, in addition to the nature or appearance of the task being performed, the underlying model of computation also decides if it comes under the class of intelligent computing. Thus, if the underlying model of computation is a low-level biological process, the current trend is to classify or categorize the computing as artificially intelligent computing or biology-inspired computing. Under this classification scheme, genetic and evolutionary computing, genetic algorithm, swarm or collective intelligence guided methods, working artificial neural network, etc., represent intelligent computing. The definition of soft computing we have alluded to earlier automatically brings all these methods under its fold and soft computing can also be categorized as a kind of AI method. The method of simulated annealing, according to our definition, is also a soft computing method, but does not qualify as an AI method as the underlying model has no biological component or representation.

It is a nature-inspired soft computing method. The major components of soft computing discussed in this book include:

1. Genetic algorithms
2. Evolutionary and genetic computing
3. Swarm/collective intelligence methods
4. Artificial neural networks
5. Simulated annealing
6. Fuzzy set and fuzzy-logic-based methods

Additionally, the possibility of interfacing soft computing with yet another emerging paradigm, quantum computing, is included as a special, futuristic topic. The switchover from traditional computing to soft computing marks a shift in the computing paradigm or computing doctrine chemists and physicists have been wedded to and the question naturally arises whether there is really a need to do so, or whether there is a dividend to be gained from such a shift.

1.3 Why Soft Computing?

The majority of the interesting and important theoretical problems that chemists and physicists deal with are not analytically tractable. Numerically solving such problems is often the only way to make progress and has been popular. Traditional numerical procedures are varied and the specific method invoked depends upon the nature of the problem. Thus, the problem at hand may require solving an eigenvalue equation, a differential or a partial differential equation, linear or nonlinear algebraic equations, and so on. The numerical recipe for each is different. The soft computing methods (SCMs) on the other hand are metaheuristics and are system independent. They are all-purpose optimizers and can be invoked with equal facility irrespective of the nature of the problem. The generality makes the SCM an attractive paradigm—just reduce the problem at hand to an equivalent optimization problem and invoke an SCM. It must be mentioned that the SCMs may have several control parameters and optimal choice of the control parameters may be problem specific, and may require adjustments which could be external or adaptive.

Traditional numerical procedures may require a good deal of precise information about the system e.g., precise initial conditions, boundary conditions, a good initial guess for starting the procedure, the nature and number of constraints, if any, which may not be available. The lack of tolerance of the numerical procedure to imprecision or uncertainty in any of the above information may cripple the numerical procedure. Even if such information is available, a smooth execution of the procedure in a reasonable time can be challenging when the system is complex and large. The ability of the SCMs to provide at least approximate but robust solutions even in the presence of all the incompleteness, imprecision, or uncertainties makes them attractive options for problems that are computationally hard. Different soft computing techniques may have complementary features which can be and have been exploited by researchers to come up with hybrid intelligent problem solvers. It is equally possible to hybridize an SCM with a traditional numerical procedure at some stage of the search for the optimal solution or fine tune the approximate solution already discovered by the SCM being used.

SCMs being derivative free become the automatic choice when the search variables are discrete (e.g., integer, 1, and so on); that is when one is looking for the maximum (minimum) of a fitness function (objective function) the argument of which are integral numbers.

A second case arises when the problem at hand admits of a combinatorically explosive number of possible choices for the solution. Direct enumeration of all the possibilities is not a viable option for locating the optimum under either a deterministic or a stochastic (random) search technique. The inbuilt biasing mechanism present in an SCM can guide the apparently random search to the important region of the search space and eventually to the optimum, even though the algorithm explores only a tiny fraction of the possibilities that exist. The computationally intelligent behavior reduces the cost of search for the global optima and makes the SCM a strikingly viable option for handling search problems (e.g., the traveling salesman problem). The exploration of the search space and exploitation of the information already present in the search space is usually achieved efficiently by the SCMs in general (although one may perform more efficiently than another in a given application). It must be stressed that the SCMs do not, in general, guarantee that the solution discovered is the global optimum. One chooses an SCM usually to discover a robust solution, however approximate, relatively quickly using a reasonable amount of computational labor. That reasonable computational labor is generally much lower than what would have been required in a random, function-only search. We will examine these features more closely in later chapters. Let us now summarize the main reasons for choosing the soft computing (SC) option in handling difficult problems:

1. System-independent structure
2. Ease of implementation
3. Function-only character
4. Ability to home in on the important region of the search space quickly
5. Workability in continuous and discrete optimization
6. Possibility of constructing hybrid methods that exploit the complementary capabilities of different SCMs

1.4 Overview of the Topics Covered

In the following chapters, we have discussed at length several soft computing options for solving problems encountered in physics and chemistry. The coverage has been broad, but we do not claim to have been exhaustive. Only methods that have already found some use in physical and chemical research have been dealt with. Some emerging topics that we anticipate may derive benefit from an interface with SCM are also discussed.

1.4.1 Genetic Algorithms

Originally developed by Holland (1965) [5] as a computer-executable mathematical model for adaptation in living systems, genetic algorithms (GAs) have proved to be powerful all-purpose problem solvers. GAs work with a population of plausible solutions which evolve on a carefully defined fitness landscape under the combined action of genetic operators such as selection (e.g., the Darwinian principle of natural selection), recombination or crossover, mutation, and inversion. Each individual is given a chromosomal representation, which encodes the genetic information (search variables) in binary language. As generations elapse, the algorithm discovers progressively better solutions in terms of fitness and stops when no further improvement of fitness takes place, or a preset value of the maximum number of generations has elapsed. Usually by that time the desired solution has already appeared in the population or at least a solution close to the desired one has been discovered. There are still many unsolved questions about GA (rate of convergence optimal choice of parameters, premature convergence, etc.) and possibly yet unexplored facets of it.

1.4.2 Evolutionary Algorithms and Genetic Computing

Evolutionary algorithms (EAs) work with fewer genetic operators. Mutations play the dominant role in the evolutionary process in EA. In genetic computing (GC), the solutions of the problem are themselves an evolving set of computer programs that undergo changes under the action of a genetic operator like crossover and mutation. The need of specific programming language (like Lisp) for writing GC or genetic programming (GP) codes has been an impediment to their acceptability outside the domain of computer science. GCs and GPs are powerful tools for solving complex problems and need exploration.

1.4.3 Random Mutation Hill Climbing and Simulated Annealing

Random mutation hill climbing (RMHC) is a single parent evolutionary strategy, mutation being the only genetic operator to act on the parent string. No downhill movement on the fitness landscape is allowed. Like RMHC, simulated annealing (SA) [6] also works with a single solution string which undergoes random thermal fluctuations in a simulated thermodynamic environment forcing it to move both up and down hill on the fitness landscape. In the limit T→0 the global minimum is reached as a result of very slow cooling. Both are natural problem solvers and have found many applications in scientific research.

1.4.4 Artificial Neural Networks

Supposed to work like the human brain, the artificial neural networks (ANNs) [7,8] are made up of many artificial neurons interconnected by edges (synapses). The neurons receive inputs (patterns/signals), activation takes place, and the ANN produces output by adjusting the weights (synaptic strength) or state of the neurons, depending upon the nature of the network. These computer models of the brain can either be trained (supervised learning) or they can learn by themselves (unsupervised learning) and can accomplish specific tasks such as classification, associative memory formation, optimization, and pattern recognition. ANNs have great potential for use in scientific research.

1.4.5 Swarm Intelligence Method

Based on a computer model of flocking behavior of birds or foraging behavior of bees, ants, or herding patterns of animals, these are powerful distributed computing models [9]. Each individual is a simple agent, but the collective ability of the swarm to perform a task far exceeds the sum of the abilities of the individuals to do the same. Interaction among the simple agents leads to the emergence of "collective" or "swarm" intelligence, and the significantly superior ability of the swarm to accomplish a difficult task.

1.4.6 Fuzzy Logic and Fuzzy-Set-Based Systems

In ordinary or "crisp logic" a proposition is either true or false. Fuzzy logic [10,11] allows a proposition to have various degrees of truthfulness. Similarly, in ordinary sets an element is either a member of a given set or not—there is nothing in-between the two possibilities. Fuzzy sets allow an element to have a degree of membership. The concept of truthfulness or membership is extended from a discrete to a continuous paradigm in fuzzy systems. Fuzzy control has found important practical applications. Fuzzy neural network and fuzzy GAs have proved to be powerful hybrid intelligent tools for problem solving, and the application of fuzzy logic in the domain of scientific research is growing.

1.4.7 Quantum Computing

Still in its infancy, the idea of quantum computing (QC) [12] has already provided at least two very powerful algorithms for executing difficult tasks like prime factoring or indexed database search, with spectacular sendups compared to classical algorithms. The superior power of QC is derived from unique features like quantum superposition and quantum parallelism. Interfacing QC with SC may be profitable. Quantum GAs and quantum annealing are some of the possibilities explored.

1.5 Interfacing Soft Computing with Problem Solving in Physics and Chemistry

The interfacing of problems of chemistry and physics with SC for obtaining solutions is rather simple once we have chosen a particular SCM for the problem. Being metaheuristics, SCMs are not problem specific and can be adapted to handle problems of any kind. The first key is to construct an objective (or fitness) function incorporating the relevant and necessary constraints, the minimization (or maximization) of which becomes the goal of the problem-solving enterprise. An important second key is to provide a suitable representation of the solution being sought, e.g., strings of numbers in binary mode, arrays, strings of integers or floating point numbers, vectors, matrices, or amplitude distribution. In many of the SCMs an initial population of plausible or potential solutions must be provided. Adequate care must be taken to choose individuals of the population so that the physical characteristics of the system and mathematical requirements for acceptable solutions are reflected (characteristics of bound states, real numbers, normalizability, orthogonality, etc.). They are usually chosen randomly within limits, but educated guesses can be exploited to seed the population without destroying the diversity or variability. The third step is to allow the members of the initial population to explore the objective surface or the fitness landscape and move from one region to another. The exact mechanism and extent of the exploration depends on the specific SCM being used (such as GA/EA, swarm intelligence, hill climbing, annealing). The fourth step is to choose an appropriate stopping criterion for the search to terminate, depending on the nature of the problem. One has to make a judicious choice as otherwise the search will consume time without any tangible improvement in the quality of the solution. Finally, one may further refine the solutions obtained by exploiting the output as inputs to another method that is known to be endowed with good depth-search quality. Some of the SCMs have a superior ability to explore the search space well (breadth search quality) and quickly arrive at good-quality approximate solutions that are close to the global optimum. It could be profitable to couple the SCM with another—this may even be a traditional method of optimization to refine the solution already discovered by the SCM. Thus, close to the global optima, a quadratic search (e.g., Newton's or Newton-like methods) could cut down the search time. The availability of analytical expressions for gradients and the Hessian of the objective (fitness) function can greatly facilitate the search.

There is practically no restriction on the type of problems that can be interfaced with SCM. We will describe in Chapters 6 and 7 some of the problems of physics and chemistry that have been tackled by SCM. They include diagonalizing a Hamiltonian matrix or solving a differential equation to searching out saddle points or global minimum energy structures on potential energy surfaces (PES), and direct calculation of electronic charge distribution in large molecules.

References

1. Kecman, V. 2001. *Learning and Soft Computing: Support Vector Machines, Neural Networks, and Fuzzy Logic Models.* Cambridge, MA: MIT Press.
2. Munakata, T. 2007. *Fundamentals of the New Artificial Intelligence.* T. Munakata (Ed.). Texts in Computer Science. London: Springer.
3. Russell, S.J., and P. Norvig. 2010. *Artificial Intelligence: A Modern Approach.* Prentice Hall Series in Artificial Intelligence. Prentice Hall.
4. Luger, G.F. 2008. *Artificial Intelligence: Structures and Strategies for Complex Problem Solving.* 6th ed. Addison-Wesley.
5. Holland, J.H. 1975. *Adaptation in Natural and Artificial Systems: An Introductory Analysis with Applications to Biology, Control, and Artificial Intelligence.* U Michigan Press.
6. Kirkpatrick, S., C.D. Gelatt, and M.P. Vecchi. 1983. Optimization by Simulated Annealing. *Science* 220, no. 4598: 671–680.
7. Takefuji, Y. 1992. *Neural Network Parallel Computing.* Norwell, MA: Kluwer Academic Publishers.
8. Hertz, J.A., A.S. Krogh, and R.G. Palmer. 1991. *Introduction to the Theory of Neural Computation.* Addison-Wesley Computation and Neural Systems Series. Avalon Publishing.
9. Kennedy, J., and R.C. Eberhart. 2001. *Swarm Intelligence.* San Francisco, CA: Morgan Kaufmann Publishers Inc.
10. Zadeh, L.A. 1965. Fuzzy Sets. *Information and Control* 8, no. 3: 338–353.
11. Zadeh, L.A. 1968. Fuzzy Algorithms. *Information and Control* 12, no. 2: 94–102.
12. Nielsen, M.A., and I.L. Chuang. 2000. *Quantum Computation and Quantum Information.* Cambridge University Press India.

2

Genetic Algorithms

2.1 Introduction

The emergence of living organisms on our earth millions of years ago was in the form of unicellular species that were able to replicate themselves. Out of purely inanimate matter, the creation of life was perhaps the outcome of a marvelous transmutation of chance encounters among simple molecules which brought order out of disorder, structure out of randomness, and complexity in an otherwise simple world. Gradually, the forces of evolution took over and millions of new species appeared and disappeared; higher and more complex forms of life started emerging and growing. The process of evolution was driven by a complex interplay of chance and necessity—the necessity to survive and the chance production of genetic materials that enhanced the beings' ability to do so in the face of the onslaught of an ever-changing environment [1]. The survival required what is now called adaptation to harsh, hostile, and changing surroundings, and that in turn required better and better designing of the genetic material. How did it happen in nature? Are there mathematical models that capture the essential elements of the long-drawn processes of biological evolution? The answers to these questions are important in two ways: first, they lead to better understanding of the deeper and more complex issues of evolutionary biology, and second, they allow us to exploit the understanding in constructing algorithms for solving practical problems that are too difficult otherwise to deal with, within the framework of standard problem-solving techniques. A number of biology-inspired models of computing evolved out of the enterprise. One such model is genetic algorithms (GAs) [2]. GA was proposed by G.S. Holland (1965) as a computer-executable mathematical model of natural adaptation and genetics of evolution in living systems. The model captures the essential features of the evolutionary process of nature, a major component of which is the principle of natural selection or the rule of the survival of the fittest as articulated by Charles Darwin [1]. The model was simple but realistic enough to be viewed as nature's own approach to solving complex designing problems in living systems by gradual adaptation to the environment to which they are exposed. A complex interplay of chance and necessity forced organisms into adaptively learning how best to survive in a given environment. The forces of adaptation led to a progressive redesigning of the organism's genetic material such that appropriate physical attributes for better survival were produced. The better adapted species survived and grew as they successfully reproduced while the inferior ones died out. The mathematical model suggested by Holland translates into a powerful problem-solving tool that works well when executed on a computer for finding solutions of problems—not of genetics, but of any field, especially problems that are computationally hard (e.g., the "traveling salesman" problem, or the "prisoners dilemma" problem). Thus, a powerful all-purpose problem solver—the GAs—came into existence and flourished.

Before describing what GAs are let us take a brief look into the essential elements of genetics in real life with special reference to ourselves. The life journey begins when the egg cell from the mother is fertilized by a sperm cell from the father [3]. Cell divisions take place until the process attains maturity, the entire ontogenetic process culminating in the organism beginning to reproduce itself thereby establishing a self-sustaining chain that perpetuates life. The whole process is scripted down to every detail by the structure of that one cell—the fertilized egg cell that life begins with. The process of cell divisions, mitosis and meiosis, is controlled by the cell nucleus wherein a number of fiber-shaped or rod-like particles called chromosomes encoding the entire genetic information reside. Each chromosome comprises a large number of blocks of twisted strands of DNA, called genes, composed of four bases—adenine (A), cytosine (C), guanine (G), and thymine (T)—with each gene encoding a specific protein. The twisted strands of DNA can be viewed as a kind of software, a biological computer program that instructs the cells what to do, and how and when to do it. We can perhaps liken it to a new type of computer that works not on a base two but on a base four method. The data and instructions here are coded by the ordering of the four bases, A, C, G, and T rather than 0 and 1. The complete collection of all chromosomes—the entire genetic material—is called the "genome." Note that each gene is located at a particular position or locus on the chromosome. In humans, the number of chromosomes in the fertilized egg cell is 2×23, one set of 23 coming from each parent. Each complete set of chromosomes (23) carries the entire code—the blueprint that scripts the entire development process starting from the fertilized egg cell to the full-blown individual. During the growth of the body by the process of mitotic cell divisions, both sets of chromosomes are duplicated. A group of cells is, however, preserved to produce the sperm cells (in males) and egg cells (in females) required for reproduction much later in life. These cells suffer fewer mitotic divisions, but undergo reductive division (meiosis) at an appropriate stage of development, producing "gametes." The point to note is that in meiosis the double chromosome set of the parent cell separates into two single (but complete) sets of chromosomes, one set each going to the two daughter cells (gametes). The doubling of chromosomes (2×23) in humans does not take place in meiosis, each gamete (daughter cells) getting only one full copy of the code (23 chromosomes). These cells (gametes) are called haploids, the ordinary cells in the human body (the somatic cells) being diploid in character. The millions of genes in each chromosome from the father and the corresponding genes in the same chromosome from the mother constitute the gene pair in the offspring. Each pair or a number of such pairs of parental chromosomes give rise to a specific physical or mental attributes of the child such as the color of the eyes, hair color, blood group, taste for music, height, and intelligence. These characteristics define the "phenotypes" [4]. All possible gene-pair contributions producing a specific attribute (eye color, blood group, etc.) are called "genotypes." The genes responsible for a particular attribute may have different values, e.g., 0 or 1 or perhaps 0, 1 and 2, called alleles. Considering genes that determine the blood group, the alleles can have values 0, 1, and 2. The possible genotypes and the corresponding phenotypes (blood group) are summarized in Table 2.1. The order of appearance of alleles in the genotype may be ignored. Even then, the mapping from geno- to phenotype remains many to one.

Now whatever the phenotype is, when the child reaches maturity and produces sperms (if male) and eggs (if female) by meiosis, one complete set of 23 chromosomes from the parental pair (2×23) is selected for the sperm or the egg cells, chance deciding which ones constitute the complete set. The key to the inheritance of certain phenotypes from the parents to the child and then to the grandchild lies in this chance event. However, pure chance and chance alone creates an additional flexibility in mixing grandparental inheritance in the offspring by a process called chromosomal crossover. Let us have a look into the process as it constitutes a very important element of the algorithm, the genetic algorithm that we will describe later in this chapter.

TABLE 2.1

The Possible Genotypes and the Corresponding Phenotypes (Blood Group)

Genotype (Allele Value Combination)	Phenotype (Blood Group)
0 0	O
1 0 (or 0 1)	A
1 1	
2 0 (or 0 2)	B
2 2	
2 1 (or 1 2)	AB

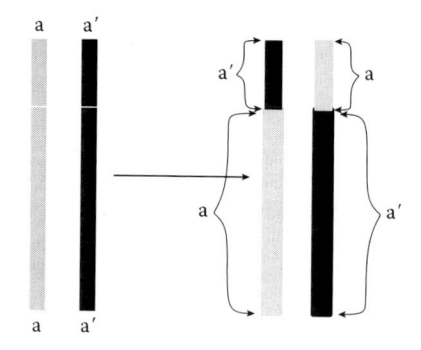

FIGURE 2.1
A schematic illustration of chromosomal crossover.

In meiosis, when the chromosome pairs separate into two complete sets of 23 chromosomes, there is no guarantee that a particular chromosome will be passed on as a whole. Before being separated in meiosis, two homologous chromosomes which come into close contact with each other may exchange a portion or portions of each other as revealed in Figure 2.1.

The physicists or chemists will note that it resembles an avoided curve crossing event in molecular physics [3]. But what is the net outcome of the process? The net outcome is that two attributes located in the respective parts of the chromosomes (a, a') will be separated in the grandchild, mere chance deciding the separation.

There is yet another chance event of monumental importance in the entire evolutionary scheme of biology. In this rather rare event, a specific gene or gene value (single nucleotide) in a particular parental chromosome becomes altered, and the altered gene is passed on to the offspring. The process takes place randomly at a low rate but can be accelerated by external factors such as exposure to radiation (X-rays, gamma rays, etc.). It is called gene mutation or simply mutation. Mutation is a low-probability event, but the probability of a mutation taking place may sometimes increase over and above what may be called the basic or inherent mutability of the gene due to changing environmental factors. If the mutated gene value imparts attributes that enhance the ability of the organisms to survive (fitness-enhancing mutation) in the given environment, it is more likely that it will live to reproduce, then the altered gene will be passed on to the progeny in reproduction and will perpetuate in succeeding generations. We note here that the "fitness" of an organism in the evolutionary scheme of things is measured typically by the probability that it lives long enough to reproduce (called the viability of the organism) or by a function of the number of offspring produced by the organism (called virility). These properties determine whether a mutation will survive through succeeding generations or be rejected. We have so far referred only to humans as a diploid species. However, the majority of organisms in nature that sexually reproduce are diploid in character. Hence the preceding discussion on chromosomal crossover and mutation remains generally valid for all such species [5].

2.2 What Are Genetic Algorithms?

Natural genetics, as the rudimentary discussion in Section 2.1 suggests, is far too complicated for a computer implementation. The GAs that we will discuss in this and subsequent sections are much simplified, abstract mathematical models of natural genetics and the process of biological evolution—models that are capable of solving difficult problems. The problem is first reduced to an equivalent problem of computing the absolute maximum of a function of the search variables. The function is called a fitness function for the problem, and then an algorithm that mimics, however weakly, the natural genetics and evolutionary biology is invoked to trace out the maximum on the fitness landscape. The mimicking algorithm has become known as GA—a simplified mathematical model of what nature does in genetics and evolution [2].

The first simplification comes in the form of introducing a finite (in fact, small) number of individuals making up a population of trial or plausible solutions. In most cases, each individual is a haploid individual carrying a single chromosome, the genes of which are numbers 0 and 1 in binary, integers (+1, −1, for example), floating point (decimal) numbers, characters, graphs, chemical structures, or parts thereof. Since individuals carrying a single chromosome are used, we will use the term "individual" and "chromosome" interchangeably, instead of using longer expressions like a chromosome representing an individual or an individual carrying the particular chromosome. The genotype of an individual in the population is just the collection of "bits" (0, 1 in binary) making up the chromosome. In general, no use is made of the phenotype, although scientists have experimented with both genotype and phenotype representations of the potential or trial solutions. The initial population is chosen randomly (with an educated guess, probably). Care is taken to ensure that they are physically meaningful in the context of the problem at hand and that enough variability is present in the chosen population. Each individual is evaluated, and given a fitness value that measures the degree or extent of acceptability of the individual as a solution of the problem under investigation. Individuals with higher fitness values are selected with a higher probability (the Darwinian principle of natural selection) to build up the "mating pool" where "crossover breeding" produces offspring (new chromosomes/individuals). Occasional mutations in the offspring are allowed to produce still newer individuals. The new individuals are evaluated for their fitness values and a new population emerges, completing what is called one generation. The entire ontogenetic process is repeated until no better individuals are produced. One hopes that the higher probability or frequency of selection of an individual with relatively higher fitness value together with crossover and mutation somehow ensures that progressively better and better solutions are discovered as generations elapse. Eventually, one hopes, either an exact solution or close approximations to it are found by the evolutionary process implemented on the machine (the computer). The basic operational philosophy of GAs as a problem-solving enterprise is now clear (see Figure 2.2)—a randomly generated population of individuals (chromosomes) evolve under the action of genetic operators such as selection, crossover, and mutation (in that order) discovering progressively newer and better solutions. The search terminates on fitness saturation, i.e., when no better individuals are produced signaling that scope for further improvement has been exhausted. The basic philosophy of GAs has clearly the underpinnings of some

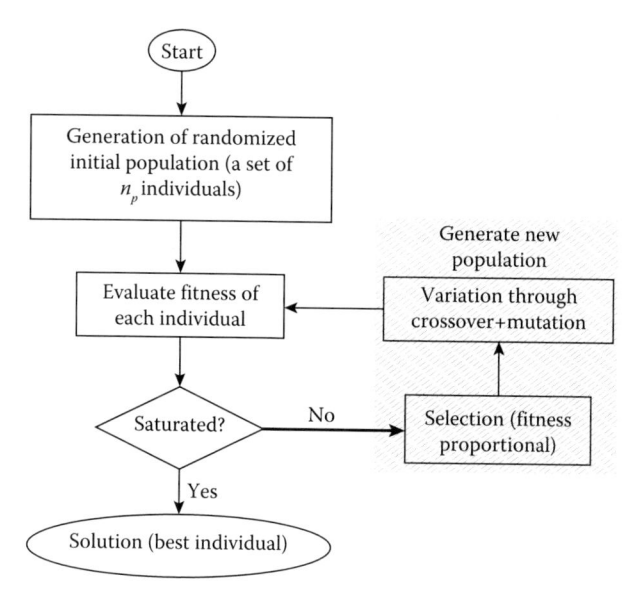

FIGURE 2.2
Basic operational philosophy of a genetic algorithm.

of the basic principles of genetics and evolutionary biology, but the goal now is not so much to achieve a better understanding of the finer details of evolution, adaptation, and genetics as it is to solve a specific problem. We have defined only the broad principles of GAs, many realizations of which are possible. There is thus not one GA but many GAs.

2.3 How Do the GAs Work?

We are now in a position to describe the organization of a GA as a problem-solving tool. GAs [2,5–10] work in a well-defined set of steps each of which can be implemented in many different ways that are to be described later. The sequence of steps is as follows:

1. *Problem definition:* The problem to be solved is reduced or transformed into a problem of maximizing a fitness function $f(X)$ where the argument X of the function represents a potential solution, say a vector with n real components $(x_1, x_2, \ldots, x_n)$. For any such input vector X_i, $f(X_i)$ returns a real number f_i where f_i is the value of the fitness of the potential solution X_i.

2. *Representing solutions:* The potential solutions $\{X_i\}$ are given a chromosomal representation—that is, each X_i is represented as a collection of genes, the genes representing the search variables of the problem each being a component of X_i. A collection of n_p number of such potential solutions is called a population. The initial population $\{X_i\}$, $i = 1$, n_p is generated randomly, subject to relevant constraints.

3. *Evaluation of potential solutions:* Each individual X_i of the population is evaluated for its fitness value f_i, and the average fitness f_{av} of the population is calculated.

4. *Selection:* A "selection" operator now acts on the population such that individuals with higher fitness values relative to the average fitness of the population have a

greater chance to be selected into the mating pool. Once exactly n_p individuals have been selected, the mating pool is complete. One hopes that the postselection population of n_p individuals emphasizes fitness—superior individuals, and undermines the inferior ones. As a result, the average fitness of the population increases.

5. *Breeding new solutions:* The individuals in the mating pool now breed new solutions by chromosomal crossover (crossover breeding) randomly with a probability p_c and the offspring undergo random mutation with a probability p_m (p_c and p_m are user defined).

6. *Screening:* New solutions generated in step (5) are evaluated and screened into a new population of n_p individuals and with that done, one generation is complete. The control returns to step (4) and the iterations continue until a satisfactory solution has been discovered or a maximum number of generations has elapsed.

Steps (2) to (6) are well defined mathematically, but they differ from steps in a standard numerical method of problem solving in that stochastic elements have been introduced in these steps. As a result, different runs may lead to slightly different outcomes. An average over many runs is all that matters. The alert reader must not have missed another point—the algorithm creates a condition in which the solution emerges.

With the working steps of a GA defined it is now time to elaborate on each of the working steps involved with a view to gaining a better understanding of the whole scheme of things in a GA, its flexibilities, powers, and pitfalls.

2.3.1 Representation of Solutions

As already stated, the solutions being sought are given a chromosomal representation. A solution here refers to any value or set of values of the unknowns that can be considered as a potential candidate for the correct solution of the problem at hand. Now which mode of representation to use is dictated by the nature of the problem, type of solution, convenience, personal preference, etc. In fact, an appropriate choice of the representation almost amounts to solving the problem. The most common representation used is to represent the potential or plausible solutions in the form of a string of characters. The "alphabets" for representing these characters are the binary {0, 1}, although alternatives are possible and available (see Section 2.6). Let us consider, for example, a linear system:

$$a_{11}x_1 + a_{12}x_2 + a_{13}x_3 = b_1$$

$$a_{21}x_1 + a_{22}x_2 + a_{23}x_3 = b_2 \qquad (2.1)$$

$$a_{31}x_1 + a_{32}x_2 + a_{33}x_3 = b_3,$$

where a_{ij}s ($i = 1, 2, 3$ and $j = 1, 2, 3$) and b_1, b_2, b_3 are known real constants. Being a linear system, there is only one solution vector $X^* = (x_1^*, x_2^*, x_3^*)^t$ that satisfies Equation (2.1), assuming that the matrix A of the coefficients is invertible ($Det\ A \neq 0$). Any other three-component vector of $X = (x_1, x_2, x_3)^t$ is a potential solution to the problem. If we use four binary bits to represent each component of the potential solution vectors (X), the solutions can be represented by binary strings of 12 bits obtained by concatenating the 4 bits representing the components x_1^1, x_2^1, x_3^1. If we call the string S_1, we have, for example,

$$S_1 = (\underline{1001}\ \underline{0011}\ \underline{1010}),$$

and we can use a higher number of bits for each component if the problem so demands. For each component with 8 bits the potential solutions will be represented as 24-bit binary strings. The bits may be regarded as genes and the string as single chromosome individual or the chromosome itself. For $n \times n$ linear systems, the potential solutions at this level of representation will be $8n$ binary bit strings. It is not difficult to guess that for large linear systems and with higher demands of accuracy these binary bit-string representations may quickly become unwieldy. It could be more convenient to represent the potential solutions as strings or "arrays" of floating point numbers, each component of the solution vector representing a decimal number such as

$$S = (1.20, -2.30, \ldots , 1.04)^t,$$

where $S(1) = 1.20$, $S(2) = -2.30$, ..., $S(n) = 1.04$.

Let us now consider a very different kind of problem say, the problem of representing a one-dimensional Ising chain of n interacting spins. Since these spins can assume only two values (1 or −1) we can represent the system of say, 100 such spins as one-dimensional arrays of signed integers 1 or −1. If the spins interact randomly there are 2^{100} such possible arrays out of which one will represent the lowest energy (ground state) of the system assuming nondegenerate ground state configuration. The interactions can be represented by a symmetric matrix $J(i, j)$ of signed real numbers, each element of the J-matrix denoting the strength and nature (attractive/repulsive) of each interaction. One such randomly chosen solution, for example, could be $S_i(100)$ where $S_i(1) = -1$, $S_i(2) = 1$, $S_i(3) = 1$, ..., $S_i(100) = -1$, the problem being to find the global minimum energy array or the string $S_g(100)$ (if the ground state is nondegenerate or a pair of strings $S_g(100)$ and $S'_g(100)$, if the ground state is doubly degenerate, and so on). A binary bit-string representation of the solutions is equally possible, but it may be more convenient to work with integer array representation $S_i(n)$ of the potential solution for n-spin systems while searching for the ground state.

For some graph problems, the solution being sought could be a graph. Such a graph can be simply represented by what is called an adjacency matrix $A(i, j)$, where $A(i, j) = 1$ if the vertices i and j are connected, else $A(i, j) = 0$. If we also want to specify the strength of the connection between vertices i and j we can introduce a weight matrix $W(i, j)$, where the elements of the W matrix are numbers representing the strength of the connection represented by the edge $(i \leftrightarrow j)$. If the vertices represent atoms, the weight matrix may be taken to represent the order or the strength of the bond formed between the atom pair i and j. We may conclude the discussion about representation by noting that there could be different ways of representing the targeted solution—but one must take care to choose the most convenient one. It is always desirable to work with a representation that is closest to the "space" of the solutions.

2.3.2 Fitness and Fitness Landscapes

For solving any problem, GA starts by defining a function that takes a string S_i as an input and returns a value $f_i = f(S_i)$ that represents a measure of the acceptability of the solution encoded by the string S_i. The value f_i is called the fitness value or simply the fitness of the solution. Higher the value of f_i, better is the quality of the solution encoded by S_i. There could be many ways of constructing a fitness function. It may be so designed as to produce fitness values (f_i) in the closed range $0 \le f_i \le 1$; we will call it a scaled fitness. If unscaled, the fitness function can produce fitness values $f_i \ge 0$. For function optimization, the fitness function should be carefully defined, considering all the relevant constraints. In this context, the term "fitness landscape" so often used and encountered in GA literature perhaps needs a little clarification. It is an important concept in biology where fitness landscape represents the space spanned by all possible genotypes and their fitness values [5]. Let us assume that a binary bit-string based representation has been adopted in GA. If m–bit strings are used, each genotype is a binary bit string of length m. The distance between two genotypes can be measured by their Hamming distance while each genotype can be assumed to have a real fitness value. The fitness landscape can then be visualized as a plot in a space of $(m + 1)$ dimensions, each genotype occupying a point in the m dimensional space, their fitness values being displayed along the $(m + 1)$th axis. Such plots of fitness values may indeed reveal structures like hills, peaks, valleys, and other topographical features present in a real physical landscape, so naturally one describes such plots as fitness landscapes. The genetic operators of GA offer simple means of moving a population of potential solutions from one region of the fitness landscape to another in search of the highest peak on the landscape and the genotype (the solution string) associated with it. However, appealing the above picture of GA dynamics might appear, the very notion of moving a population from one region of a static or unchanging landscape to another is sure to attract criticism as it appears to be biological rather artificial and far from reality. In the real-life scenario, the fitness landscape and the organisms thriving on it are inseparable [8]. The idea of a static fitness landscape is therefore artificial and flawed. However, this description based on unchanging fitness landscape has occupied a rather central place in the GA literature and therefore we will continue to use it, with its limitations always being at the back of our minds.

A second clarification concerning the notion of fitness measures in GA must be made at this point. The "fitness" in GA carries a sense that is somewhat different from the notion of reproductive fitness that biologists make use of. The fitness of an individual in GA refers to the value of the individual with respect to the particular problem-solving goal. Even then the idea of reproductive fitness, so central in evolutionary biology, does not completely lose relevance in GA. However, it does not apply to the goal-specific fitness alone but to the collective notion comprising goal-relative fitness, survival under selection, and variability of the population [11].

A third point on the theme of fitness also demands attention. The fitness in GA is assessed for each individual in the evolving population independently without any reference to the other members of the population. However, there are coevolutionary and ecological exceptions where fitness has to be evaluated in a different manner [5].

Having clarified the notion of fitness in the context of problem solving by GA, let us now turn to the specific case of solving the linear system of Equation 2.1 and explicate how fitness can be defined and evaluated for the linear system $Ax = b$. There is, as we know, only one vector $x^* = (x_1^*, x_2^*, x_3^*)^t$ with real-valued components for which $Ax^* = b$ and the corresponding residue or error vector $R^* = Ax^* - b$ is a null vector, i.e., $R^* = (0, 0, 0)^t$. For any other vector $x = (x_1, x_2, x_3)^t$ the residue vector $R = Ax - b$ is characterized by three real-valued components r_1, r_2, r_3, and the length l_r of the residue vector is given by

$$l_r = \sqrt{r_1^2 + r_2^2 + r_3^2} = (R^t R)^{0.5} \tag{2.2}$$

Clearly $l_r > 0$, unless r_1, r_2, r_3 are each equal to zero, i.e., unless the R vector is a null vector.

Thus, in a GA-based approach to solving the problem of Equation 2.1, the goal of the search is to find a vector $X \equiv (x_1, x_2, x_3)^t$ for which $l_r = 0$. A fitness function (f) specific to the goal can be defined by taking

$$f = e^{-\lambda l_r} \tag{2.3}$$

where λ being a real scalar and $l_r = (R^t R)^{0.5}$ and the residue vector.

$$R = Ax - b \tag{2.4}$$

if $l_r = 0, f = 1$ and the vector X is the exact solution of the linear system. For any other three-dimensional vector "X" with real components, $l_r > 0$, and $f < 1$. So Equation 2.3 defines a scaled fitness function in the range $0 \le f \le 1$ for the problem at hand. If the linear system is n-dimensional, a scaled fitness function can be similarly defined and evaluated. Supposing now that we are using 12-bit binary strings to represent potential solutions of Equation 2.1 in our effort to search out the "true" solution vector (four bits for each of the three components of X), the search space is the space spanned by all possible 12-bit binary strings which is clearly 2^{12} dimensional. GA seeks to discover the one string in this space for which "f" is equal to 1, without having to enumerate all the 2^{12} possibilities.

The evolution of a population of a small number (n_p) of randomly selected 12-bit strings under the action of genetic operators move the population from one part of the search space to another (or from one region of the fitness landscape to another) and in the process discovers the true solution vector X^* for which the goal-specific fitness attains the maximum value of 1. Instead of using the scaled fitness measure given by Equation 2.3, we could have used an unscaled fitness function $f'(X)$ where

$$f'(X) = \frac{1}{(R^t R)^{0.5} + \delta} \tag{2.5}$$

where δ is a small real constant added to the denominator to prevent the right-hand side from blowing up when $l_r = (R^t R)^{0.5}$ tends to become zero as the search for the true solution vector homes in to X^*. The maximum fitness value $f' \ne 1$, but is equal to $\frac{1}{\delta}$ while the minimum is still zero. This new fitness function f' is there, bounded in the range $0 \le f' \le \frac{1}{\delta}$. We must emphasize here that there could be multiple choices for the fitness function of a particular problem; the actual choice should be made with care considering convenience and benefits. The fitness function defined here is however representation independent— neither f of Equation 2.3 nor f' of Equation 2.5 depends on whether binary bit strings or strings of floating point numbers are used to represents the solutions. The representation is, however, important for the operational designs of some of the genetic operators. As already discussed, the central theme of GA is to move an initial population of potential solutions across the fitness landscape and quickly reach the region where the true solution is located. That means GA must be able to generate new potential solutions with higher goal-specific fitness values as the search proceeds. The new potential solutions (bit strings, say) are produced by a crossover (recombination) operator and a mutation operator. The designs of crossover and mutation operators depend upon the representation used for the potential solution. In what follows we explain their actions with reference to representation by binary bit strings. The designs of such operators when a different representation is used will be addressed separately at a later stage.

2.3.3 The Crossover Operator

The idea is borrowed from real-life genetics and is used to breed new solutions from existing ones. It is a binary operator and works on a pair of strings. It has been implemented in many different ways. For binary bit strings, the simplest form of it is called single-point crossover and operates in the following manner. Let us suppose that two individuals (or chromosomes) S_k and S_l of the evolving population have been randomly chosen for the crossover operation where S_k and S_l are the two 12-bit binary bit strings with, say

$$S_k = (1\ 0\ 0\ 1\ 0\ 0\ 1\ 0\ 1\ 0\ 1\ 0)$$

$$S_l = (0\ 1\ 1\ 1\ 1\ 0\ 1\ 1\ 0\ 1\ 0\ 1)$$

Once the individuals have been chosen to undergo crossover, the next step is to select the crossover site. Suppose that the crossover probability is p_c. The crossover site is then chosen randomly with a probability p_c; let us assume that the site so chosen is the fifth bit position from the right (marked with $\downarrow$). Crossover forces interchange of the entire genetic material between S_k and S_l located on the right of the chosen crossover site. That means all the bits on the first four positions from the right are swapped between S_k and S_l producing two new strings, S'_k and S'_l, the offspring, where

$$\downarrow$$
$$S_k = \left(1\ 0\ 0\ 1\ 0\ 0\ 1\ 0\ \boxed{1\ 0\ 1\ 0}\right)$$

$$\downarrow$$
$$S_l = \left(0\ 1\ 1\ 1\ 1\ 0\ 1\ 1\ \boxed{0\ 1\ 0\ 1}\right)$$

$$\downarrow crossover$$

$$S'_k = \left(1\ 0\ 0\ 1\ 0\ 0\ 1\ 0\ \boxed{0\ 1\ 0\ 1}\right)$$

$$S'_l = \left(0\ 1\ 1\ 1\ 1\ 0\ 1\ 1\ \boxed{1\ 0\ 1\ 0}\right)$$

The new strings S'_k, S'_l are expected to have fitness values different from those of S_k, S_l and will thus be located in a different region of the search space. The process of crossover is applied to different pairs of individuals selected randomly until an adequate number of offspring are produced. The process of crossover breeding exploits information already present in the search space to create new strings (potential solutions) hopefully with better goal-specific fitness. The crossover probability used is neither too low nor too high ($p_c \sim 0.8$ is common). The GA literature abounds with many different methods of introducing crossover. We will return to these possibilities later in this chapter.

2.3.4 Mutation Operator

Mutation is unary operator. It acts on one string at a time. Mutation operation once again mimics a random biological event in which a gene randomly becomes altered. It is a low intensity or low-probability event which is of paramount importance in a GA. The importance is rooted in the fact that mutation can create new genetic material that was not available in the initial population of a finite, usually small number of individuals. It can also bring back genetic material lost in the (natural) selection process introduced in a GA. We will illustrate how mutation is induced in a GA working with binary bit strings. Let us take the postcrossover string S'_k, for example and assume that the mutation probability $p_m = 0.1$. One of the 12-bits of the string S'_k is chosen at random with probability $p_m = 0.1$ and the bit is flipped to "1" if it is "0" and to "0" if it is "1." Suppose that the third bit (from the left) of S'_k was successfully picked up for mutation. The mutated string or the chromosome, let us call it S''_k, is

$$S''_k = 1\ 0\ \boxed{1}\ 1\ 0\ 0\ 1\ 0\ 0\ 1\ 0\ 1$$

The genetic makeup of S''_k is different from either S_k (parent) or S''_k (offspring before mutation). The mutated string S''_k is therefore expected to have a fitness value different from the fitness of either S_k or S'_k. Every individual in the postcrossover population is exposed to the process of mutation. Needless to mention, many realizations of the mutation operation are available and the operational details of these operators are representation dependent. Some of them will be further explored later in this chapter.

2.3.5 The Selection Operator

Unlike the binary crossover operator or the unary mutation operator, the selection operator acts on the entire population of n_p individuals to produce a breeding pool of exactly n_p individuals in a way that ensures a higher frequency of occurrence of individuals with relatively higher fitness values—that is why it has been called a fitness-proportional selection process. In a way, the process mimics the Darwinian rule of the survival of the fittest that appears to operate in the form of natural selection in biological evolution. In the context of biology, individuals of above-average fitness tend to live longer, produce more offspring, and therefore dominate the population. Low fitness individuals are naturally snuffed out of the game of survival and reproduction. A simple mechanism of enforcing fitness-proportional selection on a population of n_p individuals is provided by the traditional roulette wheel selection [2,7]. It operates in the following manner.

Imagine a slotted wheel (Figure 2.3) with slot widths proportional to fitness values available in the population. The wheel is fitted with an arrow, the arrowhead pointing to a fixed direction (say NE). The wheel is given a random spin and allowed to halt naturally. The particular slot to which the arrowhead of the indicator points when the spinning wheel halts decides which individual makes it to the breeding pool. The wheel is spun exactly n_p number of times, selecting one individual as indicated by the arrowhead on each occasion. Thus, after n_p spins a new population of exactly n_p individuals, many of them hopefully and probably with relatively higher fitness values is created. The fitness-proportional slots in the wheel create a biasing mechanism in spite of the randomness of the spins. It must be mentioned that some individuals with less than average fitness may also be selected in the process. The probability of such selection is, however, rather low so that it remains highly improbable that all the n_p individuals selected by the roulette wheel procedure have less than average fitness. A mathematical model for operating the slotted roulette wheel selection process will be taken up separately in the following subsection.

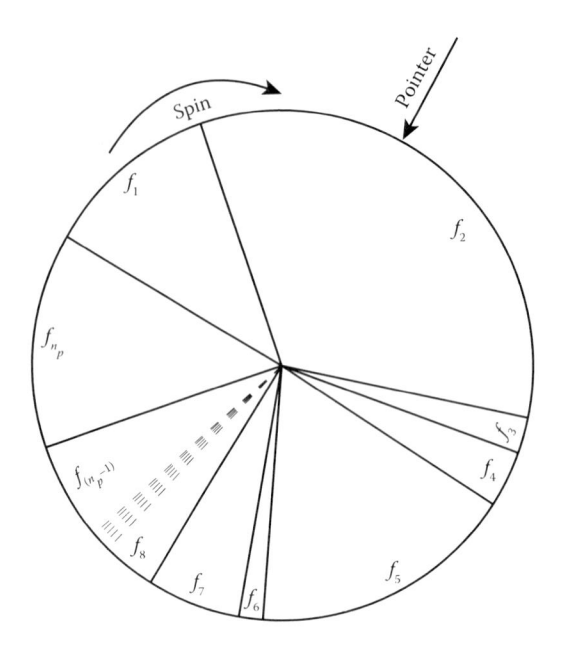

FIGURE 2.3
Roulette wheel.

2.3.6 A Specific Choice

Let us return to the matrix form of the linear system defined in Equation 2.1 and setup the equation for a specific choice of the A matrix (with elements a_{ij}) and the b vector (with elements b_1, b_2, and b_3).

$$\begin{pmatrix} a_{11} & a_{12} & a_{13} \\ a_{21} & a_{22} & a_{23} \\ a_{31} & a_{32} & a_{33} \end{pmatrix} \begin{pmatrix} x_1 \\ x_2 \\ x_3 \end{pmatrix} = \begin{pmatrix} b_1 \\ b_2 \\ b_3 \end{pmatrix} \tag{2.6}$$

For a specific choice of the elements (a_{ij}) of the A matrix and the b vector, we have, for example, an equation

$$\begin{pmatrix} 1 & -1 & 2 \\ 2 & 1 & -2 \\ 1 & 2 & 1 \end{pmatrix} \begin{pmatrix} x_1 \\ x_2 \\ x_3 \end{pmatrix} = \begin{pmatrix} 5 \\ -2 \\ 8 \end{pmatrix} \tag{2.7}$$

How does one proceed to find the solution vector $(x_1^*, x_2^*, x_3^*)^t$ in this case, by invoking a genetic algorithm? As already outlined, we start by creating a population of say, six 3-component potential solution vectors $(S_1, S_2, ..., S_6)$ each of which is represented as randomly generated 12-bit binary strings. Each such string (S_i) is mapped into a three-component vector $(x_1^i, x_2^i, x_3^i)^t$, and the residue vector $R^i = (r_1^i, r_2^i, r_3^i)^t$ is computed by computing the difference $R^i = (Ax^i - b)$, so that we have for $i = 1, 2, ..., 6$.

$$r_1^i = a_{11}x_1^i + a_{12}x_2^i + a_{13}x_3^i - b_1$$

$$r_2^i = a_{21}x_1^i + a_{22}x_2^i + a_{23}x_3^i - b_2 \qquad (2.8)$$

$$r_3^i = a_{31}x_1^i + a_{32}x_2^i + a_{33}x_3^i - b_3$$

The length of the residue vector l_i corresponding to the string S_i is computed easily as

$$l_i = \left\{ \left(r_1^i\right)^2 + \left(r_2^i\right)^2 + \left(r_3^i\right)^2 \right\}^{\frac{1}{2}} \text{ for } i = 1, 2, ..., 6 \qquad (2.9)$$

The fitness values of the six randomly chosen strings are then computed by evaluating the expression

$$f_i = e^{-\lambda l_i} \ (\lambda = 0.1, \text{ say}), \text{ for } i = 1, 2, ..., 6.$$

The n_p (= 6, here) fitness values $f_1, f_2, ..., f_{n_p}$ will be distributed in the range $0 \le f \le 1$. Now we are in a position to allow the roulette wheel operator to act on this population of six randomly chosen individuals and create a breeding pool of exactly six individuals for further processing by the GA.

The selection process involves executing the following mathematical steps:

1. Evaluate the sum F of the fitness of n_p (= 6) individuals in the population: $F = \sum_{i=1}^{n_p} f_i$ and also the average fitness $F_{av} = \dfrac{1}{n_p} \sum_{i=1}^{n_p} f_i$.

2. Calculate the probability of selecting the individual S_i as $p_i = \dfrac{f_i}{F_{av}}$. Note that $\sum_{i=1}^{n_p} p_i = 1$.

3. Find the cumulative probability $q_i = \sum_{j=1}^{i} p_i$. Note $q_1 = p_1; q_2 = p_1 + p_2 + ...; q_{n_p} = \sum_{i=1}^{n_p} p_i = 1$.

4. Generate a random number r normally distributed in the range $0 \le r \le 1$ and execute the following checks:

   a. If $r < q_1$, select the first string S_1 with the fitness value f_1.

   b. If $r \not< q_1$, search for a string S_i in the population for which $q_{i-1} < r < q_i$ ($i = 2, 3, ..., n_p$). If the condition is satisfied for $i = k$ (say), the string S_k is selected and moved into the mating pool (with replacement), i.e., the same string may be selected again. Reselection is more probable if the fitness value f_k for the string S_k is relatively high (relative to the average fitness of the population). However, such reselection or multiple selection cannot be guaranteed as the process is not a deterministic one and the population size(n_p) is small. Once n_p strings have been selected, the formation of the mating pool is complete.

5. Let n_i be the frequency with which the string S_i occurs in the postselection population (i.e., in the breeding or mating pool). n_i values may be 0, 1, 2, etc. Then the average fitness of the population in the mating pool is given by $F_{av}^M = \dfrac{1}{n_p} \sum_i^{n_p} n_i f_i$.

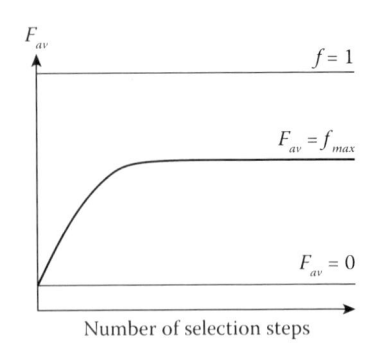

FIGURE 2.4
Growth and saturation of the average fitness when only the selection operator acts on the population.

Since strings with higher fitness values are expected to occur more frequently in the postselection pool, $F_{av}^M \geq F_{av}$. (In the highly unlikely event of all of the n_p number of random spins of the wheel selecting individuals with fitness less than the current average fitness of the population the above-mentioned inequality may not hold). Assuming that $F_{av}^M \geq F_{av}$, we note that increase in the average fitness of the postselection population is only due to a change in the structure of the population. No new strings have been added to the population. So selection alone cannot discover the true solution had it not been already present in the randomly generated initial population by chance. So, repeated applications of the selection operator will quickly lead to saturation of the average fitness, the saturation value being the fitness value (f_{max}) of the best string in the initial population (Figure 2.4).

There has to be a systematism for generating new strings so that the GA can move the population from one region of the fitness landscape to another. In GA, new strings are brought into the postselection population by the action of two genetic operators, namely operators for crossover and mutation. Let us consider crossover operation first.

6. The crossover operation: Let p_c be the crossover probability. We assume that the n_p number of strings have been ordered according to their fitness values. Let us take the first string (S_1) in the mating pool (the first parent) and randomly choose its partner string (the second parent) from the rest of the ($n_p - 1$) individuals in the population. Suppose that the third string (S_3) has been chosen in the process as the partner. The crossover site has to be chosen now with a probability p_c (= 0.8, say). One simple way to execute this step is to loop over the (12 − 1) = 11 sites of the chosen 12-bit string (the rightmost bit is excluded), generating a random number in the range between 0 and 1 ($0 \leq r \leq 1$), for each of the 11 sites. The site for which $r < p_c$ is chosen as the crossover site. Let us suppose that the condition has been satisfied for the fourth site from the left. The single-point binary crossover (the simplest crossover) between the strings S_1 and S_3 at the chosen crossover site is executed by interchanging all the bits beyond the fourth site, between the strings S_1 and S_3. Let the two new strings produced by the process be called S_1' and S_3'. Formally, we can write ($\oplus$ is the crossover operator)

$$S_1 \oplus S_3 = S_1' + S_3'$$

and explicitly in terms of the bits exchange we have

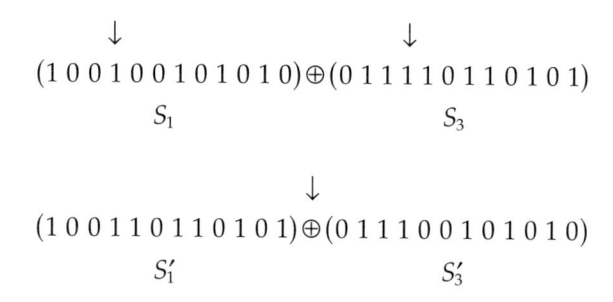

$$\downarrow \qquad\qquad\qquad \downarrow$$
$$(1\ 0\ 0\ 1\ 0\ 0\ 1\ 0\ 1\ 0\ 1\ 0) \oplus (0\ 1\ 1\ 1\ 1\ 0\ 1\ 1\ 0\ 1\ 0\ 1)$$
$$\quad S_1 \qquad\qquad\qquad\qquad\qquad S_3$$

$$\downarrow$$
$$(1\ 0\ 0\ 1\ 1\ 0\ 1\ 1\ 0\ 1\ 0\ 1) \oplus (0\ 1\ 1\ 1\ 0\ 0\ 1\ 0\ 1\ 0\ 1\ 0)$$
$$\quad S_1' \qquad\qquad\qquad\qquad\qquad S_3'$$

S_1' and S_3' have bit patterns somewhat different from the parent strings S_1 and S_3—they are regarded as the new individuals, the offspring, produced by parents S_1 and S_3. The process is repeated with the other strings ($S_2, S_3, ..., S_6$) sequentially chosen as the first parent, and the partner chosen randomly from remaining ones. Clearly, a string selected as partner may be selected again. The probability of such selection will be higher if the particular string has high fitness value as we expect it to have multiple copies in the population—thanks to the fitness-proportional selection process used to create the mating pool population.

7. The mutation operator: The postcrossover population of n_p (= 6) individuals or strings $S_1', S_2', ..., S_6'$ (offspring) is now ready for undergoing mutation. As in the real world, mutation in our GA is an event with a low probability of occurrence (a rare event). Let the mutation probability $p_m = 0.1$, so that roughly 1 bit out of 10 can mutate. To effect mutation one can loop across each of the six (n_p) strings $S_1', S_2', ..., S_6'$, generating uniformly distributed random numbers $r(0, 1)$ for each site. The site for which $r < p_m$ is selected for undergoing mutation—that is, the particular bit is flipped to 1 if it is 0 and to 0 if it is 1. Let us suppose that the third bit position on the string S_3' from the left has been chosen for mutation. The mutated string (call it S_3'') then becomes (□ marked bit is to be flipped)

$$S_3'' = S_3'^{*} = (11\boxed{0}000101010) \Rightarrow (111000101010)$$

8. The screening: The population after the crossover and mutation operations has six strings again $S_1', S_2', S_3'', S_4', S_5', S_6'$. Their fitness values are evaluated. It is now time to create a new population for the next generation; a multiplicity of choice is available. Let us examine some of the possibilities.

   a. We replace the entire population of six strings ($S_1, S_2, ..., S_6$) by the six new strings (offspring) generated by crossover and mutation ($S_1', S_2', S_3'', S_4', S_5', S_6'$) and return to step (i). This strategy may be called nonelitist screening.

   b. We may compare the fitness values of six parents and six offspring and choose the best set of the six strings (individuals) out of the combined pool of all the parents and the offspring and return to step (i). We may call it the completely elitist strategy of screening.

   c. We may adopt a kind of middle-of-the-road approach and choose n_p' number of best strings from the combined pool ($n_p' \ll n_p$). The remaining ($n_p - n_p'$) numbers of strings are selected randomly from the same pool and control is returned to step (i).

The procedure (c) has the facility of varying the selection pressure by choosing the value of n'_p. If $n'_p = 0$, the selection pressure is very low and we have the completely nonelitist strategy of screening. If $n'_p = n_p$, the selection pressure is very high and we have the completely elitist screening strategy. For intermediate values of n'_p, we have GAs with varying degrees of selection pressure.

The execution of steps (1) to (8) once completes one generation. The iterations are continued until some exit or termination criterion has been fulfilled. Most commonly, the exit criterion involves checking if the goal-specific fitness value of the best string in the population is not improving any further or a maximum number of iterations have been already executed. In the present example, if the best string has achieved a fitness value of $f_{max} = 1$ or $|1 - f_{max}| \leq \epsilon$ where ϵ is a preset small number ($= 10^{-6}$ or so), the exit condition has been met. The string with the highest fitness value then represents the solution of the linear system. A typical profile for the evolution of the fitness of the best string in the current example is displayed in Figure 2.5—where f_{max} is plotted against n_g—the number of generations elapsed (a completely elitist strategy is enforced in the screening).

The staircase-like structure of the $f_{max} - n_g$ plot is typical. The small horizontal steps are indicative of the number of generations that the search has to wait for a beneficial mutation that enhances the fitness of the best string to take place. While the appearance of the fitness evolution profile displayed in Figure 2.5 is typical, the detailed structure of the plot may depend upon the parameters of the GA employed or even with the same set of parameters; it may vary from run to run reflecting the stochastic nature of the search for a solution. It is important to realize that the average performance over many runs is important. A particular run may display performance below or above the average level of performance [5].

The GA described in this section has a number of adjustable parameters, such as population size (n_p), screening control parameter or the parameter for controlling selection pressure (n'_p), crossover probability (p_c), or mutation probability (p_m). These parameters can be held fixed over generations (static GA) or can be dynamically adjusted as generations elapse (dynamic GA). If the parameters are dynamically adjusted on the basis of analysis of the performance of the search over the past generations, we may call it an adaptively controlled GA. The interest in different aspects of adaptively controlled GA has been steadily growing.

Having seen how GAs with binary bit-string representation work in a specific problem (i.e., how the different steps are executed) we are now in a position to ask perhaps the most important question: why does GA work the way it does?

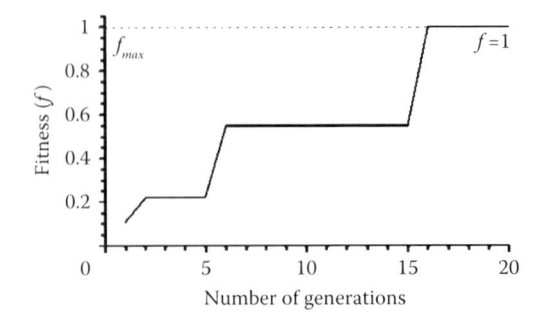

FIGURE 2.5
A typical profile for the evolution of the fitness of the super individual in a GA run.

2.4 Why Does GA Work?

The unique solution for the specific toy problem described in the preceding section is one string or chromosome with 12 binary bits. Since each bit can have two values, 0 or 1, the search space for the problem has $N = 2^{12}$ (= 4096) dimensions. A sequential search for the solution would require us to evaluate ~4096 strings (the maximum number of such evaluations) for their fitness values. The simple GA described in Section 2.2, if it quickly converges, would discover the true solution string after much fewer string evaluations. Suppose that the GA hits the solution in $n_g = 100$ generations. With a population size of $n_p = 6$, that would have required only $6 \times 100 = 600 \ll 4096$ string evaluations. Suppose now that we are dealing with a linear system of 100 unknowns $(x_1, x_2, ..., x_{50}, ..., x_{100})$. With four bits representing each unknown, the strings (chromosomes) would be made up of 400 bits (genes), the search space now having $N = 2^{400}$ dimensions. If the GA-based search converges to the correct solution in 200 generations with a population size $n_p = 10$, the total number of string evaluations (the costliest step) made would be just a modest $10 \times 200 = 2000$ evaluations—far less than the astronomically large figure of $N = 2^{400}$ number of evaluations in a direct enumeration method. The basic question "how does a GA work?" concerns how the GA homes in on the correct solution string so efficiently or quickly, i.e., with so fewer string evaluations. A complete answer to the question is not available yet. Several theoretical models have been proposed from time to time in partial response to the very basic question about GA efficiency [5]. These models have undoubtedly sharpened our understanding about the functioning of a GA. The first and until now the most popular among such theoretical models has been known as Holland's schema theorem.

The pivotal idea of Holland's schema theory [2,12,13] of GA is that the GA explores the search space by implicitly discovering building blocks called schemas or schematas (the schemas are partial structures or templates of l-bit strings with some undefined bits) that contribute significantly to the goal-specific fitness value of a string and bring them onto the same string by the action of genetic operators. The most important point to note here is that the GA, as we have seen in the last section, actually never explicitly evaluates any of the building blocks; it evaluates only the n_p number of l-bit full strings present in the population at each generation. Holland's conjecture was that the GA, through an inbuilt implicit parallelism that it is endowed with, keeps track of a huge number of building blocks and emphasizes the ones that can be stitched together to shape the correct solution strings as generations elapse and evolution saturates. The schema theorem provides the mathematical underpinning of the building block hypothesis. Before describing the model let us first explain what the building blocks are, and the attributes of consequence they have.

2.4.1 The Schemas (Schematas) and Their Attributes

Let us consider the search space spanned by l-bit binary strings. A schema in this search space is any l-bit string in which zero, one, or more bit positions are undefined. By undefined, we mean that we do not know whether the specific bit values are 0 or 1. The particular bit positions are marked by a special symbol, the asterisk (*), called the "don't care symbol" or the wild card. Let us take for example the case of 3-bit strings ($l = 3$ case). An example of a 3-bit string with one undefined bit may be (* 0.1). It represents a valid schema, let us call it H_1. Likewise, a 3-bit string with two undefined bits could be a schema $H_2 = * 1*$, for example. The schema H_1 has two defined bits and is said to be of order 2, while the schema H_2 has only one defined bit and is of order 1. In fact, the number of defined bits in a schema H defines its order and is represented by the symbol O_H (the symbol H for schemas is due to Goldberg who took the view that schemas represent hyperplanes in the relevant search space). The order O_H of the schema, as we will see later, is a rather important attribute of a schema. In the preceding examples $O_{H_1} = 2$ and $O_{H_2} = 1$. A second important attribute of a schema is called the defining length, represented by the symbol δ_H which is the distance between the first and the last defined bit positions in the schema concerned. With reference to the previous examples, $\delta_{H_1} = 1$ and $\delta_{H_2} = 0$. How many schemas can be constructed for l-bit strings? Since there are l-distinct bit positions each of which can be either * or 0 or 1, the total number of schemas that can be constructed is 3^l. The total number of l- bit strings as we already know is 2^l. So the schemas are more numerous and more flexible units for exploring the search space.

In addition to order O_H and defining length δ_H, there is a third important attribute of the schema H which is called the fitness of a schema (the same goal-specific fitness value that is used to characterize a potential solution string) f_H. f_H cannot however be evaluated directly as we do not know all the bit values in a schema H (there are undefined bits marked by "*"). f_H can only be measured indirectly by alluding to it the average fitness of all the strings in the population which are instances of the schema H at that particular time (generation). By the term "instances of the schema H" we mean the strings $\{S_k\}$ in the population in which the particular template H can be found. Let $\eta_H(t)$ be the number of instances of a schema H at the tth generation—$\eta_H(t)$ representing the number of strings in the population which matches with H at all its defined bit positions. Suppose in a population of ten 4-bit strings, $S_1, S_2, \ldots, S_{10}$ with fitness values $f_1, f_2, \ldots, f_{10}$, respectively, the strings S_3, S_7 and S_{10} are instances of the schema $H \equiv * 1\, 0\, *$. The order O_H of the schema is 2, the defining length δ_H is 1, and $\eta_H = 3$. The fitness f_H of the schema H is the average fitness of S_3, S_7 and S_{10}. Formally, we can write

$$f_H = \frac{1}{\eta_H}\left(f_3 + f_7 + f_{10}\right) = \frac{1}{3}\left(f_3 + f_7 + f_{10}\right)$$

As the population evolves under the action of the genetic operators, η_H is also expected to evolve. The question is in what manner $\eta_H(t)$ would change from generation to generation and how would the dynamics of schema evolution depend on the different schema attributes like O_H, δ_H, and f_H. We have thus come to pose the very important question in our quest for understanding why GAs work. The question is how does the number of instances of a particular schema of fitness f_H, order O_H, and defining length δ_H evolve in response to the combined action of genetic operators like selection, crossover, and mutation on a population of n_p individuals (strings), and what does it imply for the working of a GA? The answer to this question contributes what has come to be known as the fundamental theorem of GAs and the basic building block hypothesis [7,14].

It must be emphasized again that the GA does not explicitly represent and process any of the schemas or their attributes—it explicitly represents only the n_p number of l-bit strings in the population—nor does it evaluate any schema for its fitness value, nor does it store any information about them explicitly during a run. The schema theorem, as we will shortly see, describes the dynamics of the growth or decay of the number of instances of a schema in the evolving population. This dynamics, in turn, may be interpreted to imply that the GA is actually calculating, storing, and keeping track of the fitness of a huge number of schemas in the form of the averages of fitness values of their instances in the population. The interpretation creates the notion of an implicit parallelism operating in a GA.

2.4.2 The Schema Theorem

Suppose that we have a randomly chosen population of n_p number of binary l-bit strings $(S_1, S_2, ..., S_k, ..., S_{n_p})$ each of which represents a potential solution of the targeted problem, their fitness values being $f_1, f_2, ..., f_k, ..., f_{n_p}$, respectively. Let $m_H(t)$ be the number of instances of a particular schema H in the evolving population at time t—at the tth step of the evolutionary process, and $f_H(t)$ be the fitness (average fitness of the schema at the tth generation). By the definition of schema fitness, we have

$$f_H(t) = \frac{1}{m_H} \sum_{S_i \in H} f(S_i) \tag{2.10}$$

where $f(S_i)$ represents the fitness value of the string S_i present in the population and $S_i \in H$ implies that the summation is carried out over all the strings (S_i) which are instances of the schema H in the current population. The average fitness $f_{av}(t)$ of the current population of n_p number of strings is

$$f_{av}(t) = \frac{1}{n_p} \sum_{i=1}^{n_p} f(S_i) \tag{2.11}$$

The number of copies of the string S_i in the new mating pool created by the fitness-proportional roulette wheel procedure at time $t + 1$ as described in Sections 2.3.5 and 2.3.6 is $n_p \times p_i$, where

$$p_i = \frac{f(S_i)}{\sum_{i=1}^{n_p} f(S_i, t)} \tag{2.12}$$

is the probability of the string S_i being selected and n_p is the total number of strings present in the population. $f(S_i, t)$ is the fitness value of the string S_i at the tth generation. We use $f(S_i, t)$ instead of $f(S_i)$ to stress that $f(S_i)$ is a generation-dependent quantity. The fitness f_H of a schema H as we have already defined is the average fitness of all the strings currently in the population which are also instances of the schema, the number of which is $m_H(t)$. Now consider the probability p_H that one string with fitness equal to f_H occurs in the population. Clearly,

$$p_H = \frac{f_H}{\sum_{i=1}^{n_p} f(S_i, t)} \tag{2.13}$$

The total probability of occurrence of the group of $m_H(t)$ strings which are currently instances of the schema (H) in the population is then given by

$$P_H = m_H(t)p_H = m_H \frac{f_H}{\sum_{i=1}^{n_p} f(S_i, t)} \tag{2.14}$$

The number of instances of the schema H in the population of size n_p at the $(t+1)$th generation is therefore,

$$m_H(t+1) = n_p \cdot P_H = n_p \cdot m_H(t) p_H = n_p \cdot m_H(t) \frac{f_H}{\sum_{i=1}^{n_p} f(S_i, t)} = m_H(t) \frac{f_H}{\frac{1}{n_p} \sum_{i=1}^{n_p} f(S_i, t)} \tag{2.15}$$

i.e., $m_H(t+1) = m_H(t) \dfrac{f_H}{f_{av}(t)}$

Suppose now that the fitness f_H of the schema H is slightly above $f_{av}(t)$, i.e., above the average fitness of the strings (individuals) in the current population. Then we can set f_H $(1+\epsilon) f_{av}(t)$ with $\epsilon > 0$ and have [5]

$$m_H(t+1) = m_H(t) \cdot (1+\epsilon) \tag{2.16}$$

Starting with $t = 0$, and iterating we have

$$m_H(1) = m_H(0) \cdot (1+\epsilon)$$

$$m_H(2) = m_H(1) \cdot (1+\epsilon) = m_H(0) \cdot (1+\epsilon)^2$$

$$\vdots \tag{2.17}$$

$$m_H(t) = m_H(0) \cdot (1+\epsilon)^t$$

$$m_H(t+1) = m_H(0) \cdot (1+\epsilon)^{t+1}$$

Clearly, Equation 2.15 in this case predicts an exponential growth in the number of instances of the schema H with fitness higher than the average fitness of the current population as $t \to \infty$. What if f_H were slightly lower than $f_{av}(t)$? Proceeding in same manner and noting that $f_H(1-\epsilon) f_{av}$, $\epsilon > 0$ we have

$$m_H(1) = m_H(0) \cdot (1-\epsilon)$$

$$m_H(2) = m_H(1) \cdot (1-\epsilon) = m_H(0) \cdot (1-\epsilon)^2$$

$$\vdots \tag{2.18}$$

$$m_H(t) = m_H(0) \cdot (1-\epsilon)^t$$

$$m_H(t+1) = m_H(0) \cdot (1-\epsilon)^{t+1}$$

Clearly, there is now exponential decay in the number of instances of the schema H with fitness lower than the average fitness of the population as $t \to \infty$. Equation 2.15 therefore describes the general dynamics of evolution of the number of instances of a schema H of fitness f_H in the population under the action of a fitness-proportional selection scheme alone. The message is quite clear, and important. Fitness-proportional selection tends to exponentially emphasize schemas with above population-average fitness and undermine all those schemas that have less than population-average fitness in succeeding generations. We must note here that selection alone cannot generate totally new strings and therefore cannot produce new schemas with fitness values better or worse than the best or the worst schema already present in the population. New strings, as we have already described, are produced in a GA by the process called crossover and mutation. It is therefore necessary to carefully consider the effects of crossover and mutation on the schemas growth/decay in Equation 2.15. We will do the analysis in two steps. Let us consider first how crossover affects Equation 2.15.

2.4.3 Effects of Crossover

Let us consider a valid single-point crossover between strings S_k and S_l present in the population and ask whether the crossover would destroy the schema H of which S_k is an instance, ignoring the role of its partner S_l, if any. As an example, let us consider the string S_k which is assumed to be an 8-bit binary string and two schemas H_1 and H_2 of which the chosen S_k is an instance. Let the crossover probability be p_c and suppose that the second bit from the left (marked by $\downarrow$) has been chosen as the crossover site as depicted below:

$$\downarrow$$
$$S_k = 0\ 1\ 0\ 1\ 0\ 0\ 0\ 0$$
$$H_1 = *\ 1\ *\ *\ *\ *\ *\ 0$$
$$H_2 = *\ *\ *\ 1\ 0\ *\ *\ *$$

In the crossover, the schema H_1 may be lost as the two defined bits in this schema, 1 in position 2 and 0 in position 8, will be placed on different offspring, S'_k and S'_l, following crossover. The schema H_2 will, however, survive this crossover because the defined bits, 1 in position 4 and 0 in position 5 (from left) will be passed on to one of the offspring together without disruption and thus at least one string which is an instance of H_2 will be there. One may construct many such examples and conclude that the crossover site chosen and the structure of the template—the schema H—determines the probability (p_c^d) of destruction of a schema under single-point crossover. Since we are dealing with l-bit strings, there are $(l-1)$ nontrivial crossover sites on the string S_k. If the first and last defined bits on the schema H are widely separated, i.e., the defining length δ_H of the schema is large, the probability of randomly choosing a crossover site (with probability p_c) falling between the first and the last defined bits of the l bit schema H, with defining length δ_H, under single-point crossover p_c^d) is

$$p_c^d = \text{crossover probability} \times \text{fraction of available crossover sites falling between}$$

$$\text{the first and last defined bits of } H = p_c \times \frac{\delta_H}{(l-1)} \tag{2.19}$$

We may now recall that crossover is a binary operator and acts on a pair of strings, so the role of the string S_l, the partner of S_k in crossover, so far neglected in assessing p_c^d, must be factored in to provide a better estimate of p_c^d. It is not difficult to see that the schema H will survive crossover at the given site if S_l has the appropriate bit structure to compensate for the disruption of the bit pattern caused to S_k by single-point crossover and the defined bits in H will appear on the same offspring (S'_k, say) produced in the process. The schema H will survive in such a situation.

When such possibilities are accommodated in the theoretical description, the equality in Equation 2.19 for p_c^d has to be replaced by an inequality ($\leq$) and the corrected expression for p_c^d reads

$$p_c^d \leq p_c \times \frac{\delta_H}{(l-1)} \tag{2.20}$$

This is the probability of destruction of H under crossover. The probability of survival of H under single-point crossover (p_c^s) is therefore given by

$$p_c^s = (1 - p_c^d) \geq 1 - p_c \times \frac{\delta_H}{(l-1)} \tag{2.21}$$

We note here that selection and crossover are mutually independent processes; so the dynamics of growth and decay of a schema H under the combined action of selection and crossover can be obtained by multiplying the right-hand side of Equation 2.15 for the schema growth equation under selection by the survival probability (p_c^s) of the designated schema. The modified equation becomes an inequation

$$m_H(t+1) = m_H(t) \frac{f(H)}{f_{av}(t)} (1 - p_c^d) \geq m_H(t) \frac{f(H)}{f_{av}(t)} (1 - p_c \times \frac{\delta_H}{(l-1)}) \tag{2.22}$$

The message from Equation 2.22 is clear: the number of instances of a schema H of short defining length (δ_H small) and high or above-average fitness (f_H above $f_{av}(t)$) tend to grow exponentially as the population evolves under the combined action of selection (fitness proportional) and crossover (single-point). The question is: what happens to this message when mutation is taken into consideration?

2.4.4 Effects of Mutation

The effects of mutation on the schema growth/decay equation (Equation 2.22) can be introduced following a similar line of argument and multiplying the right-hand side of the inequality 2.21 with p_m^s—the probability of survival of the schema H under mutation, assuming that it has survived the combined action of selection and crossover. The attribute of schema H that determines the probability of destruction of H under mutation is its order O_H which is just the number of defined bits in H. Let p_m be the probability of mutation—it means that the chance that a single bit at any position of an l-bit string (S_k', say) gets flipped is p_m. The probability that the bit survives mutation—that is, it does not get flipped—is ($1 - p_m$). Supposing that the postcrossover string S_k' is an instance of a schema H having the order O_H. As we know, O_H is just the number of defined bits in H, each having a survival probability ($1 - p_m$). In mutation, the mutated string S_k'' will still continue to be an instance of H after mutation if all of the O_H number of defined bits of H contained in S_k'' survive mutation—remain unchanged. The probability of such an event occurring can be obtained by multiplying ($1 - p_m$) onto itself O_H number of times, each flipping event being independent of others. Thus the survival probability of a schema H under mutation ($p_m^s(H)$) is predicted to be

$$p_m^s = (1 - p_m)^{O_H} \tag{2.23}$$

The rate of growth or decay of the number of instances of the schema H in the evolving population under the combined action of selection, crossover, and mutation can be obtained by multiplying the right-hand side of the inequality 2.23 by $p_m^s(H)$ yielding

$$m_H(t+1) \geq m_H(t) \frac{f(H)}{f_{av}(t)} (1 - p_c \times \frac{\delta_H}{(l-1)})(1 - p_m)^{O_H} \tag{2.24}$$

For $p_m \ll 1$, a condition that is generally applicable in a GA, we can approximate $\left((1-p_m)^{O_H} \approx 1-p_m O_H \right)$ so that Equation 2.24 under the approximation takes a simpler form:

$$m_H(t+1) \geq m_H(t) \frac{f(H)}{f_{av}(t)} (1 - p_c \times \frac{\delta_H}{(l-1)})(1 - p_m O_H) \qquad (2.25)$$

Multiplying out the terms in parentheses on the right-hand side of Equation 2.25 and dropping cross terms involving p_c and p_m we get a further approximated form of the basic schema growth/decay equation:

$$m_H(t+1) \geq m_H(t) \frac{f(H)}{f_{av}(t)} (1 - p_c \times \frac{\delta_H}{(l-1)} - p_m O_H) \qquad (2.26)$$

Equation 2.24 or its approximations 2.25 and 2.26 have been called the fundamental theorem of GAs. The theorem provides a lower bound estimate of the growth and decay rates of a schema H as only the disruptive effects of crossover and mutation have been taken into account. Both crossover and mutation have important beneficial effects on the schema growth/decay which are not taken into account [4].

Equations 2.24 to 2.26 formalize the building block hypothesis of GA introduced earlier, taking into account the effects of selection, single-point crossover, and mutation, i.e., all the basic ingredients of GA. The message is that the GA explores the search space not just by randomly selecting and evaluating a finite number strings in the population, but by implicitly tracking the fitness values of an explosively large number of schemas, the building blocks which include the strings in the population. While doing so, GA emphasizes on short, low-order schema of better than population-average fitness—such schemas receive exponentially increasing numbers of samplings as they are not disrupted by the GA operators, rather GA tries to bring important templates onto the same string by crossover. Mutation contributes by regenerating important genetic material lost in the selection process and by creating it if it was not available in the initial population at all. That is how the GA appears to navigate through a complex search space discovering progressively better and better solutions and finally homing on to the "true" solution string for the problem concerned.

The schema theorem just described has a more general character than it seems and applies equally well to many other possible subsets of strings (templates) in the search space [5]. The schemas as defined provide a rather simple and convenient representation of the building blocks—they are not to be taken as unique, but are just elegant choices. Numerous practical applications [15,16] of GA have confirmed basic validity of the schema theorem and the building block hypothesis although there are some counter examples [17,18] questioning the infallibility of the theorem.

2.5 Other Theoretical Models of GA

2.5.1 GA as a Dynamical System

One of the more interesting mathematical models developed for gaining insight into the working of the GA is due to Vose and Liepinsl [19,20]. They considered a particularly simple GA (to be defined later) which is mapped onto an equivalent mathematical problem of finding composite genetic operators (matrices) G_p or G that act on an initial population represented by a population vector $\vec{p}(t)$ or equivalently, a selection vector $\vec{s}(t)$, to produce the population or the selection vectors for the next generation, $\vec{p}(t+1)$ or $\vec{s}(t+1)$ as the case may be.

Formally then, the task is to look for a G or G_p matrix such that

$$G\vec{s}(t) = \vec{s}(t+1) \qquad (2.27)$$

and

$$G_p\,\vec{p}(t) = \vec{p}(t+1) \qquad (2.28)$$

Assuming that such operators or matrices have been found, they can be viewed as dynamical systems describing the GA evolution process. The collection of all possible $\vec{s}(t)$ vectors (or $\vec{p}(t)$ vectors) form a surface S (or P) on which G (G_p) act producing movement from one point on the relevant surface to another. Starting from an initial selection vector $\vec{s}(0)$, for example, iterations of Equation 2.27 will produce a trajectory $\vec{s}(t)$ on S while iterations of Equation 2.28 will similarly generate a trajectory $\vec{p}(t)$ on P, starting off from $\vec{p}(0)$. These trajectories trace the dynamics of evolution in GA. If it were possible to find fixed points on S (or P) and determine their stabilities, we would know how an initially chosen population or selection vector evolves into one associated with the fixed point, or return to it when slightly perturbed signifying the existence of a stable fixed point, or move away from it signifying an unstable fixed point. In what follows, we outline Vose–Liepinsl's treatment of the problem (see reference [5] for a full description).

Let us first explicate the simple GA that Vose and Liepinsl exploited for building up their mathematical model. The GA consists of the following steps:

1. Create a random population of n_p number of l-bit binary strings (x_i).
2. Define and calculate the fitness $f(x_i)$ for all the x_is.
3. Select a pair of strings (x_k and x_l, say) with probability proportional to their fitness in the population.
4. x_k and x_l undergo crossover at a single site chosen with probability p_c. Out of the offspring pair x_k' and x_l' only one is selected randomly (unlike in a standard GA).
5. Each bit in the selected offspring (x_k' or x_l') is exposed to mutation with a probability p_m and the mutated string (x_k'' or x_l'') is retained as an individual for the next generation.
6. Return to step 3 until the population attains the desired size n_p.
7. Once the population is complete return to step 2.

In any generation Vose and Liepinsl [21], as already stated, represent the population by two vectors $\vec{p}(t)$ and $\vec{s}(t)$ each having the same length $L = 2^l$. Both the vectors are real valued. Let $p_i(t)$ and $s_i(t)$ denote the ith component of the vectors $\vec{p}(t)$ and $\vec{s}(t)$, respectively. $p_i(t)$ is the fraction of the population made up of the string x_i in the tth generation while $s_i(t)$ is the probability that x_i is selected for crossover as defined in step 2 of the simple GA already described. Thus, the vector $\vec{p}(t)$ defines the composition of the population at the t^h generation and $\vec{s}(t)$ specifies the probabilities of selection of individuals in the population at the same instant, under a selection scheme operating with the fitness-based selection operator F. On the basis of the individuals $\{x_i\}$, F can be regarded as a diagonal matrix with elements defined as follows:

$$F_{i,i} = f(x_i)$$

$$F_{i,j} = 0, i \neq j$$

$$(2.29)$$

Noting that $\vec{p}(t)$ completely specifies the composition of the population at the tth generation, it can be easily mapped onto the selection vector $\vec{s}(t)$ via the fitness matrix F as

$$\vec{s}(t) = \frac{F\vec{P}(t)}{\sum_{k=0}^{L} Fp_k(t)}$$

$$(2.30)$$

where $L = 2^l - 1$, and a fitness-proportional selection is assumed. Now the problem is to find a composite operator G which acting on $\vec{s}(t)$ will produce $\vec{s}(t+1)$ just as the simple GA working on the population at the generation t is expected to produce $\vec{p}(t+1)$ and hence $\vec{s}(t+1)$, as is evident from Equation 2.30. The targeted operator (or matrix) therefore must satisfy the relation

$$G\,\vec{s}(t) = \vec{s}(t+1)$$

$$(2.31)$$

if we take only selection into account $G = F$. That means we have already got the required relation in the form of Equation 2.30.

The problem, therefore, boils down to defining G as a composite of matrices F and M, where M encodes the effect of crossover and mutation on the evolving population. Vose and Liepinsl constructed the matrix M in two steps. The first step involved defining a matrix M' the elements M'_{ij} of which represented the probabilities $R_{ij}(O)$ that the composite recombination event involving strings i and j (supposedly selected to mate) led to the production of the string O—the string with all the l-bits set to 0. $R_{ij}(O)$ is shaped by two terms:

1. First, the probability P_1 that no crossover takes place between strings i and j, and the selected offspring is mutated to the string O.

2. Second, the probability P_2 that crossover does take place between the ith and jth strings with the probability p_c and the selected offspring is mutated to O; therefore, we have $R_{ij}(O) = P_1 + P_2$.

Let p_m be the probability that one bit of the selected string is mutated so that the probability of mutation not taking place at the same position is $(1 - p_m)$. Likewise, the probability of crossover not taking place between the selected strings i and j is $(1 - p_c)$, p_c being the probability that crossover does take place. It is straightforward then to write

$$P_1 = \frac{1}{2}(1 - p_c)\{p_m^{i_1}(1 - p_m)^{l-i_1} + p_m^{j_1}(1 - p_m)^{l-j_1}\}$$

$$(2.32)$$

where i_1 is the number of ones (1) in the string i while j_1 is the number of ones (1) in the string j. The factor of $\frac{1}{2}$ occurring in the right-hand side of expression 2.33 for P_1 takes care of the fact that each of the two offspring produced by the crossover between strings i and j has only a 50% chance of being chosen for the next generation (one of the two is chosen in the GA used).

Let us now consider the second possibility that crossover takes place, say at the site c counted from the left producing the children string h and k. The number of available crossover sites in a string of l bits is $l - 1$ and the probability that any one of the available sites is chosen is $\dfrac{1}{l-1}$. The possible values of c are 1, 2, ..., $l - 1$. The probability P_2 then becomes

$$P_2 = \frac{1}{2}\frac{p_c}{l-1}\sum_{c=1}^{l-1}\{p_m^{h_1}(1-p_m)^{l-h_1} + p_m^{k_1}(1-p_m)^{l-k_1}\} \qquad (2.33)$$

where h_1 and k_1 are the number of ones (1) in the strings h and k, respectively, produced by crossover between the strings i and j already present in the population. What values can h_1 and k_1 assume? Once this question is answered the quantity P_2 of Equation 2.33 becomes completely defined and together with P_1 of Equation 2.32 leads to an analytical expression for $R_{ij}(O)$.

Let i' be the substring of i containing $l - c$ bits lying to the left of the crossover site. i and i'' be the substring containing the c –bits lying to the right of c. Let j' and j'' be the counterparts of i' and i'' on the string j. It is easy to establish that

$$h_1 = i_1 - i_1'' + j_1''$$

$$k_1 = j_1 - j_1'' + i_1''$$

where i'' is the number of ones (1) in the substring i'' and j'' represents the same quantity for the substring j'. We note here that h_1 or k_1 will obviously depend on the location of the site c for crossover and introduce the entity $\Delta_{i,j,c}$ as

$$\Delta_{i,j,c} = i_1'' - j_1''.$$

With $\Delta_{i,j,c}$ thus defined we are now able to represent h_1 and k_1 for a specific choice of c as h_1^c or k_1^c with

$$h_1^c = i_1 - \Delta_{i,j,c}$$

$$k_1^c = j_1 + \Delta_{i,j,c}$$

The expression of P_2 is now completely defined:

$$P_2 = \frac{1}{2}\frac{p_c}{l-1}\sum_{c=1}^{l-1}\{p_m^{h_1^c}(1-p_m)^{l-h_1^c} + p_m^{k_1^c}(1-p_m)^{l-k_1^c}\} \qquad (2.34)$$

Combining P_1 and P_2 and doing some simple algebra they arrived at an explicit expression for $R_{ij}(O)$ as revealed below:

$$R_{ij}(O) =$$

$$\frac{(1-p_m)^l}{2}\left[\left(\frac{p_m}{1-p_m}\right)^{i_1}\left\{1-p_c+\frac{p_c}{l-1}\sum_{c=1}^{l-1}\left(\frac{p_m}{1-p_m}\right)^{-\Delta_{i,j,c}}\right\}+\left(\frac{p_m}{1-p_m}\right)^{j_1}\left\{1-p_c+\frac{p_c}{l-1}\sum_{c=1}^{l-1}\left(\frac{p_m}{1-p_m}\right)^{+\Delta_{i,j,c}}\right\}\right]$$

$$(2.35)$$

The general recombination matrix M was then expressed in terms of $M'(=R(O))$ following sophisticated permutational arguments and clever use of logical operators (see Vose and Liepinsl [19] for details). Without sketching the details, we may come straight to final results of the analysis, which established that there is an operator $G(\vec{X})$ which in the limit of infinite population satisfies

$$G\left(\vec{S}(t)\right) \sim \vec{S}(t+1) \tag{2.36}$$

where for vectors $\vec{x}$

$$G(\vec{x}) = F \odot M(\vec{X}) \tag{2.37}$$

with $\odot$ denoting composition operator. Similarly, they also demonstrated that

$$G_p\left(\vec{p}(t)\right) \sim \vec{p}(t+1)$$

$$\text{with } G_p\left(\vec{X}\right) = M \odot \left(\frac{F(\vec{X})}{|F(\vec{X})|}\right) \tag{2.38}$$

Vose and Liepins then went on to find the fixed points of F and M separately and analyzed the dynamics of GA evolution. The main results assuming infinite population size can be summarized as follows:

1. When selection operator (F) alone acts on the population, only stable fixed points correspond to the population with maximally fit strings. A slight change in the fitness distribution only temporarily disturbs the dynamics which returns to the same point after some time (generations).
2. When M alone is acting, there is only one fixed point on S defined by the selection vector $\vec{s}$ which has equal probability for all strings in the search space. The corresponding population vector $\vec{p}$ defines a fixed point where all strings are present in equal proportion.

The important question concerning the possible effects of interaction between the F and M operators on the GA dynamics was largely left unexplored although some conjectures were made by the authors. The finite population situation was probed by Nix and Vose [22]. They succeeded in drawing some more general conclusions. For example,

1. As $n \to \infty$, the probability that the Markov-chain trajectories converge to iterates of G (or G_p) may grow arbitrarily close to 1.

2. If G_p has a single fixed point, the time spent by the GA at the fixed point becomes asymptotically large as $n \to \infty$.

3. If more than one fixed point exists, the time spent by the GA in regions away from the fixed points asymptotically approaches 0 as $n \to \infty$.

Both the finite and infinite population models were further extended by [20] who demonstrated that the short-time behavior of GA is determined by the initial population while the long-time behavior is shaped by the structure of what the authors call the GA surface on which population trajectories are traced. The GA in the limit of infinite population homes in on the fixed points having the largest basins of attraction. The work is noteworthy because of the dynamical view of the evolution process that it offers with special emphasis on the dynamical stability of the evolving system in GA.

2.5.2 GA: A Statistical Mechanical View

The schema theorem, as we have already described, analyzes the GA behavior in terms of the dynamics of growth and decay of smaller subunits of individuals or strings called schemas. This leads to the emergence of an important and a very popular idea of building blocks and the building block hypothesis. The Vose and Liepins theory, on the other hand, examines the dynamics of evolution of the population in a GA under the action of genetic operators on strings in order to understand how the GA moves the population from one point on the relevant surface to another, and ascertains when the GA dynamics remains stable. The analysis was based on individuals or the whole strings (not subunits). A third interesting view to the macroscopic GA behavior has been through the window of statistical mechanics of interacting individuals or strings. Central to the theme is the idea of defining a macroscopic property and following its evolution driven by GA operators, with the help of tools and concepts borrowed from statistical mechanics. Prügel-Bennet and Shapiro [23] chose a one-dimensional chain of N Ising spins $\{S_i = \pm 1\}$ with random interaction (J_i) between pairs of the nearest neighbors $(i, i + 1)$. The total energy of the system $(\vec{S})$ of $N + 1$ spins is

$$E(\vec{S}) = -\sum_{i=1}^{N} J_i S_i S_{i+1} \tag{2.39}$$

where the spin vector $\vec{S} = (s_1, s_2, ..., s_i, ..., s_N, s_{N+1})$ with the components s_i capable of taking only two values, +1 or −1. The coupling constants J_is are assigned fixed values of +1 or −1 randomly. The problem is to find the spin orientations (values of s_is) in the ground state, i.e., to discover which set of spin orientations or spin values will globally minimize the energy of the spin system $(\vec{S})$.

GA is eminently suitable for the problem, and choices of several features almost suggest themselves. Thus an obvious representation for the system may be in the form of integer strings: $\vec{S}_k = (s_1^k, s_2^k, ..., s_i^k, ..., s_{N+1}^k) = (1, 1, -1, ..., 1, ..., -1, 1)$ is one possible choice. The $N + 1$ components $(s_i^k)_{i=1,2,...,N+1}$ can be chosen randomly. A collection of n such randomly generated strings constitutes the initial population. The fitness of the string S^k is just the negative of its energy, i.e., $f_k = - E(S^k)$. The single-point crossover with a preset probability p_c can operate in just the same manner as it does in the case of binary bit strings (the alphabet here is +1 and −1, instead of 0 and 1) and mutation involves flipping a component of $\overline{S^k}$ chosen randomly with the probability p_m, to 1 if it was −1 and to −1 if it was 1. For selection, Prügel-Bennet and Shapiro [23] introduced what may be called a Boltzmann selection in place of the traditional roulette-wheel-based procedure already described in detail. Thus the probability p_k of selecting the kth individual of the population into the "breeding pool" was estimated by evaluating p_k where

$$p_k = \frac{e^{-\beta E_k}}{\sum_{k=1}^{n} e^{-\beta E_k}} \tag{2.40}$$

A controlled selection pressure could be applied to the population by varying β which plays the role of the system temperature (high value of $\beta \to$ low temperature higher $\to$ selection pressure, for example). Having defined the system, the population size (n), the goal-specific fitness (f_k), selection probability (for a given β), and crossover and mutation probabilities, the task was to theoretically predict how the fitness or the energy distribution $P(E)$ in the population evolved as GA worked on it through many generations and assess how the predictions compared with what was observed in actual computer runs. In the GA runs, mutation was not used—the only GA operators used were selection and crossover. The evolving distribution in different generations averaged over many independent distribution runs was symmetric about $E = 0$ to start with, became skewed, and approached a δ distribution with peaks centered at progressively lower values of energy.

The question is: could this behavior be predicted using statistical mechanics as a tool? Prügel-Bennet and Shapiro [23] invoked the random energy model of Derida [24] and derived analytical expressions for various cumulants ($c_1, c_2, ..., c_N$) of the evolving energy distribution in the full GA. They also derived expressions for the cumulants when either the selection operator or the crossover operator alone worked on the population. The most fascinating aspect of the work is that the behavior of cumulants observed in actual GA runs on the spin system and those predicted by the theoretical model turned out to be remarkably similar, lending credence to the huge possibility of a statistical mechanical approach in deciphering the detailed macrobehavior of GAs and its response to changes in important GA parameters like population size (n), selection pressure (β), or crossover probabilities (p_m), and eventually designing an optimal GA for the problem. These possibilities have hitherto remained unexplored in the more general context such as in the case of two- or three-dimensional Ising models, spin glasses, or Coulomb glasses where GA can certainly be profitably used to discover the ground state. The work nevertheless remains important in opening up a new direction of research in GAs.

2.6 Variants of GA

What we have described and analyzed so far are the simplest realizations of GA in terms of representations, and construction of GA operators. The potential solutions have been represented as binary bit strings (most generally) or integer bit strings (in the special case of Ising spin systems). In either case, the crossover and mutation operators have been designed similarly. The selection has been based on goal-specific fitness realized through the roulette wheel method or by Boltzmann selection. However, that is not all. For many problems, it is convenient to work with floating point strings—that means representing the potential solution in the form of arrays containing floating point numbers as genes. For example, the potential solutions of the linear system of Equation 2.1 which are vectors $(x_1, x_2, x_3)^t$ can be represented as arrays of decimal numbers, for example:

$$S1(1) = x_1 \,;\; S1(2) = x_2 \,;\; S1(3) = x_3$$

The fitness calculation can now be carried out by forming the error vector $(AX - b)$ directly using the array. Unlike with binary strings, there is no need first to convert the binary strings into decimal numbers and then evaluate fitness. However, floating point strings require specially designed crossover and mutation operators for generating new strings. In this section we will review many possibilities of implementing crossover and mutation while working with floating point representation of potential solutions. In addition, we will also describe a host of methods of enforcing selection on a population (independent of representation) along with reflections on their merits and demerits.

2.6.1 Variants of Crossover Operators

These operators are binary and designed for sharing information between a randomly chosen pair of individuals (parents) and swapping information between them or mixing up the genetic materials encoded by the individual strings, producing new individuals—the offspring. Let the parents P_j and P_k be represented as strings (arrays) of cardinality d:

$$P_j \equiv (p_1^j, p_2^j, ..., p_d^j)$$

$$P_k \equiv (p_1^k, p_2^k, ..., p_d^k)$$

where p_i^j and p_i^k represent the ith gene of the parents P_j and P_k. Let the offspring produced be denoted by similar strings O_j and O_k of the same cardinality with genes denoted by $\{O_i^j\}$ or $\{O_i^k\}$

$$O_j \equiv (o_1^j, o_2^j, ..., o_d^j)$$

$$O_k \equiv (o_1^k, o_2^k, ..., o_d^k)$$

2.6.1.1 Simple Crossover

A single crossover point i is chosen randomly with $i \in \{1, 2, ..., d - 1\}$. The interacting chromosomes, individuals, or strings are cut at the site i and the sections after the cut are swapped, creating two new individuals O_j and O_k in which the parental genes have been swapped (exchanged).

$$O_j \equiv (p_1^j, p_2^j, ..., p_i^j, \; p_{i+1}^k, ..., p_{d-1}^k, p_d^k)$$

$$O_k \equiv (p_1^k, p_2^k, ..., p_i^k, \; p_{i+1}^j, ..., p_{d-1}^j, p_d^j)$$

It is possible to design N-point simple crossovers by generalizing the definition to N-different segments. The crossover operation just defined works with floating point strings as well as with binary strings.

2.6.1.2 Uniform Crossover

Descendants are created by swapping the genes [25] of the two parent chromosomes P_j and P_k with a probability of 0.5. Thus

$$o_i^j = \begin{cases} p_i^j, \text{ if } r_i > 0.5 \\ p_i^k, \text{ otherwise} \end{cases} \tag{2.41}$$

where r_i is a uniform random number drawn from the range [0,1]. The genetic compositions of the two descendants O_j and O_k may be, for example,

$$O_j \equiv (o_1^j, o_2^j, \dots, o_d^j) \equiv (p_1^j, \boxed{p_2^k}, \boxed{p_3^k}, \dots, p_i^j, p_{i+1}^k, \dots, p_{d-1}^j, \boxed{p_d^k})$$

$$\tag{2.42}$$

$$O_k \equiv (o_1^k, o_2^k, \dots, o_d^k) \equiv (p_1^k, \boxed{p_2^j}, \boxed{p_3^j}, \dots, p_i^k, p_{i+1}^j, \dots, p_{d-1}^k, \boxed{p_d^j})$$

2.6.1.3 Heuristic Crossover

A single offspring is generated by linear extrapolation [26] of two individuals (parents), P_j and P_k. The choice of parents must satisfy the condition that $f(P_j) \geq f(P_k)$ where $f(P)$ denotes the fitness of the individual P. The elements of the offspring (vector) are given by

$$\{O_i\} = \left\{ p_i^j + r\,(p_i^j - p_i^k) \right\} \tag{2.43}$$

where r is a uniformly distributed random number lying in the interval [0,1]. In case the extrapolation produces an infeasible offspring vector, it is discarded, and the new vector is generated with a different choice of r.

2.6.1.4 Linear Crossover

Three offspring are generated out of the recombination of two parents P_j and P_k, one of which is in the exploitation zone and second in the exploration zone [26]. Two of the most promising in terms of goal-specific fitness are chosen as offspring. The three offspring are

$$O_1 = 1/2\,P_j + 1/2\,P_k$$

$$O_2 = 3/2\,P_j - 1/2\,P_k$$

$$O_3 = 3/2\,P_k - 1/2\,P_j$$

2.6.1.5 Arithmetic Crossover

Complementary linear combinations of the parents are generated [8] on the basis of arithmetic mean:

$$O_1 = rP_j + (1 - r) P_k$$

$$O_2 = rP_k + (1 - r) P_j$$

r being a random number drawn from a normal distribution in the range [0,1]. If m is the chosen crossover site, only components lying to the right of the crossover site are combined.

2.6.1.6 Simplex Crossover

A simplex is constructed with randomly selected N-parents $\{P_i\}$ from the population. Centroid C of the simplex is calculated by evaluating

$$C = \sum_{i=1}^{N} P_i / N$$

The simplex is then expanded by a small degree ϵ, with the expanded simplex points given by

$$O_j = C + \epsilon \,(P_j - C), \; j \,\epsilon \; \{1, 2, \,...,\, N\}$$

One or two individuals are then sampled from the expanded simplex to become offspring [27].

2.6.1.7 BLX-α Crossover

To expand the range of arithmetic crossover Eshelman and Schaffer [28] suggested the method of "blend crossover," which generates a single offspring by blending two floating point parent vectors P_j and P_k with elements $\{p_i^j\}$ and $\{p_i^k\}$. The elements of the offspring vector O_1 are given by

$$o_i^j = R\left(\left(L_i + \propto.I\right),\left(U_i + \propto.I\right)\right) \; i = 1, 2, \; ..., \; N$$

where, $U_i = \max(p_i^j, p_i^k)$, $L_i = \max(p_i^j, p_i^k)$, $I = U_i - L_i$

and $R(a, b)$ is a function generating a random number uniformly distributed in the range (a, b). The parameter $\propto$ is user defined [29,30], the most commonly used value being 0.5.

2.6.1.8 Similarity Crossover

A variant of the arithmetic crossover scheme obtained by replacing the random number r used in arithmetic crossover by a similarity measure [31] ($S = e^{-EN}$) where

$$EN = \frac{1}{\sqrt{d}}.\{(p_1^l - p_1^m)^2 + (p_2^l - p_2^m)^2 + ... + (p_d^l - p_d^m)^2\}$$

P_k and P_m are the parent vectors undergoing crossover while $\left\{ p_i^k \right\}$ and $\left\{ p_i^m \right\}$ are the d real-valued components of the parent vectors.

2.6.1.9 Cell Crossover

A cell is a set of genes. In cell crossover [32], cells of two randomly chosen parent individuals are swapped with a preset probability, producing two offspring as shown in the example below.

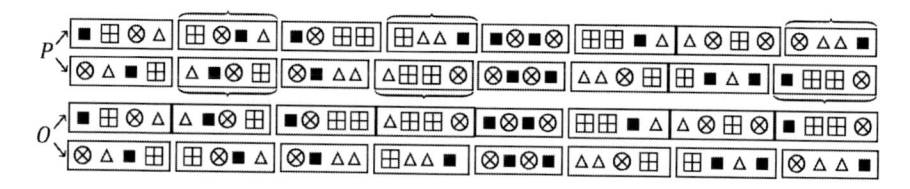

GAs have been extensively and most successfully exploited in the calculation of energy and prediction of structures of atomic/molecular clusters. For N-atom clusters the individuals (chromosomes) in the populations are 3N-component vectors, the components being the Cartesian coordinates of the constituent atoms. Crossover operation in such clusters has often been specifically designed to meet the requirements of exploitation. We present here two crossover operators that have been widely used in cluster physics, namely average offspring method, and cut and splice crossover.

2.6.1.10 Average Offspring Method

The average offspring method (AOM) [33] was designed for prediction of structures of cluster by GA and showed promise in systems having rather compact global minimum energy structures. The operative steps are the following.

1. Two individual parent vectors representing two alternative cluster geometries are randomly selected.
2. For each atom of the first individual (a cluster geometry), the closest lying atom of the second individual (another cluster geometry) is identified.
3. An offspring (a new cluster geometry) is generated by placing the atoms randomly on the line connecting the corresponding pairs of atoms of the parent clusters. The random placement ensures that not too many similar offspring are generated.

2.6.1.11 Cut and Splice Crossover

This operator is an extension of the simple crossover method already described and has been designed for cluster geometry optimization problems. It works extremely well and involves the following steps:

1. Choose two parent geometries randomly.
2. The first parent is cut randomly into two complementary halves by a randomly chosen horizontal plane.

3. The second parent is also cut similarly with the rider that the number of the atoms above and beneath the cutting plane match the number of atoms in the two complementary halves of the first parent.

4. 'Splice' then joins the 'head' of one string to the 'tail' of the other.

A modified form of cut and splice crossover was suggested by Pereira et al. [34]. The modification comes in the form of an alternative strategy of determining which subclusters would be exchanged. In the modified cut and splice method, an atom of the first parent (N-atom cluster geometry) is randomly selected to act as an atom of the offspring. A random integer ($M \in [1, N - 1]$) is generated and M atoms of the first parent (cluster) closest to the selected first atom are added to the offspring cluster. The remaining ($N - M - 1$) are taken from the second parent (cluster geometry) on the basis of their closeness to the first selected atoms (skipping atoms which have already been accommodated in the offspring).

Many other forms of crossover operators have been experimented with. Notable among them are fuzzy recombination [35], SBX [36], PCX [37], XLM [38], Laplace crossover [39], differential evolution [40], partition [41,42], linear BGA [43], UNDX [44], fuzzy connectives-based [45], direction-based [46], multiple crossover [47], ring crossover [48]. Several hybrid crossover methods have also appeared.

2.6.2 Variants of Mutation Operators

The mutation operator for binary bit strings (chromosomes) or integer arrays have already been defined. The definition needs to be extended to handle more general cases, for example, when floating point strings or N-component vectors with real components are used to represent the potential solutions. The design of a general mutation operator takes into account the basic fact that mutation is a rare event and when it happens it perturbs one or more components (genes) of the selected individual with either a preset or a dynamically determined mutation probability p_m. The event raises the entropy and therefore reduces the mutual information in the population. Mutation plays an extremely important role—it can restore genetic material lost in the selection process and bring in fresh genetic information in the population thereby helping the GA distribute potential solutions widely across the search space so that exploration continues unhindered by premature convergence. A schematic view of biological mutation and a "structural mutation" are displayed in Figure 2.6a and b. Such structural mutations can play an important role in discovering lowest energy structures in a GA-based structural optimization protocol.

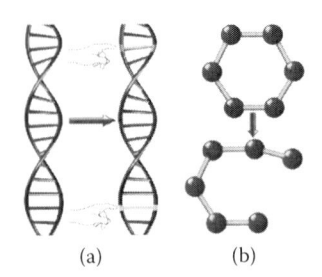

(a)   (b)

FIGURE 2.6
(a, b) Schematic representation of biological mutation and structural mutation.

Let us now consider different forms of mutation operator that researchers have experimented with. In what follows we will use a uniform notational convention for representing an individual or chromosome before and after mutation as described below: $\{x_i\}$ and $\{y_j\}$ are genes

$$X = (x_1, x_2, ..., x_i, ..., x_{d-1}, x_d)$$

(before mutation)

$$\Downarrow$$

$$Y = (y_1, y_2, ..., y_i, ..., y_{d-1}, y_d)$$

(after mutation)

2.6.2.1 Uniform Mutation

Genes in an individual $\{x_i\}$ are selected with a predefined probability p_m and replaced with a random value $\{y_j\}$ uniformly distributed in a given range [8] according to the following scheme:

$$y_i = \begin{cases} R(L_i, U_i), & \text{if } p_m > r_i, \ r_i \in [0,1] \\ x_i, & \text{otherwise} \end{cases}$$

$R\,(a, b)$ is a function generating a uniform random number in the range (a, b). U_i and L_i are the upper and lower bounds to the gene value at the ith location. Uniform mutation tends to forestall premature convergence. However, if the search has already moved into the neighborhood of the optimal region, the process of fine-tuning the evolving solution to the optimum may become disrupted as uniform mutation induces random oscillations which must be damped.

2.6.2.2 Nonuniform Mutation

This is essentially a dynamic form of mutation [8]. The mutation intensity varies from generation to generation. It is one of the most commonly used mutation schemes in real coded GAs. As in uniform mutation a gene x_i in an individual (X) is randomly selected with a predetermined mutation probability p_m and mutated to y_i where

$$y_i = \begin{cases} x_i + \Delta(t, U_i - x_i), & \text{if } r > 0.5 \\ x_i - \Delta(t, x_i - L_i), & \text{otherwise} \end{cases}$$

$$\Delta(t, z) = z \left(1 - r^{(1-\frac{t}{t_{max}})^b} \right)$$

where $r \in [0,1]$ is a random number, and t is the generation number; b is a user-defined number that controls nonuniformity.

2.6.2.3 Gaussian Type Mutation

A noise with zero mean and a standard deviation that is 0.1 times the maximum value of the gene being mutated is added to the gene value. A new (mutated) individual is obtained by adding a random value to each element of the selected parent (a vector with real components). There are a number of possibilities of which the Gaussian and Cauchy mutations [8,9,49] have found extensive use in evolutionary programming and evolutionary strategy-based problem solving. The mutation functions that transform an element of d-dimensional vector X into a new vector Y are

$$
G = \begin{cases}
\displaystyle\prod_{i-1}^{d} \frac{1}{\sqrt{2\pi}\sigma_i} e^{\left(-\frac{(\bar{y}_i-\bar{x}_i)^2}{2\sigma_i^2}\right)} & \to \text{Gaussian} \\[2em]
\displaystyle\pi^{-d}\prod_{i-1}^{d} \frac{\tau}{\tau^2+(\bar{y}_i-\bar{x}_i)^2} & \to \text{Cauchy}
\end{cases}
$$

The mutated components in the two cases are given

$$
y_i = \begin{cases}
x_i + N(0,\sigma_i), & \to \text{Gaussian} \\
x_i + C(0,\tau_i), & \to \text{Cauchy}
\end{cases}
$$

Many other forms have also been explored. Lévy-type mutation [50,51], real number creep [52] are just two of the popular variants of Gaussian mutation. The real number creep operates on a d-dimensional vector as follows:

$$
y_i = \begin{cases}
x_i + N\left(0,\sigma_i^2\right), & \text{if } i=k,\ k=\text{Uniform}\left[1,d\right] \\
x_i, & \text{otherwise}
\end{cases}
$$

σ_i is set as $\dfrac{U_i - L_i}{1000}$

2.6.2.4 Dynamic Random Mutation

The mutation step size or mutation intensity is dynamically controlled [53] in every generation t. The component-wise change in the parent vector leading to the components of the mutated vector is produced as follows:

$$
y_i = x_i + s.\phi_{0i}(UB_i - LB_i)
$$

where ϕ_0 is a random perturbation vector with elements in the range (0,1). The dynamic adjustment of the step size s is induced through the formula

$$
s = (1 - t/t_{max})^b,\ b > 0
$$

Changing the value of b can control the decay rate of step size s. The idea is borrowed from metallurgy where a similar strategy is adopted in constructing an annealing schedule.

2.6.2.5 Arithmetic Mutation

This is possibly the simplest of all the mutation operators acting on vectors with real-valued components. The operation is defined as follows: Choose one component (x_i) of a parent vector (X) randomly with a preset probability p_m. The corresponding component of mutated vector Y is given by

$$y_i = \begin{cases} x_i + (-1)^L \Delta s \ r, & \text{if } r_i < p_m \\ x_i, & \text{otherwise} \end{cases}$$

Δ_s is the mutation step size or intensity (static), L is a randomly generated integer, and r is a random number in the range $(0,1)$ drawn from a normal distribution.

This has been used extensively in quantum mechanical wave function as well as nuclear geometry optimization with success.

2.6.2.6 Directed Random Mutation

Closely related to the arithmetic mutation scheme, this adjusts mutation step size Δ on the basis of an analysis of the degree of acceptability of mutation (i.e., the percentage of fitness-enhancing mutations occurring during the n number of past generations). The value of a randomly chosen gene in a parent is altered according to the following scheme [54,55]:

$$y_i = \begin{cases} x_i \pm \Delta_m \ r, & \text{if } p_m < r(0,1) \\ x_i, & \text{otherwise} \end{cases}$$

The + or − sign is chosen with equal probability. The Δ_m value is dynamically adjusted following the schedule given below

$$\Delta_m = \begin{cases} \Delta_m/(1+r_g), & \text{if } N_{accept} < N_{lower} \\ \Delta_m \cdot (1+r_g), & \text{if } N_{accept} > N_{lower} \end{cases}$$

N_{accept} is the number of fitness-enhancing mutations during the last n-generation. N_{upper} and N_{lower} are user-defined upper and lower limits of N_{accept}.

2.6.2.7 Wavelet Mutation

A continuous function $\psi(w)$ is called a "mother wavelet" if it has the following properties:

$$\int_{-\infty}^{+\infty} \varphi(\omega) \, d\omega = 0$$

$$\int_{-\infty}^{+\infty} |\varphi(\omega)|^2 \, d\omega < \infty$$

A Morlet wavelet defined below is an example of a mother wavelet:

$$\varphi(\omega) = e^{-\frac{\omega^2}{2}} \cos(5\omega)$$

A more flexible mother wavelet $\psi_{a,b}(w)$ can be produced by introducing a shift (or translation) parameter b and dilation parameter a where

$$\varphi_{a,b}(\omega) = \frac{1}{\sqrt{a}} \varphi\left(\frac{\omega - b}{a}\right)$$

Ling et al. [56] proposed wavelet mutation in which each gene (a component) in a chromosome (a real-valued vector) is picked up to undergo mutation as usual with a predefined mutation probability $p_m \in [0, 1]$, and the mutated gene value is calculated as follows:

$$y_i = \begin{cases} x_i + \varphi_{a,0}(U - x_i) \text{ if } \varphi_{a,0}(\omega) > 0 \\ x_i + \varphi_{a,0}(x_i - L) \text{ if } \varphi_{a,0}(\omega) \le 0 \end{cases}$$

The value of w is randomly generated in the range $w \in [-2.5, 2.5]$.

2.6.2.8 BGA Mutation

The mutated gene (y_i) is generated [30] from the parent gene (x_i) of a d-component vector X by taking

$$y_i = x_i \pm \text{range}_i \cdot \delta$$

The + or − sign is chosen with a probability of 0.5, range_i is normally set to 0.1, and δ is computed as follows:

$$\delta = \sum_{i=1}^{d} A_i 2^{-i}, \; A_i \; \varepsilon \; 0,1$$

A_i values are randomly generated and take on the value equal to one with a probability $p = 1/d$. Other means of calculating the crucial parameter δ have been proposed leading to more general forms of BGA mutation in continuous as well as discrete spaces.

2.6.2.9 Soft Mutation

Specially designed for GA-driven crystal structure prediction or calculation of the structure of clusters or soft matter, soft mutation [57,58] discards complete randomness in choosing possible mutation sites. It replaces it by a directed event in which concerted mutation takes place at several sites simultaneously, chosen on the basis of fitness-improving potential. It is achieved by moving the atoms along the low-frequency (soft) eigenvalues of the system (crystal or cluster) that correspond to a low curvature of the energy surface. Since eigenvectors corresponding to the smallest nonzero eigenvalues (softest mode) determine the direction of mutation, it has been called soft mutation. The determination of the soft modes, however, requires calculation of the eigenvalues of the dynamic matrix (force constant) which can be costly, if accurate ab initio calculations are targeted. Instead, a much cheaper route involving the construction and diagonalization bond-stiffness matrix serves the purpose in most cases.

2.6.3 Variants of Selection Operators

We have already described the roulette wheel method in detail. However, many other forms of selection have been proposed and tested in the GA literature [2,10,59]. The central idea has been to design a fitness sensitive operator that acts on the entire population and ensures that the more promising individuals (potential solutions with higher fitness) enjoy a higher probability of being selected. To what extent it is done determines the selection pressure. A high selection pressure could lead to faster but premature convergence to a suboptimal solution. A low selection pressure will inevitably increase the search time, but reduce the chance of premature convergence if the variability (diversity) in the population is not compromised. The selection operator has two components—one, the calculation of selection probability, and two, a sampling mechanism. Together, they provide the basic biasing mechanism. A few well-tried selection mechanisms are described next.

2.6.3.1 Generational Replacement

The entire set of parents is replaced by their descendants or the *n*-worst parents (in terms of goal-specific fitness) are substituted for *n*-best offspring.

2.6.3.2 Truncation Replacement

A fraction (*p*) of the population is selected on the basis of fitness values of individuals and are copied *p*-times so that the population size remains constant. The less-fit individuals are denied the opportunity to evolve and are ruthlessly discarded.

2.6.3.3 Stochastic Universal Sampling

The population (size N) is mapped into a pie chart on a wheel, in which zones are allocated for individuals in the population with zone areas proportional to the fitness of the representative individuals, exactly as in the roulette wheel selection. Only one random spin is given to the wheel. There are N equally spaced pointers with arrowheads pointing to different segments of the chart. The position of the first pointer is chosen randomly in the interval $(0, 1/N)$. Individuals are selected by generating N pointers, starting with the randomly generated first one, spaced by $1/N$ by checking out the individuals whose fitness spans the positions of the N-pointers. Stochastic universal sampling is preferred to the roulette wheel selection as it tends to preserve diversity better, provides low spreading over the desired distribution of individuals and prevents the "best individual" from dominating the population. But the major defect of any fitness-proportional selection is not completely eliminated.

2.6.3.4 Rank-Based Selection

Rank-based selection offers a means of preventing too-quick convergence. N = individuals in the population are first ranked on the basis of their goal-specific fitness values in ascending order, assigning rank = 1 to the least fit individual and rank = N to the fittest individual. The selection probability of an individual depends on its rank and not on the absolute value of its fitness. The function that maps ranking indices of individuals onto their selectibilities has a crucial role to play.

For linear rank-based selection, the selection bias is controlled by introducing a selection pressure (SP) such that $1 \leq SP \leq 2$. The expected sampling rate for the best individual is SP, while that of the worst individual is $(2 - SP)$. The expected rate for other individuals (i) in the population can be obtained by a linear interpolation

$$F(i) = (2 - SP) + 2(SP - 1)\frac{r_i - 1}{N - 1}$$

where r_i is the rank of the ith individual in the population of size N.

Rank-based selection schemes are computationally costly as the population has to be sorted in every generation, but has the benefit of increased preservation of diversity leading to better exploration and avoiding premature convergence due to the preservation of the super individual (the best individual is always assigned the same rank no matter what its fitness value is). The convergence can, however, be slow.

2.6.3.5 *Tournament Selection*

A tournament is held among n-competitors selected randomly from the population of N individuals, and the winner of the tournament is selected on the basis of total score [59]. The most commonly used tournament size is $n = 2$, but the SP can be increased by increasing the size of the tournament. The implementation is simple: two individuals i (fitness f_i) and "j" (fitness f_j) are randomly chosen from the current population and a uniformly distributed random number r is drawn from the interval (0,1). Let c be a preset parameter with $c = 0.75$ (say). If $r < c$, and $f_i > f_j$, individual i makes it to the new population; if $f_i < f_j$, then the individual j is chosen. If $r \not< c$, the less-fit individual is selected. Both are then returned to the current population and may be selected again. Binary tournament selection and linear rank-based selection are expected to perform similarly but the former is preferred for its ease of implementation, superior time complexity, and easy parallelizability.

2.6.3.6 *Clonal Selection*

A fraction of the more fit individuals are selected for cloning [60]. Each of these individuals receives a number of copies (clones) proportional to the individual's position in the ranking. The clones are allowed to undergo a process of maturation. A given individual together with its matured clones forms a subpopulation. The "best" of each subpopulation is passed onto the next generation.

2.6.3.7 *Sexual Selection*

The sex (F or M) of an individual in the current population is assigned either randomly or on the basis of some well-defined attribute. Chromosomal crossover is allowed only between F and M [61,62]. The females are given the chance to reproduce only once irrespective of their fitness values, facilitating exploration of the search space. The male selection is fitness-biased, but the feminine selection allows a female to choose one of the males based on a problem-specific attraction function. Sexual selection works in a way different from that of natural selection. Natural selection emphasizes traits that increase survival probability of the organism while sexual selection stresses traits that help the organism attract potential mates. They do not, however, act at cross-purposes—in the end, both increase the probability that an organism will reproduce and transmit the genes for the preferred traits to the offspring. Many mathematical models have been developed to understand the nature and consequences of sexual selection in the presence of natural selection, but this model of selection has found little use so far in solving problems of physics and chemistry.

References

1. Darwin, C. 1859. *On the Origin of the Species by Means of Natural Selection: Or, the Preservation of Favoured Races in the Struggle for Life*. New York: D. Appleton and Company.
2. Holland, J.H. 1975. *Adaptation in Natural and Artificial Systems: An Introductory Analysis with Applications to Biology, Control, and Artificial Intelligence*. Ann Arbor, MI: U Michigan Press.
3. Schrödinger, E. 1992. *What Is Life? With Mind and Matter and Autobiographical Sketches*. Cambridge Paperback Library. Cambridge: Cambridge University Press.
4. Munakata, T. 2008. *Fundamentals of the New Artificial Intelligence*. Texts in Computer Science. London: Springer-Verlag.
5. Mitchell, M. 1998. *An Introduction to Genetic Algorithms*. Cambridge, MA: MIT Press.
6. Holland, J. 1966. Universal spaces: A basis for studies of adaptation. In *Automata Theory*, ed. E. Caianiello, pp. 218–231. New York: Academic Press.
7. Goldberg, D.E. 1989. *Genetic Algorithms in Search, Optimization, and Machine Learning*. Boston, MA: Addison-Wesley.
8. Goldberg, D.E. 2006. *Genetic Algorithms*. Pearson Education.
9. Michalewicz, Z. 1996. *Genetic Algorithms + Data Structures = Evolution Programs* (3rd ed.) New York: Springer-Verlag.
10. Baeck, T., D.B. Fogel, and Z. Michalewicz. 1997. *Handbook of Evolutionary Computation*. London: Institute of Physics Publishing.
11. Spector, L. 2004. *Automatic Quantum Computer Programming*. Vol. 7. Genetic Programming. Boston, MA: Kluwer Academic.
12. Bridges, C.L., and D.E. Goldberg. 1987. An Analysis of Reproduction and Crossover in a Binary-Coded Genetic Algorithm. In *Proceedings of the Second International Conference on Genetic Algorithms on Genetic Algorithms and Their Application*, pp. 9–13. Hillsdale, NJ: L. Erlbaum Associates.
13. Holland, J.H. 1995. *Hidden Order: How Adaptation Builds Complexity*. Redwood City, CA: Addison Wesley Longman.
14. Stephens, C., and H. Waelbroeck. 1999. Schemata Evolution and Building Blocks. *Evolutionary Computation* 7, no. 2: 109–124.
15. Harik, G.R., F.G. Lobo, and K. Sastry. 2006. Linkage Learning via Probabilistic Modeling in the Extended Compact Genetic Algorithm (ECGA). In *Scalable Optimization via Probabilistic Modeling*, eds. M. Pelikan, K. Sastry, and E. CantúPaz, pp. 39–61. Berlin: Springer.
16. Pelikan, M., D.E. Goldberg, and E. Cantú-Paz. 1999. BOA: The Bayesian Optimization Algorithm. In *Proceedings of the 1st Annual Conference on Genetic and Evolutionary Computation— Volume 1*, pp. 525–532. GECCO'99. San Francisco, CA: Morgan Kaufmann.
17. Echegoyen, C., A. Mendiburu, R. Santana, and J.A. Lozano. 2013. On the Taxonomy of Optimization Problems Under Estimation of Distribution Algorithms. *Evolutionary Computation* 21, no. 3: 471–495.
18. Martins, J.P., and A.C.B. Delbem. 2016. Pairwise Independence and Its Impact on Estimation of Distribution Algorithms. *Swarm and Evolutionary Computation*, 27: 80–96.

19. Vose, M.D., and G.E. Liepinsl. 1991. Punctuated Equilibria in Genetic Search. *Complex Systems* 5: 31–44.

20. Vose, M.D. 1995. Modeling Simple Genetic Algorithms. *Evolutionary Computation* 3, no. 4: 453–472.

21. Vose, M.D.1991. Generalizing the Notion of Schema in Genetic Algorithms. *Artificial Intelligence* 50, no. 3: 385–396.

22. Nix, A.E., and M.D. Vose. 1992. Modeling Genetic Algorithms with Markov Chains. *Annals of Mathematics and Artificial Intelligence* 5, no. 1: 79–88.

23. Prügel-Bennett, A., and J.L. Shapiro. 1994. Analysis of Genetic Algorithms Using Statistical Mechanics. *Physical Review Letters* 72, no.9:1305–1309.

24. Derrida, B. 1981. Random-energy model: An exactly solvable model of disordered systems. *Physical Review B* 24, no. 5: 2613–2626.

25. Syswerda, G. 1989. Uniform Crossover in Genetic Algorithms. In *Proceedings of the 3rd International Conference on Genetic Algorithms*, pp. 2–9. San Francisco, CA: Morgan Kaufmann.

26. Wright, A.H. 1991. Genetic Algorithms for Real Parameter Optimization. In *Foundations of Genetic Algorithms*, ed. G.J.E. Rawlins, pp. 205–218. San Mateo, CA: Morgan Kaufmann.

27. Tsutsui, S., M.Yamamura, and T. Higuchi. 1999. Multi-Parent Recombination with Simplex Crossover in Real Coded Genetic Algorithms. *Proceedings of the Genetic and Evolutionary Computation Conference* 1: 657–664.

28. Eshelman, L.J., and J.D. Schaffer. 1992. Real-Coded Genetic Algorithms and Interval-Schemata. In *FOGA*, ed. L.D. Whitley, pp. 187–202. San Mateo, CA: Morgan Kaufmann.

29. Radcliffe, N.J. 1991. Equivalence Class Analysis of Genetic Algorithms. *Complex Systems* 5, no. 2: 183–205.

30. Mühlenbein, H., and D. Schlierkamp-Voosen. 1993. Predictive Models for the Breeder Genetic Algorithm I. Continuous Parameter Optimization. *Evolutionary Computation* 1, no. 1: 25–49.

31. Guvenc, U. 2010. Combined Economic Emission Dispatch Solution Using Genetic Algorithm Based on Similarity Crossover. *Scientific Research and Essays* 5, no. 17: 2451–2456.

32. Bao, Z., and T. Watanabe. 2009. A Novel Genetic Algorithm with Cell Crossover for Circuit Design Optimization. In *IEEE International Symposium on Circuits and Systems*, pp. 2982–2985. Taipei, Taiwan: IEEE Conference.

33. Schönborn, S.E., S. Goedecker, S. Roy, and A.R. Oganov. 2009. The Performance of Minima Hopping and Evolutionary Algorithms for Cluster Structure Prediction. *The Journal of Chemical Physics* 130, no. 14: 144108.

34. Pereira, F.B., and J.M.C. Marques. 2011. Analysis of Crossover Operators for Cluster Geometry Optimization. In *Computational Intelligence for Engineering Systems*, eds. A. Madureira, J. Ferreira, and Z. Vale, Vol. 46, pp. 77–89. Intelligent Systems, Control and Automation: Science and Engineering. Dordrecht: Springer.

35. Voigt, H.-M., H. Mühlenbein, and D. Cvetkovic. 1995. Fuzzy Recombination for the Breeder Genetic Algorithm. In *Proceedings of the Sixth International Conference on Genetic Algorithms*, pp. 104–111. San Francisco, CA: Morgan Kaufmann Publishers.

36. Deb, K., and R.B. Agrawal. 1994. Simulated Binary Crossover for Continuous Search Space. *Complex Systems* 9, no. 3: 1–15.

37. Deb, K., A. Anand, and D. Joshi. 2002. A Computationally Efficient Evolutionary Algorithm for Real-Parameter Optimization. *Evolutionary Computation* 10, no. 4: 371–395.

38. Takahashi, O., and S. Kobayashi. 2001. An Adaptive Neighboring Search Using Crossover-like Mutation for Multi-Modal Function Optimization. *IEEE International Conference on Systems, Man, and Cybernetics* 1: 261–267.

39. Deep, K., and M. Thakur. 2007. A New Crossover Operator for Real Coded Genetic Algorithms. *Applied Mathematics and Computation* 188, no. 1: 895–911.

40. Storn, R., and K. Price. 1997. Differential Evolution—A Simple and Efficient Heuristic for Global Optimization over Continuous Spaces. *Journal of Global Optimization* 11, no. 4: 341–359.

41. Whitley, D., D. Hains, and A. Howe. 2009. Tunneling between Optima: Partition Crossover for the Traveling Salesman Problem. In *Proceedings of the 11th Annual Conference on Genetic and Evolutionary Computation*, pp. 915–922. GECCO '09. New York: ACM.

42. Whitley, D., D. Hains, and A. Howe. 2010. A Hybrid Genetic Algorithm for the Traveling Salesman Problem Using Generalized Partition Crossover. In *Parallel Problem Solving from Nature, PPSN XI*, eds. R. Schaefer, C. Cotta, J. Kolodziej, and G. Rudolph, Vol. 6238, pp. 566–575. Lecture Notes in Computer Science. Berlin: Springer.

43. Schlierkamp-voosen, D., and H. Mühlenbein. 1994. Strategy Adaptation by Competing Subpopulations. In *Parallel Problem Solving from Nature (PPSN III)*, pp. 199–208. Berlin: Springer.

44. Ono, I., H. Kita, and S. Kobayashi. 2003. A Real-Coded Genetic Algorithm Using the Unimodal Normal Distribution Crossover. In *Advances in Evolutionary Computing*, eds. A. Ghosh and S. Tsutsui, pp. 213–237. Natural Computing Series. Berlin: Springer.

45. Herrera, F., M. Lozano, and J.L. Verdegay. 1998. Tackling Real-Coded Genetic Algorithms: Operators and Tools for Behavioural Analysis. *Artificial Intelligence Review* 12, no. 4: 265–319.

46. Arumugam, M.S., M.V.C. Rao, and R. Palaniappan. 2005. New Hybrid Genetic Operators for Real Coded Genetic Algorithm to Compute Optimal Control of a Class of Hybrid Systems. *Applied Soft Computing* 6, no. 1: 38–52.

47. Chang, W.-D. 2006. Coefficient Estimation of IIR Filter by a Multiple Crossover Genetic Algorithm. *Computers & Mathematics with Applications* 51, no. 9–10: 1437–1444.

48. Kaya, Y., M. Uyar, and R. Tekin. 2011. A Novel Crossover Operator for Genetic Algorithms: Ring Crossover. *CoRR abs/1105.0*, https://arxiv.org/pdf/1105.0355.pdf

49. Dorronsoro, B., and P. Bouvry. 2011. Improving Classical and Decentralized Differential Evolution With New Mutation Operator and Population Topologies. *IEEE Transactions on Evolutionary Computation* 15, no. 1: 67–98.

50. Iwamatsu, M. 2002. Generalized Evolutionary Programming with Lévy-Type Mutation. *Computer Physics Communications* 147, no. 1–2: 729–732.

51. Tinós, R., and S. Yang. 2011. Use of the Q-Gaussian Mutation in Evolutionary Algorithms. *Soft Computing* 15, no. 8: 1523–1549.

52. Davis, L. 1996. *Handbook of Genetic Algorithms*. eds. Davis L., Van Nostrand Reinhold (1991). New York: International Thomson Computer Press.

53. Chuang, Y.-C., and C.-T. Chen. 2011. A Study on Real-Coded Genetic Algorithm for Process Optimization Using Ranking Selection, Direction-Based Crossover and Dynamic Mutation. In *IEEE Congress on Evolutionary Computation (CEC)*, pp. 2488–2495. New Orleans, LA: IEEE Conference Location.

54. Sharma, R., and S.P. Bhattacharyya. 2007. Direct Search for Wave Operator by a Genetic Algorithm (GA): Route to Few Eigenvalues of a Hamiltonian. In *IEEE Congress on Evolutionary Computation (CEC)*, 25–28 September, pp. 3812–3817. Singapore: IEEE.

55. Sarkar, K., R. Sharma, and S.P. Bhattacharyya. 2010. Blending Determinism with Evolutionary Computing: Applications to the Calculation of the Molecular Electronic Structure of Polythiophene. *Journal of Chemical Theory and Computation* 6, no. 3: 718–726.

56. Ling, S.H., and F.H.F. Leung. 2007. An Improved Genetic Algorithm with Average-Bound Crossover and Wavelet Mutation Operations. *Soft Computing* 11, no. 1: 7–31.

57. Lyakhov, A.O., A.R. Oganov, and M. Valle. 2010. How to Predict Very Large and Complex Crystal Structures. *Computer Physics Communications* 181, no. 9: 1623–1632.

58. Lyakhov, A.O., A.R. Oganov, H.T. Stokes, and Q. Zhu. 2013. New Developments in Evolutionary Structure Prediction Algorithm USPEX. *Computer Physics Communications* 184, no. 4: 1172–1182.

59. Goldberg, D.E., and K. Deb. 1991. A Comparative Analysis of Selection Schemes Used in Genetic Algorithms. In *Foundations of Genetic Algorithms*, ed. G.J.E. Rawlins. pp. 69–93. San Mateo, CA: Morgan Kaufmann.

60. Campelo, F., F.G. Guimaraes, H. Igarashi, and J.A. Ramirez. 2005. A Clonal Selection Algorithm for Optimization in Electromagnetics. *Magnetics, IEEE Transactions on* 41, no. 5: 1736–1739.

61. Goh, K.S., A. Lim, and B. Rodrigues. 2003. Sexual Selection for Genetic Algorithms. *Artificial Intelligence Review*. 19, no. 2: 123–152.

62. Agrawal, A.F. 2001. Sexual Selection and the Maintenance of Sexual Reproduction. *Nature* 411, no. 6838: 692–695.

3

Evolutionary Computing

3.1 Introduction

Genetic algorithms developed by Holland and his group of coworkers (notably K. A. De Jong, David E. Goldberg) are a class of algorithms that exploit the idea of biological evolution. The same idea of evolution has also been utilized in a number of algorithms which are similar to the GAs but are not identical. Collectively, all these techniques of problem solving including GAs belong to what has now become recognized as a new computing paradigm—the paradigm of evolutionary computing (EC). The subtle but important differences among different components of EC, their operational details and their basic philosophies need elaboration. This chapter addresses some issues including those relating to parallel implementation of EC.

Let us recall that GAs work by creating and evaluating fixed size populations of individuals (chromosomes) represented by binary bit strings of fixed lengths and selecting individuals randomly with a bias provided by the goal-specific fitness. Survivors (presumably the fittest ones) are paired for crossover breeding and the offspring produced then undergo mutation in which some genes (bits) flip. Sexual recombination of genetic materials in crossover offers a rather powerful strategy of adaptation and is a characteristic feature of genetic algorithms. As a class of algorithms, GAs are often regarded as "all-purpose optimizers/problem solvers." They indeed are, but a fundamental and distinctive feature of all GAs often goes unnoticed—in GAs, we are not directly solving a problem. We are merely creating a condition in which optimization takes place naturally. The feature is common to all the different realizations of EC. Then, where do they differ? As we have emphasized in the preceding chapter, GAs employ what is called genotype or genotypic representation of the potential solutions and work directly in the genotype space. That means encoding and processing of information is done in the genotype space. It is possible to think of algorithms similar to GAs in many respects but differing in a crucial one. These algorithms stress the phenotype space of features and observables, and seek to evolve behavior; the solution of the problem at hand emerges in the process. One such algorithm has now become known as evolutionary programming (EP) [1,2]—one of the members of the family of EC. Let us consider briefly how EP differs from GAs in philosophy, architecture, and operational details.

3.2 Evolutionary Programming

Evolutionary programming derives its operational strength from its core concept of simulating the adaptive behavior noted (or rooted) in the evolutionary process of nature. Just like GAs, EP makes use of a population of randomly chosen potential solutions for the specific problem at hand. The population is made to evolve under the action of EP operators which tend to mimic the very basic top-down feature of adaptive behavior noted in nature. Thus, working in the phenotypic space of observed behaviors, EP aims at evolving behavior (without worrying about the underlying genotypic space) and solving the problem in the process. The emphasis in EP is on simulating the adaptive behavior in the evolutionary scheme instead of simulating the genetics of evolution as practiced in GA. The way evolution is handled in EP is much more flexible. The operators have no fixed forms; they are freely adapted to suit the problem being solved. The central role in EP evolution is played by the process of mutation, i.e., by the action of a mutation operator which produces offspring, there being virtually no role played by sexual recombination in producing progeny. EP is thus a single-parent strategy. The number of offspring produced by mutation of the population of single parents is usually equal to the number of parents who were selected into the "breeding pool," the selection being effected by a tournament selection (instead of a roulette wheel-based scheme). Alternative selection operators based on elitism, proportional selection, or selection based on nonlinear ranking can be used. The selected parents have their features mutated to produce the desired number of offspring. The better half of the combined population of parents and offspring is retained and passed on to make the population for the next generation. In addition to the normal mutation, EP makes use of yet another process for producing offspring—self-adaptation [3]. Self-adaptation offers the strategy parameters of EP an opportunity to evolve themselves which in turn may nudge the mutations to explore more promising areas of the search space.

3.2.1 The Basic Architecture

The EP algorithm just introduced may be broken down into a sequence of steps which are executed repeatedly until an exit criterion is fulfilled. The steps are summarized as follows:

1. Choose a population size (n_p) and create the initial population of size n_p by choosing a set of n_p strings randomly.
2. Allow the population to interact with the environment. Random forces now come into play affecting the strings in the long run. A fitness function is defined to measure fitness. Strategy parameters are determined.
3. Evaluate each individual member of the population for its fitness.
4. Allow each parent to undergo a random mutation induced by "noise"—the mutated strings become the offspring.
5. Evaluate the offspring for their fitness values.
6. Select the better half of the combined pool of parents and offspring on the basis of fitness or some function of fitness.
7. Loop back to step (B), and iterate until some stopping criterion is satisfied.

A flowchart is displayed in Figure 3.1.

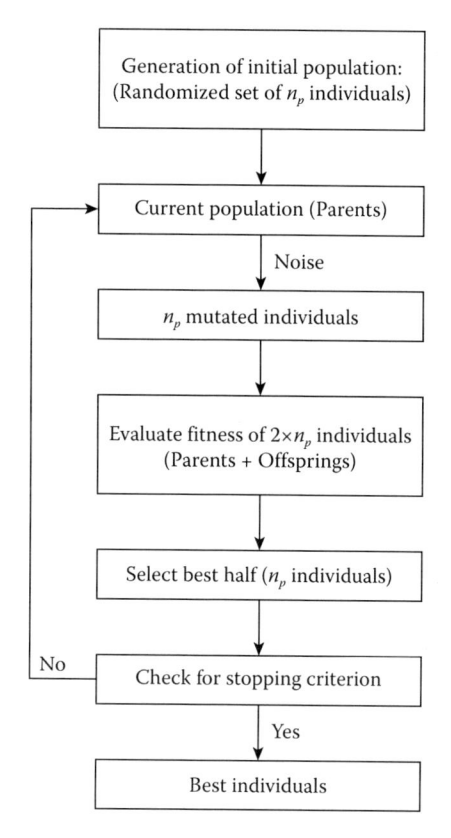

FIGURE 3.1
A flowchart for evolutionary programming.

NOTES:

1. In step (1), we have to choose a representation for the potential solutions. If the problem at hand admits solutions in the real computable space, the strings chosen are vectors with N-components, each component having a "real value" chosen randomly from a specified range (say from $-a$ to $+a$).

2. The population size n_p chosen is dictated by the problem. Generally, n_p varies between 50 and 500.

3. There is no mutation probability—every string is allowed to mutate.

We can now illustrate the operational details of EP as applied to, say, the problem of globally optimizing a real-valued function $F(x, y, z)$ in three dimensions as defined below:

$$F(x, y, z) = k(x^2 + y^2 + z^2), k > 0. \tag{3.1}$$

The extremum at $x = y = z = 0$ is the global minimum. Let the population size n_p be fixed at 25. The initial population of 25 vectors $V_i = (x_i, y_i, z_i)^t$ for $i = 1, 2, 3, \ldots, 25$ can be created by choosing the three components of $V_i(v_{i1}, v_{i2}, v_{i3})$ randomly from a range $[-a, a]$. The fitness value f_i of each individual V_i may be defined as

$$f_i = \frac{1}{d_i + \delta} \tag{3.2}$$

where d_i is the Euclidean distance of the point (x_i, y_i, z_i) from the origin, δ is a small constant (0.01, say) added to the denominator to prevent exponential overflow if and when $d_i \to 0$. The next step is to mutate each of the 25 vectors to produce offspring. The mutation, as advocated by Fogel, is brought about by adding a Gaussian random variable $N(\mu, \sigma^2)$ to each component of the parent vector V_i that has been chosen for mutation. N has mean $\mu = 0$ and variance σ^2 which is just the error value of the parent V_i as measured in this case by the Euclidean distance from the origin. The jth component of the mutated vector V_i (call it V_i') then reads

$$V_{i,j}' = V_{i,j} + N(O, \beta_j, f_i, \epsilon_j) \tag{3.3}$$

where f_i is the fitness value of the ith parent (V_i), β_j is a scaling parameter, and ϵ_j is an offset. After each n_p-parent (the complete set of $\{V_i S\}$) has been mutated into n_p offspring (the set of vectors $\{V_i'\}$, the combined pool of $2n_p$ vectors $\{V_i, V_i'\}_{i=1,n_p}$ is created and the fitness of each individual is evaluated. Half of the augmented population of size $2n_p$ is then selected into the next generation by a "tournament selection" (TS). In TS, each member is made to play a tournament with m others in the population, and the total score is counted. The "winning half," selected on the basis of the total score, makes it to the next generation. In addition to the mutation defined in Equation 3.3, EP often makes use of mutation by self-adaptation. In the preceding example, the standard deviations can be modified based on their current values so that the search adapts itself to the contour of the error surface. EP with local search involving hill climbing steepest descent can be useful.

The EP method appears to be very suitable for locating the global minima on complex potential energy surfaces or for optimizing parameters in the wave function so that energy is minimized.

3.3 Evolutionary Strategy

In both GAs and EP, the population of individuals moves from one region of fitness landscape to another, the movement being engineered by evolution. In nature, evolution is believed to have optimized the biological processes over millions of years. If we interpret evolution itself as one of such processes, we are free to assume that evolution has also optimized itself. Evolutionary strategy (ES) is a computing paradigm which exploits the ability of a population to evolve its innate evolvability, perhaps even optimize it and in the process solve a problem [4].

The basic steps involved in an ES application are as follows:

1. Generate an initial population of μ individuals. An individual is represented by a set of n-features—a feature vector, and the components $x_1, x_2, ..., x_n$ are values of the decision variables and corresponding set of strategy parameters—generally variances or standard deviations [5,6].

2. Allow recombination (crossover) using μ numbers of parents to produce λ number of children.

3. Use the strategy parameters to mutate the feature vectors of the offspring.

4. Evaluate either λ or $(\mu + \lambda)$ number of individuals of the population for their fitness values—the former being called (μ, λ) ES, and the latter termed $(\mu + \lambda)$ ES. The (λ/μ) ratio, in either case, is commonly kept around 7.

5. μ-individuals are selected as parents for the next generation.

6. If termination criterion is not met, move to step (2) and iterate; terminate otherwise.

Recombination in ES modifies the entire set of variable values (features) by using either a local or a global method by means of a discrete or an intermediate mode of recombination. In the local method, the features or the components of the offspring vector are generated from the components of the two randomly selected parents. The global method, as the name signifies, exploits the entire population for generating the features of the children vectors. The most widely used version of recombination in ES is the intermediate mode of recombination in which each component of the child vector has a value lying between values of the corresponding parent vectors. Let the two parents be identified as P_1 and P_2 and the particular component of the feature vectors (X) be marked by the index i. Then the intermediate recombination process works as follows:

$$X_{child,i} = X_{P_1,i} + \alpha(X_{P_2,i} - X_{P_1,i}). \tag{3.4}$$

The scalar α in Equation 3.4 is usually set at a value of 0.5 but can be adjusted further, based on experience.

In discrete recombination, $X_{child,i}$ can be selected from the corresponding components of the feature vector $X_{P_1,i}$ or $X_{P_2,i}$ (on an either/or basis). It is important to note that the strategy parameters are also modified by an appropriate recombination process.

Mutation of features in ES is induced and controlled indirectly. Let us recall that an individual in ES is represented by a set of features (the components of a feature vector) and the corresponding sets of strategy parameters (usually standard deviations). The features of the parents' feature vectors are slightly perturbed (mutated) by adding random numbers generated from a probability distribution that has mean zero and a standard deviation defined by the strategy parameters. Note here that the strategy parameters themselves undergo evolutionary changes (recall the recombination step). So, the mutation intensity and variability of changes to be manifested in the next generation also evolve along with features, the latter being the actual target of the optimization process. If the variance is high, the steps (step lengths) that alter the features on the fitness landscapes are larger, signaling the onset of exploration. Low variance, on the other hand, leads to small step lengths in mutation, indicating an exploitative phase of the search (depth search, as opposed to breadth search in the high variance situation).

A novel feature of mutation in ES has been the idea of an "evolution window." Rechenberg [7] showed that the probability of a successful mutation (fitness-enhancing mutation) taking place is approximately 0.2 if mutations are carried out with an optimal standard deviation (i.e., within a window). The window can be made dynamic, in which case the mutation step lengths are adapted to a dynamic window. These mutations bring in the benefit of what has been called metaevolution.

Selection in ES has most often been through the process of TS. Tournament selection exploits local competition to select survivors, and several realizations of TS have been successfully exploited. In the simplest form, pairs of individuals are chosen at random and the better one of each pair (the winner) is selected. The process is repeated until the next generation is adequately populated. In another version, individuals are paired to play in a preset number of competitions. Each time an individual wins a point, it is added to the score. An individual with a total or cumulative score that is higher than a given threshold is selected. One can use bigger tournament sizes, but the practice has been to restrict the size to only two.

3.4 Genetic Programming

The fourth member of the quadruplet collectively called evolutionary computing is called genetic programming (GP) [8]. Just as in GAs, EP, and ES, GP too works with a population of potential solutions which evolve under the action recombination (i.e., crossover), mutation, and fitness-sensitive or fitness-guided selection operators with the objective of maximizing fitness value. These similarities notwithstanding, GP fundamentally differs from the three other constituents of EC in two major ways: (i) the way of representing potential solutions, and (ii) the way fitness is evaluated. These key differences are summarized below.

1. *Representation:* The potential solution (chromosomes) are syntactically valid and evolvable computer programs in a suitable language.
2. *Fitness:* These programs must be executed to calculate the goal-specific fitness of an individual in the population on the basis of the output produced.

The programs representing individuals can be coded in the form of parse trees (visual representation) of varying shapes (branching patterns), sizes, and complexities and run on a machine with specific compilers (such as Lisp). The goal of the other three members of the EC family has been to discover a string of bits or numbers that produces the best possible fitness value, the corresponding string representing the most acceptable solution to the problem. The goal in GP has been shifted to discover the computer program that yields the maximum fitness value when run with a set of input parameters specific to the problem [9,10]. The search takes place in the space of all possible syntactically valid computer programs, for the particular problem at hand. The parse tree representation of these executable programs has internal points where functions defined for the problem appear while all constants and variables appear at the leaves of the tree, also called the external points. These differences are, however, only superficial. At a more fundamental level of analysis, no sharp demarcation line can be drawn between the normal chromosomal encoding used in EC and the representation in terms of executable structures, i.e., programs used in GP. Thus whatever form of encoding is used, the elements of it can be regarded as commands in a computer language. The interpreter or the compiler then serves as the chromosome decoder. On the other hand, chromosomes in GP being computer programs or equivalent parse trees, can vary in length as opposed to fixed length strings normally used in other forms of EC. The program length variation is a common feature of computer programs, in general. However, the programs must conform absolutely to the semantics and syntax of the specific computer language being used. At the same time, these chromosomes (programs or parse trees) have rather strong hierarchical and compositional structures making it possible to accommodate in them problem-specific, special purpose functions like *SIN, COS, SQRT,* along with general purpose instructions like $+$, $-$, $\times$, $/$ and these structures evolve. Viewed in this light GP is not just a problem-solving enterprise; it has another very important dimension—it also represents an automatic programing or code development system. How does it happen? The automatic programing clue lies in the fitness function designed for the problem to be solved. It communicates to the system what we expect it to accomplish. A successful run of the system then produces by genetic evolution an output program that has been designed to solve the problem at hand to the desired level of accuracy. The very idea of an automatically evolving computer program has great appeal to many. The appeal increases manifoldly especially when it is difficult to develop particular types of code, for example, codes for massively parallel computers, or codes for a quantum computer. The difficulty in the first case stems from the fact that our ideas are rooted in serial computing while in the second the problem is rooted in the inability of our thought processes to leave the classical mooring. In either case, an automatic code development system can ease the bottleneck. However appealing the idea is, automatic programming is still considered a theoretical curiosity or an object of interest to a few specialists in computer science. As articulated by John Koza in a number of publications [11,12] automatic programming as envisaged in GP is endowed with quite versatile and general capability that can be exploited rather easily in many areas of science and technology, and probably in arts as well. An important, perhaps even the most important, aspect of the automatic programming technology lies in its ability to open new vistas for explorations in computer science. A question of paramount interest in computer science concerns whether a computer program with a special capability or priority exists or not. Although logical, mathematical, or analytical arguments enable us to make progress, the prospect of using the computer itself to search through the space of all computer programs and find whether such a program indeed exists, opens up a new avenue to move forward.

3.4.1 The Parse Tree and Lisp Program

John Koza was possibly the first to invoke the genetic algorithm to evolve a computer program automatically, using a special high-level computer language called Lisp. Lisp has a long and rich history in computer science and has sophisticated facilities. What Koza used in his work was only a few of the significant features of Lisp. Koza's work led to the development of what has been hailed as genetic programming—a shorter form of genetic algorithms in automatic programming. The technique used is quite general and has both the sophistication and robustness needed for automatic general purpose code development.

In Koza's GP a candidate solution is encoded as a parse tree with nodes and leaves. Consider for example the parse trees in Figure 3.2a through d, and the corresponding algebraic expressions and their Lisp transcriptions, assuming that we are using only common functions like $+, -, \times, /, \sqrt{}$ and the terminal set consists of two constants (variables) A and B.

As the reader will easily note, the operators in Lisp precede the argument—$A * B$ is written as $(*AB)$. The subset of Lisp syntax used in GP is generally the following:

Program::= terminal | (function program)

Note that a "terminal" is a program in Lisp. More often than not, terminals are constants. (say "8," or, "1.414"), although they can be variables or functions with zero argument.

Consider the following parse tree (Figure 3.2c) for an expression that can be algebraically represented as $AB - \{C + (y - 5)\}$. Its Lisp transcription reads as follows:

$$(-(*AB)) (+C(y - 5)).$$

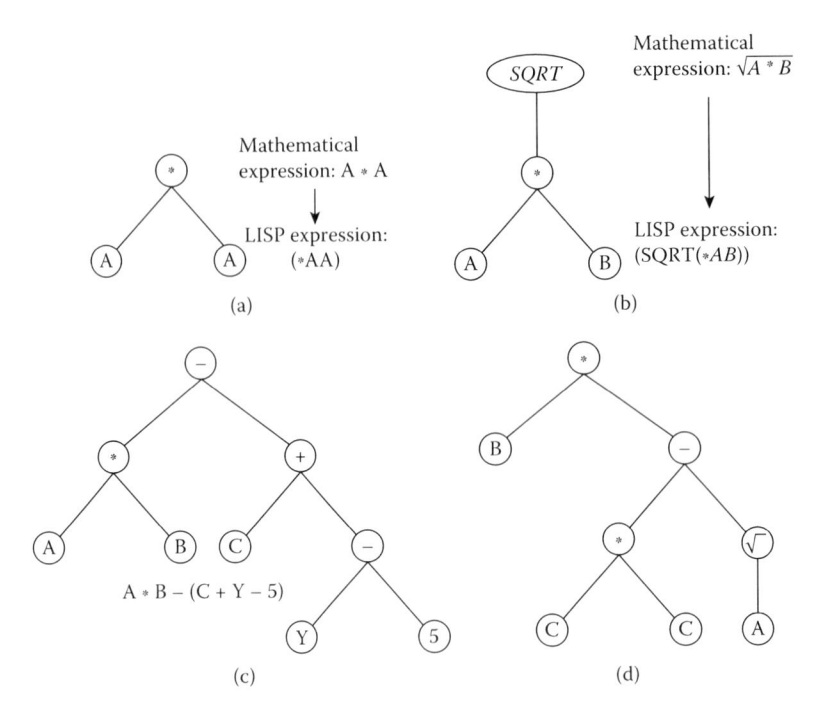

FIGURE 3.2
(a–d) Parse tree structures representing some algebraic expressions.

Each program in the expression above returns a value to the enclosing context. The value is then passed on as an argument to the function that appears at the head of the enclosing expression. Thus, in the preceding expression the program returns the difference between two values—the first being the product of A with B, and the second being the sum of C and the difference between y and 5. We assume that A, B, C, y are variables containing values. Such values can be assigned by the self-operator as follows:

$$(Self\ C\ 2.14).$$

In the expression displayed above, the self-operator has two arguments and self-assigns, the second argument (which is the value 2.14) to the first argument (which is variable C). Let us consider yet another parse tree. The tree 3.4d translates into an algebraic expression $B(C^2 - \sqrt{A})$. In Lisp it takes the following form:

$$(*B(*CC(-\sqrt{A}))).$$

The preceding example confirms the assertion made earlier that each expression in Lisp can be interpreted functionally. The primary task of "call to a function" is to return a value to the enclosing context. Such an interpretation is not mandatory, but certainly useful. What is the most important attribute of the Lisp program representation we have described so far? It is the syntactic uniformity (SU). This SU is highly significant in the context of GP where we wish to evolve the programs. The SU allows one to substitute any subprogram by another subprogram within a program without affecting the syntactic validity of the program. That enables the programmer to design genetic operators that can act rather blindly on programs producing new, syntactically valid programs that would run without difficulty. Thus, we have at our disposal, at least in principle, a system for automatic program evolution. However, certain additional preventive measures may have to adopted to ensure that arbitrary programs generated from a population of programs remain semantically valid—that is, they will run without flagging any error signal and produce output that can be interpreted and used in fitness calculation. With the brief background given, we are now in a position to address the most important question: How to define traditional genetic operators (mutation and recombination) in the context of GP?

1. *Mutation in GP:* Mutation in GP involves the replacement of an arbitrarily (randomly) chosen subprogram in a program (a subtree in a parse tree representation) with a newly generated subprogram: $(SQRT(^{*}AB))$ which can be mutated to $(SQRT\,(/AB))$ or perhaps to $(Cos(^{*}AB))$. In general, the practice has been to replace an entire function call in the mutation process rather than replace only the terminals. It is also possible to put additional safeguards in place so that mutation replaces a subprogram (subtree) only by another subprogram (subtree) of similar size. Such mutations have been called size-fair or no-bloating mutations (see, for example, Langdon 1998 [13]).

2. *Recombination (crossover) in GP:* This works by swapping subprograms (subtrees) between two chosen parent programs (trees). Let P_1 and P_2 be the two parent programs selected for undergoing crossover. The chosen subprograms on P_1 and P_2 for exchange are marked by enclosing them in boxes.

$$P_1 = \left(-\boxed{(^{*}AB)}(+C(-y5))\right)$$

$$P_2 = \left(-(^{*}5.4(+3B))\left(^{*}\boxed{(+(^{*}2A)C)}\right)\right).$$

The crossover produces two child subprograms C_1 and C_2 by the exchange where

$$C_1 = (-(+(^{*}2A)C)(+C(-y5)))$$

$$C_2 = (-(^{*}5.4(+3B))(^{*}AB)).$$

Many variations on the recombination theme have been proposed and tested from time to time. The basic motivation has been either to enhance the probability that children of fit parents remain fit, or to control the statistics of shape and size distribution of the individual trees in the evolving population.

Selection in GP is fitness guided or fitness sensitive. For implementing selection, each subprogram in the population is run. The output is used to calculate fitness and the fitness values dictate selection. John Koza has given the following guidelines for writing a GP code.

Step 1: At the outset one must choose a set of candidate functions and terminals to be used in the programs. The user is required to make an intelligent or a reasonable guess as to the terminals and functions that could be needed.

Step 2: An initial population of randomly constructed parse tree, or programs using the candidate functions and terminals chosen in step 1 are generated; the only requirement here is that the program must be syntactically correct and must run producing an interpretable output. The random parse trees can vary in size and shape.

Step 3: Calculate the fitness value of each program in the population, by running it on a set of test cases—on a set of inputs for which exact or very high quality results are known. The fitness of a program is a function of the number of test cases for which it produces the correct answer. Selection is biased by the fitness values. The usual population size is a few hundred. The initially chosen programs must be able to provide sufficient fitness differential so that natural selection can operate with success.

Step 4: Apply selection, crossover, and mutation for generating a new population. Usually 10–20% of the best individuals are copied unchanged into the new population while the remaining 80–90% are generated by crossover and mutation operations.

Step 5: Check for the satisfaction of termination criterion. If not satisfied go to step 3 and iterate until the termination condition is met with. Exit, otherwise.

In the realm of physics and chemistry, GP has not been exploited to a significant extent. As an automatic programming system GP has found interesting use in producing quantum-computing codes (see Chapter 10). Some intelligent use of GP for producing an analytical formula of energy eigenfunction of a Hamiltonian (for which no exact solution is available) has been reported (see Chapter 6). The most common use of GP that one can think of is in analytically representing data for which random guesses refuse to work. We will explore such possibilities in Chapter 7.

3.5 Differential Evolution

Differential evolution (DE) is a variant of evolutionary computing technique operationally different from the other EC techniques discussed so far. Proposed by Rainer Storn and Kenneth Price (1977) DE explores the basic theme of function optimization in continuous spaces with the help of a novel population-based search, with the population wandering across the function surface or an equivalent fitness landscape [14,15]. The main emphasis in DE is on mutation which is treated in a novel manner and serves as the exploratory step. Selection is fitness guided and is the bias provider. Crossover operation is carried out on each of the postmutation individuals and exploits the information already present in the search space in order to move the population from one part of the fitness surface to another. We summarize below the sequence of steps DE works with:

1. Create an initial population of n_p-potential solution vectors $\{\vec{X}_i\}_{i=1,\ldots,n_p}$ randomly. Each $\vec{X}_i$ is an N-dimensional vector with N real number components, each component representing one of the N-search variables.

2. Define a fitness function $f(\vec{X}_i)$, the maximization of which can be identified as the goal of the search. For each $\vec{X}_i$, $f(\vec{X}_i)$ returns a real number called the fitness value f_i of the individual ($\vec{X}_i$).

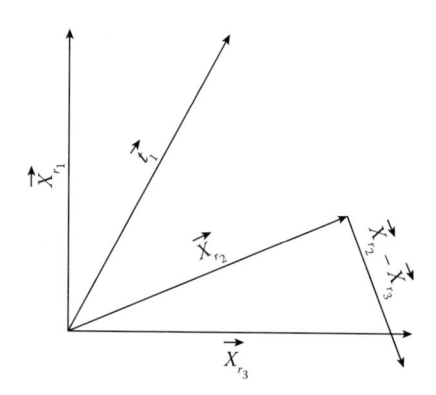

FIGURE 3.3
Trial vector formation in DE.

3. For each of the n_p individuals $\vec{X}_i$, execute steps 2–5. Randomly select three individuals from the population and call them $\vec{X}_{r1}, \vec{X}_{r2}, \vec{X}_{r3}$.

4. Create a new trial vector $\vec{t}_i$ by adding the vector difference $(\vec{X}_{r2} - \vec{X}_{r3})$ weighted by a mutation intensity parameter W_m to $\vec{X}_{r1}$, i.e.,

$$\vec{t}_i = \vec{X}_{r1} + W_m(\vec{X}_{r2} - \vec{X}_{r3}) \tag{3.5}$$

We note here that the distances between vectors (individuals) provide a very good estimate of the diversity of the population and rough estimate of the step size W_m. See Figure 3.3 for a pictorial explanation of this step.

5. Perform a crossover operation between $\vec{X}_i$ and $\vec{t}_i$ with a preset probability p_c to generate a new individual $\vec{U}_i$ (offspring). The N-components of $\vec{U}_i$ are determined as follows:

$$\vec{U}_{ij} = t_{ij}, \text{ if } r(j) \leq p_c \text{ or } j - j_r$$
$$= X_{ij}, \text{ otherwise} \tag{3.6}$$

$r(j)$ is a random number in the range (0,1) for the jth gene (jth component of U and t), and j_r is a random integer in the range [1, N].

6. Evaluate the new candidate's ($\vec{U}_i$) solution for its fitness. If $f(\vec{U}_i) > f(\vec{X}_i)$, select $\vec{U}_i$ into the population for the next generation.

7. Return to step 3 until the termination criterion is satisfied, exit otherwise. The individual with the highest fitness value is the best solution discovered. DE has been used with impressive success for locating the global minimum on complex potential energy surfaces and in dynamic optimization problems, where both locating and following the trajectory of the global minimum are important.

NOTE: The chance of being selected as a parent in DE is fitness independent (as opposed to fitness proportional selection in GA). The offspring produced by a self-adjusting mutation operation and a crossover process compete with the parents, with the better one winning the competition. That is, a greedy scheme dictates the selection. All these features make DE a fast converging search that is simple and flexible and free from any Hamming cliff problem.

3.6 Parallel GAs

GAs or other EC techniques including the technique of DE make use of a population of evolving solutions. Usually, the computation of fitness of n_R individuals is the costliest computing step in GA although the selection and crossover process could also be time-consuming. If we concentrate only on the fitness calculation of n_p individuals, parallelization seems to be the automatic choice for enhancing performance. The central idea behind parallel program development has been to divide a large computing problem into a number of smaller independent tasks and then attempt to solve the subtasks simultaneously using multiple processors. As readers will immediately notice, this is a divide and conquer strategy and can be applied to GAs in more than one way. GA literature is replete with reports of successful implementation of parallel GAs [16,17]. Some of these implementations work with a single population with a fitness calculation of n_p individuals distributed over the available processors while some others divide the entire population into isolated subpopulation with occasional interactions among them. There are parallel GA methods that exploit massively parallel computing architectures. There are also others, which are better adapted to systems with fewer but more powerful processing elements operating in a slower network.

Parallel GAs can be classified in a number of ways. We will follow the most commonly used classification which recognizes four major types of parallel GAs:

1. Master–slave GA (MSGA) working with a single population of n_p individuals
2. GAs working with multiple populations (MPGA) with occasional migration of individuals from one subpopulation to another
3. Fine-grained GAs (FGGA)
4. Hierarchical hybrid GAs (HHGA)

MSGA uses a single population with one master node managing the basic GA steps like selections, crossover, and mutation. The evaluation of fitness of individuals is distributed among the available N-slave nodes which evaluate the fitness of individuals sent to them by the master node and return the output (fitness value) to the master. The selection, crossover, and mutation processes are executed by the master nodes on the entire population. Master slave GAs are therefore called global parallel GAs.

MPGAs use several subpopulations with the occasional exchange of individuals between subpopulations which is called migration. The multiple population GAs with migration are the most popular and sophisticated realizations of parallel GAs. They are, however, rather poorly understood and hence difficult to control. Among the controlling factors are the number and the size of the subpopulations, how frequently the migration should take place, the number of migrants and their destinations, how to select the migrants, etc.

MPGAs are also known as distributed GAs in as much as they are frequently implemented on distributed memory MIMD computers. Since communications between subpopulations are few and far between, the computation to communication ratio is high—that is why they are also called coarse-grained GAs. In view of the similarity between MPGA and the island model of population genetics in which the population is divided into geographically separated subpopulations with the possibility of occasional migration of individuals between any two subpopulations, MPGAs are also called island parallel GAs. In many ways, MPGAs bear closer analogy with natural population genetics compared to single population GAs, and phenomena observed in natural populations are also observed in MPGAs. One such observation concerns the similarity between the evolution process in a parallel GA and the theory of punctuated equilibria. Elredge and Gould suggested [18] that significant changes in a population occur rarely—most of the time the population is practically in equilibrium. Rapid evolutionary changes that occur in a population rarely appear to be triggered by external events, such as migration from one subpopulation to another. Assuming that the subpopulations have discovered locally optimal solutions, one effect of migration would be to nudge the other subpopulations to explore further and eventually discover better solutions.

Fine-grained GAs (FGGA) also called cellular GAs (CGA) employ a single but spatially structured population. The population structure resembles a two-dimensional rectangular grid with one individual assigned to one grid point, and if possible one processor being made available for each individual. In such a case, fitness evaluation of all the individuals of the population can be performed simultaneously. Selection and crossover breeding are confined to a small overlapping neighborhood around each individual. The overlapping neighborhood ensures that the good attributes of a superior individual spread through the entire population mimicking the process of random diffusion of particles in a fluid—that is why this particular class of parallel GA is also called diffusion model-based GA. The FGGA are particularly suitable for implementation on parallel SIMD computers, although it is perfectly feasible to implement them efficiently on coarse-grain MIMD computers.

A hierarchical parallel GA combines multiple populations with either master–slave or fine-grained GA enjoying the benefits of both, and has the potential to perform better than either of the components. They are categorized as hierarchical parallel GA because they behave as a single population parallel GA at a lower level while at a higher level they behave as multiple population GAs. Parallel GAs can be efficiently implemented on symmetric multiprocessors with ease. Before closing this section, let us note a few points briefly: (i) The master–slave GA does not usually affect the behavior of the genetic algorithm itself—but the FGGAs and their hierarchical hybrids can strongly affect the way GAs work, and a lot more needs to be done to understand their behavior completely. (ii) Contrary to widespread belief, no specialized hardware is needed to implement parallel GAs—simple inexpensive hardware such as Beowulf clusters or web-based computations can be used profitably. (iii) Each type of parallel GA can be of synchronous or asynchronous types, the synchronous type having all the processors (irrespective of speed) dealing with individuals of the same generation all the time. That essentially means a faster processor has to wait until the slower ones have completed their tasks. In the asynchronous parallel GAs, the idle processor time is remarkably reduced as the processors continue working at their own speed and they may not all be processing individuals of the same generation. Parallel GAs are easy to construct and implement, but mathematical analysis of their performance with a view to designing all optimal parallel GA is not a simple task. In what follows, we outline such an analysis of the simplest of the parallel GAs, namely the master–slave GA working with a single population [16].

3.6.1 Designing an Optimal Master Slave GA

In the master–slave single population parallel GA, there are several aspects to be considered while designing and optimizing it. There are cost elements coming from (a) fitness evaluations, (b) basic GA operations like selection, crossover, and mutation, and (c) cost of communication between the master and slave nodes. Early work by Bethke [19], Grefenstette [20], Fogarty and Huang [21], and Abramson and Abela [22] tended to conclude that parallelization of the fitness evaluation step (the costliest step) alone can increase the efficiency of the capacity utilization of the available processors and produce great speedups. Practical experience, however, indicated that parallel efficiency decreased as more processors were added, signaling that communication overhead in master–slave (MS) architecture has an important role to play. It is therefore necessary to identify first the primary trade-off between computation and communication overheads, and arrive at the MS-configuration that minimizes the execution time for code generation and then proceed to analyze the parallel efficiency of the algorithm.

Having identified the task, we may proceed with the analysis under the assumptions that (a) the time taken by the basic GA operations $\ll$ the fitness evaluation time as well as the communication time; (b) each processor evaluates a fixed number of individuals (c) the evaluation time is independent of the individual; and (d) the master executes not only all the basic GA steps in a negligible time, but also evaluates a fraction of the population which consumes time. Let t_f be the time required to evaluate the fitness of one individual, n_p the size of the population, n_s the number of slave nodes while P is the total number of processors including that of the master ($p = n_s + 1$). The master starts by sending a fraction of the population to each of the n_s slaves consuming a time t_c to communicate with each slave node. The master evaluates a fraction ns/p of the population at the cost of time T_{cal} where

$$T_{cal} = \frac{n_s}{p} t_f.$$

(3.7)

The time consumed in communication

$$T_{com} = n_s t_c + t_c = p t_c.$$

(3.8)

The elapsed time for one generation is

$$T_p = T_{com} + T_{cal}$$

$$= p t_c + \frac{n_s}{p} t_f.$$

(3.9)

It is therefore clear that as more processors are added (p increases) the computation time diminishes while the communication time increases and the trade-off between the two leads to an optimal value of p (= p_0, say) that minimizes the execution time for one generation. p_0 is found easily by minimizing T_p with respect to p (see Cantú-Paz [16]) which leads to

$$p_0 = \sqrt{n_p \left(t_f / t_c \right)}.$$

(3.10)

If we let the ratio (t_f/t_c) be represented by r we have $p_0 = \sqrt{r n_p}$ and $n_s = p_0 - 1$.

If $r \approx 1$, $p_0 = \sqrt{n_p}$ so the simple thumb rule for the optimal design of a parallel GA in such a case is to have $n_s = \left(\sqrt{n_p} - 1 \right)$—for a population size of 100 only nine slave processors would make the optimal choice. In general, $r \neq 1$, and therefore the optimal number of processors in a master slave parallel GA would depend nontrivially on the ratio of t_f and t_c. When considering the benefits of parallelization, a pertinent question concerns how to measure parallel speedups. In the present context of master–slave parallel GAs, the parallel speedup can be measured by the ratio of the time spent in evaluating the n_p individuals per generation in a serial implementation of the GA (t_{ser}) and T_p given by Equation 3.9 for the master–slave GA. It is easy to see that the time spent in evaluating n_p individuals in a serial calculation is

$$t_{ser} = n_p t_f. \tag{3.11}$$

In order that the parallel implementation has an edge over the serial GA, we must have

$$\frac{t_{ser}}{T_p} = \frac{n_p t_f}{\dfrac{n_p t_f}{p} + p t_c} > 1. \tag{3.12}$$

Recalling that $\dfrac{t_f}{t_c} = r$, we have the simple looking condition

$$\frac{nr}{\dfrac{nr}{p} + p} > 1 \tag{3.13}$$

for achieving positive parallelization benefits.

It is easy to find that the above inequality puts a bound on the kind of $r = \dfrac{t_f}{t_c}$ ratio needed in order that the parallel master–slave GA performs better than the serial implementation. Thus solving Equation 3.13 for r, we have

$$r > \frac{p^2}{n_p (p-1)} \tag{3.14}$$

Let us consider the case of four processors in a parallel GA working with a single population of size $n_p = 5$. Then Equation 3.13 suggests that for earning parallelization dividend

$$\frac{t_f}{t_c} = r > \frac{16}{15}. \tag{3.15}$$

If we know t_f, t_c, it is immediately possible to check whether the designed MS parallel GA would surpass the serial version of the same GA in performance.

An important concept in this contest is the idea of parallel efficiency (ϵ_p) which is just the ratio of t_{ser} to T_p divided by the number of processors (p): that means

$$\epsilon_p = \frac{t_{ser}}{p \, T_p}. \tag{3.16}$$

If the ideal situation prevails ϵ_p should always remain 1 and parallel speedup would increase linearly with the number of processors. But that does not happen due to the presence of the communication overhead in T_p. It has been argued that when ϵ_p falls below 0.5, the parallelization becomes meaningless. It can be easily verified that the larger the value of r the larger the magnitude of p for which the linear speedup can be realized. Let us assume that we would like to maintain a particular efficiency level—ϵ_p^*, say. What is the critical number of processors p_c that would be needed to maintain this desired level of efficiency? It is straightforward to make use of Equation 3.16 with $\in_p$ replaced by $\in_p^*$ and Equation 3.12, and solve for p to get

$$p_c = \left(\frac{1 - \in_p^*}{\in_p^*} \right)^{1/2} \left(n_p \frac{t_f}{t_c} \right)^{1/2}. \tag{3.17}$$

Recalling the expression for the optimal number of processors (p_c) as given in Equation 3.10, we can immediately conclude that $p_c < p_0$ except for the case where $\in_p^* = 1/2$. At the optimal configuration the simple master–slave GAs, therefore, have an efficiency of 50%. That means 50% of the time the processors are either idle (waiting for the master to assign them tasks) or busy in communication. The master cannot do any better as it has to wait for all the slaves to complete their evaluation of individuals, before it can select parents for the next generation and assign individuals to different nodes. The synchronous parallel GA has its pitfall in the requirement to maintain generational synchronization. An asynchronous GA can do better. For further details we refer our readers to the excellent book by Cantú-Paz [16].

References

1. Fogel, L.J. 1962. Autonomous Automata. *Industrial Research* 4: 14–19 March 1–3, 1995, San Diego, California.
2. Fogel, L.J. 1964. *On the Organization of Intellect*. Los Angeles, CA: University of California.
3. McDonnell, J.R., R.G. Reynolds, and D.B. Fogel. 1995. An Evolutionary Programming Approach to Self-Adaptation on Finite State Machines. In *Evolutionary Programming IV: Proceedings of the Fourth Annual Conference on Evolutionary Programming*, March 1–3, pp. 355–365. San Diego, CA: MIT Press.
4. Zurada, J.M., R.J. Marks, and C.J. Robinson. 1994. *Computational Intelligence: Imitating Life*. New York: Institute of Electrical & Electronics Engineers.
5. Engelbrecht, A.P. 2007. *Computational Intelligence*. Computational Intelligence. Chichester, UK: Wiley.
6. Kennedy, J., and R.C. Eberhart. 2001. *Swarm Intelligence*. San Francisco, CA: Morgan Kaufmann Publishers Inc.
7. Rechenberg, I. 1994. Evolution Strategy. In *Computational Intelligence Imitating Life*, Zurada, J.M., Marks, R.J., and Robinson, C.J. (eds.), pp. 147–159. New York: IEEE Press.
8. Koza, J.R., D. Andre, F.H. Bennett, and M.A. Keane. 1999. *Genetic Programming III: Darwinian Invention & Problem Solving*. 1st ed. San Francisco, CA: Morgan Kaufmann Publishers Inc.
9. Koza, J.R. 1989. Hierarchical Genetic Algorithms Operating on Populations of Computer Programs. In *Proceedings of the 11th International Joint Conference on Artificial Intelligence – Volume 1*, August 20–25, pp. 768–774. IJCAI'89. San Francisco, CA: Morgan Kaufmann Publishers Inc.
10. Koza, J.R. 1992. *Genetic Programming: On the Programming of Computers by Means of Natural Selection*. August 20–25 1989. Cambridge, MA: MIT Press.
11. Koza, J.R. 1990. *Genetic Programming: A Paradigm for Genetically Breeding Populations of Computer Programs to Solve Problems*. Technical report. Stanford, CA: Stanford University.
12. Koza, J.R. 1994. *Genetic Programming II: Automatic Discovery of Reusable Programs*. Cambridge, MA: MIT Press.
13. Langdon, W.B., T. Soule, R. Poli, and J.A. Foster. 1999. The Evolution of Size and Shape. In *Advances in Genetic Programming*. Spector, L., Langdon, W.B., O'Reilly, U.-M., and Angeline, P.J. (eds.), Cambridge, MA: MIT Press. pp. 163–190.

14. Price, K.V. 1997. Differential Evolution vs. the Functions of the 2/sup nd/ICEO. In *IEEE International Conference on Evolutionary Computation,* April 13–16, pp. 153–157. Indianapolis, IN: University Place Hotel Indianapolis.

15. Price, K.V., R.N. Storn, and J.A. Lampinen. 2005. *Differential Evolution: A Practical Approach to Global Optimization.* Natural Computing Series. Springer-Verlag, Berlin, Printed in Germany.

16. Cantú-Paz, E. 2000. *Efficient and Accurate Parallel Genetic Algorithms.* Genetic Algorithms and Evolutionary Computation. New York: Springer.

17. Alba, E., and J.M. Troya. 1999. A Survey of Parallel Distributed Genetic Algorithms. *Complexity* 4, no. 4: 31–52.

18. Eldredge, N., and S.J. Gould. 1972. Punctuated Equilibria: An Alternative to Phyletic Gradualism. *Models in Paleobiology.* Schopf, T. J. M. (ed.), San Francisco: Freeman, Cooper and Co. pp. 82–115

19. Bethke, A.D. 1976. *Comparison of Genetic Algorithms and Gradient-Based Optimizers on Parallel Processors: Efficiency of Use of Processing Capacity.* Technical Report (University of Michigan. Logic of Computers Group). Ann Arbor, MI: Arts, Computer and Communication Sciences Department, College of Literature, Science, University of Michigan.

20. Grefenstette, J.J. 1981. *Parallel Adaptive Algorithms for Function Optimization: (Preliminary Report).* Technical Report (Vanderbilt University. Department of Computer Science). Computer Science Department, Vanderbilt University. Nashville, Tennessee.

21. Fogarty, T.C., and R. Huang. 1991. Implementing the Genetic Algorithm on Transputer Based Parallel Processing Systems. In *Parallel Problem Solving from Nature,* Schwefel, H.-P. and Männer, R. (eds.), Berlin: Springer-Verlag. pp. 145–149.

22. Abramson, D., and J. Abela. 1992. A Parallel Genetic Algorithm for Solving the School Timetabling Problem. In *International Joint Conference on Artificial Intelligence,* August 24–30, pp. 1–11. Sydney, NSW, Australia: Division of Information Technology, C.S.I.R.O.

4

Random Mutation Hill Climbing and Simulated Annealing Methods

4.1 Introduction

So far we have discussed a number of problem-solving techniques that have their roots in biology or more specifically in low-level biological processes. One common feature shared by all these methods is that they make use of a population of plausible or potential solutions which interact, share information, undergo random low-intensity changes, and slowly evolve into better and still better solutions. In all these methods, a condition is created in which the solution emerges naturally. The use of a population of n_p individuals (potential solutions) means roughly a n_p-fold increase in the computational labor compared to traditional methods that work with a single guessed solution, which is iteratively modified into a better solution following some well-defined recipe based on analysis. The solution does not emerge in such cases but is hammered out of the initial guess. The problem of enhanced computational requirement in the population-based methods, however, can be taken care of by invoking the parallel versions of the evolutionary algorithms discussed in Chapter 3. Could there be alternatives which retain global search efficiency without using a population-based strategy and therefore avoiding the need to parallelize? It is in this context that one can think of developing nature-inspired computational strategies which use a single potential solution as the starting point and nudge it slowly to develop into the globally optimal solution by following low-level biological or natural processes. In recent years, such methods have gained importance in view of their extreme simplicity and the ability to explore the search space well, navigating around or avoiding local traps. In the context of evolutionary algorithms, such a method can be called "a single parent" method of exploring the search space, while in a more general context we can categorize these methods as single string-based natural search algorithms [1–5]. The use of a single string or individual reduces the exploitable amount of information already present in the search space but avoids the complex problem of finding an optimal, or near-optimal, population size for a population-based method to perform well. There is also no need to design an efficient and appropriate mechanism for the exchange of information among the individual members of the population or enforce a viable selection mechanism for choosing individuals for breeding new solutions. All these make the single string-based search algorithms extremely simple to work with. There is a bonus in that these methods may still retain the ability to reach the global optima. In the following, we will first consider variants of a single-parent evolutionary algorithm in which "mutation" alone decides how evolution would proceed.

4.2 The Random Mutation Hill Climbing Method (RMHCM)

Let us suppose that our objective is to find the global minimum of an N-dimensional function $f(\vec{X}) \equiv f(x_1, x_2, ...x_k, ..., x_N)$, where x_1, x_2,..., x_N are the real-valued components of the vector $\vec{X}$. Suppose further that the problem has been reduced to an equivalent problem of finding the global maximum of a fitness function $F(\vec{X}, F_L) = F(x_1, x_2, ...x_k, ..., x_N; F_L)$ where

$$F(\vec{X}, F_L) = \{(f(\vec{X}) - F_L)^2 + \delta\}^{-1}. \tag{4.1}$$

F_L being an updatable lower bound to the values of the function $f(\vec{X})$. $\vec{X}$ is a trial solution vector with N real-valued components. As $f(\vec{X}) \to F_L$, the denominator tends to vanish and the fitness value rises uncontrollably, becoming infinite (∞) when $f(\vec{X}) = F_L$. When such a situation arises, the small positive number δ in the denominator of the fitness function takes care of the blow-up situation and the fitness value remains finite. If we represent $\vec{X}$ by a randomly generated binary bit string S_0 (the chromosome) with m bits assigned for each of the N components of $\vec{X}$, we have a single bit string of length $l = mN$ as the trial solution vector $\vec{X}(0)$ to start with, having the fitness value of, say, F_0. Random mutations are then allowed to take place at one or more sites with the probability p_m (the mutation probability), producing a new solution string S_1 (say) which is evaluated for its fitness value F_1. If the mutation has been beneficial (i.e., $F_1 > F_0$) or at least has not been harmful (i.e., $F_1 \not< F_0$), the mutated string (S_1) is accepted as a new solution string replacing S_0. Otherwise, S_1 is rejected and S_0 is allowed to undergo fresh mutations. In either case, one generation of fitness evolution is counted to have elapsed.

The RMHCM in its simplest form works on this principle of a single chromosome undergoing mutations producing better chromosomes and the cumulative effects of mutations over many generations, finally succeeding in discovering the correct solution vector or a good approximation to it. The RMHC algorithm can be represented as a pseudocode as described in the next section.

4.2.1 RMHC Pseudocode

1. Define p_m (mutation probability) and the fitness function $F(\vec{X})$.

2. Choose an $N{\cdot}m$ bit binary string randomly, map it onto the vector $(\vec{X}) = (x_1, x_2, ..., x_N)$, and evaluate its fitness. Call it the current best $\vec{X}_{bestc}$ with fitness value f_{bestc}.

3. Identify a locus on the string at random with a probability p_m and flip the bit producing a new bit string. Map it onto the vector $\vec{X}_{new}$ and calculate its fitness value f_{new}.

4. If $f_{new} \geq f_{bestc}$, then reset

   $\vec{X}_{bestc} = \vec{X}_{new}$

   $f_{bestc} = f_{new}$

   Otherwise, reject $\vec{X}_{new}$

   Go back to step 3.

5. Continue the iteration until a preset number of trials (n_{max}) has been executed or until no better string has been generated during the last M number of trials (generation) consecutively.

6. Print the $\bar{X}_{bestc}$ and f_{bestc} as the closest approximation to the solution vector and its fitness value.

In many problems of interest to physicists and chemists, it could be more convenient to work with a single string (a vector X) with N real-valued components: $S(x_1, x_2,..., x_N)$. The RMHC procedure remains exactly the same except that the mutation is not done by bit-flipping but by adopting an appropriate floating-point mutation scheme that works on the components of the vector directly. Supposing that the kth component of the string $S(x_1, x_2,..., \overset{\downarrow}{x_k},..., x_N)$ has been randomly selected with the probability p_m to undergo mutation, the mutated string is generated by mutating the value x_k to x'_k,

$$x'_k = x_k + (1 - 2r)r_1\Delta_m \tag{4.2}$$

where Δ_m represents the mutation intensity (which may be static, i.e., generation independent, or dynamic, i.e., changing with generations), r is a uniformly distributed random number between 0 and 1, while r_1 is another random number in the same range drawn from a Gaussian distribution.

The product $r_1\Delta_m$ on the right-hand side of Equation 4.2 defines the actual magnitude of the random mutation, while the term $(1-2r)$ fixes the sign of the mutation (whether the mutation increases or decreases the value of x_k over its current value, i.e., whether $x'_k > x_k$ or $x'_k < x_k$) with equal probability. The acceptance or rejection of the mutated string is decided by the fitness value of the mutated string as described in the pseudocode. The algorithm just described can be viewed as a single-parent evolutionary algorithm based on random mutations working on floating-point strings and is a variant of RMHCM. The simple method, however, presents some interesting possibilities for further improvement.

4.2.2 Adaptive RMHC

Let us note that the method just described has two adjustable parameters, namely p_m (the mutation probability) and Δ_m (the mutation intensity). One of the possibilities of practically realizing dynamic adjustment of p_m and Δ_m as generations elapse is to allow the algorithm to determine the suitable values of p_m and Δ_m completely adaptively based on continuous assessment of the success and failure rates of random mutation during the span of the last M generation (M: user-defined, a typical value could be 100). The adaptive control on p_m and Δ_m can be enforced as follows:

Suppose that we have made a reconfiguring (mutation) move and produced a new solution vector $\bar{X}_{new}(t+1)$ from $\bar{X}_{old}(t)$, t being the index counting the generations that have elapsed. Let $f_{new}(t+1)$ and $f_{old}(t)$ be the fitness values of the new and the old strings, respectively. If $f_{new}(t+1) \geq f_{old}$, the mutated string $\bar{X}_{new}(t+1)$ is accepted and it replaces $\bar{X}_{old}(t)$; otherwise, $\bar{X}_{new}(t+1)$ is rejected. Each time the mutated string is accepted (successful mutation), a counter $icount$ is incremented by 1 ($icount + 1$). The fraction f_r of the successful moves during the last M generation is therefore given by $f_r = icount/M$. After M generations, $icount$ is reset to its initial value $icount = 0$.

If $f_r < \varepsilon_l$ (ε_l, user-defined lower limit of success rate), the mutation intensity Δ_m is reduced randomly, while if $f_r > \varepsilon_u$ (ε_u, user-defined upper limit of success rate) Δ_m is increased randomly as follows:

$$\text{If } f_r < \varepsilon_l, \ \Delta_m \leftarrow \frac{\Delta_m}{1+r}$$

$$\text{If } f_r > \varepsilon_u, \ \Delta_m \leftarrow \Delta_m(1+r) \tag{4.3}$$

where r is a random number in the range (0,1) drawn from a Gaussian distribution. The logic behind the adaptive adjustment of Δ_m displayed in Equation 4.2 can be understood more easily through a concrete example. Let us suppose that we have set $\varepsilon_l = 0.01$, $\varepsilon_u = 0.1$ and $M = 100$. Then Equation 4.3 states that if during the last 100 generations not even a single mutation with intensity Δ_m was accepted, it is time to reduce Δ_m by a factor $\frac{1}{1+r}$ and use the new $\Delta_m \left(\Delta'_m = \frac{\Delta_m}{1+r} \right)$ value for further exploration of the searches space. The reduction in Δ_m value essentially means that the algorithm now emphasizes on the "depth search" aspect of exploration. On the other hand, if more than 10% of the mutations were successful during the last 100 generations, it would be prudent now to increase the Δ_m value by a factor $(1 + r)$ (i.e., to set $\Delta'_m = \Delta_m(1+r)$) which would amount to laying emphasis on the "breadth-search" aspect of the exploration. Not only Δ_m but also the mutation probability p_m can be adaptively determined by following a strategy similar to what has been described for adaptively controlling mutation intensity. The only change needed is in the values of δ_l and δ_u, the lower and upper limit of success/failure rates, respectively, based on which p_m values are adjusted. The adaptive strategy can be represented as follows:

$$\text{If } f_r < \delta_l, \ p_m \leftarrow \frac{p_m}{1+r}$$

$$\text{If } f_r > \delta_u, \ p_m \leftarrow p_m(1+r). \tag{4.4}$$

The rationale almost suggests itself—if too few mutations were getting accepted during the last M generation, it makes sense to reduce the mutation probability randomly by a factor $\frac{1}{1+r}$. If the percentage of beneficial mutations during the last M generation has been greater than the threshold ($\delta_u \times 100\%$), it would be prudent to carry out mutation with a randomly enhanced mutation probability (enhanced by a factor $(1+ r)$). The adaptive determination of p_m and Δ_m makes RMHC more efficient and has been called the completely adaptive RMHCM [6]. The choice of threshold value for ε_l, ε_u, δ_l, and δ_u may be problem dependent and some initial experiments may be necessary to fix their values. We will describe some interesting application of completely adaptive random mutation hill climbing (CARMHC) in quantum chemical calculations in Chapter 7.

4.2.3 Variants of RMHC

The RMHC procedure can be invoked easily when the solution of the problem is represented as an integer string. Take, for example, the problem of finding the ground state of a one-dimensional Ising chain of N spins (S_i), S_is taking on only one of the two values

+1 or −1 and interact only with its immediate neighbor S_{i+1}, the strength of the interaction being fixed by the value of the coupling constant (J). Formally, the energy of Ising chain can be written as

$$E(S) = -\frac{1}{2} \sum_{i,j=1}^{N} J_{ij} S_i S_j \tag{4.5}$$

where S_i, S_j can take on values +1 or −1 randomly, and $J_{ij} = J$ (>0) Only if $|i - j| = 1$. The problem is to find the ground state configuration of the spins—to find the spin values (+1 or −1) at N different sites that minimize the energy $E(S)$. The initial string S chosen randomly is a collection of N spin values (±1), say, S_0 (1 − 1 − 1 − 11... −1). Let the energy be E_0. One of N sites in S_0 is randomly chosen with a probability p_m and the spin is flipped to −1 if it was +1 or to +1 if it was −1. Let the mutated string be called S_1 and the energy of the mutate string be E_1. If $E_1 \leq E_0$, the mutated string is accepted and replaces the string S_0 ($S_0 = S_1$ and $E_0 = E_1$) which then undergoes another mutation, and the procedure is continued until no further lowering of energy takes place. The string S_0 then hopefully defines the ground state configuration of the N-spins, and E_0 the ground state energy. Note here that we have used the RMHCM directly for energy minimization as opposed to indirectly minimizing the energy through maximization of a suitably defined fitness function.

A little reflection suggests that the system being probed has a pair of degenerate ground states. Actually, the states with all spins up $S_i = 1$, $i = 1, 2, ..., N$ or all spins down $S_i = -1$, $i = 1, 2, ..., N$ has the same energy $E_0 = - (N - 1)J$ which is the lowest possible energy that the system can have and are therefore degenerate ground states. We may be curious to know which of the degenerate pair of ground states does the RMHC search converges to. The answer is that the RMHCM will discover any one of the two degenerate ground states with equal probability—that means if RMHC is run for a large number of times, half the runs will discover the all +1 spin ground state while the remaining half will lead to the all −1 spin ground state. Being a stochastic search, in a particular run the RMHC will lead to one of the two degenerated states randomly with equal probability, as long as the energy of the chain is the only quantity guiding the search. It is, however, possible to tweak the search so that it converges to only one of the two degenerate states. The tweaking comes in the form of a modified fitness function or objective function. Suppose that we are looking for the all +1 spin ground states and E_L is an improvable lower bound to the ground state energy. Then our objective would be to minimize the energy $E(S)$ and maximize the number of +1 spins in the chain. Let n_+ be the number of +1 spins at any particular state of the search and $E(S)$ be the energy of the chain. A fitness function consistent with our objective can, for example, be written as

$$f_t(E(S), E_L) = \frac{1}{(E(S) - E_L)^2 + \lambda\left(1 - \dfrac{n_+}{N}\right)^2 + \gamma_s} \tag{4.6}$$

where γ_S is a small shift that takes care of the blow-up situation when the other two terms in the denominators approach and become zero, and λ is a penalty factor (>0). As energy is minimized $E(S) \to E_L$, $(E(S) - E_L)^2$ approaching zero. That contributes to increases in the fitness value f_+. However, the increase will be damped unless the term $\lambda\left(1 - \dfrac{n_+}{N}\right)^2$ also tends to zero simultaneously as $E(S) \to E_L$. This is possible only if $n_+ \to N$ as $E(S) \to E_L$. Thus, the

RMHC search guided by the fitness function f_+ will seek a solution (spin configuration) that reduces both the terms in the denominator maximally for a particular choice of γ_S and $\lambda(> 0)$ and converges to the all +1 spin ground state in any run. Similarly, we may define another fitness function $f_-(E(S), E_L)$ where

$$f\_(E(S), E_L) = \frac{1}{(E(S) - E_L)^2 + \lambda\left(1 - \dfrac{n_-}{N}\right)^2 + \gamma_s} \qquad (4.7)$$

n_- being the number of -1 spins in the chain at a particular stage of the search. By following the same logic as used for the maximization of $f_+(E(S), E_L)$, we can conclude that maximization of $f_-(E(S), E_L)$ by the RMHC procedure will lead to the all -1 spin in ground state. One can explore many such modified fitness functions for handling the problem of degenerate ground states and pick up one of them exclusively, although the search becomes more complex and time-consuming.

The RMHC, a simple optimization algorithm, has been viewed as a single-parent evolutionary method. It appears to work well as evidenced by reported applications (see Chapters 6–7) to complex problems. In view of its successes, it would be worthwhile to examine its expected time complexity theoretically.

4.2.4 Analysis of RMHCM

The building block hypothesis discussed in Chapter 2 emphasizes the idea that short, low-order schemas with above-average fitness are gradually brought onto the same string by the recombination (crossover) operator creating intermediate-order schemas of higher fitness which are combined to produce schemas of even higher fitness, and so on. The role of the recombination process in creating progressively higher-order schemas of higher fitness is central to the operation of the building block hypothesis. RMHC being a single-parent evolutionary algorithm does not have the recombination operator and its benefits. That appears to seriously limit the efficacy of the search (compared to genetic algorithms). The practical experience has, however, been otherwise. The remarkable effectiveness of RMHC or its adaptive variants in a number of diverse applications has been sought to be rationalized by appealing to biology. Let us note that the biological evolution of a species is marked by minor changes in a small portion of the "genetic code" taking place over a span of many generations. The small but beneficial changes tend to accumulate, leading to the evolution of a superior species. More extensive changes in the code brought in by a major mutation or by combining genes of different species seldom produce viable descendants. The RMHC procedure succeeds probably because it is designed to allow only rather small changes in the "chromosome" at a time, which accumulate over generations to shape the ultimate solution string. It would be interesting to analyze how it happens and understand the kind of time complexity the emergence of the correct solution in RMHC is associated with, even if it is in the context of a model problem, however artificial.

Consider a problem that has an exact solution represented by a binary string of M bits each of which is one (1). We imagine that the M bits are subdivided into N adjacent blocks, each block comprising L ones (1). The blocks represent the schemas (S_i) of order L to be discovered. The problem is to find N adjacent blocks of L ones, starting from a randomly generated binary bit string of length $M (= NL)$ comprising zeros and ones. The fitness function of the problem can be defined in terms of schemas $\{S_i\}$ (see Royal Road function [7]) as follows:

$$F(x) = \sum_i C_i \delta_i(x) \tag{4.8}$$

where

$$\delta_i(x) = \begin{cases} 1 \text{ if } x \in S_i \\ 0, \text{otherwise.} \end{cases}$$

The string x is an instance of the schema S_i. All the C_is have the same value (say, equal to 4). Assuming that $M = 16$, $L = 4$, there are $N = 4$ blocks each containing 4×1)—these are the order-4 schemas to be discovered. They are displayed in Table 4.1. The exact solution of the problem is the string S_0 reported in the same table which has the maximum fitness.

The value of f_0 is 16. Clearly, S_0 is formed by juxtaposing the four order-4 schemas S_1, S_2, S_3, S_4. The task of the RMHC algorithm is to discover S_0. In the general case, S_0 will be comprised of M blocks of L ones.

Suppose that the algorithm discovers the first block of L ones in time $T(L, 1)$ through random mutations of the $N = ML$ bits of a randomly chosen string. The question now is to assess the time that algorithm will consume to discover the second block of L ones. Let the time required be $T(2, L)$. A little reflection will suggest that $T(2, L)$ is going to be longer than $T(1, L)$—why? The reason is that as the $N = ML$ bits on the evolving string S are sampled, a fraction of the trials will inevitably be allocated to the L bits in the already discovered first block of L ones. These trials will be wasted as any mutation within this block will work only to lower the fitness value defined for the problem. The fraction q of the $N = ML$ trials that could be useful in discovering the second block is $\frac{ML - L}{ML} = \frac{M - 1}{M}$. Therefore, the time required to find the second block would be $\frac{M}{M - 1}$ times higher than the time consumed in discovering the first block as the algorithm spends only a fraction $\frac{M}{M - 1}$ of the trials for achieving beneficial mutations. So the time to the discovery of the second block is

$$T(L, 2) = T(L, 1) + T(L, 1)\frac{M}{M - 1}. \tag{4.9}$$

TABLE 4.1

The Schemas and Their Fitness Values

S_i	Schemas	C_i	Fitness
(S_1)	111************	$C_1 = 4$	$f_1 = 4$
(S_2)	****111********	$C_2 = 4$	$f_2 = 4$
(S_3)	********1111****	$C_3 = 4$	$f_3 = 4$
(S_4)	************1111	$C_4 = 4$	$f_4 = 4$
(S_0)	1111111111111111		$f_0 = 16$

Using a similar argument, the time to the third block is

$$T(L,3) = T(L,2) + T(L,1)\frac{M}{M-2}. \tag{4.10}$$

Continuing the process until the last block of L ones is (the Mth block) found, we have an estimate of the total expected time

$$T(L,M) = T(L,M-1) + T(L,1)\frac{M}{M-(M-1)}$$

$$= T(L,M-2) + T(L,1)\frac{M}{M-(M-2)} + T(L,1)\frac{M}{M-(M-1)} \tag{4.11}$$

$$= MT(L,1)\left\{\frac{1}{M} + \frac{1}{M-1} + \frac{1}{M-2} + \cdots + \frac{1}{3} + \frac{1}{2} + \frac{1}{1}\right\}.$$

Rewriting the series within the curly bracket, we have

$$T(L,M) = MT(L,1)\left\{1 + \frac{1}{2} + \frac{1}{3} + \cdots + \frac{1}{M}\right\}$$

$$= MT(L,1)\{\ln M + \gamma\} \tag{4.12}$$

where γ is the well-known Euler's constant and lnM is the natural algorithm of M. A conservative estimate of this is $T(1, L) = \beta 2^L$ ($\beta > 1$) so that the total expected time to find the M blocks of L ones—the time required to find the exact solution—is

$$T(M,L) = \beta 2^L M(lnM + \gamma) \ (\beta > 1, \ \gamma = 0.5170). \tag{4.13}$$

It is not an astronomically long time even when ML is large as $T(M, L)$ grows as $M \, lnM$. It has been mathematically demonstrated that compared to an ideal genetic algorithm (IGA) the expected time in RMHC is only M times higher (see Mitchell et al. [4]). That is

$$T_{IGA}(M,L) = \beta 2^L (lnM + \gamma). \tag{4.14}$$

It is not unlikely therefore that RMHC or its adaptive analogues (CARMHC) may work well in many cases and, considering its simplicity, may be the method of choice (see Chapter 7) for solving complex problems (nondeterministic polynomial time hard (NP-hard) problems, for example).

4.3 The Simulated Annealing Method

The RMHCM of finding the optima of a multidimensional function of continuous or discrete variables described and analyzed in the previous sections has an analogue in what has become well known as the classical simulated annealing method (SAM). The method was proposed [1] for locating the global optima in problems that are known to be NP-hard (e.g., traveling salesman problem) or problems with the requirement to satisfy conflicting conditions (i.e., problems with frustrations).

Unlike the RMHC, the SAM has its roots in a natural but non-biological process. Imagine how molten metals cool, freeze, and anneal. If the molten metal is cooled very slowly, it is likely to freeze into a perfectly crystalline form, that is, the form or structure with the lowest free energy. If the liquid is cooled rapidly (quenched), the result may not be the perfectly crystalline species with the lowest free energy, but a solid with some degree of quenched-in randomness or disorder. The solid does not occupy the global minimum free energy well on the free energy hypersurface but represents a higher free energy structure trapped in a local minimum—a metastable state so to say on the free energy hypersurface. Because of rapid cooling, the atoms could not escape from the local minimum and reorganize themselves in a more crystalline arrangement of lower free energy. The problem has been well known in metallurgy and metallurgists try to solve this problem by a rather simple recipe—alternately heat and cool the sample obtained by rapid quenching. The heating provides adequate thermal stimulus that nudges the system out of the confines of the local minimum or the trap and thereby explores the free energy hypersurface further in search of the global minimum. After several heating–cooling cycles, the system hopefully attains the global minimum free energy structure. The heating–cooling cycles, called the annealing schedule, are predesigned on the basis of experience.

The rapid cooling and the consequent problem of getting the system trapped in a local free energy minimum has a parallel in the optimization of functions having multiple minima—a quick-descent on the function hypersurface from an arbitrarily chosen starting point on the function surface will generally reach a local minimum, the global minimum remaining elusive. What do we mean by quick-descent on the function surface? Let us try to visualize and analyze the problem. Suppose $f(x)$ is a nonlinear function of a single real variable x characterized by multiple minima and maxima. Speaking theoretically, all the optima that the function $f(x)$ admits of can be found by differentiating $f(x)$ with respect to x, equating the first derivative $f'(x)$ to zero and solving the resulting equation $f'(x) = 0$ for x. The equation $f'(x) = 0$ will be generally nonlinear, admitting multiple solutions $\{x_i^*\}_{i=1,2,...}$ for which $f'(x_i^*) = 0$. Unless the values of the function $f(x_i^*)$ are evaluated at all such points $\{x_i^*\}$, it is not possible to identify the global minimum (global minima, if it is degenerate) as there is no *a priori* means of knowing which of the optima represents the global minimum. There is the additional problem that $f'(x_i^*) = 0$ also at the local or global maxima of the function so that the values of the second derivation $f''(x_i^*)$ must also be calculated to ascertain the nature of the optimum represented by x_i^*. Figure 4.1 displays the profile of a hypothetical function $f(x)$ supporting multiple (5) minima at $x_1^*, x_2^*, x_3^*, x_4^*$, and x_5^*, respectively, and maxima (4) at $x_1^\#, x_2^\#, x_3^\#$, and $x_4^\#$. Visual inspection of the graph identifies x_3^* as the global minimum and $x_2^\#$ as the global maximum. The straightforward method outlined above is workable only when $f'(x) = 0$ is analytically solvable and the values of the second derivative $f''(x)$ at all the designated points are available. This severely restricts the viability of the straightforward method. A more practical and workable route to the optima could be to solve the equation $f'(x) = 0$ by well-established numerical

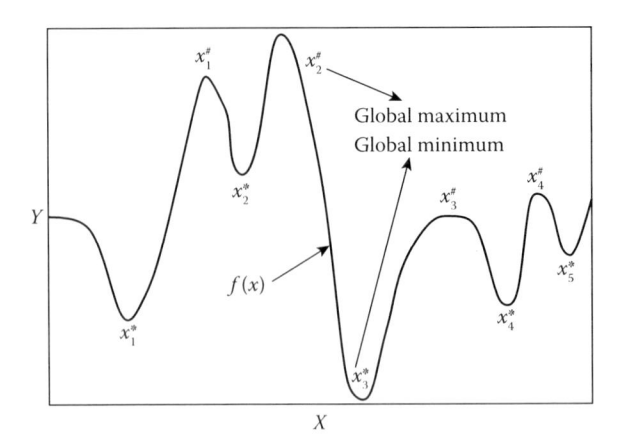

FIGURE 4.1
A hypothetical function $f(x)$ supporting multiple optima.

procedures such as the Newton–Raphson method. Let us examine the suggestion in detail. Suppose we write $f'(x) = g(x)$ so that the equation to be solved is

$$g(x) = 0.$$

Let us start from a point x_i where $g(x_i) \neq 0$, but which is in the neighborhood of a solution of the equation $g(x) = 0$. Taylor-expanding $g(x)$ to order "h" about x_i, and assuming that at the new point $x_i + h$ the function $g(x)$ is zero, we have

$$g(x_i + h) \approx g(x_i) + hg'(x_i) = 0 \tag{4.15}$$

which immediately suggests that

$$h = -\frac{g(x_i)}{g'(x_i)} \tag{4.16}$$

This means that the new point is

$$x_i + h = x_i - \frac{g(x_i)}{g'(x_i)}. \tag{4.17}$$

Calling the new point x_{i+1}, we can set up an iteration scheme for finding the optimum lying closest to the starting point (x_0).

$$x_{i+1} = x_i - \frac{g(x_i)}{g'(x_i)}, \quad i = 0, 1, 2, \ldots. \tag{4.18}$$

The method just described is the celebrated Newton–Raphson method. Under certain conditions, the iterations converge quadratically to one of the solutions of the equation $g(x) = f'(x) = 0$, depending on the starting point of the iterations. Thus, there is no *a priori* guarantee that the iterations will converge to the global minimum of $f(x)$—we can only claim that the iterations would converge to one of the optima, that is, one of the solutions of the equation $f'(x) = 0$—not even necessarily a minimum. Once the solution is reached, the iterations remain stuck there as $g(x) = 0$ in Equation 4.18. To seek other solutions (optima), we must reset the value of x_0 and start afresh.

As an alternative to identifying the minima by numerically solving the equation $f'(x) = 0$ for optima and then categorizing the nature of the optima, it is possible to seek them by directly minimizing the function $f(x)$ by following the descent path suggested by the local gradient $g(x)$ of the function $f(x)$ $(g(x) = f'(x))$. Suppose that we are in the neighborhood of a minimum. Taylor expansion of $f(x)$ about the designated point x_i gives us, for a small value of h,

$$f(x_i + h) \approx f(x_i) + hf'(x_i) + \frac{h^2}{2} f''(x_i). \tag{4.19}$$

If the expansion is convergent, $\frac{h^2}{2} f''(x_i) < hf'(x_i)$. In that event, $f(x_i + h) < f(x_i)$ if $hf'(x_i) < 0$, a condition that is guaranteed to be satisfied if we choose $h = -\lambda f'(x_i)$, $(\lambda > 0$, but small). This means that the function value is guaranteed to decrease if we move from the point x_i in the direction opposite to that of the local gradient at x_i in a small step of length λ. The decrease continues until we reach a point x^* where $f'(x^*) = 0$—that is, a local minimum has been found. This strategy of minimizing a function $f(x)$ is known as the method of steepest descent. Clearly, the same strategy can also be exploited to maximize a function by choosing $h = \lambda f'(x_i)$, $(\lambda > 0$, but small) as the choice ensures that $f(x + h) + f(x)$ the method being known as the method of steepest ascent. Thus, the steepest descent (ascent) method provides a direct numerical means of minimizing (maximizing) a function $f(x)$ so that one class of optima can be targeted. The problem of getting stuck in a local optimum, however, remains. As soon as the steepest descent (SD) method hits a point x^* where $f'(x^*) = 0$, there is no way to escape from the minimum as the gradient vanishes locally. There is an additional worry that unless the magnitude of λ (> 0) is carefully chosen, the iterations may overshoot the minimum being approached. On the other hand, if λ is too small, the descent becomes too slow. We can get around the overshooting problem and the problem of too-slow descent by incorporating more information about the local geometry of the function surface into the algorithm. Thus, by minimizing $f(x_i + h)$ of Equation 4.19 with respect to h, we have

$$f'(x_i) + hf''(x_i) = 0 \tag{4.20}$$

which leads to

$$h = -\frac{f'(x_i)}{f''(x_i)} \tag{4.21}$$

The updating formula for the local minimizing moves then becomes

$$x_{i+1} = x_i - \frac{f'(x_i)}{f''(x_i)}. \tag{4.22}$$

The moves are still in the direction opposite to that of the local gradient, but the step lengths are modulated by the local curvature—if the local curvature (>0 in the neighborhood of a minimum) is large, the step length (λ) is small—if the curvature is small, the step being taken is longer. This strategy of direct minimization is known in the literature as Newton's method.

The strategy is simple—starting from the chosen point x_i to move in a direction opposite to that of the local gradient ($f'(x_i)$) in steps of length inversely proportional to the local curvature ($f''(x_i)$). The movement stops as one reaches a local minimum at x^* where $f'(x^*) = 0$. The method works well, but requires calculation of both the first and the second derivatives of the function being minimized. It avoids the problem of overshooting the minimum being approached but there is no inbuilt mechanism in Newton's method (NM) to escape from a local minimum once it reaches the minimum. The only option left is to choose a new starting point (x_{new}^0) and redo the calculations to see if a deeper minimum can be found. The global minimum may thus elude the NM-based search just as it happens in the SD method. Let us note here that both the SD and NM can be easily extended to handle optimization problem in many dimensions. We will describe these extensions briefly here.

Let $f(x_1, x_2, \ldots, x_k, \ldots, x_n) = f(\vec{X})$ be an *n*-dimensional function. Let the starting point be $\vec{X}_i \equiv (x_1^i, x_2^i, \ldots, x_k^i, \ldots, x_n^i)$.

1. *The SD method in many dimensions*
   a. Compute the function value $f(\vec{X})$ at $\vec{X} = \vec{X}_i$.
   b. Compute local gradient vector $\vec{g}_i = \vec{\nabla} f(\vec{X}_i)$.
   c. Move downhill to find a new point $\vec{X}_{i+1}$
      where $\vec{X}_{i+1} = \vec{X}_i - \lambda \vec{\nabla} f(x_i)$, ($\lambda > 0$, small).
      Set $\vec{X}_i = \vec{X}_{i+1}$.
   d. Go back to step (a) and iterate until the norm of the gradient vector becomes vanishingly small or $|f(\vec{X}_{i+1}) - f(\vec{X}_i)| < \epsilon$.

2. *The NM in many dimensions*
   a. Starting point: $\vec{X}_i$, compute starting function value $f(\vec{X}_i)$.
   b. Compute the local gradient vector $\vec{g}_i$ and the matrix H_i of the second derivatives of the function with respect to the search variable $x_1, x_2, x_k, \ldots, x_n$.
   c. Find a new point $\vec{X}_{i+1}$

   where
   $$\vec{X}_{i+1} = \vec{X}_i - H_i^{-1} \vec{g}_i \tag{4.23}$$

   Set $\vec{X}_i = \vec{X}_{i+1}$

   d. Go back to step (a) and iterate until the norm of the gradient vector becomes vanishingly small or there is no significant lowering of the value of the function ($|f(\vec{X}_{i+1}) - f(\vec{X}_i)| < \epsilon$).

The problem of getting stuck at a local minimum by following the local gradient and curvature of the function being minimized is reminiscent of what happens when a molten metal is allowed to freeze quickly—it solidifies not necessarily into the perfectly crystalline state of the lowest free energy, but more probably into a metastable structure with

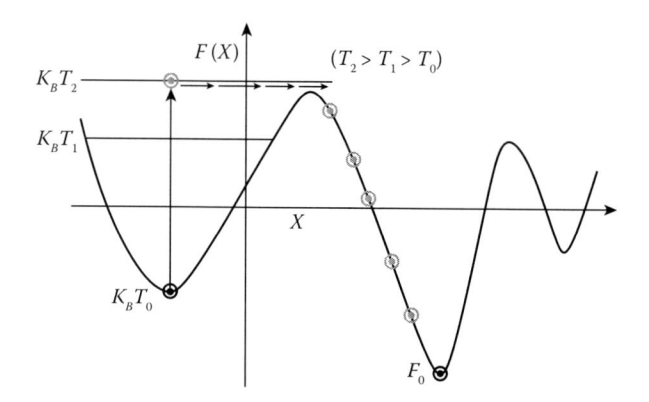

FIGURE 4.2
Escape from the local minimum by means of thermal excitation.

higher free energy. One way to enable the system to escape from the local minimum is to provide it with enough thermal energy (by heating the system up) so that thermal fluctuations (the scale determined by the temperature) are strong enough to dislodge the system from the local minimum and push it up over the local barrier (Figure 4.2) and explore the free energy landscape further in search of a deeper minimum and eventually find it as the system is slowly cooled down again to a very low temperature. At low enough temperature, the thermal fluctuations are small so that the system remains confined to the potential well until the system is heated up again sufficiently.

The message from the analogy is that an algorithmic search for the global minimum of a function must have an inbuilt mechanism that allows both downhill and uphill movements across the function surface in the exploration phase guided by a control parameter. The idea was exploited and very successfully implemented by Kirkpatrick et al. [1] and independently by Černý [2] in what has become known as the SAM. The algorithm is described below.

1. Let the n dimensional function to be minimized globally be $f(x_1, x_2, \ldots, x_k, \ldots, x_n)$, which is assumed to be in a thermodynamic environment characterized by a temperature "T_0." $f(X)$ is assumed to be akin to the thermodynamic free energy function and $x_1, x_2, \ldots, x_k, \ldots, x_n$ to various thermodynamic coordinates, which are to be optimized. T acts as the control parameter.

2. Because of thermal noise, the search variables $x_1, x_2, \ldots, x_k, \ldots, x_n$ are assumed to be undergoing random changes as the system approaches equilibrium at the given temperature (T_0). Let us pick up one of the variables, say, x_k, randomly as the one that suffers a random fluctuation at the current instant and assumes a new value, x'_k (say) where

$$x'_K = x_k + \Delta x_k (1 - 2r)$$

r being a uniformly distributed random number in the range (0, 1). Δx_k, the size of the fluctuation can be broken down into a product of two terms as ΔS and r_1, where ΔS is a user-defined small step length and r_1 is a random number in the range (0,1) drawn from a normal distribution. The reconfiguring move then assumes the form

$$x'_k = x_k + (1 - 2r)\,\Delta S\,r_1. \tag{4.24}$$

The term $(1 - 2r)$ fixes the sign ($\pm$) of the fluctuation Δx_k (i.e., whether there is increment or decrement of x_k following the fluctuation) with equal probability $p_t = \dfrac{1}{2}$.

3. The question now is to determine if the system accepts or rejects the change. Let the value of the function before the fluctuation be F, that is, $f(x_1, x_2, \ldots x_k, \ldots, x_n) = F$, and the post-fluctuation value be F' so that $f(x_1, x_2, \ldots x_k, \ldots, x_n) = F'$.

    The change in the function value is $\Delta F = F' - F$. To determine the acceptability of the change, we appeal to classical statistical mechanics. Let us assume that in the simulated thermodynamic environment F is akin to the thermodynamic free energy. According to the Boltzmann distribution, the probability of realizing the state with the function value F' from the state with the function value F by fluctuation at a temperature T is $P(\Delta F) = e^{-\Delta F}/k_B T$. Two possibilities arise.

   a. If $\Delta F < 0$, that is, $F' < F$, $P(\Delta F) > 1$ so that such a move is always accepted by the system.

   b. If $\Delta F > 0$, that is, $F' > F$, $P(\Delta F) < 1$, and we must determine if such a reconfiguring move is thermally allowed or acceptable. Let r be a random number between 0 and 1 drawn from a normal distribution.

     i. If $r < P(\Delta F)$, the move is accepted as uphill movement appears to be more probable than any random event at the given temperature, that is, the system experiences thermally assisted movement over a barrier of height ΔF.

     ii. If $r > P(\Delta F)$, the move is rejected—the scale of thermal fluctuation is inadequate in this case to enable the system to surmount the barrier.

4. Go back to step 3 and choose a new variable x_l (say) and iterate for a preset maximum number of times or until the simulated thermodynamic system hopefully attains thermal equilibrium at the current temperature T_i.

5. Reduce the temperature by following a predesigned cooling schedule (e.g., in a linear schedule $T_{i+1} = \eta T_i$, with, say, $\eta = 0.9$, and $i = 0, 1, 2, \ldots, N$). Go back to steps 2–4 as before. The cooling is continued until T_i becomes too small, that is, $T_i \to 0$ and no further lowering of the function value takes place. Hopefully, we have found the optimal set of parameters $(x_1^*, \ldots, x_n^*)$ and the global minimum of the function $f^*\left(x_1^*, x_2^*, \ldots, x_n^*\right) = F^*$. Store the information.

6. To ensure that we have indeed reached the global minimum, the system may be heated up again to a temperature T_0' and cooled following a predesigned annealing schedule executing steps 2–5. Check if a lower minimum has been discovered. If yes, that is accepted as the current global minimum. If no, the previously discovered minimum is accepted. (Note: the alternate heating and cooling cycle may be repeated a number of times as dictated by the predesigned annealing schedule.) The success of the annealing method in locating the ground state depends critically on the annealing schedule $(T(t))$. If it is fast, the search may get trapped in a local minimum. Geman and Geman [5] have shown that if we choose $T(t) \geq N/\ln t$, the system eventually reaches the ground state.

As we have seen, SAM only requires the evaluation of the function to be optimized and calculation of transition probabilities from the current state to a randomly generated state. No information about the local gradient vector or the second derivative matrix is used in the search. Being thus a function-only method endowed with the built-in ability to escape from local traps and reach the global minimum, SAM has been extensively used to handle wide varieties of problems which can be cast in the form of a constrained or unconstrained minimization problem. Both continuous and discrete optimizations have been tackled by the SAM. Some of these applications will be considered in Chapters 6 and 7. It is important at this stage to try to understand how the SAM works. In the following, we present a thermodynamic argument that explains the function of the SAM.

Suppose that we wish to locate the global minimum of the function $f(x_1, x_2, ..., x_k, ..., x_n)$. We may think of the function f as something akin to the thermodynamic internal energy function f which is distorted into a free energy function $F = U - TS$ as the system is placed in a simulated thermodynamic environment in which the system is in contact with a bath at temperature T_0. The bath delivers thermal noise to the system ($f \equiv U$) and the optimization variables ($x_1, x_2, ..., x_n$) fluctuate. When the simulated thermodynamic system comes to equilibrium (after many reconfiguring steps induced by thermal fluctuations), the system attains the free energy minimum state of $F = U = T_0 S \equiv f - T_0 S$ (free energy is minimum at equilibrium) at the current temperature T_0. Call it F_0. As the system is cooled to a temperature $T_1 < T_0$, the fluctuations get diminished in scale and a new equilibrium state characterized by the minimum free energy, say, F_1, emerges. If the cooling is slow enough and is continued to very low temperature ($T_n \to 0$), the system will pass through a continuous succession of equilibrium states characterized by the equilibrium free energies $F_0, F_1, F_2, ..., F_n$, each of which is the global minimum of the simulated free energy function at the corresponding temperature. As $T_n \to 0$, the sequence of minima $F_0, F_1, F_2, ..., F_n$ converges to the global minimum of $F = F_*$ which is just the global minimum of $U \equiv f$ as the distortion brought in by thermal perturbation vanishes at $T = 0$. The central theme of the SAM is thus to distort the function (f) to be minimized globally into a free energy-like function by thermal perturbation and then slowly reduce the temperature to $T = 0K$—so slowly that at every temperature the system attains the equilibrium state of minimum free energy—the so-called ground state. In the limit of $T = 0K$, the succession of global minima of the distorted function (F) smoothly passes over into the global minimum of the undistorted original function as the distortion asymptotically vanishes at $T = 0$. In a way, this is akin to what is done in the homotopy method. Let us have a look at the homotopy method.

Suppose that we wish to solve nonlinear algebraic equation $f(x) = 0$. Let $g(x) = 0$ be another equation for which a solution is known (say, $x = x_0$). Then we may construct a linear convex homotopy function $h(x, t)$ where

$$h(x,t) = (1-t)f(x) + tg(x) \tag{4.25}$$

t being a parameter that can assume value in the range $(0,1)$ continuously. Let t be equal to 1 to start with. Then $h(x, 1) = g(x)$. So starting with the vector $x = x_0$ which solves $h(x, 1) = 0 = g(x)$ we slowly decrease the value of t and solve $h(x, t) = 0$ for x (say, by NM) at each t value always starting the iterations with the solution obtained in the previous step. Let the solutions be $x_0(t)$. As t is continuously and slowly decreased to $t = 0$, we obtain a sequence of solution vectors $x_0(t)$ that are solutions of the equation $h(x, t) = 0$. In the limit $t \to 0$, this sequence converges to the vector $x_0(0)$ which is a solution of the original equation $h(x, 0) = f(x) = 0$. The connection with global minimization by homotopy method now becomes intuitively clear. If we are looking for the global minimum of $F(x)$, and we have at our disposal a function $G(x)$ with the known global minimum at $x = x^*$, we may seek the global minimum of the homotopy function $H(x, t) = (1 - t) F(x) + tG(x)$ for a series of decreasing values of t, starting from $t = 1$ for which known global minimum of $H(x, t)$ coincides with the global minimum of $G(x)$ at $x_{min}(t = 1) = x^*$. As the value of t is slowly diminished to $t = 0$, the sequence of global minima of $H(x, t)$ at different value of t pass over smoothly into the global minimum of the undistorted original function $F(x)$. The key, of course, is the rate at which t is switched off. A homotopic path may be defined by plotting $x_{min}(t)$ as a function of the distortion parameter t as shown in Figure 4.3.

The required slowness of the rate at which the distortion is removed reminds us of the quantum adiabatic theorem which asserts the following:

Let $H(x)$ be the Hamiltonian of our interest, and $H_0(x)$ be a simpler Hamiltonian the exact ground state of which is known. Then to arrive at the ground state of $H(x)$, we construct a time-dependent Hamiltonian $H'(x, t)$, where

$$H'(x,t) = H_0(x) + \lambda(t)\{H(x) - H_0(x)\} \tag{4.26}$$

and the switching function $\lambda(t) = 0$ at $t = 0$ and $\lambda(t) = 1$ at $t = T$, T being the time (large) where the switching concludes. The quantum adiabatic theorem asserts that the ground state of H_0 evolves into the ground state of $H(x)$ at the conclusion of switching provided $\lambda(t)$ is a very slowly evolving function of t $\left(\dfrac{d\lambda}{dt} \sim 0\right)$ and satisfies Kato condition including the boundary behavior at $t = 0$ and $t = T$. Once $\lambda(t)$ has been properly chosen, the integration of the time-dependent Schrodinger equation from $t = 0, T$ by a suitable algorithm can be expected to provide the exact ground state of $H(x)$. We will see later in Chapter 10 how this theorem paves the way to quantum adiabatic computing or quantum adiabatic search for the ground state of complex combinatorial explosive problems.

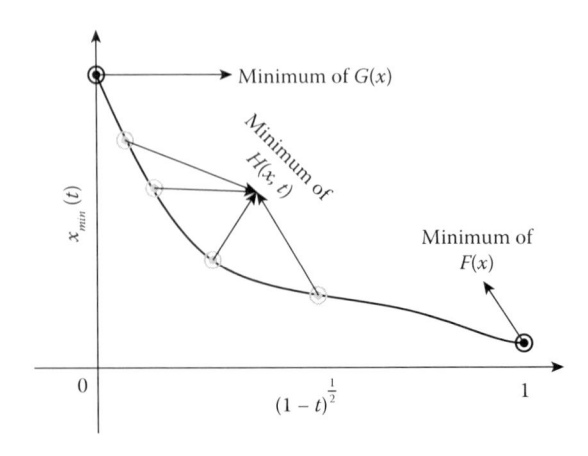

FIGURE 4.3
Homotopic path to the global minimum of $F(x)$.

4.4 The Method of Quantum Annealing (QA)

We have already seen that the SA is a powerful tool to locate the global minimum (or minima, in case there is degeneracy) in problems that support many local minima in the search space (the configuration space) of the problem. If there are N-independent variables to be optimized, the problem often turns out to have a computational complexity class known as NP-hard. In such problems, the computational time required to search out the global minimum (minima) cannot be represented by any polynomial in N, as N increases, the required search time diverges exponentially. The SAM may turn out to be inadequate to handle some of these problems, for example, the problem of locating the global minimum of a nonergodic system like spin glasses (N randomly interacting spins each of which can assume one of the two orientations, up (+1) or down (−1)). These systems are typically characterized by many local energy minima separated by barriers of heights that are of the order of N. Thermal dynamics at finite temperatures fail to sample the entire phase space resulting in the system getting trapped in one of the many local minima as $N \rightarrow \infty$. In this limit, finite thermal fluctuations are unable to help the system surmount the barriers separating the local minima and therefore fail to make the system ergodic [1,2,8]. The idea of quantum annealing provides a means of handling the situation in a more efficient manner [9,10]. In quantum annealing, thermal fluctuations are replaced by quantum fluctuations (nonzero even at $T = 0K$) and controlled by using a tunneling field Γ (an external magnetic field) in place of temperature (T). The basic idea of quantum in annealing can be illustrated clearly with the help of diagrams (Figure 4.4) and compared with that of SA (thermal annealing).

In thermal annealing, the escape probability by thermally induced over barrier transition from the local minimum (L) to the global minimum (G) is given by $P \sim e^{\Delta E / k_B T}$. On the other hand, the probability of escape through the same barrier by quantum mechanical tunneling is given by $P \sim e^{-\Delta W \sqrt{\Delta E} / \Gamma}$, where ΔW is the width of the barrier and Γ is the fluctuation field strength. Obviously, if ΔW is small compared to ΔE, the quantum passage through

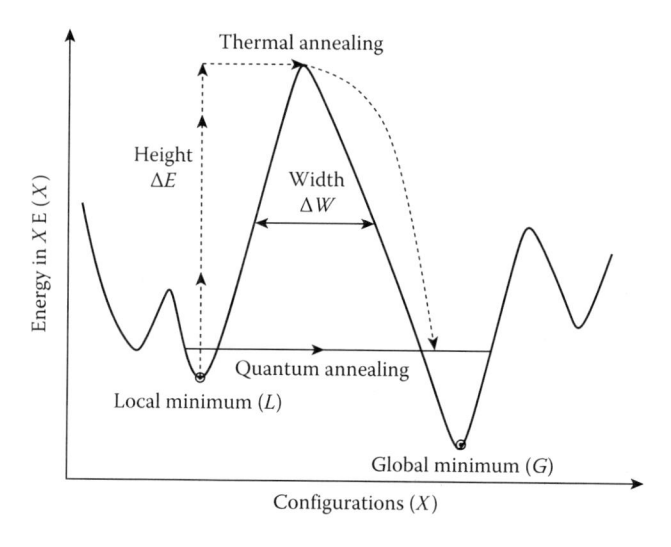

FIGURE 4.4
The schematic diagram of quantum annealing.

the barrier will be more probable than the classical thermal passage over the same barrier within a given time interval. We can therefore anticipate better efficiency for a quantum annealing-based search for the global minimum in such cases. Just like T in SAM, the control parameter Γ in quantum annealing (QA) must be reduced very slowly over time, that is, the conditions for quantum adiabatic passage from one state to another must be satisfied. We will review some application of quantum annealing in studying spin glass in Chapter 10. In spite of the promising nature of the quantum annealing method, it has not been explored that extensively in the context of quantum chemical optimization problem. We will examine some of the possibilities in the context of quantum soft computing options in the last chapter.

4.5 Elementary Applications of SAM in Quantum Mechanics

It would be useful to have a look at some of the early applications of classical annealing technique in the context of the search for solutions to quantum mechanical problems. The literature has grown and diversified enormously. Here, we will only review some of the simple applications of the SAM that illustrate the workability and power of the method for solving standard quantum mechanical problems and point to the possibilities that exist.

Dutta and Bhattacharyya [11], for example, experimented with an orthogonality-constrained variational calculation of excited state wave functions and energies by invoking the SAM for directly searching through the relevant parameter space. These authors stressed that their method does not require any information about the derivatives of the energy functional and has manageable computational complexity. The method has the built-in capability to escape from a local minimum along with the ability to handle multiple equality constraints. Being one of the very first calculations of wave functions and energies using the Metropolis SAM, it would be worthwhile to examine the method proposed by these authors in some detail.

Suppose that $\psi_1, \psi_2, ..., \psi_{n-1}$ are the $(n-1)$ lowest bound eigenstates supported by Hamiltonian H with energies $E_1 \leq E_2 \leq \cdots \leq E_{n-1}$. Let ψ be the trial wave function for the nth eigenstate of H which is constrained to remain orthogonal to the $(n-1)$ lower eigenstates $\psi_1, \psi_2, ..., \psi_{n-1}$. The objective function of the problem $f(a_1, a_2, ..., a_m)$ including the relevant constraints ($a_1, a_2, ..., a_m$ are parameters in the wave functions) was set as

$$F(a) = \left(\in (a) - E_L \right)^2 + \sum_{k=1}^{n-1} \lambda_k g_k \left(\langle \tilde{\psi} | \phi_k \rangle \right). \tag{4.27}$$

The constraint functions were so chosen that $g_k \left(\langle \tilde{\psi} | \phi_k \rangle \right) = 0$ only if $\langle \tilde{\psi} | \phi_k \rangle = 0$ and $g_k \left(\langle \tilde{\psi} | \phi_k \rangle \right)$ becomes large if $\langle \tilde{\psi} | \phi_k \rangle \neq 0$. Needless perhaps is to mention that λ_k is the penalty weight factor for the kth constraint and has appropriate dimension (energy squared). A simple choice for g_ks could be to take, for example

$$g_k \left(\langle \tilde{\psi} | \phi_k \rangle \right) = \left| \langle \tilde{\psi} | \phi_k \rangle \right|^2, \tag{4.28}$$

although many alternative forms are possible. For example, we could have chosen

$$g_k\left(\langle\tilde{\psi}|\,\phi_k\,\rangle\right)=\left[\left\{1-\left|\langle\tilde{\psi}|\,\phi_k\,\rangle\right|^2\right\}^{-1}-1\right]^2 \tag{4.29}$$

and that could be a very effective choice for the penalty function. With the choice of the objective function defined in Equation 4.27 and the constraint function as defined in Equations 4.28 and 4.29, Dutta and Bhattacharyya [11] applied their SAM-based method for the calculation of the first excited states (singlet 1s2s) of a number of two-electron atoms and ions. The ground state and the first excited state wave functions were chosen in the Eckert–Hylleraas form.

First, the ground state wave function was calculated by minimizing $F(\alpha_0, \beta_0)$

$$F\left(\alpha_0,\beta_0\right)=\left(\frac{\langle\psi_0|\,H|\,\psi_0\,\rangle}{\langle\psi_0|\,\psi_0\,\rangle}-E_L\right)^2=\left(\epsilon_0\left(\alpha_0,\beta_0\right)-E_L\right)^2$$

where α_0 and β_0 are the optimization variables (no orthogonality constraint was required for the ground state—therefore, g_{ks} were set $= 0$) and the trial wave function is chosen as

$$\psi_0\left(r_1,r_2\right)=N_0\left[exp\left\{-\left(\alpha_0 r_1+\beta_0 r_2\right)\right\}+exp\left\{-\left(\beta_0 r_1+\alpha_0 r_2\right)\right\}\right] \tag{4.30}$$

E_L is an updatable lower bound to the ground state energy. In the second step, the first excited singlet ($^1s_1$) state wave function and energy were calculated by minimizing the constrained functional

$$F\left(\alpha',\,\beta',\,\alpha_0,\,\beta_0\right)=\left(\epsilon_0\left(\alpha',\beta'\right)-E_L\right)^2+\lambda\left\{S\left(\alpha',\,\beta',\,\alpha_0,\,\beta_0\right)\right\}^2 \tag{4.31}$$

where

$$\epsilon\left(\alpha',\beta'\right)=\left(\frac{\langle\psi_1\left(r_1,r_2\right)|H|\,\psi_1\left(r_1,r_2\right)\rangle}{\langle\psi_1|\,\psi_1\,\rangle}-E_L\right)^2 \tag{4.32}$$

and

$$\psi_1\left(r_1,r_2\right)=N_1\left[u_{1s}\left(\alpha',r_1\right)u_{2s}\left(\beta',r_2\right)+u_{1s}\left(\beta',r_1\right)u_{2s}\left(\alpha',r_2\right)\right] \tag{4.33}$$

where

$$u_{1s}\left(\alpha',r\right)=N_{1s}\exp\left(-\alpha'r\right)$$

$$u_{2s}\left(\beta',r\right)=N_{2s}\exp\left(-\frac{1}{2}\beta'r\right)\left\{1+\frac{1}{2}\beta'r\right\} \tag{4.34}$$

$S(\alpha', \beta', \alpha_0, \beta_0)$ represents the value of the overlap integral between ψ_1 and ψ_0 (preoptimized). These authors showed that the calculated excited state energies had strict upper-bound character and the excited state wave function remained orthogonal to the preoptimized approximate ground state. The residual overlap, if any, could easily be taken care of by carrying out a 2×2 configuration interaction (CI) calculation [12] with $\psi_0(r_1, r_2)$ and $\psi_1(r_1, r_2)$ as nonorthogonal basis functions. They also demonstrated that the SAM-based strategy always succeeded in discovering a better constrained minimum compared to a gradient-based method like the SD. (In Table 4.2, the overlap with the ground state was uniformly small in the SAM-based calculation.)

Dutta and Bhattacharyya [13] in a sequel to the work just described considered the possibility of exploiting the SAM-based orthogonality constrained direct search technique to compute the excited state wave functions and energies of a system when the trial space is linear. Suppose that we are interested in the nth excited state when all $n - 1$ lower lying state including the ground state are already known. In a fixed basis set $\{\chi_k\}$ spanning an N-dimensional linear space, the trial wave function for the nth excited state can be expressed as

$$\tilde{\psi}_n = \sum_{k=1}^{N} a_k^{(n)} \chi_k; \qquad \left(\langle \chi_k | \chi_l \rangle = S_{kl} \right). \tag{4.35}$$

Suppose further that all the energies and wave functions of the $n - 1$ lower eigenstates of H are known $\left(\phi_1(E_1), \phi_2(E_2), \ldots, \phi_{n-1}(E_{n-1}) \right)$ with $E_1 \le E_2 \le \cdots \le E_{n-1}$. Our problem is to find an optimal set of the linear expansion coefficients $\left(a_k^{(n)}, k = 1, 2, \ldots, N \right)$ that minimize the energy (E_n) of the nth excited state and keeps ψ_n orthogonal to all the lower states. It means that the following constraint conditions must be satisfied during the optimization:

$$\langle \tilde{\psi}_n | \phi_k \rangle = 0, \quad k = 1, 2, \ldots, n-1. \tag{4.36}$$

In addition, $\tilde{\psi}_n$ must stay normalized $\left(\langle \tilde{\psi}_n | \tilde{\psi}_n \rangle = 1 \right)$. Dutta and Bhattacharyya [13] constructed an objective function $\tilde{F}(a)$ whose minimization was the goal of the search:

$$\tilde{F}(a) = \left(\tilde{E}_n(a) - E_L^n \right)^2 + \sum_{k=1}^{n-1} \lambda_k \left| \langle \tilde{\psi}_n | \phi_k \rangle \right|^2 + \beta \left\{ 1 - \langle \tilde{\psi}_n | \tilde{\psi}_n \rangle \right\}^2. \tag{4.37}$$

In Equation 4.37,

$$\tilde{E}_n(a) = \sum_{k,l}^{N} a_k^{(n)} a_l^{(n)} H_{kl} \Bigg/ \sum_k a_k a_l s_{ij}. \tag{4.38}$$

TABLE 4.2

Comparison of the Constrained Minima Obtained from Steepest Descent (SD) and SAM-Based Calculations on the First Excited States ($^1s_1$: 1s2s) of He and Helium-Like Atoms/Ions

Atoms/Ions (z)	Energy (au) (SD)	Energy (au) (SAM)	Overlap with Ground State
2	−2.058769	−2.058770	-2.1×10^{-10}
6	−21.133733	−21.133733	5.9×10^{-8}
10	−60.207140	−60.206670	1.1×10^{-4}

Note: au, atomic unit; z, atomic number.

E_L^n is an estimated lower bound to the energy of the nth excited state. λ_k is the penalty weight factor for the kth orthogonality constraint and β is the penalty weight factor for preventing violation of the normalization constraint. The search variables were the linear expansion coefficients $a_1^{(n)}$, $a_2^{(n)}$, ..., $a_N^{(n)}$ for the wave function in the nth state (i.e., the $(n-1)$th excited state of the system). The search proceeds in the following steps.

1. Set the initial temperature T_0 and an initial set of expansion coefficients $\left\{a_1^{(n)}(0)\right\}$; calculate energy $E_n(0)$ and value of the objective function F(0). Fix λ_{ks} and β. Set $istep = 0$, $T_{new} = T$, and define maxstep (the maximum number of samplings to be undertaken at a given temperature).

2. Set $istep = istep + 1$, $T = T_{new}$, choose one of the N-expansion coefficients randomly (suppose $a_k^{(n)}$ has been chosen), and check if $T_{new} < T_0$; if yes, exit (through step 6).

3. Modify $a_k^{(n)}(0)$ to $a_k^{(n)}(1)$, where $a_k^{(n)}(1) = a_k^{(n)}(0) + (1 - 2\pi)\Delta r_1$, where r is a random number lying in the range $0 \le r \le 1$. Δ is a small step length (user specified), and r_1 is another random number between 0 and 1.

4. Evaluate energy and constraint (orthogonality as well as normalization) values and hence the new value of the objective function of $F(1)$.

   a. If $F(1) < F(0)$, accept the change and reset $a_k^{(n)}(0)$ to $a_k^{(n)}(1)$.

   b. If $F(1) > F(0)$, generate a random number r (0,1) and check if $r < e^{-\Delta F/kT}$

   c. where $\Delta F = F(1) - F(0)$.

      i. If yes, accept the change, reset $a_k^{(n)}(0) \leftarrow a_k^{(n)}(1)$.

      ii. If not, reject the change.

5. In either case, check if $istep > maxtep$. If yes, set $T_{new} = T * \eta$ (η is a user-defined fraction), reset $istep$ to 0; go back to step 2 and report latest values of expansion coefficients, energies, overlaps, etc., along with T_{new}, $istep$.

A similar scheme was implemented by the same authors calculating the wave functions and energies of the two-electron atoms or ions in the first excited singlet state (1s2s $\equiv$ ^{1}s$_1$). First, the ground state ($1S_0 \equiv 1S^2$) was calculated by expressing $\psi_0(r_1, r_2)$ as a superposition of 10 configuration functions (1s^2, 1s 2s, 1s 2s 2s^2, 3s^2, 4s^2, 5s^2). The ns wave functions were represented by Slater type of orbitals with

$$\phi_{ns} = \exp\left(-\xi_n r\right) r^{n-1}$$

with fixed exponents

$$\xi_n = z/n$$

z being the nuclear charge of the atom or ion being considered. The trial ground state wave function then reads

$$\psi_0\left(r_1, r_2\right) = \sum_{k=1}^{10} a_k^0 \psi_k\left(r_1, r_2\right)$$

$$(4.39)$$

where $\psi_k(r_1, r_2)$ represents the kth singlet configuration function $\phi_{ns}(r_1)\phi_{n's'}(r_2)$. The objective function reads

$$F_0\left(a_1^0 \ldots a_{10}^0\right) = \left(E_0\left(a_1^0 \ldots a_{10}^0\right) - E_L\right)^2 + \beta\left\{1 - \langle\tilde{\psi}_0|\tilde{\psi}_0\rangle\right\}^2. \qquad (4.40)$$

The search variables are the 10 linear expansion coefficients to be optimized by the SAM. Once the ground state wave function (approximate) was obtained, the excited singlet (1s2s) state wave function was found by invoking the SAM to optimize the expansion coefficients $a_1^1, \ldots, a_{10}^1$ of the same 10 configuration functions used to express the ground state wave functions, that is, $\psi_1(r_1, r_2) = \sum_{k=1}^{10} a_k^1 \psi_k(r_1, r_2)$. The objective functions incorporating both the orthogonality and normalization constraint read

$$F_0\left(a_1^1 \ldots a_{10}^1\right) = \left(E_1\left(a_1^1 \ldots a_{10}^1\right) - E_L\right)^2 + \lambda_1\left(\sum_{i=1}^{10}\sum_{j=1}^{10} a_i^0 a_j^1 s_{ij}\right)^2$$

$$+\beta\left(1 - \sum_{i=1}^{10}\sum_{j=1}^{10} a_i^0 a_j^1 s_{ij}\right)^2. \qquad (4.41)$$

The authors showed that the excited state energy and wave function could be obtained rather smoothly for $z = 2 - 8$ and the results matched nicely with the results of direct diagonalization of the nonorthogonal CI matrix. It was claimed that the orthogonality constrained calculations on the excited state could be computationally cost-effective as the computational complexity of SAM-based optimization in a search space of N dimensions is expected to grow as N^p (p—a small positive constant between 1 and 2). One drawback of the calculations is the use of basis functions $\phi_{n3}(r)$ with fixed exponents that limit the flexibility of the trial wave function considerably. It would be natural to experiment with compact, short CI expansions in which the single-electron basis functions are allowed to vary, that is, the exponents ξ_ks on the basis functions $\phi_{ns}(r, \xi)$ are optimized simultaneously with the linear parameters (a_i^ks) of the CI wave function.

The authors experimented with such a possibility within the framework of their SAM-based scheme of optimization of wave functions. Two important points must be noted in the context of variational calculations of excited state wave functions by simultaneously optimizing both linear and nonlinear parameters present in the trial wave function. First, the process has to be sequential, that is, the ground state wave function is optimized first and then the relevant excited state wave function is optimized under appropriate orthogonality constraints. Short, compact CI expansion for the ground state of a many-electron system will not provide us with the exact ground state wave functions, but only with approximate ones. The excited state trial wave function ($\tilde{\psi}$) for a state having the same symmetry must not only have to be kept orthogonal to the approximate ground state ($\langle\tilde{\psi}|\tilde{\psi}_0\rangle = 0$) and normalized ($\langle\tilde{\psi}|\tilde{\psi}\rangle = 1$); it must also have to be kept decoupled with the approximate ground state wave function. By decoupling we mean that the matrix elements of H between $\tilde{\psi}_0$ and $\tilde{\psi}$ must be strictly zero. The satisfaction of these conditions ensures that the variational estimate of the excited state energy retains its upper-bound character and no variational collapse takes place during the optimization of both linear and nonlinear parameters in $\tilde{\psi}$.

Let $\tilde{\psi}(a, \xi)$ be the trial wave function for the $(n-1)$th excited state (or the nth state) and $\tilde{\psi}_k(a^k, \xi^k)$ (k = 1, 2, ..., $n-1$) be the already optimized wave functions of the $n-1$ lower states of the system including the ground state. The a^k s and ξ^k are the collections of the optimized sets of linear and nonlinear parameters of the wave function for the kth state of H. The trial wave function for the nth state is represented as

$$\tilde{\psi}\left(a^n, \xi^n\right) = \sum_{i=1}^{N} a_i^n \phi_i\left(\xi_i^n\right) \tag{4.42}$$

where the index i runs over all the basis states involved in the CI expansion. The following constraint conditions are to be enforced during the optimization of a_i^n and ξ_i^n:

1. Normalization of $\tilde{\psi}$: $\left\langle \tilde{\psi}\left(a^n, \xi^n\right) \middle| \tilde{\psi}\left(a^n, \xi^n\right) \right\rangle = 1$
2. Orthogonality to $n-1$ lower states

   $$\left\langle \tilde{\psi}\left(a^n, \xi^n\right) \middle| \tilde{\psi}_k\left(a^k, \xi^k\right) \right\rangle = \left\langle \tilde{\psi}_k\left(a^k, \xi^k\right) \middle| \tilde{\psi}\left(a^n, \xi^n\right) \right\rangle = 0$$
   for $k = 1, 2, ..., n-1$

3. Decoupling constraints: $\left\langle \tilde{\psi}\left(a^n, \xi^n\right) \middle| H \middle| \tilde{\psi}_k\left(a^k, \xi^k\right) \right\rangle = \left\langle \tilde{\psi}_k\left(a^k, \xi^k\right) \middle| H \middle| \tilde{\psi}\left(a^n, \xi^n\right) \right\rangle = 0$

Dutta and Bhattacharyya [13] chose the following objective function which incorporated all the necessary constraints; the minimization of the objective function was carried out by the SAM.

$$F_c^n\left(a^n, \xi^n\right) = \left(\epsilon_c^n - E_L^n\right)^2 + \alpha\left(1 - \left\langle \tilde{\psi}\left(a^n, \xi^n\right) \middle| \tilde{\psi}\left(a^n, \xi^n\right) \right\rangle\right)^2$$

$$+ \sum_{k=1}^{n-1} \beta_k \left| \left\langle \tilde{\psi}\left(a^n, \xi^n\right) \middle| \tilde{\psi}\left(a^n, \xi^n\right) \right\rangle \right|^2$$

$$+ \sum_{k=1}^{n-1} \lambda_k \left| \left\langle \tilde{\psi}\left(a^n, \xi^n\right) \middle| H \middle| \tilde{\psi}_k\left(a^k, \xi^k\right) \right\rangle \right|^2 \tag{4.43}$$

where ϵ_c^n is the expectation value of energy in the excited state $\tilde{\psi}\left(a^n, \xi^n\right)$ (the $(n-1)$th excited state), and E_L^n is an estimate lower bound to the energy of the nth state (updatable). The second term on the right-hand side of Equation 4.43 takes care of the normalization constraint in $\tilde{\psi}$, α representing the penalty weight factor for violating the normalization constraint. Similarly, the third group of terms take care of the orthogonality constraints ($n-1$ of them), the penalty weight factor for the kth orthogonality constraint being β_k, while the fourth group of terms enforce the decoupling constraints with λ_k denoting the penalty weight factor for violating the kth decoupling constraint. The SAM-based algorithm was applied for sequential calculation of the ground ($^1s_0$) and the first excited 1s2s ($^1s_1$) states of some two electrons, atoms, or ions. Two sets of calculations were performed on the ground state. In the calculations of set I, a 55-configuration CI function involving the ground singlet configuration ($1s^2$) and all the singly and doubly excited configurations generated by excitations into the space spanned by 2s, 3s, ..., 10s orbitals represented by STOs ($\chi_n = N_n \, exp[-z\pi\xi/n]r^{n-1}$) were used. All the linear expansion coefficients and the scale factor (ξ) were fully optimized by SAM.

In the second set of calculations, only five configuration functions (spin singlet representing $1s^2$, $1s\,2s$, $2s^2$, $3s^2$, and $4s^2$) were used for the CI expansion (i.e., five linear parameters were optimized), the ns orbitals were replaced by STOs ($\chi_n = N_n\,exp[-\xi_n r]r^{n-1}$), and all the nonlinear parameters were optimized. The results were uniformly better for the five configuration calculations (Table 4.3) with complete optimization of all the nonlinear parameters on the basis function (ns orbitals) and approached the s-limit energies.

Dutta and Bhattacharyya [13] also reported results of calculation on the excited ($^1s_1$; $1s\,2s$) state energies of a two-electron atom by simultaneously optimizing all the linear and nonlinear parameters in a three-term CI function (configurations included were $1s^2$, $1s2s$, and $2s^2$) with the inclusion of the relevant orthogonality and decoupling constraints (the ground state ($^1s_0$; $1s^2$) was represented by the same three-term CI function). The results were encouraging considering the limited variational freedom provided by such a simple trial function. Some of the results are reported in Table 4.4 along with available benchmark results [14].

The calculations performed by Dutta et al. are rather small-scale calculations. Nevertheless, they prove a point. It is indeed possible to optimize the linear parameters as well as nonlinear ones globally in a given analytical form of the wave function of two-electron atoms in the excited states by a function-only method like the SAM. That demands the relevant orthogonality and decoupling constraints are incorporated in the objective function. The approximate nature of the ground state wave function then does not vitiate the upper boundedness of the computed excited state energies. Thus, the energies and the wave functions of the excited states remain physically meaningful.

It is a pity that further extension of these calculations to larger and more flexible basis sets is unavailable. Analytical two-electron atomic wave functions obeying the nuclear and interelectronic cusp conditions in both the ground and excited states are especially difficult to optimize by traditional means. The SAM would have been a highly viable technique to construct such wave functions.

TABLE 4.3

Predicted Ground State Energies of Two-Electron Atoms by SAM-Based Optimization of Wavefunction Parameters

Atoms/Ions (z)	Set I ($E(^1s_0)$) au	Set II ($E(^1s_0)$) au
2	−2.878261	−2.869396
4	−13.626003	−13.614137
8	−59.124871	−59.112306

TABLE 4.4

Comparison of the Energies of the 1s2s State of the He Isoelectronic Sequence Predicted by SAM with Exact Energies

Atoms/Ions (z)	Energy ($^1s_1$; 1s2s) au Predicted by SAM	Exact Results au
2	−2.117190	−2.1477294
4	−21.191027	−21.221361
8	−38.227088	−38.257883

Source: Sharma, C.S., and Coulson, C.A., *Proc. Phys. Soc.,* 80(1), 81–96, 1962.

4.5.1 The Generator Coordinate Method

The SAM-based strategy of calculations for good quality wave functions for both ground and excited states of a system was explored by Chaudhury et al. [15] in a slightly different variational context. These authors noted that the generator coordinate (GC) method provides an easy route to optimally dispose of a particular class of nonlinear variational parameters, but that leads to the problem of solving a set of integral equations that may not be so tractable. Discretization then converts the integral equations into linear equations with built-in dependence of the linear parameters on the nonlinear ones. These equations can be solved efficiently by invoking the SAM. The authors further explored the possibility of exploiting the optimized energy eigenvalues and eigenvectors for constructing a good unperturbed (zeroth order) Hamiltonian $(H^{(0)})$ and a perturbation $(H^{(1)})$ that enabled them to obtain a perturbative expansion that converged faster. Let us consider their proposal and its rudimentary implementation briefly.

Let α represent a class of nonlinear parameters in the trial wave function $\psi(x)$ of a one-dimensional system represented by $H(x)$. The GC method provides a compact way of representing $\psi(x)$ as

$$\psi(x) = \int_a^b c(\alpha)\phi(x|\alpha)d\alpha \tag{4.44}$$

where a, b define the range of the parameter α. The energy $E[c(\alpha)]$ can be minimized with respect to $c(\alpha)$ leading to a Griffin–Wheeler GC equation:

$$\int_a^b d\beta\big[H(\alpha,\beta) - E\Delta(\alpha,\beta)\big]c(\beta) = 0. \tag{4.45}$$

The solutions of the integral equation provide strict upper bounds to the corresponding exact energy eigenvalues. The parameter α that serves to generate the trial wave function acts as an extra coordinate—the so-called GC—but does not appear in the final wave function. The integral equations are, however, not always tractable. It is possible, in such situations, to choose a simpler alternative involving discretization of the GC which converts Equation 4.45 into the familiar secular equations in a nonorthogonal basis set $\{\phi(x|\alpha_i)\}$ with α_i restricted to the range $[a, b]$:

$$\sum_j \big(H_{ij}^\alpha - E\Delta_{ij}^\alpha\big)c_j^\alpha = 0. \tag{4.46}$$

That the c_js depend on the nonlinear parameter α_is make the simultaneous determination of the optimal set of the discrete generate coordinates $(\alpha_1, \alpha_2, \alpha_3, \dots)$ and the corresponding linear expansion coefficients $\{c_j^\alpha\}$ a tricky task. SAM appears to provide an easy recipe for solving the problem. Thus, the authors define a cost functional $F[\alpha]$:

$$F[\alpha] = \big\{\in\big[c(\alpha)\big] - E_L\big\} \tag{4.47}$$

where

$$\in \left[c(\alpha) \right] = \sum_i \sum_j c_i^*(\alpha_i) H_{ij} c_j(\alpha_j) \, / \, \sum_i \sum_j c_i^*(\alpha_i) \Delta_{ij} c_j(\alpha_j)$$

$$H_{ij} = \left\langle \phi\left(x \vert \alpha_i\right) \vert H \vert \phi\left(x \vert \alpha_j\right) \right\rangle$$

$$\Delta_{ij} = \left\langle \phi\left(x \vert \alpha_i\right) \vert \phi\left(x \vert \alpha_j\right) \right\rangle. \tag{4.48}$$

E_L is an approximate (updatable) lower bound to the energy eigenvalue being sought (say, $\in_0$). A temperature (T) is set (high to start with). One of the discretized GCs (say, α_k) is randomly chosen and given a random value in the range $[a, b]$, the H_α matrix is constructed and diagonalized leading to the eigenvalues ($\in_0, \in_1, \in_2, \ldots$). The targeted eigenvalue ($\in_0$) is chosen and $F(\alpha)$ of Equation 4.47 is computed. The new value of the kth GC is accepted or rejected based on the Metropolis algorithm at the given temperature T. Sufficient number of samplings are done at a particular temperature before lowering T according to a pre-defined cooling schedule. Usually as $T \to 0$, the search converges to the ground eigenvalue. After the ground eigenvalue has been obtained, the next higher eigenvalue ($\in_1$) is targeted by resetting E_L to $\in_0$ and so on for higher eigenvalues. Instead of adopting the sequential search for a set of energy eigenvalues and eigenfunctions, the SAM-guided discretized generator coordinate (DGC) method was also explored for estimating several energy eigenvalues simultaneously by invoking what may be called a manifold minimization principle. Suppose that we are interested in M number of lowest eigenvalues including the ground state. Then an objective function for the manifold of M states can be defined as:

$$F_M = \frac{1}{M} \sum_{i=1}^{M} \left(\in_i (t) - E_L^i \right)^2 \tag{4.49}$$

and the method of annealing can be used to obtain a set of discrete values of the GCs that are optimal for the manifold of states in a least-squares sense. All the energy eigenvalues obtained through the manifold minimization principle retain their strict upper-bound character.

Both the methods (the sequential and the manifold) were applied to estimate the ground and excited states of one- and two-dimensional anharmonic (quartic) oscillators. It was interesting that the manifold minimization principle worked well in both the cases even when the anharmonic strength was large. In either case, the basis sets used were harmonic oscillator energy eigenfunctions of even and odd parities (three even and three odd). Considering the very limited size of the basis set, the predicted energies were surprisingly accurate. The authors also succeeded in constructing a new zeroth-order Hamiltonian (H^0) by analyzing the optimized DGC wavefunctions and identifying which GC carried the maximum weights in the DGC wavefunction. The new H^0 provided a perturbative series with better convergence properties.

4.5.2 (Multiconfiguration) Self-Consistent Field Calculation and the SAM

The method of calculating ground and excited state wavefunctions by invoking the SAM described in the preceding section hinges on the construction of an appropriately constrained energy functional and its global minimization by SAM with respect to the parameters (linear as well as nonlinear) in the trial wave function. The idea was carried forward further by Dutta and Bhattacharyya [16] to the realm of the self-consistent field (SCF) and multiconfiguration self-consistent field (MCSCF) calculations of wave functions and energies of a many-electron system (atoms/molecules).

Suppose that we are considering the closed-shell ground state wave function of a $2n$-electron atom or molecules which can be well represented (approximately though) by an antisymmetrized product of $2n$ spin orbitals ψ_0, where

$$\psi_0 = \frac{\hat{A}}{\sqrt{2n}}\left\{\phi_1\overline{\phi_1}\dots\phi_k\overline{\phi_k}\dots\phi_n\overline{\phi_n}\right\} \tag{4.50}$$

The space parts of the spin orbital are expressed as linear combinations of a fixed set of M basis functions (usually atom-centered single electron normalizable functions, but can also be more general) (χ). Hence, we can write:

$$\phi_i = \sum_{p=1}^{M}\chi_p T_{pi}, \ i = 1, 2, \ \dots, \ n. \tag{4.51}$$

In matrix form, we have

$$\Phi = \chi T, \ \ T^\dagger S T = 1 \tag{4.52}$$

S being the metric of the basis (χ).

We note here that the space parts of ϕ_k and $\overline{\phi_k}$ are the same, but their spin parts differ. The SCF problem for the ground state concerns finding the optimal set of ϕ_ks so that the ground state energy $\left(\widetilde{E}_0 = \langle\psi_0|H|\psi_0\rangle\right)$ is minimized subject to the orthonormality of the orbital set. The optimization problem can be cast as the problem of globally minimizing a constrained functional $F[\psi_0]$ with respect to the expansion coefficients $\{T_{pk}\}$.
where

$$F[\psi_0] = \left(\widetilde{E}_0 - E_L\right)^2 + \beta \ Tr\left[\left(T^\dagger S T - \mathbb{1}\right)^\dagger\left(T^\dagger S T - \mathbb{1}\right)\right] \tag{4.53}$$

and

$$\widetilde{E}_0 = \sum_{i=1}^{n}\sum_{p}^{M}\sum_{q}^{M}2T_{pi}^*T_{qi}\left(h_{pq}\right) + \sum_{i,j=1}^{n}\sum_{pq}^{M}\sum_{rs}^{M}2T_{pi}^*T_{qj}\left(g_{pqrs}\right)T_{rj}T_{sj}. \tag{4.54}$$

E_L is, as usual, an estimated updatable lower bound to the ground state energy; h_{pq}s represent the matrix elements of the one electron part $h(1)$ of the atomic or molecular Hamiltonian involving the basis function χ_p

$$h_{pq} = \left\langle\left.\left\|\chi_p^{(1)}\right|T(1) - \sum_I\frac{z_I e^2}{r_{I1}}\right|\chi_q^{(1)}\right\rangle. \tag{4.55}$$

g_{pqrs} are the matrix elements of the two-electron part of the Hamiltonian over the basis functions and are defined as follows:

$$g_{pqrs} = \left\langle\chi_p(1)\chi_q(2)\left|\frac{e^2}{r_{12}}\right|\chi_r(1)\chi_s(2)\right\rangle. \tag{4.56}$$

β is a penalty weight factor for any violation of the orthonormality constraint on the orbital $\{\phi_i\}$ If $T^\dagger S T \neq \mathbb{1}$, constraints are violated and minimization by SAM tends to orient the search variables $\{Tp_i\}$ so that $\tilde{E}_0 \rightarrow E_L$ as closely as possible and $T^\dagger ST \rightarrow \mathbb{1}$ as the temperature $\rightarrow 0$. The SCF procedure can thus be posed as a global minimization problem for the functional $F[\psi_0]$.

The MCSCF problem can be similarly mapped into a constrained global minimization problem of a more general nature. The MCSCF trial wave function ψ is not a single determinant wave function, but is a many determinant one:

$$\psi = \sum_{I=1}^{N} C_I \psi_I (\phi) \tag{4.57}$$

where ψ_Is are configuration functions that are each antisymmetrized product of $2n$ orthonormal spin orbitals $\{\phi_i\}$ where, as before

$$\phi = \chi T, \quad T^\dagger ST = \mathbb{1}$$

$$\sum_{I=1}^{N} C_I^2 = 1. \tag{4.58}$$

It is convenient to choose the configuration function ψ_I as eigenfunction of $\hat{S}^2$ (spin). The standard approach in MCSCF is to make the energy $E(\psi)$ stationary with respect to variations in $\{C_I\}$s and $\{T_{pi}\}$s. The authors of reference, on the other hand, cast the problem as a global minimization problem of the functional $F[\psi]$, where

$$F[\psi] = (E[\psi] - E_L)^2 + \beta \left(1 - \sum_I C_I^2\right)^2 + \gamma \ Tr\left[\left(T^\dagger ST - \mathbb{1}\right)^\dagger \left(T^\dagger ST - \mathbb{1}\right)\right]. \tag{4.59}$$

In Equation 4.59

$$E[\psi] = Tr\left(hTP_1T^\dagger\right) + \frac{1}{2} Tr[zT^\dagger]. \tag{4.60}$$

The two-electron part of the interaction is coded by the $z(N \times M)$ matrix (cf. Equation 4.60)

$$z_{pi} = \sum_{jkl}^{occupied \ MOs} T_{js}^* g_{pqrs} T_{kq} T_{lr} (P_2)_{ijkl} \tag{4.61}$$

P_1 and P_2 are the one- and two-electron density matrices in the $M0(\phi)$ basis and are determined by the linear expansion coefficients $\{C_I\}$. Tr means matrix trace operation. The first term on the right-hand side of Equation 4.59 directs the energy minimization, the second term takes care of the normalization condition on ψ while the third group of terms takes into account the constraint that the orbital $\{\phi_I\}$ (MOs) must be an orthonormal set. β and γ are the penalty weight factors for the relevant constraints. Three groups of search variable on which the functional $F[\psi]$ depends can be identified:

1. $\{C_I\}$ or the linear expansion coefficients appearing in the CI expansion of ψ

2. $\{T_{pi}\}$ or the linear expansion coefficients that determine the MOs (ϕ_is) appearing in the different configuration functions $\{\psi\}$

3. The parameters $\{R_\alpha\}$ that define the coordinates of various nuclei in the molecule or the nuclear framework defining variable (like bond lengths, bond angles, dihedral angles).

Thus, $F[\psi] = F\left[\{C_I\}, \{T_{pi}\}, \{R_\alpha\}, E_L, \beta, \gamma\right].$

Once the lower bound to the constrained minimum energy (E_L) and the penalty weight factors β, γ are fixed, SAM can be invoked to minimize F globally as a function of the parameters $\{C_I^0\}, \{T_{pi}^0\}$ and $\{R_\alpha^0\}$ simultaneously. It is important to note that all the parameters are to be optimized simultaneously by randomly sampling them many times at a given temperature (T) and slowly cooling the system to lower temperatures $T_1, T_2, ..., T_n \to 0$. When the $T_n \to 0$ limit is reached, the current sets of parameters $\{C_I\}, \{T_{pi}\}$ and $\{R_\alpha\}$ are accepted as solutions.

$$\psi_{opt} = \sum C_I^0 \psi_I\left(\phi^0\right) = \psi^0$$

$$\phi^0 = \chi T^0 \tag{4.62}$$

$$E\left[\psi_{opt}\right] = \left\langle \psi_{opt} | H | \psi_{opt} \right\rangle = \left\langle \psi^0 | H | \psi^0 \right\rangle.$$

The corresponding energy $E\left[\psi^0\right]$ gives the desired energy of the state, and $\{R_\alpha^0\}$ defines the equilibrium geometry of the molecule in the state ψ_{opt}. The authors demonstrated the workability of their Metropolis simulated annealing-based MCSCF strategy in several model applications. It is surprising that the strategy was not explored any further. We note here that the method appears to be appealing especially when the linear expansion coefficients $\{C_I\}$ and the orbital expansion coefficients $\{T_{pi}\}$ are strongly coupled and a deterministic route like the quadratically convergent MCSCF would require accurate analytical calculations of the first and second derivatives of energy with respect to the linear expansion coefficients and the orbital expansion coefficients at each nuclear configuration. In contrast, the method of Dutta and Bhattacharyya does not require calculation of any derivative—first or second. It is a function-only method so that the dominant computational cost comes from the computation of the energy $E[\psi]$, which, in turn, is dominated by the time consumed in computing the two-electron integrals in the MO basis. It has been claimed that in the worst-case scenario, the computational complexity in a SAM-based optimization of N parameters grows as N^p $(1 < p < 2)$ which is highly acceptable. Efficient parallelization can provide further scope of comfort. Sadly though, this alternative soft computing approach has not been explored or exploited so far. It is our belief and assessment that there is enough scope for further work in the development of a Metropolis simulated annealing-based strategy of performing multiconfiguration, SCF calculations. It remains to be seen if quantum annealing could be exploited to accelerate convergence in MCSCF calculations within the framework of the model described.

Before concluding this section, we must point out that the exploitations of the SAM as a global minimizer in the calculation of atomic and molecular structure discussed in this chapter can be regarded as a Monte Carlo-based-approach to solve constrained function(al) minimization problems where the constraints (equality) are imposed in the form of penalty functions. The constraints considered are normalization constraint on ψ, mutual orthogonality constraints of wavefunctions representing different states of the system, decoupling constraints, etc. Another annealing-driven strategy developed by different groups of researchers for molecular electronic structure calculation can be linked to the molecular dynamics-based formulation originally proposed and popularized by Car and Parrinello [17,18]. In the Car–Parrinello (CP) method, the essential step is to treat the electronic wave function ψ_e as a dynamical variable and define a fictitious Lagrangian, called the Car–Parrinello Lagrangian, for the electronic system and exploit it to derive classical equations of motion which are solved for generating the Born–Oppenheimer surface of a many-electron system. A key step here is to slowly withdraw kinetic energy from the electronic degree of freedom, that is, cool the system slowly so that as $T \to 0$ the system ends up in the ground electronic state.

4.6 Elementary Applications of RMHC in Cluster Physics

Random mutation hill climbing, as has been already described, is a single-parent (or string)-based method of stochastic search for the global maximum on a fitness landscape. Being a single-parent method, the binary crossover operation is redundant in RMHC; only the unary random mutation operator acts on the string to move it from one location on the fitness landscape to another. It is natural noise that drives the evolution in RMHC. A selection pressure is applied by requiring the mutated string to have fitness higher than or equal to the fitness of the parent string. Therefore, the RMHC technically looks like the method of simulated annealing at zero temperature $(T = 0K)$ so that thermal fluctuations are absent to start with and gradual cooling to $T = 0$ is not required to ensure that the system reaches the maximum point on the fitness landscape (or a minimal point on the objective function landscape). It would be inherently interesting therefore to compare the method of simulated annealing and random mutation hill climbing in the search for the global maximum on complex fitness landscapes. In either case, we propose to work with floating-point solution strings as representation in the form of binary strings would immediately lead to loss of accuracy. In addition, the use of floating-point strings in the RMHCM leaves us with the option of two control parameters, namely mutation intensity (Δm) and mutation probability (p_m), which can be leveraged to make the search more effective. Let us note that in the floating-point string-based implementation of the SAM, we have at our disposal two parameters—the step length (ΔS) (which can be likened to the mutation intensity (Δm)) and the temperature (T) that fixes the probability of realizing a particular thermal fluctuation in the system configuration (which can be likened to the mutation probability (p_m)). Having satisfied ourselves that we are indeed comparing the comparables, it is necessary to choose a problem that can bring both the methods to a fairly harsh test. Search for the global minimum energy configuration of a three-dimensional (3D) confined Coulomb cluster offers such a testing ground. Apart from being computationally hard, the problem can be linked to models for many important physical systems like electrons in quantum dots, charged particles in ion traps, or charged particles in plasma traps. Highly charged dust grains of micro- or sub-micrometer dimension are a common occurrence in low-temperature laboratory plasmas of various kinds. It has been known that the dust particles embedded in an electrical discharge can form regular patterns known as plasma crystal, even at room temperature. In fact, homogeneous 3D spherical clouds of monodisperse charged dust particles have been experimentally realized rather recently [6,19]. The strong correlation among particles in such system leads to the emergence of liquid-like or even crystal-like behavior as a result of competition between columbic repulsion and forces of confinement. The repulsion among like charges tends to keep the particles maximally separated while the forces of confinement try to bring them closer, resulting in arrangements of the charged particles in concentric, nested shells ultimately leading to the appearance of 3D plasma crystals.

Let us consider a simplified model in which N point charges are interacting via Coulomb forces, embedded in a homogeneous neutralizing background, and look for the spatial arrangement of the charges that produce the global minimum of energy. No analytical solution of the problem for an arbitrary N is available, but numerical procedures have suggested what may be called the apparent global minimum energy configurations for some values of $N < 112$. The number of local minima (L_N) has been shown to be an exponentially increasing function of N with [20]:

$$L_N = 0.382 \times \exp\left(0.0497N\right). \tag{4.63}$$

The computational complexity therefore makes the problem of identifying the global minimum in such systems (one-component plasma) a hard one. Suppose now that N identically charged point particles interacting via Coulomb forces are confined in a 3D spherical (harmonic) potential, and we try to identify the charge configurations that minimize the energy in a global sense. The computational complexity of the problem would be similar to what has been found for the one-component plasma problem. We propose to put both RMHC and the SAM to test in the context of the problem posed.

4.6.1 The Problem and the Method

A system containing N identical point charges $\{q_i\}_{i=1,N}$ confined in a parabolic trap is characterized by potential energy

$$\in\left(\vec{r_1},\vec{r_2},\ldots,\vec{r_N}\right) = \frac{1}{2}\sum_{i=1}^{N}\sum_{j=1\,(j\neq i)}^{N}\frac{q_i q_j}{\left|\vec{r_1}-\vec{r_2}\right|} + \frac{1}{2}K\sum_i q_i r_i^2 \tag{4.64}$$

where $\vec{r_i}$ represents the position vector locating the ith charge from the origin (the center of the confining potential). In terms of the Cartesian coordinates, we have

$$\in\left(\vec{r}\right) = \frac{1}{2}\sum_i^N\sum_{j\neq i}^N\frac{q_i q_j}{\sqrt{\left(x_i-x_j\right)^2+\left(y_i-y_j\right)^2+\left(z_i-z_j\right)^2}} + \frac{1}{2}K\sum_i q_i\left(x_i^2+y_i^2+z_i^2\right). \tag{4.65}$$

The configuration of the system of N confined point charges is completely specified by a single string S containing $3N$ real coded Cartesian coordinates of N point charges. Random mutation affects the coordinates making the string S evolve on the fitness landscape F. Let the configuration of point charges at the pth stage of evolution be coded by $S\left(X^{(p)}\right)$ where $X^{(p)}$ is a $3N$-dimensional vector of real numbers representing the Cartesian coordinates of the charges.

$$X^{(p)} = \left(x_1^p,y_1^p,z_1^p,\ \ldots,\ x_k^p,y_k^p,z_k^p,\ \ldots,\ x_N^p,y_N^p,z_N^p\right)$$

$$\equiv\left(\xi_1^p,\xi_2^p,\xi_3^p,\ \ldots,\ \xi_{3N-2}^p,\xi_{3N-1}^p,\xi_{3N}^p\right) \tag{4.66}$$

The fitness of the string $S\left(\xi^{(p)}\right)$ is then defined as

$$F^{(p)} = \frac{1}{\left(V^p - V_L^p\right)^2 + \delta_p^2}$$

(4.67)

where

$$V^p = \frac{1}{2}\sum_i^N\sum_{j\neq i}^N \frac{q_i q_j}{\sqrt{\left(x_i^p - x_j^p\right)^2 + \left(y_i^p - y_j^p\right)^2 + \left(z_i^p - z_j^p\right)^2}} + \frac{1}{2}K\sum_i q_i\left(\left(x_i^p\right)^2 + \left(y_i^p\right)^2 + \left(z_i^p\right)^2\right).$$

(4.68)

V_L^p is an estimated lower bound to the potential energy of the system at the pth stage of evolution and δ_p is a small number that takes care of exponential overflow if $V^p = V_L^p$ accidentally. In the standard RMHCM, one chooses one or more of the charge coordinates randomly with a probability p_m and mutates the chosen coordinates randomly with mutation intensity Δ_m^p. The mutated coordinate (say, ξ_k^p) of the new configuration, i.e., $(p+1)$th configuration reads

$$\xi_k^{p+1} = \xi_k^p \pm \Delta_m^p r$$

(4.69)

where r is a random number from $0 < r < 1$, and sign is chosen randomly with a probability of 0.5. If $F^{(p+1)} \geq F^{(p)}$, the new configuration is accepted. If F_{max}^p is the best fitness value found up to the pth stage of evolution, the procedure ensures that a strongly elitist strategy is enforced. Sarkar and Bhattacharyya proposed a modified RMHC strategy in which the mutation probability (p_m) and intensity (Δ_m) were adaptively adjusted based on the accumulated experience of the success or failure rates of random mutations over the past M trials. The algorithm has been called the CARMHC. Thus, with $M = 100$, the mutation intensity is adjusted as follows:

1. If the fraction f_R of the past 100 trials that were successful exceeds a preset upper bound (say, 0.1), then Δ_m is stretched, i.e., $\Delta_m^{p+1} = \Delta_m^{p+1} \times (1 + r)$, where r is a Gaussian random number in the range (0,1).
2. If f_R over the same number of past trials turns out to be less than a preset lower bound (say 0.01), then Δ_m is contracted by setting $\Delta_m^{p+1} = \Delta_m^p \times (1 + r)^{-1}$.

A similar strategy was adapted to stretch or contract the mutation probability. Thus:

1. If f_R turned out to be larger than a preset upper bound (say, f_u), p_m was scaled up by setting $p_m^{new} = p_m^{old} \times (1 + r)$.
2. If f_R was lower than a preset lower bound (say, f_l), p_m was scaled down by setting $p_m^{new} = p_m^{old} \times (1 + r)^{-1}$.

The adaptive control requires us to define the initial and final values of $\Delta_m (= \Delta_m^0, \Delta_m^f)$ as well as $p_m (= p_m^0, p_m^f)$. The initial value of p_m was set equal to $p_m^0 = \dfrac{1}{a}$, where $3 \le a \le 4$ (the lower and upper bounds are adjustable). The final value of p_m beyond which dynamic adjustment of mutation probability freezes out was $p_m^f = \dfrac{1}{3N}$, that is, when uniform probability of mutation was reached. Δ_m^0 was set at 1.0 and Δ_m^f was set at 10^{-12}. Sarkar and Bhattacharyya claim that the adaptive control on mutation probability and mutation intensity imparts enhanced search ability to RMHC which performed better than the SAM, at least in the search for the global minimum energy configurations of the confined monodisperse Coulomb clusters. A word or two about the basic philosophy of the adaptive control proposed would perhaps be in order at this point. If the current mutation probability has been leading to many successful reconfiguring moves over the last M trials, it would be prudent to work with a higher value of mutation probability. On the other hand, if it led to too many unsuccessful trials, resetting p_m to a lower value would be a judicious adaptive move. Similarly, if the current value of Δ_m has been producing too many successful mutations, it would be wise to resort to using a larger mutation step length to broaden the search perspective. Similarly, if the current value of Δ_m has been resulting in too many unsuccessful mutations, it is a signal to adopt smaller mutation steps and thus emphasize the "depth search" aspect. It would be possible, it seems, to adaptively control the cooling and heating schedule (the overall annealing schedule) in a simulated annealing-guided search following similar logic. It remains to be seen whether SAM with adaptively controlled annealing schedule would perform as well as the CARMHC procedure of Sarkar and Bhattacharyya. So far, no such comparative data are available. We may mention that the adaptive control mechanism followed in CARMHC is not unique and there is enough scope for improvement. The results available seem to indicate that adaptive control on p_m and Δ_m enforced in CARMHC allows it to avoid getting trapped in a local optimum, at least in the present context. We refer the readers to reference for further details.

It is interesting that starting with a random initial distribution of point charges, a random mutation process with a strong elitist criterion for successful mutation (biasing) fairly quickly leads to the emergence of regular shell structures in a confined monodisperse system of point charges. Table 4.5 displays the shell occupancy levels of fitness maximal structures in an isotropically and harmonically confined system of N identical point charges ($N = 151, 300, 500, 1000$) discovered by the CARMHC algorithm.

We will further explore the CARMHC strategy of search for the global minima in molecular systems in Chapter 7.

TABLE 4.5

Predicted Shell Occupancy Levels of Clusters of Different Sizes

		Shell Occupancy						
N	Energy	K	L	M	N	O	P	Q
151	3719.563	12	43	96	–	–	–	–
300	11833.04	10	40	90	160	–	–	–
500	27903.25	7	34	79	147	233	–	–
1000	89109.08	1	15	50	105	177	270	382

4.7 Search for the Minima (Local and Global) in Model Potentials and Spin Systems

It is often necessary to have complete information about the local minima lying close to the global minimum on a potential energy or a free energy surface. At nonzero temperatures, the thermally averaged properties of the system are the actual observable and theoretical prediction of such properties that require averaging over all the thermally accessible higher energy structures lying in the neighborhood of the global minimum. Methods like the SAM or the CARMHC are stochastic global minimizers (optimizers) that can be tailored to trace out local minimum energy structures in the close neighborhood of the global minimum. We illustrate a procedure for doing so within the framework of the method of simulated annealing already discussed in the current chapter at length.

Let $V(X) = V(x_1, x_1, \ldots, x_n)$ be the n-dimensional function and V_L^0 be an estimated lower bound to the global minimum of V. X is an n-dimensional configuration vector with real components $x_1, x_1, \ldots, x_n$. An objective function of the global minimization, which can be identified as the target of the search in this context, can be defined (see Dutta and Bhattacharyya).

$$F\left(X, V_L^0\right) = \left(V(X) - V_L^0\right)^2 + \lambda \sum_{i=1}^{n}\left(\frac{\partial V}{\partial x_i}\right)^2, \quad \lambda > 0. \tag{4.70}$$

At a minimum, global or otherwise, the $\left(\dfrac{\partial V}{\partial x_i}\right)$ terms are all zero; so the minimization of $F\left(X, V_L^0\right)$ is dictated by how close $V(X)$ is to the estimated lower bound V_L^0 provided by the user. It appears therefore that V_L^0 can be leveraged to fine-tune the search to a minimum lying close to V_L^0. However, $\left(\dfrac{\partial V}{\partial x_i}\right)_{i=1,n}$ terms are zero at any stationary point (minima, maxima, and saddle points), so it is mandatory to check the nature of the stationary point reached by diagonalizing the matrix H $\left(H_{ij} = \dfrac{\partial^2 V}{\partial x_i \partial x_j}\right)$ and ascertaining the nature of the eigenvalues (at a minimum all the $n-6$ eigenvalues are >0, the six out of n eigenvalues left out are the ones representing three rotational and three translational modes of the center of the mass of the system which do not encounter any restoring force and can therefore be zero). It is convenient to target the global minimum first and then sequentially look for higher minima by using V_L^0 as the tuning or directing parameter. The search may be executed in the following stepwise manner.

1. First choose a low enough value of V_L^0 and high enough value of initial temperature T^0.

2. Search for the global minimum of the objective function of Equation 4.70 with $T \to 0$.

3. Check if $\left(V(X) - V_L^0\right)^2$ is still $\gg 0$. If so, reset the temperature to a high value and reset V_L^0 dynamically at every temperature step

$$V_L^0(i+1) = V_L^0(i) + a\left(\Delta F_i\right)^{\frac{1}{2}} \tag{4.71}$$

where a is a user-specified shift parameter (>0) and F_i is the value of the objective function at the conclusion of samplings at the ith temperature step.

4. Check if a lower minimum has been reached as $T \to 0$. If not, accept the current minimum as the global one.

   If a lower minimum has been reached, reset V_L^0 to a lower value by setting

$$V_L^0 (new) = V_L^0 (old) - \sigma \qquad (4.72)$$

where σ is a shift parameter. The temperature is reset to a high value and minimization is carried out again to ascertain the stability of the global minimum discovered so far. Let the global minimum discovered be V_G and the corresponding configuration be $X_G (x_1^G, x_1^G, \ldots, x_n^G)$.

5. Reset the temperature to a high value, set a new value of V_L^0 (say, V_L^1) where

$$V_L^1 = V_G + n\sigma. \qquad (4.73)$$

   A new starting configuration may be created by applying random perturbations to the parameters defining the global minimum.

$$X_{new} = X_G \pm r_i \quad (0 < r_i < 1) \qquad (4.74)$$

where r_i is the random shift to the ith coordinate in X_G, and $\pm$ is chosen with equal probability. The minimization by SAM is carried out to ascertain if a minimum close to V_L^1 exists.

6. Stability of the higher minimum is checked as was done in the case of search for the global minimum.

7. Higher minima can be located similarly by resetting the lower bound to V_L^2, V_L^3 in the same manner.

8. Chaudhury et al. [21] taking the cue from an earlier work by Dutta et al. [22] demonstrated the viability of a similar method in the context of search for the global as well as local minima of two-dimensional Müller–Brown potential. They demonstrated that

   a. The global minimum could be reached irrespective of the starting point, provided that the cooling rate was slow. The total simulation time is however different for different starting points.

   b. The local minima could be discovered sequentially without difficulty.

We anticipate that the search strategy would be useful in locating the global as well as minimum energy configurations in atomic or molecular clusters. We will examine these and other possibilities further in Chapter 7. Needless perhaps it is to mention that the same problems could also be easily solved by involving the CARMHC algorithm.

Chaudhury et al. also tested their code on the problem of locating the global minimum energy configuration in a one-dimensional spin glass with random nearest-neighbor interactions or couplings. Let there be N spins $\{S_i\}$. The chain with $S_i = \pm 1$ and the nearest neighbor coupling constants $\{J_i\}_{i=1,}^{N-1}$ are assumed to have random positive values in the range $0 < J_i < 1$. The energy of the chain $E(S)$ is given by

$$E(S) = -\sum_{i=1}^{N-1} J_i(S_i S_{i+1}).$$ (4.75)

For all J_is > 0, the global minimum is doubly degenerate, that is, S_+^0 or S_-^0, the S_+^0 referring to the configuration in which all the spins are +1 while S_-^0 represents the spin configuration with all spins −1. The authors used a chain of 25 spins and an objective function

$$F_{obj}^i = \left(E(S) - E_L^i\right)^2$$ (4.76)

where E_L^i is the estimated lower bound to the ground state energy at the ith temperature state. The iterations started with $E_L^i = -13.0$ which was updated at the onset of a new temperature step by adopting the Morrison function technique.

$$E_L^{i+1} = E_L^i + const. \sqrt{F_{obj}^i} \quad (const. > 0).$$ (4.77)

The spins were randomly chosen and flipped, and new configuration was accepted or rejected based on the Metropolis criteria. As $T \to 0$, their search converged to the global minimum energy configuration with all spins assuming +1 value and energy $E_G = -10.9902$. If the initial spin configuration (chosen randomly) is different, the search may also lead to the other component of the twofold degenerate ground state (all the spins having the value −1). Even in such a simple system, the search may encounter difficulty as many local minima exist on the search landscape and the search may get stuck at a local minimum for a long time before a strong thermal fluctuation dislodges it from the trap. It would be interesting to explore whether the idea of quantum annealing could come to our rescue and a suitable choice of tunneling field would accelerate escape from local minimum (see Section 10.9). As to tuning the search to discover a particular component of the doubly degenerate ground state with certainty, we have two alternatives: either add an additional term in the objective function with an appropriate penalty weight factor or use a magnetic field that forces all spin to align along the field direction. The first option can be implemented by using, for example, an objective function F_+ where

$$F_+ = \left(E(S) - E_L^i\right)^2 + \lambda\left(1 - \frac{n_+}{n}\right)^2, \quad (\lambda > 0).$$ (4.78)

The global minimum of F_+ would be reached on cooling as $E(S) \to E_L^i$ and $\frac{n_+}{n}$ (i.e., the fraction of spin having +1 value) $\to 1$. The use of a magnetic field H_z modifies the energy expression,

$$E(S, H_z) = -\sum_{i=1}^{N-1} J_i(S_i S_{i+1}) - \sum_{i=1}^{N}(S_i H_z).$$ (4.79)

The field-modified energy expression can be used to construct the objective function without any additional constraint:

$$F_H = \left(E(S, H_z) - E_L^i \right)^2.$$ (4.80)

The magnetic field plays the role of the constraint used in defining F_+. If F_+ is replaced with another objective function F_-, then

$$F_- = \left(E(S) - E_L^i \right)^2 + \lambda \left(1 - \frac{n_-}{n} \right)^2, \quad (\lambda > 0)$$ (4.81)

where $\dfrac{n_-}{n}$ represents the fraction of spins having the value –1, and the search would home in to the other component of the doubly degenerate ground state (all spins having the value –1). Thus, with properly constructed objective functions, the annealing may be nudged to discover a specific component of a degenerate ground state as $T \to 0$. We would advise curious readers to write a code to test the alternatives.

References

1. Kirkpatrick, S., C.D. Gelatt, and M.P. Vecchi. 1983. Optimization by Simulated Annealing. *Science* 220, no. 4598:671–680.
2. Černý, V. 1985. Thermodynamical Approach to the Traveling Salesman Problem: An Efficient Simulation Algorithm. *Journal of Optimization Theory and Applications* 45, no. 1: 41–51.
3. Rata, I., A.A. Shvartsburg, M. Horoi, T. Frauenheim, K.W.M. Siu, and K.A. Jackson. 2000. Single-Parent Evolution Algorithm and the Optimization of Si Clusters. *Physical Review Letters* 85, no. 3: 546–549.
4. Mitchell, M., J.H. Holland, and S. Forrest. 1994. When Will a Genetic Algorithm Outperform Hill Climbing. In *Advances in Neural Information Processing Systems,* vol. 6, ed. J.D. Cowan, G. Tesauro, and J. Alspector, pp. 51–58. San Francisco, CA: Morgan-Kaufmann.
5. Geman, S., and D. Geman. 1984. Stochastic Relaxation, Gibbs Distributions, and the Bayesian Restoration of Images. *IEEE Transactions on Pattern Analysis and Machine Intelligence* PAMI-6, no. 6: 721–741.
6. Sarkar, K., and S.P. Bhattacharyya. 2013. Single String Based Global Optimizer for Geometry Optimization in Strongly Coupled Finite Clusters: An Adaptive Mutation-Driven Strategy. *The Journal of Chemical Physics* 139, no. 7:74106.
7. Mitchell, M., S. Forrest, and J.H. Holland. 1992. The Royal Road for Genetic Algorithms: Fitness Landscapes and GA Performance. In *Proceedings of the First European Conference on Artificial Life,* pp. 245–254. Cambridge, MA: MIT Press. https://www.cse.unr.edu/~sushil/class/gas/papers/mitchell91royal.pdf (Accessed on August 8, 2016).
8. van Laarhoven, P.J.M., and E.H.L. Aarts. 1987. Simulated Annealing. Inbook. *In Simulated Annealing: Theory and Applications,* pp. 7–15. Dordrecht: Springer Netherlands. https://books.google.com/books?id=-IgUab6Dp_IC&printsec=frontcover&source=gbs_ge_summary_r&cad=0#v=onepage&q&f=false (Accessed on August 8, 2016).
9. Ray, P., B.K. Chakrabarti, and A. Chakrabarti. 1989. Sherrington-Kirkpatrick Model in a Transverse Field: Absence of Replica Symmetry Breaking due to Quantum Fluctuations. Article. *Physical Review B* 39, no. 16: 11828–11832.

10. Kadowaki, T., and H. Nishimori. 1998. Quantum Annealing in the Transverse Ising Model. *Physical Review E 58*, no. 5: 5355–5363.

11. Dutta, P., and S.-P. Bhattacharyya. 1990. Orthogonality-Constrained Variational Calculation of Excited State Wavefunctions and Energies. A New Strategy Based on the Method of Simulated Annealing. *Chemical Physics Letters 167*, no. 4: 309–314.

12. Dutta, P., and S.-P. Bhattacharyya. 1990. A New Strategy for the Calculation of Configuration Interaction Wavefunctions: Direct Search Involving Metropolis Simulated Annealing. *Physics Letters A 148*, no. 6–7: 331–337.

13. Dutta, P., and S.P. Bhattacharyya. 1991. Simultaneous Optimization of Linear and Non-Linear Parameters of a Trial Wavefunction: The Efficacy of the Method of Simulated Annealing. *Chemical Physics Letters 184*, no. 4: 330–334.

14. Sharma, C.S., and C.A. Coulson. 1962. Hartree-Fock and Correlation Energies for 1s2s ^{3}S and ^{1}S States of Helium-like Ions. *Proceedings of the Physical Society 80*, no. 1: 81–96.

15. Chaudhury, P., P. Bandyopadhyay, and S.P. Bhattacharyya. 1996. Exploring a Discretized Generator Coordinate Route to Approximate Eigenstates of One- and Two-Dimensional Anharmonic Oscillators within the Framework of the Simulated Annealing Method. *Chemical Physics Letters 262*, no. 6: 764–770.

16. Dutta, P., and S.P. Bhattacharyya. 1992. The MC SCF Problem and Simulated Annealing Strategy of Global Minimization. *Chemical Physics Letters 199*, no. 1–2: 169–175.

17. Car, R., and M. Parrinello. 1985. Unified Approach for Molecular Dynamics and Density-Functional Theory. *Physical Review Letters 55*, no. 22: 2471–2474.

18. Car, R., and M. Parrinello. 1988. Structural, Dymanical, and Electronic Properties of Amorphous Silicon: An Ab Initio Molecular-Dynamics Study. *Physical Review Letters 60*, no. 3: 204–207.

19. Apolinario, S.W.S., and F.M. Peeters. 2011. Binary Dusty Plasma Coulomb Balls. *Physical Review E 83*, no. 4: 41136.

20. Erber, T., and G.M. Hockney. 1995. Comment on "Method of Constrained Global Optimization." *Physical Review Letters 74*, no. 8: 1482–1482.

21. Chaudhury, P., P. Dutta, P. Bandyopadhyay, P. Sarkar, and S.P. Bhattacharyya. 1996. A Random Walk to Local Minima and Saddle Points on a Potential Energy Surface. A Strategy Based on Simulated Annealing. *Chemical Physics Letters 250*, no. 2: 238–246.

22. Dutta, P., D. Majumdar, and S.P. Bhattacharyya. 1991. Global optimization of molecular geometry: A new avenue involving the use of Metropolis simulated annealing. *Chemical Physics Letters 181*, no. 4: 293–297. doi: 10.1016/0009-2614(91)80073-7.

5

Swarm Intelligence

5.1 Introduction

We are living in an already complex world which is getting even more complex by the day. We are overburdened with unending streams of information, the mere abundance of which is intimidating. It is impossible for a human mind to comprehend the huge volume of available information, categorize it, discover structure or pattern embedded in it, let alone make efficient and productive use of it. We are therefore helplessly becoming dependent on hardware or machines (computers) and software (computer programs) to make sense of a myriad of information and process it for practical use. In order to handle complex problems involving processing of enormous volumes of information, the software systems are increasingly getting so much more complex and intractable that they can no longer be effectively controlled by a single human mind. It is in this context that "swarm intelligence" guided computing emerges as a refreshingly novel alternative to the more traditional computing paradigm involving control, programing, and centralization. The swarm intelligence-based computing paradigm leads to a natural way of designing intelligent systems in which autonomy, emergence, and distributed functioning replaces control, programing, and centralization. It is pertinent to ask what benefits the change of paradigm brings. The benefits are indeed enormous as such systems can solve problems that are computationally very hard (*NP* hard). The biological inspiration behind this paradigm shift emanated from studies of behavioral patterns of foraging flocks of birds, herds of animals, swarms of bees, colonies of ants, or fireflies glowing in unison, etc. Diverse as these entities are, there are certain unifying features among them which prompt us to describe them by the general term—swarm—and the phenomenon of collective motion noted among them by swarming. Swarming is the most visibly and easily observed phenomenon in flocks of birds. The individuals here move collectively in a well-coordinated manner as a single unit—the swarm displays a clear structural order. The movement of the individuals in a swarm becomes so integrated that despite occasional changes in shape and direction (Figure 5.1), the collection of individuals continues to move as a coherent single entity.

As a flock of birds flies by it may appear so much like a formation of aircraft participating in a fly-past. The aircraft are executing a drill, well planned with every detail charted out beforehand; there is a leader of the formation and there is a ground controller constantly monitoring the flying machines and communicating with the individual fliers who in turn communicate with each other. The flocks or the swarms on the other hand do not have a preprogrammed flight path, or a leader of the flock (though a leader may emerge temporarily as the swarm flies), nor is there a ground controller giving instructions. Indeed, the motion of each individual of the swarm could be completely random;

FIGURE 5.1
A flock of birds changing the shape of a flock.

yet the formation—the swarm—moves, maintaining a coherent structure. Understanding how the collective behavior of the swarms (call it swarm organization) emerges and what is achieved in the process is important as the understanding can then be translated into a new computing paradigm in which a self-organizing assembly of simple computing elements (agents) cooperate to solve problems that are computationally hard.

Extensive computer modeling experiments have revealed the main set of principles that lead to the emergence of the collective behavior of swarms. They are

1. *Homogeneity:* the principle implies that every individual in the flock behaves similarly—the flock moves without a definite and permanent leader.
2. *Locality:* the movement of each bird (individual) is directly influenced only by the motion of its nearest flock mates; that is, the flock members interact only locally, the interaction diffuses across the flock leading to integration and self-organization.

3. *Collision avoidance:* each bird tries to avoid collision with its neighboring flock mates; thus, there appears to be a force of repulsion when two flockmates come close to each other in space.

4. *Velocity matching:* each bird tries to match its velocity with those of its neighbors; i.e., they interact only locally.

5. *Flock centering:* each bird tries to stay close to its immediate neighbors; that means except during maneuvering for avoiding a collision, as if there is a force of attraction operating among flockmates, each trying to move closer to the flock center.

The readers may note that the collective or self-organizing behavior of the flock is a consequence of a set of rather simple rules of interactions among flockmates that are strictly local. What does the swarm achieve in the process? An individual in a swarm is a simple agent with rather modest capability to execute a task. When these simple agents are integrated into a swarm which is nothing but a large group of simple agents interacting and cooperating to perform a task or achieving an objective, a collective intelligence (swarm intelligence) that is superior to the sum of the intelligence of all the individuals, emerges and helps the swarm accomplish a task that may be too hard for an individual to do. The emergence of swarm intelligence appears as an automatic consequence of interactions among the simple agents (birds) through simple local rules as already described. It turns out that this biological model of collective behavior can be translated into a distributed computing paradigm for performing hard computing tasks and has led to the development of swarm-based computing techniques like the artificial bee colony algorithm, ant colony optimization algorithm, firefly algorithm, particle swarm optimization (PSO) algorithm, and so on. We propose to consider some of these paradigms in the subsequent sections. The first swarm algorithm to be considered is the PSO proposed by Kennedy and Eberhart [1,2].

5.2 Particle Swarm Optimization

Kennedy and Eberhart proposed a problem-solving technique apparently exploiting the sociocognitive learning elements of a foraging flock of birds. However, they argued that the PSO algorithm almost suggests itself as a simple-minded extension of the cultural model of Axelrod [3]. Axelrod hypothesized and later demonstrated by means of computer simulations that agents who are similar to each other are more likely to interact and then become even more similar. Similarity among individuals can therefore result in the spread of a culture. In Axelrod's theory, the similarity is causally responsible for interaction and exchange of cultural features and is an inevitable consequence of cognitive optimization. The analysis of Kennedy and Eberhart, however, rejected the causal position of similarity in Axelrod's theory after careful examination and analysis and suggested an adaptive culture model. Here, an individual strives to adopt features of randomly chosen neighbors conditionally (not necessarily on the basis of similarity) and the spread of culture leads to optimization of other functions that produce similarity among neighboring individuals. Similarity thus becomes an effect of cultural adaptation and not the cause of it. The adaptive culture model (ACM) was shown to be capable of navigating through complex search spaces and locating optima in several combinatorial optimization problems. The socio-cognitive underpinnings of the ACM are simple and universal—an individual in a society evaluates its own performance, compares it with the performance of neighbors, and imitates the better ones to improve its position in that society. It turns out that these simple rules translate into a powerful problem-solving enterprise, namely the PSO algorithm.

5.2.1 The Algorithm

Let us consider the problem of finding the global minimum of a multivariate function $f(x_1, x_2, ..., x_i, ..., x_N)$ where $x_1, x_2, ..., x_N$ are continuous decision variables with values distributed in the range $(-L, L)$. Suppose now that a group of n-particles (the swarm) are randomly flying through the N-dimensional search space of our problem. At any instant of time t the position of the ith particle in this space can be represented by an N-component (real-valued) vector $\vec{X}_i(t) \equiv (x_{i1}(t), x_{i2}(t), ..., x_{iN}(t))$ and its dynamical state by the corresponding N-component velocity vector $\vec{V}_i(t) \equiv (v_{i1}(t), v_{i2}(t), ..., v_{iN}(t))$. At $t = 0$, the components of $\vec{X}_i(t)$ and $\vec{V}_i(t)$ are randomly chosen from preset ranges ($-a$ to a for $x_{ik}(0)$ and $-v$ to v for $v_{ik}(0)$, for $i = 1, 2, ..., n$ and $k = 1, 2, ..., N$). If the particles do not interact, they will fly randomly through the search space without forming any coherent structure. Suppose now that the individual particles are each trying to locate the global minimum of the function $f(\vec{X})$—the function representing the foraging objective. With respect to this objective, a measure of the goal-specific fitness or performance of an individual i at time t is the value $f_i(t) = f(\vec{X}_i(t))$. Let the lowest value of f up to time t achieved by the immediate neighbors of the ith particle be $f_{i,lbest}(t)$, the corresponding position vector being $\vec{X}_{i,lbest}(t)$. The subscript *lbest* refers to the local nature of the minimum as it has been discovered by a few neighborhood particles of the ith individual. The lowest value of f found so far by the ith individual itself may similarly be denoted by $f_{ibest}(t)$ and the corresponding position vector by $\vec{X}_{ibest}(t)$.

The ACM suggests that the ith particle would try to emulate the best neighborhood particle to improve or better its own position in the search space and therefore correct its velocity vector $\vec{V}_i$ for the next instant of time which is $t + 1$. How would it correct the velocity? From where would it draw the correction? A part of the correction would presumably come from the difference between its current position vector $\vec{X}_i(t)$ and the personal best $\vec{X}_{ibest}(t)$ it has achieved so far. Another part of the correction could come from the difference between $\vec{X}_i(t)$ and the neighborhood best $\vec{X}_{i,lbest}(t)$ achieved so far by its immediate neighbors. The first part of the correction can be immediately recognized as the cognitive part while the second part can be identified as the social component of it. The corrected velocity vector of the ith particle at the next instant of time $(t + 1)$ then takes the following form:

$$\vec{V}_i(t+1) = \eta \vec{V}_i(t) + c_1 r_{1i}(t)\left(\vec{X}_{ibest}(t) - \vec{X}_i(t)\right) + c_2 r_{2i}(t)\left(\vec{X}_{i,lbest}(t) - \vec{X}_i(t)\right). \qquad (5.1)$$

The updated position vector of the ith particle at $(t + 1)$th instant of time is given by

$$\vec{X}_i(t+1) = \vec{X}_i(t) + \chi \vec{V}_i(t). \qquad (5.2)$$

In Equation 5.1, η is a real-valued scalar called the inertial weight—it represents the innate tendency of the particle or individual i to remain in the same velocity state $\vec{V}_i(t)$, a measure of its inertia. c_1 and c_2 are called acceleration coefficients corresponding to the cognitive and social components of velocity correction or learning. $r_{1i}(t)$ and $r_{2i}(t)$ are two random numbers drawn from a normal distribution in the range. $(0,1)$ χ in Equation 5.2 is a real scalar that acts like a damping constant, limiting the size of the correction for updating the position vector so that no runaway situation develops and threatens disintegration of the swarm. In the swarm literature, χ has been called a constriction coefficient, introduced to ensure convergence of the swarm-based search. Why are random numbers $r_{1i}(t)$ and $r_{2i}(t)$ introduced in Equation 5.1? The answer becomes clear if we note that we do not have any means to assess beforehand the relative weights of cognitive and social components of velocity correction or learning. The acceleration coefficients c_1 and c_2 are therefore randomly modulated allowing the particle to oscillate between the two attractors ($\vec{X}_{ibest}$ and $\vec{X}_{i,lbest}$). As the search converges to the best optimum, the two attractors tend to merge into one and the random oscillations subside.

Along with the velocity and position-updating Equations 5.1 and 5.2, we must also have equations for updating $\vec{X}_{ibest}(t)$ and $\vec{X}_{i,lbest}(t)$, the positions of two attractors for completing the description of the present version of the PSO algorithm for solving problems with continuous (nondiscrete) search variables. The relevant equations are

$$\vec{X}_{ibest}(t+1) = \vec{X}_i(t), \quad \text{if} \quad f\left(\vec{X}_i(t+1)\right) \geq f\left(\vec{X}_i(t)\right)$$

$$= \vec{X}_i(t+1), \quad \text{if} \quad f\left(\vec{X}_i(t+1)\right) < f\left(\vec{X}_i(t)\right). \tag{5.3}$$

For determining $\vec{X}_{ibest}(t)$—the position of best particle in the immediate neighborhood of the ith particle, the swarm is divided into several overlapping groups of particles; the group M_i defining the neighborhood of the particle i, $\vec{X}_{i,lbest}(t)$ is then determined as follows:

$$\vec{X}_{i,lbest}(t) = Min\left\{f\left(\vec{X}_i(t)\right)\right\} \ \forall X_i \in M_i. \tag{5.4}$$

The neighborhood topologies can be constructed in several ways. One simple way is to assume a ring lattice (see Figure 5.2a) and define

$$M_i = \left\{\vec{X}_{i-l}(t), \vec{X}_{i-l+1}(t), ..., \vec{X}_i(t), \vec{X}_{i+1}(t), ..., \vec{X}_{i+l-1}(t), \vec{X}_{i+l}(t)\right\}.$$

If $l = 1$, we have a three-particle neighborhood topology with particle numbers $(i-1, i, i+1)$ which is the most commonly used topology. With larger values of l, we will have larger neighborhoods. In the *lbest* swarm with a three-particle neighborhood, an individual in the swarm is influenced by only two of its neighbors, (particle with cardinality $i-1$ and $i+1$), one on the left and one on the right. The interaction among particles is therefore entirely "local" (cf. the velocity-matching rule of flocks of birds). Swarms of this kind will converge slowly, but locate the global optimum with greater facility. The *lbest* swarm can navigate around local optima as the "subswarms" are able to explore multiple optima in the search space simultaneously.

An alternative topology that is global in nature is shown in Figure 5.2b. It can be clearly seen from Figure 5.2b that each individual in the *gbest* swarm is influenced by the best individual in the entire swarm at a particular instant of time. All the individuals are simultaneously attracted to the same individual—the *gbest* (as if simulating flocking) and

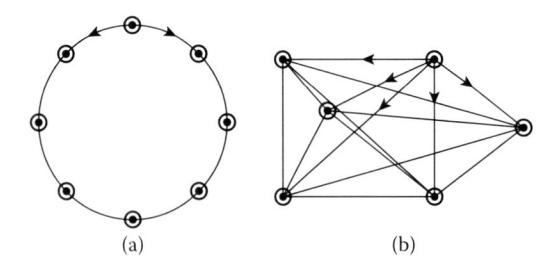

FIGURE 5.2
(a) An *lbest* swarm (b) a *gbest* swarm.

that helps the swarm converge faster. There is only one center of attraction, so there is no subswarming—and in that there is a potential source of trouble. Unless the best individual is already close to the global optimum, the entire swarm may miss the global optimum. This can happen in the absence of subswarming, as there is little exploration of promising areas of search space other than that in the neighborhood of *gbest*. The velocity updating equation for *gbest* swarms is slightly different and is obtained by replacing $\vec{X}_{i,lbest}(t)$ by $\vec{X}_{gbest}(t)$ in Equation 5.1 leading to

$$\vec{V}_i(t+1) = \eta \vec{V}_i(t) + c_1 r_{1i}(t)\left(\vec{X}_{ibest}(t) - \vec{X}_i(t)\right) + c_2 r_{2i}(t)\left(\vec{X}_{gbest}(t) - \vec{X}_i(t)\right) \tag{5.5}$$

while

$$\vec{X}_i(t+1) = \vec{X}_i(t) + \chi \vec{V}_i(t+1). \tag{5.6}$$

Many other topologies have been experimented with. The best one for a given problem can usually be decided through experiments at smaller scales (see, for example, Kennedy and Mendes). The PSO described in this section works rather well in the search for the global minimum on fixed potential energy surfaces supporting multiple minima. By modifying the objective function or the fitness function appropriately (see Chapter 7), the PSO algorithm can be profitably utilized to locate local minima on the PES as well. If the PES is not fixed, but evolving along with the electronic charge distribution, the PSO algorithm can be cleverly marshaled to calculate the equilibrium electronic charge distribution along with the distribution of the nuclear position variables at equilibrium simultaneously. We will address these issues along with other possibilities in Chapter 7.

Suppose now that the decision variables x_{ik}s are not continuous, but assume only discrete values such as 0 and 1, −1 or 1. Can the PSO algorithm be used to find the global optimum of a function of discrete variables? The answer is in the affirmative. However, such applications require some modifications in the algorithm, as described in the following section.

5.3 Particle Swarm Optimization for Discrete Search Variables

The PSO algorithm is structurally flexible enough to accommodate optimization problems where a binary decision (e.g., 0 or 1; yes or no) must be enforced on the variables of the problem. However, some crucial changes must be introduced. Suppose the search vector $\vec{X}_i(t)$ has N-component (x_k, $k = 1, N$) each of which may either 0 or 1. The search vector $\vec{X}_i(t)$ can be regarded as a bit string S_i with N bit positions (1, 2, ..., N) carrying either zero (0) or one (1). In this situation we can only talk about the probability $P(x_{ik}(t))$ that the kth bit position of the ith string is 1 or 0 at time t rather than specifying what the value of x_{ik} is. This probability, say $P(x_{ik}(t) = 1)$ will be a function of personal (cognitive) experience and social influences along with the value that x_{ik} had in the immediately previous time step ($t - 1$), and the predisposition of this component to have a value equal to 1. Formally we can write

$$P(x_{ik}(t) = 1) = function\left(x_{ik}(t-1),\ v_{ik}(t-1), \vec{X}_{k,ibest}, \vec{X}_{k,lbest}\right). \tag{5.7}$$

The quantity on the left hand side of Equation 5.7 is just the probability that the kth bit of the string i has a value equal to 1 at time t. $x_{ik}(t)$ represents that what the current state of the kth bit of string i (or S_i) is while $v_{ik}(t-1)$ measures the extent of predisposition of the ith string to take on a value equal to 1 at the kth bit position. $\vec{X}_{k,ibest}$ is the best state for the kth bit that the ith individual in the swarm has discovered so far, while $\vec{X}_{k,ibest}$ is the best state for the kth bit that the swarm has been able to find until now. Thus $\vec{X}_{k,ibest} = 1$, if the ith individual performed the best when x_{ik} was equal to 1 ($\vec{X}_{k,ibest} = 0$, otherwise). The same definition applies to $\vec{X}_{k,lbest}$ values. The required change in the basic philosophy of the PSO algorithm for continuous search variables described in Section 5.2, for handling optimization problems with discrete decision variables is now clear, and we can proceed to explicate an updating procedure for the discrete decision variables x_{ik}s following what was prescribed for the continuous case. First, consider the velocity updating problem. Since we are not in a position to guess whether the individual's cognitive or social learning will dominate, it is reasonable to assign random weights to each and write

$$v_{ik}(t+1) = v_{ik}(t) + c_1 r_1(t)\left(\vec{X}_{k,ibest}(t) - \vec{X}_{k,i}(t)\right) + c_2 r_2(t)\left(\vec{X}_{k,lbest}(t) - \vec{X}_{k,i}(t)\right)$$
$$= v'_{ik}\ (\text{say}). \tag{5.8}$$

We recall that v_{ik}s measure the predispositions to have one of the discrete values (0 or 1) at a specific bit position of a particular string (i) and are neither themselves discrete nor distributed over a fixed range. Therefore, we must first map the updated values of v_{ik}s (i.e., v'_{ik}s) onto a function that produces values within a specific range (say 0 and 1) and convert them into binary decision values. One procedure to accomplish the task is to proceed as follows:

Let us first feed the updated velocity components (v'_{ik}s) into a sigmoid function $S(v'_{ik})$ where

$$S(v'_{ik}) = 1/(1 + exp(v'_{ik}))\quad \text{with}\ \ 0 \leq S(v'_{ik}) \leq 1. \tag{5.9}$$

Consider now a random N-dimensional vector R_i with components R_{ik} randomly assigned values 0 or 1. The binary decision on the search variables x_{ik}s at the next instant can be taken by adhering to the following rule:

$$\text{If } R_{ik} \leq S(v'_{ik});\ x_{ik}(t+1) = 1$$

$$\text{else}\qquad\qquad x_{ik}(t+1) = 0 \tag{5.10}$$

The updated values of x_{ik} are used to construct the vector $\vec{X}_i(t+1)$ and fitness function defined for the problem, testing at every time step whether the fitness value has become better. If it is better, the $\vec{X}_{ibest}$ is updated.

The algorithm described here can be profitably utilized for finding the global minimum energy configuration of N randomly interacting spins on a lattice, each spin assuming a value +1 or −1. We will return to examine the possibility in Chapter 6.

5.4 Convergence of the PSO Algorithm

Let us suppose that the inertia weight η in Equation 5.1 and damping constant χ in Equation 5.2 are both set to have a value equal to 1. In the velocity updating equation there are contributions from two attractors, X_{ibest} and X_{lbest} in the *lbest* swarm case or X_{ibest} and X_{gbest}, if the swarm is the *gbest* swarm. The two contributions are weighted by the corresponding acceleration coefficients c_1 and c_2, respectively, multiplied by random numbers r_1 and r_2 in the range (0,1). Let $c_1 r_1$ be equal to ϕ_1 and $c_2 r_2$ be equal to ϕ_2. The net effect of the two contributions is to force the particle i to oscillate unevenly about a point defined as the weighted average of the *ibest* and *lbest* components [1,2]. Let us call it $X_{av}(t)$ where

$$X_{av}(t) = \frac{\left[\phi_1\left(\vec{X}_{ibest}(t) - \vec{X}_i(t)\right)\right] + \phi_2\left(\vec{X}_{lbest}(t) - \vec{X}_i(t)\right)}{\left(\phi_1 + \phi_2\right)}. \tag{5.11}$$

Since ϕ_1 and ϕ_2 are both stochastic in nature, $X_{av}(t)$ will undergo change at every iteration implying that the oscillation takes place about a continuously changing reference position. There is nothing in the model described here that forbids the oscillation from increasing in amplitude and becoming so large that the swarm disintegrates or perhaps loses the coherent structure. One way to prevent the "explosion" is to set a limiting value V_{max} and force the velocity from exceeding this limit in any of the N-dimensions. The simplest way to implement this is to introduce the following corrective actions:

$$(1)\ \textit{if}\ V_{ik} > V_{max}, \qquad \textit{then}\ V_{ik} = V_{max}$$

$$(2)\ \textit{if}\ V_{ik} < -V_{max}, \qquad \textit{then}\ V_{ik} = -V_{max}. \tag{5.12}$$

The enforcement of the conditions laid down in Equation 5.12 will ensure that the oscillations remain bounded and no explosion occurs, without, however, introducing any tendency in the swarm to collapse or converge to a point in the search space. The convergence to a point can be guided either by introducing a dynamically changing inertia weight $\eta(t)$ or by introducing a damping or a constriction coefficient χ, or both. The idea of using a constriction coefficient was introduced and experimented with by Maurice Clerc [4,5]. Let us examine his idea in the form of a simpler construct for updating velocity and position when only one attractor is present and both η and χ are equal to 1. Then the trajectory of the ith individual in the swarm is determined by the following equations:

$$V_{t+1} = V_t + \phi(X_{best} - X_i(t)) = V_t + \phi Z_t \tag{5.13}$$

$$X_{t+1} = X_t + V_{t+1} = X_t + V_t + \phi Z_t. \tag{5.14}$$

Hence,

$$X_{best} - X_{t+1} = -V_t + (1-\phi)Z_t$$

$$i.e., \quad Z_{t+1} = -V_t + (1-\phi)Z_t. \tag{5.15}$$

Combining with Equation 5.13 we can recast the velocity and position-updating equations in the form of a matrix equation (see Kennedy and Eberhart [1,2])

$$\begin{pmatrix} 1 & \phi \\ -1 & 1-\phi \end{pmatrix} \begin{pmatrix} V_t \\ z_t \end{pmatrix} = \begin{pmatrix} V_{t+1} \\ z_{t+1} \end{pmatrix}. \tag{5.16}$$

Equation 5.16 defines a dynamical system determined by the matrix M where

$$M = \begin{pmatrix} 1 & \phi \\ -1 & 1-\phi \end{pmatrix}. \tag{5.17}$$

If the two eigenvalues of M, $\in+$ and $\in-$, have moduli ≤ 1, the iterative system defined by Equation 5.16 converges starting from an initial vector $y_0 = \begin{pmatrix} V_0 \\ z_0 \end{pmatrix}$. The two eigenvalues of the matrix M are

$$\epsilon_{\pm} = \frac{(2-\phi) \pm \sqrt{\phi^2 - 4\phi}}{2}. \tag{5.18}$$

We can immediately conclude that $|\in_+| = 1$, if $\phi = 4$. The choice of ϕ is therefore very important for the swarm to converge to an optimum. Normally, one chooses the acceleration coefficients $c_1, c_2 \geq 2$ which ensures that ϕ is in the appropriate range. The analysis is a little more complex when a constriction or damping coefficient "χ" is introduced. Here, too let us consider the simplified, single attractor case of velocity and position-updating equations, with the inertial weight fixed at $\eta = 1$. Instead of introducing χ in the position-updating step, we multiply χ into the right-hand side of Equation 5.13 so that the equations determining the trajectory of a particle in the swarm become

$$\vec{V}_{t+1} = \chi(V_t + \phi(\vec{X}_{best} - \vec{X}(t))$$

$$\vec{X}_{t+1} = \vec{X}_t + \vec{V}_{t+1}. \tag{5.19}$$

The corresponding dynamical system in matrix form turns out to be

$$M_1 \begin{pmatrix} \vec{V}_t \\ \vec{Z}_t \end{pmatrix} = \begin{pmatrix} \vec{V}_{t+1} \\ \vec{Z}_{t+1} \end{pmatrix}$$

$$\text{where,} \quad M_1 = \begin{pmatrix} \chi & \chi\phi \\ -\chi & 1-\chi\phi \end{pmatrix} \tag{5.20}$$

$$\vec{Z}_t = \left(\vec{X}_{best} - \vec{X}(t)\right)$$

$$\vec{Z}_{t+1} = \left(\vec{X}_{best} - \vec{X}(t+1)\right).$$

The convergence is ensured, if the eigenvalues of the matrix $M_1(\varepsilon)$ have moduli less than or equal to 1 [1,2]. The appropriate value of the constriction coefficient χ can be determined from the eigenvalues of the M_1 matrix by taking

$$\chi = \frac{2\,\varepsilon}{\left|(2-\phi) - \sqrt{\phi^2 - 4\phi}\right|}. \tag{5.21}$$

Since $|\varepsilon| \leq 1$ assures convergence, we can estimate χ by taking $\varepsilon = 1$, and $\phi > 4$, say 4.1, where $\chi = 0.73$. In practice, one almost invariably uses $\chi = 0.7 - 0.8$ and c_1, c_2 are so chosen that $\phi > 4$.

5.5 The Fireflies Algorithm

A swarm philosophy-guided algorithm developed during 2007–2008 [6,7] that has proved its usefulness and versatility in solving very different problems become known as the "firefly algorithm." It is inspired by the flashing behavior of fireflies which are insects belonging to the family of Lampyridae. These insects have light-emitting organs under their abdomen. They breathe in oxygen which combines with a substance called luciferin present in their system with the help the enzyme luciferase producing bioluminescence that produces visible light with very little heat. The light emission is intermittent—so the fireflies appear to blink at night when there is darkness around. The intensity of the emitted light and the blinking pattern depend on the species and a host of other factors. It is believed that the blinking serves as an optical signal to potential mates, and in a humid and warm atmosphere, many small swarms of fireflies flashing in unison are quite a common sight. The flashing behavior of fireflies has been mathematically modeled leading to the development of a rather potent algorithm called the firefly algorithm (FFA) and its many variants. They can be invoked to solve complex nonlinear multimodal optimization problems with easy facility. Proposed in 2008 by Xin-She Yang [6,7], the firefly algorithm has recorded a rather impressive growth as well as diversification and has evolved into an important tool for solving difficult problems. It has so far been used mostly by engineers and computer scientists—but being a metaheuristic, the FFA is flexible enough to be adapted for solving hard computational problems of physics and chemistry.

5.5.1 The Algorithm

Suppose that we are looking for the global minimum of a multidimensional function $f(x_1, x_2, \ldots, x_k, \ldots, x_N)$. The potential solutions can be represented as a vector $\vec{X}^i$ in the N-dimensional search space with real-valued components $x_1^i, x_2^i, \ldots, x_k^i, \ldots, x_N^i$. Let us define an objective function $O(\vec{X})$, the highest maximum of which is the goal of the search. The objective function returns a real value $O_i = O(\vec{X}^i)$ for every input vector $\vec{X}^i$. As already stated, the purpose of the search is to find the vector $\vec{X}^*$ for which $O(\vec{X}^i)$ attains the global maximum. For a pure maximization problem $O(\vec{X})$ could be the function $f(\vec{X})$ itself. The FFA is designed to trace the maximum of $f(\vec{X})$ by allowing a simulated swarm of interacting fireflies (simple agents) to move on the landscape generated by the objective function in accordance with a set of simple rules as detailed below:

1. The fireflies are attracted to each other without any gender bias. The attraction potential or the attractiveness of a firefly is solely guided by the intensity of its glow as seen by other fireflies in the vicinity.

2. The intensity of emission I_i of a simulated firefly designated by the index i is a function of the value of the objective function $O_i = O(\vec{X}^i)$ where $\vec{X}^i = x_1^i, x_2^i, ..., x_k^i, ..., x_N^i$ denotes the location of the ith firefly in the N-dimensional search space of the problem. In the present case, where we are looking for the maximum of a function, $I_i \propto O_i$ or more simply $I_i = O(\vec{X}^i) = O_i$.

3. The intensity of the glow of a firefly i and therefore its attractiveness falls off with the distance (r_i) from which the glow is observed. That means, we have $I_i = \beta_0 e^{-\gamma r_i^2}$, with $r \geq 0$ where r_i is the distance separating the observer from the ith firefly. γ is a scale factor that controls the rate at which I_i falls off with the distance. β_0 is simply the intensity at the site of the ith firefly where r_i is obviously zero.

If the simulated fireflies were noninteracting (i.e., neither attracted to nor repelled by others) they would have randomly moved on the fitness landscape. There is, however, a flashing intensity-driven interaction that brings in a degree of coherence in their motion and induces self-organization to take place. The following updating rule describes the mechanism of emergence of coherent structures in the swarm of fireflies and determines how they move.

Let there be n fireflies randomly located on the objective landscape, their positions in the search space being designated by N-dimensional vectors $\vec{X}^i(t)$ ($i = 1, 2, ..., n$) at time t. Consider two fireflies, i and j, designated by the vectors $\vec{X}^i$ and $\vec{X}^j$ having flashing intensities $I_i(t)$ and $I_j(t)$ with $I_j > I_i$ (suppose) which means the fly j is at a better location than the fly i on the objective function landscape. The firefly i will therefore be attracted by the fly j and will move toward j guided by the following rule [6–8]:

$$\vec{X}^i(t+1) = \vec{X}^i(t) + \beta_0 e^{-\gamma r_{ij}^2}\left(\vec{X}^j(t) - \vec{X}^i(t)\right) + \alpha_t \vec{R}^i(t) \tag{5.22}$$

where $r_{ij}^2 = \sum_{k=1}^{N}\left(x_k^i - x_k^j\right)^2$, x_k^i being the kth component of the vector $\vec{X}^i$ and x_k^j being the same component of the vector $\vec{X}^j(t)$. r_{ij} is thus a measure of the Euclidean distance between the two fireflies in the N-dimensional search space. $\vec{R}^i(t)$ is an N-dimensional vector with N-randomly chosen components in the range $0 < r_i < 1$ drawn from a uniform, normal, or any other distribution. The coefficient α_t is chosen as $\alpha_t = \alpha_0 \theta^t$ with $0 < \theta < 1$ and $\alpha_0 = 10^{-2} \cdot L$ where L is the average scale of the search or design variables (components of the vector $\vec{X}(t)$). It is easy to recognize that γ is also linked to the scale L with $\gamma = L^{-\frac{1}{2}}$.

The updating Equation 5.22 is written in vector form. In the component form, the equations look the same with

$$x_{ki}(t+1) = x_{ki}(t) + \beta_0 e^{-\gamma r_{ij}^2}\left(x_{kj}(t) - x_{ki}(t)\right) + \alpha_t r_{ki}(t) \tag{5.23}$$

for $i = 1, 2, ..., n$; $k = 1, 2, ..., N$.

Here $x_{ki}(t + 1)$ represents the kth component of the search space position vector ($\vec{X}^i$) of the ith firefly at $(t + 1)$, $x_{kj}(t)$ is the kth component of the search space position vector $\vec{X}^j$ at time t, x_{ki} is the kth component of the ith random vector.

1. If we set $\beta_0 = 0$, then the updating Equation 5.22 (or 5.23) represents a random search on the objective function surface.

2. If $\alpha_t = 0$, $\beta_0 > 0$, and $\gamma = 0$, Equation 5.22 reverts to an *lbest* PSO-like algorithm.

3. If $\alpha_t = 0$, $\beta_0 > 0$, and $\gamma = 0$ and the jth firefly toward which the ith fly moves has the most intense emission at time t, Equation 5.22 represents something akin to the PSO algorithm of the *gbest* variety. However, if $\alpha_t = 0$, there is no exploration of the search space whereas if $\beta_t = 0$, the exploitation ceases completely. Thus, correct choices of the α_t and β_0 pair are essential to strike a proper balance of exploration and exploitation. Once that is done, FFA can become a very potent tool. α_t and β_0 are the design variables of the algorithm. One possibility is to optimize α_t and β_0 adaptively so that every problem makes its own optimal choices.

4. A more general choice $\beta(r)$ is possible, for example, we may set $\beta(r) = \beta_0 e^{-\gamma r^m}$ $(m \geq 1)$. For a fixed γ, the characteristic length for the problem is $\Gamma = \gamma^{-\frac{1}{m}}$. Clearly, $\Gamma \to 1$ as $m \to \infty$.

5. For a fixed length scale (Γ) in an optimization problem, γ can be exploited as a typical initial value of the parameters being optimized with $\gamma = (\Gamma^{-m})$.

A variant of the standard firefly algorithm that has proved to be very useful and a powerful global search tool has been called Levy-flight firefly (LFF) algorithm. The position-updating scheme in LFF algorithm [9] has been defined as

$$X_i(t+1) = X_i(t) + \beta_0 e^{-\gamma r_{ij}^2} \left(X_j(t) - X_i(t)\right) + \alpha \, sign \, \left(rand - 0.5\right) \oplus Levy. \qquad (5.24)$$

The first two terms on the right-hand side of Equation 5.24 exactly coincide with the corresponding terms in the standard firefly algorithm (cf. Equation 5.22). The third term in Equation 5.24 represents randomization of the position vector $X_i(t)$ via Levy flights, α being the randomization parameter. The term $(rand (0,1) - 0.5)$ provides a randomized sign on the direction of movement while $\oplus$ represents entry-wise multiplication. The net effect of the third term is to force the firefly to take a step the length of which is drawn from a Levy distribution ($u = t^{-\lambda}$ with $1 < \lambda < 3$) and the direction provided by the term $(rand (0,1) - 0.5)$. Thus, the steps of firefly motion in the LFF algorithm resemble a random walk on the function surface with a power law distribution of step length that has the heavy tail characteristic of Levy distribution.

In addition to the FFA with Levy flight a number of variants of the originally proposed algorithm have been proposed and tested. A discrete firefly algorithm proposed recently to deal with NP-hard problems has been claimed to be superior to the ant colony algorithm in performance [10]. In addition, a chaotic firefly algorithm and various modified firefly algorithms [11,12] have been designed to enhance the algorithms' performance in specific problems. It has, however, not found much use until now in solving hard-to-solve problems that one encounters in chemistry and physics (for example, in finding the ground states of spin glass systems, or protein conformation).

5.6 Artificial Bee Colony Algorithms

5.6.1 Introduction

A number of models have been developed to mimic the rather intelligent collective behavior of a swarm of honeybees and exploited to solve combinatorial as well as numerical optimization problems. The applications dealing with numerical function optimization had been rather scarce until recently. A virtual bee algorithm (VBA) was proposed by Yang [13] in which a swarm of virtual bees was generated and induced to move randomly in the search space. The moving bees interact when they find targets containing nectar which corresponds to the encoded values of the function being optimized. The intensity of interactions among the bees was used as the guiding parameter to trace the optimum. The algorithm was tested in a function optimization involving only two parameters. A significant development in this area took place with the advent of an artificial bee colony algorithm proposed by Akay and Karaboga [14]. To place the algorithm in proper perspective, let us first explore briefly the foraging behavior of real honeybees and try to understand how collective intelligence emerges from the actions of individual bees (simple agents) as they forage and gather honey. Three important components that play crucial roles in the activities of honeybees are the food sources (flowers containing nectar), employed honeybees (those who are already in the process of nectar gathering exercise from particular food sources), and onlooker bees (they are waiting in specific areas of the hive, called the "dance floor") gathering information about food sources from employed bees returning to the hive to unload the gathered nectar. The bees display three important modes of activities, namely recruitment to a food source, abandoning a food source, and scouting for new sources of nectar in the space surrounding the hive. The bees grade the value or importance of food source depending on several attributes such as the proximity of the source to the hive (nest), the richness of the source as to its nectar content, the ease of extracting the nectar. Collectively, these attributes can be viewed as a single determinant namely the "profitability" of the source. The employed bees—the foragers who are already associated with a particular food source they are in the process of currently exploiting—carry a lot of information about the particular source, especially the exact location of the source (distance from the hive and the direction in which it lies), how rich the source is in the nectar content, etc. They may share all this information with the onlooker bees with a certain probability. The unemployed foragers of type I, i.e., the bees waiting in the hive, are looking for information on food sources that are profitable enough to be exploited, on the basis of information that the employed bees returning to the hive share with them. The unemployed bees of type II are scouts who are ready to explore the space around the hive with a view to discovering new profitable food sources, if any. They are fewer in number. The most important element responsible for the emergence of collective intelligence in the activities of the swarm of honeybees is the process of sharing of information among the onlooker and employed bees. The exchange of information takes place on the dance floor or the "dancing area" where the onlookers watch the employed bees execute the so-called waggle dance (a coded form of communication) and learn about the location of the promising food sources already found by them, along with their profitabilities. An onlooker may watch a number of dances and then decide to assign itself perhaps to the most profitable source so far discovered by the employed bees. In this process an onlooker bee becomes an employed bee. We note here that the probability that onlookers are motivated to choose the more profitable food sources is higher because the availability of information about such sources is better. In addition, the employed bees share the information in their possession with a probability that is proportional to the profitability of the food source. The process of information sharing through the waggle dance is longer in duration in such cases, thereby increasing interaction time. All such factors conspire to make the recruitment proportional to the profitability of the food source. This discussion paves the way to a better understanding of the foraging behavior of honey bees and transforms it into an effective algorithm for optimization as detailed below:

1. The important first point to note is that not all the bees start foraging at the same time. Experimental data confirm that the rate at which new foragers join the search is proportional to the difference between the total number of bees and the number of bees foraging currently.

2. The second point is that the potential forager bees initially start off as random fliers—as unemployed foragers—they move without any prior knowledge about the food sources and their dispositions in the space around the hive. They have the option to work as scouts and look for food sources around the hive, or some become recruits after watching waggle dances executed by already-employed foragers gathering information and begin searching for food sources based on the information shared with them. Once a food source is detected, such a bee starts exploiting the source after memorizing the relevant information about it. They become employed foragers. These bees will collect nectar and return to the hive to unload the collected food.

3. That takes us to the important third point in the behavior of the bees. The returning bees have several options after delivering the nectar to the hive. The options are:

    a. The bee may abandon the food source and become an "uncommitted follower."

    b. It can execute the waggle dance, recruit new foragers, and return to the original food source.

    c. It may not recruit any new foragers but continue to forage at the same food sources.

Akay and Karaboga [14] developed their artificial bee colony (ABC) algorithm based on the behavioral features of real-life honeybees that we have just described. It turns out that the ABC algorithm proposed by them is a stochastic global optimization technique that can handle difficult multidimensional numerical function optimization problems with remarkable efficacy. The artificial bees of Karaboga are "random walkers" on the function (objective or fitness) hypersurface. They simultaneously evaluate the function at multiple points on the hypersurface and through an intelligent communication scheme among them explore the surface for optima with great efficiency. The communication scheme—the heart of the intelligent search that the ABC algorithm is endowed with—has been modeled in the mold of the communication pattern of real-life honeybees, with simplifications and modifications required for handling difficult optimization problems in science and engineering.

In a numerical function minimization (maximization) problem, the artificial bees fly over the function hypersurface. They are looking for profitable sources which are nothing but points on the function hypersurface where the function values are comparatively lower (higher). Their target is to mobilize the collective search to eventually discover the point(s) on the hypersurface where the function assumes the lowest (highest) value which is akin to finding the most profitable food source. In the ABC algorithm of Akay and Karaboga [14], the search is implemented in the following manner:

1. Choose a population of n artificial bees, half the number of which are employed bees (n_e) and the remaining ones are onlookers (n_0) Then execute the following steps indexed by t.

2. Choose randomly n_E number of points on the hypersurface and assign n_E employed bees to these points—one bee per point. The positions of the employed bees on the surface are represented by one-dimensional vectors $X_i(t)$, the components of which are real-valued quantities determining the exact location of the point in the search. Thus $X_{ki}(t)$ represents the kth coordinate of the ith employed bee at the current time t. X_{ki} s are the search variables or the optimization variables of the problem.

3. Evaluate the function values $f(x_i(t)) = f_i(t)$, $i = 1, 2, ..., n_E$—these values represent the profitability of the food sources, i.e., the sources assigned to the employed bees—the larger (higher) the value of f_i, more profitable is the source. The next step involves information sharing.

4. Assign the n_0 onlookers to food sources—this step is crucial and is probabilistic in nature. Let $\{Xi(t)\}$; $i = 1, 2, ..., n_E$ be the total number of points (food sources) randomly chosen around the hive (of these n_E are already being exploited by the employed bees).

   The probability that an onlooker chooses a particular source (j) at cycle t is determined as

   $$P_j(t) = \frac{f(x_j(t))}{\sum_{i=1}^{n_E} f(x_i(t))} \tag{5.25}$$

   A roulette wheel selection may be used to choose the source based on the probabilities $P_j(t)$. The onlookers move to the sources selected and explore the neighborhood for a more profitable source. If a better source is available, they discard the old one, keep the new coordinates in memor, and become employed bees. If a better source is not found after a predetermined maximum number of attempts, the onlooker bee abandons the source and becomes a scout. It now moves randomly in search of a profitable source.

5. After the information sharing, all the employed bees return to their respective food sources and explore the neighborhood for better sources. If such a source is found by a bee the previous source is abandoned and it memorizes the new source and shares the information with the onlookers in the next cycle ($t + 1$). If no such solution is discovered in a predetermined number of trials the bee becomes a scout after abandoning the food source.

6. Send the scout to look for good food sources, i.e, points on the hypersurface where the function values are low (high). If the scouts discover such sources (solutions), the abandoned sources are replaced by the new ones.

7. Set t to $t + 1$ and repeat steps 2–4 until termination conditions are met.

It is now necessary to explicate how the employed bees or the onlookers explore the search space around their current respective food sources (solutions vectors X_i). Let X_i be the current solution vector assigned to the ith bee. The components are real values x_{ij}

where $i = 1, 2, \ldots, n_E$ and $j = 1, 2, \ldots, 3d$. A better candidate solution (food source) for ith (scout) bee is generated by taking

$$X'_{ij} = X_{ij} + \phi_{ij}(X_{ij} - X_{kj}) \tag{5.26}$$

where ϕ_{ij} is a random number between -1 and $+1$, the index $k \neq i$, but otherwise randomly chosen from the set $1, 2, \ldots, n_E$.

A food source (a solution) that has been abandoned is replaced by a new solution (source) by the following scheme.

Let the abandoned solution be X_l. Then the new solution generated by the scout may be called X_{lj}^{new} $(j = 1, 2, \ldots, d)$ where

$$X_{lj}^{new} = X_j^{min} + rand[0,1](X_j^{max} - X_j^{min}) \tag{5.27}$$

X_j^{min} and X_j^{max} are minimum and maximum cutoff values for the jth components of the solution vector. The new candidate solutions are evaluated by the artificial bee. If the function value $f(X_i^{new})$ is equal to or lower than the previous value (corresponding to the abandoned vector), X_i^{new} replace the old one—otherwise the old solution X_i is retained. For the benefit of the readers, we present a pseudocode of the ABC algorithm [14] below:

1. Randomly generate a population of solution vectors X_i, $i = 1, 2, \ldots, n_E$.
2. Evaluate the function or objective function values of all the members of the population.
3. Let the cycle counter be set as $t = 1$.
4. Iterate.
5. Generate new solutions (X'_j) for the employed bees using Equation 5.26, replace the old one, if better one found—retain the old one, otherwise. Abandon, if needed.
6. Calculate the probabilities $P_j(t)$s for solutions $X_i(t)$ from Equation 5.25.
7. Produce new solutions for the onlookers using Equation 5.27. Retain/discard depending on the profitability of the new solutions. Abandon if needed.
8. Identify the abandoned solutions, if any. Replace them with new ones.
9. Determine the best solution discovered so far and keep it in memory.
10. Advance the cycle number $t \rightarrow t + 1$.
11. Check if termination criterion $t = t_{max}$ has been satisfied—if so, print the best solution obtained so far and exit. If not, return to step 4 to continue the iteration process.

Karboga and Akay [14] carefully compared the performance of their ABC algorithm with those of other population-based algorithms like GA, DE, PSO as well as ES algorithms on a large number of unconstrained test functions. The authors claim that their ABC algorithm performs either better than the other population-based methods or equally well with fewer control parameters. The authors note that their ABC algorithm does not use the parental crossover mechanism employed in GA or DE to spawn new candidate solutions; instead, it produces a fresh candidate solution by taking the difference between randomly selected parts of the parent and a randomly chosen solution (individual) from the current population, increasing the speed of convergence to a local minimum (optimum). Not only does the crossover mechanism get amply modified in the ABC algorithm, but also the mutation operation of GA or DE. The mutation process in GA or DE randomly alters a randomly selected part (component) of a solution and helps maintain diversity in the evolving population. The ABC algorithm has a mutation-like option to maintain population diversity as in DE and GA. In addition, it accommodates the possibility of replacing a whole solution by another discovered randomly by the scout bee. This second mode of mutation-like process enhances the global search potential of the ABC method while the first mode enables the algorithm to fine-tune the already discovered solution or enhance the local search capacity. The tendency to converge rather prematurely is thereby forestalled by the inbuilt capacity of the mutation-like process to search both locally and globally. The ABC algorithm or its variants have recently received close attention from theoretical chemists and cluster physicists who have exploited their remarkable search efficiency on rugged and complex potential energy landscapes of different types of clusters. The ABC algorithms have shown great promise in handling the problem. We will return to these applications and take a closer look at their strength and weakness as well their potentialities in the next two chapters. Right now we will explore yet another social-insect-inspired algorithm, namely the ant colony optimization (ACO) algorithms. As the name implies, these algorithms mimic the foraging behavior of real ants and build quite sophisticated mathematical models that are capable of handling hard optimization problems of both continuous and discrete varieties.

5.7 Ant Colony Optimization Algorithms

5.7.1 Introduction

Ants are ancient creatures, perhaps as ancient as the habitable earth we have inherited. They are supposed to have appeared on earth some 100 or 120 million years ago. At the present time they are far more numerous than human beings—there could be an estimated 10^{16} ants right now. It is clear therefore that ants have been able to sustain themselves and grow in a fast-changing world. The key to their survival and growth perhaps lies in their social nature. A majority of the species of ants are social insects. They live in colonies of hundreds or even millions of individuals and display rather complex collective behavior although individually they are simple agents. Their complex collective behavior has attracted serious attention and a lot of research has been carried out to understand and model their foraging behavior, brood care, labor distribution, nest construction, cemetery organization, etc. The main thrust of the research has been on algorithmic modeling of their foraging behavior and mathematically understand how different aspects of the complex pattern of their social or collective behavior emerge. Looked at individually, an ant is a rather simple entity with limited capability. To be more specific they are rather simple stimulus-response (SR) agents. They just respond to a local stimulus in a simple way, performing a rather basic action a large component of which is random in nature (not a trained reflex action), yet, their collective response indicates that they represent a highly structured social organism. The emergence of such complex behavior from a collection of simple SR agents has been intriguing and stimulated extensive research activities. The particular trait of the behavior of ants that first attracted the attention of entomologists is their ability to find the shortest route from their nest to a source of food when many possible routes exist. Marco Dorigo was the first to propose an algorithmic model of the foraging behavior of ants. Since then a number of variants of the basic algorithm advanced by Dorigo have come up [15–23]. These algorithms have been successfully applied to handle difficult to solve problems (the traveling salesman problem [24], for example). Collectively, these algorithms have come to be known as ACO algorithms. As with the swarm-intelligence methods, ACOs are also metaheuristics—rather novel and sophisticated ones.

The central question in the foraging behavior of ants is: how do the ants collectively always find or select the shortest route between the nest and a source of food? The question becomes all the more important because the ants do so even though there is no centralized mechanism of coordinating these activities. The clue to solving the puzzle came from experiments. Experimental observation indicates that the ants begin their search for food sources randomly. At this stage no coordinated activity is present. Once a food source is located by one or two individuals their activities become increasingly coordinated with progressively larger and larger numbers of ants choosing the same path between the nest and the food source. It is clear therefore that the other ants must be communicating with the ants that have already discovered a food source and the most dominant feature of the communication is to induce others to trail the same path that the discoverers of the food source have traveled. The form of communication, at least for a majority of the ant species is an indirect one mediated by a chemical substance called pheromones. It turns out that an ant, after locating the food source, carries particles of food to the nest laying a trail of pheromone on the path. Other foragers choose the path along which they sense higher concentrations of pheromone with higher probability. After some time, as more and more ants follow the same route the pheromone trail gets reinforced as many more ants deposit pheromone as they carry food particles back to the nest. The collective response of the ants resembles a kind of autocatalytic behavior in which a positive feedback from a particular path draws increasingly larger number of ants to that path. If that is the mechanism of path selection, the task now is to propose a mathematical model for it. A formal mathematical model for the path selection process was first proposed by Dorigo et al. on the basis of the findings from the famous bridge experiment by Deneubourg et al. [25,26].

5.7.2 The Bridge Experiment

In the laboratory-based experiment of Deneubourg et al., the nest (N) and the food source (S) were kept separated by a bridge with two branches (A, B) of equal length as shown in Figure 5.3.

To start with, both the branches were devoid of any ant pheromone. At these stage ants chose one branch or the other randomly and were thus scattered along the two paths more or less equally. After some time, a vast majority of the ants were found to take up only one of the two equivalent paths of equal length. It appears that once one of the two equivalent paths (A, B) was chosen randomly, the pheromone trail attracted many more onto the same path. The pheromone concentration rose along the chosen path because more and more ants laid pheromone while carrying food particles to the nest along the path. Pasteels et al. [27] established empirically that the probability $P_A(t)$ that a particular path A is chosen by the ants at the time t is well represented by the following expression:

$$P_A(t) = \frac{\left(C + n_A(t)\right)^\alpha}{\left(C + n_A(t)\right)^\alpha \left(C + n_B(t)\right)^\alpha} \tag{5.28}$$

$$P_B(t) = 1 - P_A(t). \tag{5.29}$$

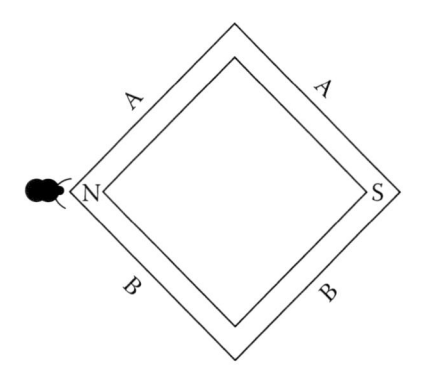

FIGURE 5.3
Bridge experiments with two equivalent paths (A,B).

In Equation 5.28, $n_A(t)$ and $n_B(t)$ are the numbers of ants on the paths A and B respectively at time t, C is a measure of the degree of attraction that the ants may have for the unexplored path while α represents a measure of the bias for pheromone-dependent decision-making by the ants. A high value of α indicates that even a slightly higher concentration of pheromone on a path will succeed in attracting ants to the path in question. On the contrary, the higher the value of C, the higher the pheromone concentration-differential needed to attract ants to the path.

We can now imagine the following scenario: an ant arrives at the bridgehead (Figure 5.3), evaluates the probability $P_A(t)$ and compares it with the probability that a random number $r(0, 1)$ generated at that instant $r(0, 1)$ is the probability of a random event. If $r(0, 1) < P_A(t)$, the ant selects path A. If not, it chooses the other path (B). While the decision-making in the case of equivalent paths is clear, the question that needs to be settled concerns what happens when there exists a path differential—that is, when path A and B are of unequal lengths. Experiments indicate that the ants will ultimately and collectively always choose the shortest path although in the beginning when there is no pheromone biasing they would pick up both the path randomly with equal probability. Why does this happen? The answer lies in the fact that shorter the path the more quickly it is traversed by the ants carrying food from the source to the nest. Therefore the pheromone concentration will build up more quickly along the shorter path, inducing more and more ants to take up the shorter path through what has been called pheromone-assisted decision-making. Experiments indicated that the probability of selecting the shortest path increased with the increase in the ratio of the longer and the shorter path lengths. The mode of communication among ants just described is based on a chemical agent deposited by the ants. The other ants sense the agent locally leading to sign-based stigmergy. There is another form of stigmergy called sematectonic stigmergy in which communication is made via changes in the physical characteristics of the environment. In either case, there is cooperation among the individual agents.

The artificial ACO algorithm constructs a model for the cooperative behavior of real ants by exploiting the stigmeric mode of communication observed in real ant colonies. In effect, the modeling is based on artificial stigmergy. An essential or central element of the modeling is the definition of the stigmergy variables for the artificial ant colony in the context of the specific problem-solving goal. We will illustrate one such artificial ant colony model capable of dealing with the problem of choosing the shortest path between two selected nodes on a graph. The connections with traveling salesman problem are obvious.

5.7.3 A Simple Ant Colony Algorithm

Let us consider the graph displayed in Figure 5.4. The graph G has n_G number of vertices which are contained in the set V and the node connectivity is represented by the matrix E. Thus $G = (V,E)$. There is a source node on G marked S and a destination node O marked D. The artificial ants are walkers (initially random) on the graph G constructing various paths between the source and the destination nodes.

In the simple ant colony algorithm (SACOA) each edge on the graph is initially given a small, but randomly chosen value $\tau_{ij}(0)$ which represents the initial pheromone concentration along the path $i \to j$ ($\tau_{ij}(0)$) which would have been zero for each i, j pair in a real-life ant problem. A number of ants identified by integers $k = 1, 2, \ldots, n_k$ are placed on the source node. Each ant will now choose an edge based on the initial (random) pheromone concentration using the simple logic: $r(0,1) < r_{ij}^k(0)$ the edge i (source node) $\leftrightarrow j$ (one vertex) will be chosen by the ant k initially. Each ant will now be incrementally constructing a path to the destination node. Thus, at each iteration, the ant (k) on the node i at the instant t takes a decision as to the node j to which it can move. It does so assuming that the node is a member of the set N_i^k which is a collection of all nodes accessible from the node i (also called feasible nodes) for the ant k. The decision is taken on the basis of the transition probability $p_{ij}^k(t)$ which the ant k computes as follows:

$$p_{ij}^k(t) = \frac{\tau_{ij}^\alpha(t)}{\displaystyle\sum_{j \in N_i^k} \tau_{ij}^\alpha(t)} \quad \text{if } j \in N_i^k$$

$$= 0, \quad \text{otherwise.} \tag{5.30}$$

If the set N_i^k turns out to be empty (i.e., $N_i^k = 0$) then the immediate predecessor to the node i is adopted as a member of the set N_i^k. This action may lead to the formation of loops in the path constructed by the ant k. These loops will have to be removed once the ant k reaches the destination node. In Equation 5.30, α is a positive number that is used to strengthen the effect of pheromone concentration on the ants' decision-making process. We must be cautious in choosing α values especially in the initial phase as a large value of α could enhance the effect of pheromone excessively and affect decision-making. It may accelerate convergence to a nonoptimal path. The process is repeated with all of the n_k number of ants. Once all the ants have completed the construction of paths from the source to the destination node, and all the loops, if formed, have been removed, each ant (k) will

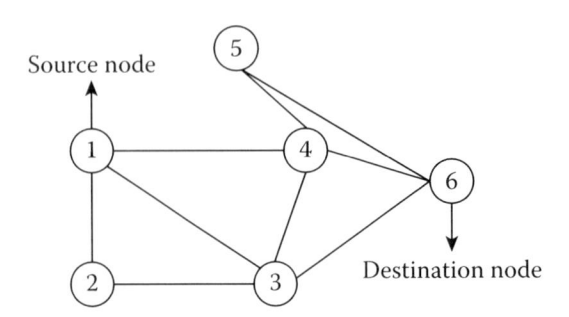

FIGURE 5.4
A graph *G(V,E)*.

retrace the path that it has constructed. While retracing the path, ants will be depositing pheromone on the path. The amount of pheromone deposited ($\Delta\tau_{ij}^{k}(t)$) on the link (or path segment) (i,j) by the ant k may be assumed to be inversely proportional to the length $L_k(t)$ of the path that it has just constructed, i.e, $\Delta\tau_{ij}^{k}(t) = \text{constant} \times \dfrac{1}{L_k(t)}$. So the pheromone concentration on the $i \leftrightarrow j$ link at $t = t + 1$, incorporating pheromone build up caused by all the ants collectively will be

$$\tau_{ij}(t+1) = \tau_{ij}(t) + \sum_{k=1}^{n_k} \Delta\tau_{ij}^{k}(t). \tag{5.31}$$

A little reflection will suggest that $\Delta\tau_{ij}^{k}(t)$ can be an indicator of the quality of the solution constructed by the ant k (the fitness indicator). The smaller the length of $L_k(t)$, the higher will be the amount of pheromone deposited on the path constructed by the kth ant and the quality of the solution is also better. This is an example of explicit evaluation of the solution quality. The question then arises: what if $\Delta\tau_{ij}^{k}(t)$ are all equal? There is then only the differential path length effect that can alter or bias the path selection toward the shortest path. For path selection in such a scenario one can use a different measure for evaluating the suitability or quality of the path $x^k(t)$. A function $f(x^k(t))$ of the constructed path $x^k(t)$ may be used to measure the quality of the solution. The choice of the function may be such that a longer path would produce a larger $f(x^k(t))$ value. In that case we can make the amount of pheromone deposited on any link in the path $x^k(t)$ inversely proportional to the $f(x^k(t))$ (i.e., $\Delta\tau_{ij}^{k}(t) = \dfrac{c}{f(x^k(t))}$). That automatically ensures that all links on the longer path become undesirable in shaping the solution being searched for. This is also an example of "explicit evaluation" where the amounts of pheromone deposited on various links by an ant are proportional to some quality measured by the path length. If the pheromone deposits are made independent of the solution quality, then the ants can still exploit differential path length effect to bias other ants toward the shortest path, exemplifying what has been called "implicit evaluation" in the literature.

Pheromone evaporation

The binary bridge experiments confirmed that ants rapidly converge to a particular solution without much exploration for alternative paths or solutions. The behavior enhances the possibilities of premature convergence to suboptimal solutions. It would be desirable therefore to force ants to explore more and more, and forestall premature convergence. One way to achieve the objective would be to allow the pheromone deposit on any link to evaporate at each iteration causing the ants to forget previous decisions on the path to some extent, depending on the rate of evaporation. For the link (i, j), we can set

$$\tau_i(t) = (1 - R)\tau_{ij}(t) \tag{5.32}$$

with $0 \le R \le 1$, R being the rate of pheromone evaporation. If $R = 1$, the ants become completely random walkers. If $R = 0$, the ants are guided completely by the pheromone deposit and no further exploration takes place. The larger the value of R (<1), the more random is the behavior of the ants, facilitating better exploration. Dorigo and Di Caro [16] observed that for complex graphs the role of evaporation of pheromone becomes crucial. If R is set equal to zero (no evaporation limit) the simple ant colony algorithms do not converge. If a large R is used (i.e., evaporation takes place too rapidly) the algorithm converges to suboptimal solutions for complex problems. Like R, the parameter α too has an important influence on the convergence. If a small α is chosen, the simple ant colony algorithm generally converges to the solution representing the shortest path while the use of larger value of α critically affects the convergence behavior. In what follows we present a pseudocode for a SACOA where the pheromone evaporation has been included.

$t = 0$
Randomly initialize $\tau_{ij}(0)$ || see Note 1
Place n_k number of ants on the source node
Do…
    For each ant $k = 1, 2, \ldots, n_k$
    Construct a path $X^k(t)$
    Set $X^k(t) = 0$
    Do
        Choose the next node based on the transition probability $p_{ij}^k(t)$ of Equation 5.29
        Add the $(i \leftrightarrow j)$ link to $X^k(t)$
        Continue until the destination node D is reached.
        Remove loops if any on $X^k(t)$
        Calculate path length/cost $L_k(t) = f(x^k(t))$ || see Note 2
    End Do
    For each link $(i \leftrightarrow j)$ Do
        (Perform pheromone evaporation)
        $\tau_{ij}(t) = (1 - R)\,\tau_{ij}(t)$ (cf. Equation 5.32)
    End Do
    For each ant $k = 1, 2, \ldots, n_k$ Do
        For each link $(i \leftrightarrow j)$ on $X^k(t)$ Do

$$\Delta\tau^k = \frac{c}{f\left(x^k(t)\right)} \quad ||\ \text{see Note 3}$$

        Reset τ_{ij} using Equation 5.31
        End Do
    End Do
Increment $t \rightarrow t +!$
Until stopping criterion is satisfied || see Note 4
Print/return $X^k(t)$ with the smallest value of $f(x^k(t))$ as the optimal solution

    Note 1: In the simple ant colony algorithm described above, there are several parameters the judicious choice of which can enhance the performance of the algorithm in a specific application. The number of ants n_k, for example, should be chosen with care. If n_k is too small, the exploration ability of the algorithm is reduced. Too many ants mean that it takes significantly longer for pheromone concentration to increase on good links than on bad links, increasing computation time. Similarly, the initial pheromone concentration τ_0 on a link needs to be chosen with care. If τ_0 is large and τ_{ij} $(t = 0)$ s are chosen randomly in the range $(0, \tau_0)$ from a uniform distribution, the starting pheromone concentrations may differ significantly causing a bias toward the links with higher initial pheromone intensities reducing the chances of proper explorations of the search space.

Notes 2 and 3: Fitness or desirability of a path can be determined by appropriately defining the function $f(x^k)$. If $f(x^k)$ is an increasing function of the path length L_k of the path x^k, then the desirability of the path is set to be inversely proportional to the value $f(x^k)$.

Note 4: The stopping criterion is generally fixed by setting a maximum number of iterations n_i. If n_i is too small, the ants do not get enough time to explore the search space and construct the best path. If n_i is too large, there would be significant increase in the computation time.

The selection of the best values of the important parameters mentioned cannot be made by following any routine procedure. Neither is there any analytical tool for doing so. The best way could be to do a preliminary optimization, or use an adaptive protocol which allows the algorithm to come up with the optimal parameter set for each problem.

Other ant algorithms (ant systems)

An ant system (AS) is an improvement over the simple ant colony algorithm that we have already described. The improvement comes in the form of a changed transition rule which now includes heuristic information for constructing the transition probability $p_{ij}^k(t)$, and by allowing the ants to have some memory capabilities through the inclusion of a tabu list. The probability that the kth ant moves from node i to node j at time t is now given by

$$p_{ij}^k(t) = \frac{\tau_{ij}^\alpha(t)\eta_{ij}^\beta(t)}{\sum_{l \in N_i^k} \tau_{il}^\alpha(t)\eta_{il}^\beta(t)} \quad \text{if } j \in N_i^k(t)$$

$$= 0, \quad \text{otherwise.} \tag{5.33}$$

The set of $N_i^k(t)$ represents the set of feasible nodes available to the kth ant when it is located on the node i at time t. The set $N_i^k(t)$ may include all nodes not yet visited by the ant k or only the immediate neighbors of the node i. In the first case, a tabu list for ant k has to be maintained. As ant k visits a new node it is added to the tabu list for k and deleted from the set N_i^k. It ensures that ant k does not visit a node more than once.

The transition probability in AS is controlled by two entities, τ_{ij} and η_{ij}, the first representing dependence on the history of past successful moves while the second takes into account the intrinsic desirability of the particular move. Together they provide a balance between exploration and exploitation provided α and β in Equation 5.33 are chosen properly.

Clearly, previous search experience is discarded if we set $\alpha = 0$, while $\beta = 0$ leads to the neglect of the potential benefit of a move completely. A careful choice of the middle path can improve the performance. If the problem requires the path length or the cost of travel on the path to be minimized, we can set

$$\eta_{ij} = \frac{c}{d_{ij}} \tag{5.34}$$

where d_{ij} is the distance between the nodes (i, j) and c is a constant.

5.8 Continuous Ant Colony Optimization

We have so far considered the ACO algorithm for handling discrete optimization problems (DOPs). The search or the solution variables in DOPs are restricted to have a fixed set of values only. We may now ask the question, can the ant algorithms be adopted to solve say a continuous function optimization problem? If yes, how does one go about it? A little reflection suggests that the extension of the ant algorithms originally developed to solve graph or connectivity problems to the continuous domain requires a systematism to map the continuous space problems onto an equivalent problem of graph search without discretizing the continuous search or solution variables. Such an extension with global search capability is available and has become known as the continuous ant colony algorithms.

The continuous ACO makes a marked departure from other ACO methods and divides the ants (n_a) into two sets of ants, n_l of which performs local search or optimization and n_g which carries out global search ($n_a = n_l + n_g$) and optimization. The entire search space is first randomly divided into n_R regions, each region representing a point (x_i) in the continuous multidimensional (m) search space of the problem. The value of the fitness of a region x_i is just the value of the function $f(x_i) = f_i$. The f_i values are used to arrange n_R regions in descending order of fitness. The n_g regions of lowest fitness are allocated to n_g ants for global search while the n_l regions of relatively higher fitness values are allocated to the n_l numbers of ants for local search. Each x_i has m components corresponding to the m-dimensions of the search space. Let $x_{min,j}$ and $x_{max,j}$ be the minimum and maximum values of the domain in the jth dimension. Thus for each region $i = 1, 2, ..., n_R$ and dimension $j = 1, 2, ..., m$, we set $x_{ij} \sim U(x_{min,j}, x_{max,j})$ initially. The pheromone concentration of each regions (i) is set to $\tau_i(0) = 1$.

Local search
The local search in the regions of better fitness is carried out by the n_l ants by adopting the probabilistic transition rule (Equation 5.33). Thus the kth ant (one of the n_l local ants) selects a region x_i with the probability $p_i^k(t)$

$$p_i^k(t) = \frac{\tau_i^\alpha(t)\eta_i^\beta(t)}{\sum_{j \in N_i^k} \tau_i^\alpha(t)\eta_i^\beta(t)} \tag{5.35}$$

η_l represents the potentiality of the ith region and is therefore set proportional to f_i (the fitness value of the ith region). N_i^k, as defined before, represents the set of feasible regions for the kth ant in the ith region. Assuming that the region selected is x_i, the ant makes a move to a new region x_i' where $x_i' = x_i + \Delta x_i$ and Δx has components $\Delta x_{ij} = c_i - \gamma_i a_i$. Both c_i and γ_i are user-defined numbers, and a_i represents the age of the ith region which is a measure of the relative weakness of the solution concerned. If $f(x_i') > f(x_i)$ then $x_i \leftarrow x_i'$, otherwise x_i is kept unaltered, but the value of a_i is incremented, and a new direction of movement is assigned to ant k. If a better region is discovered, the ant keeps moving in the same direction, and the pheromone concentration is updated.

Global search
The n_g weakest regions are explored by the n_g number of global ants, 90% of which undergo crossover and random mutation in an effort to find new regions for further explorations and 10% undergo "trail diffusion"—a euphemism for arithmetic crossover. The components x_{ij}' of an offspring produced by crossover are determined by randomly choosing a vector X_i from the weak region and setting $x_{ij}' = x_{ij}$ with probability p_c. The variables x_{ij}' defines the offspring x_i' which is allowed to undergo mutation by adding Gaussian random noise to each component x_{ij}'. Thus the mutated variables become $x_{ij}'' = x_{ij}' + N(0, \sigma^2)$

(this in fact is borrowed from evolution strategy). The mutation step size is dynamically reduced at each time step (t):

$$\sigma(t) = \sigma_{max} \left\{ 1 - r^{\left(1 - \frac{t}{n_0}\right)^{\gamma 1}} \right\} \tag{5.36}$$

where $r \sim U(0, 1)$, σ_{max} is the maximum step size for mutation, n_0 is the maximum number of iterations and γ_1 controls the degree of non-linearity.

References

1. Kennedy, J., and R. Eberhart. 1995. Particle Swarm Optimization. In *Proceedings of ICNN'95—International Conference on Neural Networks*, November 27–December 1, vol. 4, pp. 1942–1948. Perth, WA: IEEE.
2. Kennedy, J., and R.C. Eberhart. 2001. *Swarm Intelligence*. San Francisco, CA: Morgan Kaufmann Publishers Inc.
3. Axelrod, R. 1997. The Dissemination of Culture. *Journal of Conflict Resolution 41*, no. 2: 203–226.
4. Clerc, M. 1999. The Swarm and the Queen: Towards a Deterministic and Adaptive Particle Swarm Optimization. In *Proceedings of the 1999 Congress on Evolutionary Computation-CEC99 (Cat. No. 99TH8406)*, July 6–9, pp. 1951–1957. Washington, DC: IEEE.
5. Clerc, M., and J. Kennedy. 2002. The Particle Swarm—Explosion, Stability, and Convergence in a Multidimensional Complex Space. *IEEE Transactions on Evolutionary Computation 6*, no. 1: 58–73.
6. Yang, X.S. 2008. *Nature-Inspired Metaheuristic Algorithms*. 1st ed. Frome: Luniver Press.
7. Yang, X.-S. 2009. Firefly Algorithms for Multimodal Optimization. In *Proceedings of the Stochastic Algorithms: Foundations and Applications: 5th International Symposium, SAGA 2009, Sapporo, Japan, October 26–28*, Watanabe, O., and Zeugmann, T. (eds.). Berlin: Springer, pp. 169–178.
8. Fister, I., I. Fister, X.-S. Yang, and J. Brest. 2013. A Comprehensive Review of Firefly Algorithms. *Swarm and Evolutionary Computation 13*: 34–46.
9. Yang, X.-S. 2010. Firefly Algorithm, Lévy Flights and Global Optimization. In *Research and Development in Intelligent Systems XXVI*. Bramer, M., Ellis, R., and Petridis, M. (eds.). London: Springer, pp. 209–218.
10. Sayadi, M.K., R. Ramezanian, and N. Ghaffari-Nasab. 2010. A Discrete Firefly Meta-Heuristic with Local Search for Makespan Minimization in Permutation Flow Shop Scheduling Problems. *International Journal of Industrial Engineering Computations 1*, no. 1: 1–10.
11. dos Santos Coelho, L., D.L. de Andrade Bernert, and V.C. Mariani. 2011. A Chaotic Firefly Algorithm Applied to Reliability-Redundancy Optimization. In *2011 IEEE Congress of Evolutionary Computation (CEC)*, June 5–8, pp. 517–521. New Orleans: IEEE.
12. Tilahun, S.L., and H.C. Ong. 2012. Modified Firefly Algorithm. *Journal of Applied Mathematics*, no. 467631: 1–12.
13. Yang, X.-S. 2005. Engineering Optimizations via Nature-Inspired Virtual Bee Algorithms. Inbook. In *Proceedings of the Artificial Intelligence and Knowledge Engineering Applications: A Bioinspired Approach: First International Work-Conference on the Interplay Between Natural and Artificial Computation, IWINAC 2005, Las Palmas, Canary Islands, Spain, June 15–18, Part II*, Mira, J., and Álvarez, J.R. (eds.), Berlin: Springer, pp. 317–323.
14. Akay, B., and D. Karaboga. 2012. Artificial Bee Colony Algorithm for Large-Scale Problems and Engineering Design Optimization. *Journal of Intelligent Manufacturing 23*, no. 4: 1001–1014.
15. Dorigo, M., V. Maniezzo, and A. Colorni. 1996. Ant System: Optimization by a Colony of Cooperating Agents. *IEEE Transactions on Systems, Man and Cybernetics, Part B (Cybernetics) 26*, no. 1: 29–41.

16. Dorigo, M., and G. Di Caro. 1999. Ant Colony Optimization: A New Meta-Heuristic. In *Proceedings of the 1999 Congress on Evolutionary Computation-CEC99 (Cat. No. 99TH8406)*, July 6–9, vol. 2, pp. 1470–1477. Washington, DC: IEEE.

17. Dorigo, M., G. Di Caro, and L.M. Gambardella. 1999. Ant Algorithms for Discrete Optimization. *Artificial Life* 5, no. 2: 137–172.

18. Bonabeau, E., M. Dorigo, and G. Theraulaz. 2000. No Title. *Nature* 406, no. 6791: 39–42.

19. Dorigo, M., E. Bonabeau, and G. Theraulaz. 2000. Ant Algorithms and Stigmergy. *Future Generation Computer Systems* 16, no. 8: 851–871.

20. Dorigo, M., and T. Stützle. 2004. *Ant Colony Optimization*. Cambridge, MA: The MIT Press.

21. Dorigo, M., M. Birattari, and T. Stutzle. 2006. Ant Colony Optimization. *IEEE Computational Intelligence Magazine* 1, no. 4: 28–39.

22. Labella, T.H., M. Dorigo, and J.-L. Deneubourg. 2006. Division of Labor in a Group of Robots Inspired by Ants' Foraging Behavior. *ACM Transactions on Autonomous and Adaptive Systems* 1, no. 1: 4–25.

23. Engelbrecht, A.P. 2007. Ant Algorithms. In *Computational Intelligence: An Introduction*, 2nd ed., pp. 359–411. Chichester, UK: Wiley.

24. Dorigo, M., and L.M. Gambardella. 1997. Ant Colony System: A Cooperative Learning Approach to the Traveling Salesman Problem. *IEEE Transactions on Evolutionary Computation* 1, no. 1: 53–66, http://onlinelibrary.wiley.com/book/10.1002/9780470512517

25. Goss, S., S. Aron, J.L. Deneubourg, and J.M. Pasteels. 1989. Self-Organized Shortcuts in the Argentine Ant. *Naturwissenschaften* 76, no. 12: 579–581.

26. Deneubourg, J.-L., S. Aron, S. Goss, and J.M. Pasteels. 1990. The Self-Organizing Exploratory Pattern of the Argentine Ant. *Journal of Insect Behavior* 3, no. 2: 159–168.

27. Pasteels, J.M., J.L. Deneubourg, and S. Goss. 1987. Self-Organization Mechanisms in Ant Societies. I. Trail Recruitment to Newly Discovered Food Sources. In *From Individual to Collective Behavior in Social Insects: Les Treilles Workshop*, Pasteels, J.M., and Deneubourg, J.-L. (eds.). Basel: Birkhauser. Experientia Supplementum, vol. 54, p. 155–175.

6

Application of Soft Computing in Physics

6.1 Introduction

In the preceding chapters, we have seen how biology-inspired soft computing techniques have grown into potent tools for solving global optimization problems involving discrete as well as continuous search variables. It is time now that we looked into the prospect of invoking these techniques for solving some of the problems that physicists and chemists often deal with. A problem that immediately comes to mind has relevance in both physics (atomic and molecular physics) and chemistry (quantum chemistry)—the problem of computing the eigenvalues of an atomic or molecular Hamiltonian matrix (H) constructed in a suitable basis set. The relevant Hamiltonian matrix can be Hermitian. Since the basis sets can be chosen to be real for atoms and molecules the H-matrix can be assumed to be a real symmetric matrix. Formally therefore we can say that our problem is to compute the eigen spectrum of an atomic or a molecular Hamiltonian operator (the energy operator, $\hat{H}$) resolved in a suitable basis—that effectively comes down to computing the eigenvalues and the associated eigenvectors of a real symmetric matrix. Frequently, the primary interest concerns computing the lowest eigenvalues and the corresponding eigenvector of the H-matrix—finding the ground state of the system represented by the Hamiltonian $\hat{H}$. In a more general context, we may also be interested in finding a small subset of the lowest eigenvalues and eigenvectors including the ground state and compute transition energies (ΔE_{ij}), and oscillator strengths. More often than not, we will be computing the eigenvalues and vectors of the H-matrix constructed in a fixed (predesigned and preoptimized) basis set (we may call it static diagonalization). However, sometimes we may also wish to form the H-matrix in the most appropriate (the optimal) basis set for the problem at hand and diagonalize the Hamiltonian simultaneously to extract the desired eigenvalues and vectors (we may call it dynamic diagonalization). We plan to look into the prospect of invoking soft computing options for both static as well as dynamic diagonalization of the Hamiltonian matrix.

The problem of diagonalizing a Hamiltonian matrix appears quite naturally in atomic and molecular physics. That happens because the very edifice of atomic and molecular electronic structure theory has been built mainly on the idea of expanding the unknown molecular electronic wave function (ψ) in finite-dimensional (N) analytic basis sets. Quite expectedly, the problem of finding an optimally designed basis set has attracted a lot of attention of researchers in the context of calculation of atomic and molecular electronic structure. A natural choice of researchers in this area has been to exploit the variational method. The connection between optimization theory and methods of variational electronic structure calculation is rather direct. The variational recipe casts the problem of finding an optimal expansion of the unknown wave function ψ in the mold of extremalization of an appropriately constrained energy functional $E[\psi]$. In the most general case, the extremalization leads to equations that are coupled nonlinear equations and offer multiple solutions which may physically represent minima, maxima, or saddle points of the functional. The problem often lies in locating the lowest energy minimum (the ground state) and perhaps a few of the higher energy minima (the low-lying excited states). Can the soft computing methods offer viable tools for solving the problem? Can we profitably use a soft computing recipe to track down the global minimum of the relevant functional? In the following sections we will explore various issues and possibilities relating to soft computing-based search options for obtaining eigenvalues and eigenvectors of real symmetric matrices in general and the ground eigenvalue and eigenvector of such a matrix in particular.

6.2 Variational Principle, Energy Extremalization, and Diagonalization of Hamiltonian by GA

Let $\hat{H}$ be the Hamiltonian of an atomic or molecular system the spectrum of which is bounded from below. In that case the energy functional $E(\psi) = \langle \tilde{\psi}| \hat{H} |\tilde{\psi} \rangle / \langle \tilde{\psi}| \tilde{\psi} \rangle$ attains a global or an absolute minimum ($E = E_0$) for the exact ground state wave function $\tilde{\psi} = \psi_0$ [1]. For all other admissible wave functions, $E[\tilde{\psi}] \geq E_0$. In the variational context, the problem of finding the ground state wave function boils down to the search for a normalizable ψ that globally minimizes the energy E considered as a functional of ψ. For the excited states, the problem is far less simple. The functional $E[\tilde{\psi}]$ can be shown to have a stationary point $E' = E[\tilde{\psi}']$ which is upper bound to the corresponding exact excited state energy E_n if and only if $\tilde{\psi}$ is constrained to remain orthogonal to all the lower exact eigenstates (ψ_0, ψ_1, ψ_2, ..., ψ_{n-1}) of $\hat{H}$ [1]. This requirement operationally cripples the variational search for the excited states from the point of view of optimization theory since the constraint condition on the trial $\tilde{\psi}$ involves (n) lower exact eigenstates of H which will not be known to start with. The search for the excited state wave function and energies are therefore carried out by invoking a far more restricted form of the variational recipe provided by the so-called linear variational principle as embodied in the Hylleraas–Undheim and Macdonald theorem [2,3]. The only restriction imposed here is the linearity of the trial space which essentially means that the unknown wave function ψ is expanded in terms of a large (if possible, complete, but fixed) set of orthonormal basis functions $\{\phi_i\}_{i=1,N}$. Thus we write

$$\psi = \sum_{i=1}^{N} C_i \phi_i, \quad \langle \phi_i | \phi_j \rangle = \delta_{ij}. \tag{6.1}$$

The stationary condition on $E[\tilde{\psi}]$ subject to the normalization constraint on ψ ($\langle \psi | \psi \rangle = 1$) then leads to the following matrix eigenvalue problem:

$$HC = C\tilde{E} \tag{6.2}$$

where N columns of C represent N eigenvectors of H, $\tilde{E}$ being the diagonal matrix containing the N eigenvalues. Assuming that the eigenvalues have been ordered ($\tilde{E}_1 \leq \tilde{E}_2 \leq \cdots \leq \tilde{E}_N$) the first column of C contains the N expansion coefficients (C_{11}, C_{21}, ..., C_{N1}) which define the approximate linear variational ground state wave function $\tilde{\psi}_1 = \sum_{p=1}^{N} C_{p1}\phi_p$ and $\tilde{E}_1$ is the linear variational approximation to the ground state energy. Similarly, the kth column of C contains the N-expansion coefficients (C_{N1}, C_{1k}, ..., C_{Nk1}) for the wave function $\tilde{\psi}_k$ having the approximate eigen energy $\tilde{E}_k$. In the linear variational scheme the search for the ground and excited state wave functions and energies therefore boils down to the construction and diagonalization of an $N \times N$ Hamiltonian matrix H. In atomic and molecular problems $\hat{H}$ is real and Hermitian, the basis set $\{\phi_i\}$ can be chosen to be real so that the H matrix is a real symmetric matrix. The beauty of the linear variational recipe is that all the eigenvalues of the $N \times N$ Hamiltonian matrix are upper bounds to the corresponding exact energies of the system that the Hamiltonian $\hat{H}$ supports. As N increases, these eigenvalues approach the corresponding exact eigen energies ($\tilde{E}_k \rightarrow E_k$) from above and in the limit of a complete basis set ($N \rightarrow \infty$), $\lim_{N \to \infty} \tilde{E}_k = E_k$. The convergence of linear expansion is usually slow, so the question of designing the basis set gains importance. We must mention here that the condition of using an orthonormal basis can be relaxed into the condition for using a linearly independent (but nonorthogonal) basis $\{x_i\}_{i=1,N}$ with $\langle x_i | x_j \rangle = s_{ij}$ defining the metric of the basis set. The linear variation process can be carried out on such a basis. The variation, in this case, leads to a generalized matrix eigenvalue problem:

$$HC = SCE \tag{6.3}$$

where S represents the $N \times N$ matrix of the so-called overlap integral s_{ij}. We will have occasions to refer to Equation 6.3 and discuss how to handle it, and find the eigenvalues and vectors, later in this chapter. Before that let us consider the problem of a soft-diagonalizing H formed on an orthonormal basis.

6.2.1 Stochastic or Soft Diagonalization by GA

The problem of diagonalizing a real symmetric matrix (like the real symmetric Hamiltonian matrix already referred to) is an age-old problem of computational linear algebra [4,5] and has been extensively investigated. There are well-known deterministic methods of handling the problem in general and in the particular case of diagonalizing many (n) electron Hamiltonians on the basis of N orthonormal states represented by N n-electron Slater determinant (SD) $\{\phi_i\}$. The direct or brute force diagonalization route is the simplest and straightforward recipe to extract the eigen energies and wave functions of the system and compute the spectrum. On the basis of N SDs, the storage and CPU requirement of the standard deterministic direct diagonalization-based route to the spectrum of H grow as N^2 and N^3, respectively making them rather unviable for large systems. If the focus is mainly to compute the lowest eigenvalue, the Lanczos [6] and the generalized Davidson methods [7] are far better computing options since the demand on storage grows only linearly with N while the CPU requirement rises as the square of $N(\sim 0(N^2))$. More often than not the interest lies mainly in finding the ground state energy or possibly energies of a small group of low-lying excited states of the system. From the point of view of the nature of the computational problem the task here involves the search for a small subset (n) of the lowest eigenvalues and vectors of H from a search space of very large dimension (N). The stochastic global search methods—the soft computing methods—are known to be capable of exploring large search spaces well and exploit the information available in the space with great efficacy. It is in this context that

Nandy et al. [8] has sought to exploit genetic algorithm to either partially or fully diagonalize the H matrix and compute the relevant eigenvalues and eigenvectors leading to what they have termed diagonalization by GA. We will now analyze their proposals and their implementations.

Let $\{\phi_i\}$ be a set of real n orthonormal functions which is used to construct the real symmetric Hamiltonian matrix H with elements $H_{ij} = \langle \phi_i | \hat{H} | \phi_j \rangle = H_{ji}$. If a plane rotation defined by the angle θ_{ij} is applied to the pair of basis function ϕ_i and ϕ_j, we can represent its action on the H matrix by an orthogonal matrix $U(\theta_{ij})$ with real valued elements where

$$U(\theta_{ij}) = \begin{pmatrix} 1 & 0 & 0 & 0 & 0 & \cdots & \cdots & & \cdots & \cdots & 0 & 0 & 0 \\ 0 & 1 & 0 & 0 & 0 & \cdots & \cdots & & \cdots & \cdots & 0 & 0 & 0 \\ \vdots & \vdots & \vdots & \vdots & \vdots & \vdots & & \vdots & & \vdots & \vdots & \vdots & \vdots \\ 0 & 0 & 0 & 0 & 0 & \cdots & \cos\theta_{ij} & \sin\theta_{ij} & \cdots & & 0 & 0 & 0 \\ 0 & 0 & 0 & 0 & 0 & \cdots & -\sin\theta_{ij} & \cos\theta_{ij} & \cdots & & 0 & 0 & 0 \\ \vdots & \vdots & \vdots & \vdots & \vdots & \vdots & \vdots & \vdots & & \vdots & \vdots & \vdots & \vdots \\ 0 & 0 & 0 & 0 & 0 & \cdots & & & \cdots & & 0 & 0 & 1 \end{pmatrix}. \tag{6.4}$$

Suppose our aim is to design an orthogonal transformation matrix, say U_1 which annihilates all the $(n-1)$ off-diagonal elements of H on the first row and the first column—the elements $H_{12}, H_{13}, \ldots, H_{1n}$ and their Hermitian conjugate elements $H_{21}, H_{31}, \ldots, H_{n1}$ have to be reduced to zero by the transformation. The U_1 matrix for our problem can be constructed as a cumulative product of, say, m number of plane rotation matrices represented by $u(\theta)$ matrices, $u^1(\theta_{12}), u^2(\theta_{13}), \ldots, u^3(\theta_{1n})$, each designed to remove or annihilate a specific pair of off-diagonal element of the H matrix. For example, $u^2(\theta_{13})$ is designed to annihilate the element H_{13} and H_{31} of the H matrix, and so on.

The cumulative effect of m ($>n-1$) such transformations would then bring the H matrix into the following form

$$U_1^{\dagger} H U_1 \equiv \begin{pmatrix} \tilde{E} & 0 \\ 0 & \tilde{H}_{n-1} \end{pmatrix} \tag{6.5}$$

where the cumulative transformation matrix U_1 is given by

$$U_1 = u_n^m(\theta_{1n}) \cdots u_n^{m-1}(\theta_{1n-1}) \cdots u_n^2(\theta_{13}) u_n^1(\theta_{12}). \tag{6.6}$$

The subscript n on $U\theta_{1k}$ matrices denotes that we are dealing with $n \times n$ transformation matrices, while the superscript m indicates that we are considering the mth transformation matrix designed to eliminate the off-diagonals H_{1k} and H_{k1}, θ_{1k} being the angle of rotation required for the purpose. $\tilde{H}_{n-1}$ represents the $(n-1) \times (n-1)$ nondiagonal matrix block produced by the cumulative transformation U_1, assuming that it succeeds in eliminating all the off-diagonals on the first row and column of the H matrix. In view of the block structure of the transformed H matrix in Equation 6.5, it is obvious that $\tilde{E}_1$ represents one of the eigenvalues of the H matrix.

Since our primary objective here has been to extract the lowest energy eigenvalue (ground state energy) and the corresponding eigenvector of H matrix, it is necessary to ensure that $\tilde{E}_1$ is indeed the lowest eigenvalue. That requires a little preprocessing of the H matrix itself before the appropriate rotations are applied. The preprocessing involves permuting the rows and columns of H in such a way that the post rearrangement matrix (call it H') satisfies the following condition:

$$H'_{11} \leq H'_{kk} \quad (k = 2, 3, \ldots, n). \tag{6.7}$$

Once the conditions in Equation 6.7 are satisfied and the appropriate U_1 matrix bring the rearranged matrix into the same form as given in Equation 6.5, $\tilde{E}_1$ is provably the lowest eigenvalue of the H matrix.

6.2.2 The Algorithm for Calculating the Lowest Eigenvalue

Let us now examine how GA can accomplish the task of constructing the appropriate U_1 matrix and extract the lowest eigenvalue of H. The following points may be noted.

1. **Chromosomal representation of potential solution:** The basic search variables of the problem are the $(n - 1)$ Jacobi rotation angles $\{\theta_{1,j}\}_{j=2, N}$. Unfortunately, these angles can only be iteratively determined. This is because the rotation $u^k(\theta_{ij})$ that exactly annihilates the off-diagonals H_{ij} and H_{ij} can produce nonzero off-diagonals in other rows and columns even though they might have been annihilated by previous rotations or were zero to start with. So the chromosomal solutions can be represented as collection of $(n - 1)$ Jacobi rotation angles generated randomly in the appropriate range $\left(0 \leq \theta \leq \dfrac{\pi}{4}\right)$. Thus, a population of n_p number of string $S_1, S_2, \ldots, S_{np}$ is to be created to begin with where θ_{ij}^ks are floating-point numbers:

$$S_1 = \left(\theta_{12}^1 \theta_{13}^1 \cdots \theta_{1k}^1 \cdots \theta_{1n-1}^1\right)$$

$$\vdots \tag{6.8}$$

$$S_{n_p} = \left(\theta_{12}^{n_p} \theta_{13}^{n_p} \cdots \theta_{1k}^{n_p} \cdots \theta_{1n-1}^{n_p}\right).$$

We can treat the strings as arrays of $(n - 1)$ floating-point numbers.

2. **The objective and fitness functions:** The objective of GA here is to find the set of optimally conditioned Jacobi rotation angles $(\theta_{12}^0 \, \theta_{13}^0 \ldots \theta_{1n}^0)$ that minimizes the quantity $\sigma = \sum_{k=2}^{n} \left(|H_{1k}|^2 + |H_{k1}|^2\right)$ in the absolute sense when the rearranged H matrix is transformed by the orthogonal transformation matrix (U_1) generated by cumulative product of m transformation matrices generated by the angles of rotations.

$$U_1 = u_n^m(\theta_{1n}) \cdots u_n^{m-1}(\theta_{1\ n-1}) \cdots u_n^2(\theta_{13}) u_n^1(\theta_{12}). \tag{6.9}$$

The objective function σ can be mapped into a scaled fitness function f where

$$f = e^{-\lambda \sigma} \quad (\lambda > 0). \tag{6.10}$$

The condition $f = 1$ (maximum fitness) implies $\sigma = 0$ which in turn guarantees that σ, i.e., the sum of the squares of the off-diagonal elements on the first row and column of H matrix, has been reduced to zero. That is equivalent to ensuring that all off-diagonal elements on the first row and column of the transformed matrix $H_1 = U_1^\dagger H U_1$ matrix have become zero. When that is achieved, we have

$$H_1 \equiv \begin{pmatrix} \tilde{E}_1 & 0 \\ 0 & \tilde{H}_{n-1} \end{pmatrix} \qquad (6.11)$$

and $(H_1)_{11} = \tilde{E}_1$ is the lowest eigenvalue of H. The first column of U_1 on the other hand yields the corresponding eigenvector.

3. **GA-based search for the optimally conditioned rotation angles:** Having defined a chromosomal representation of the potential solution strings and the fitness function for the problem at hand the task now is to exploit GA for finding the (n − 1) optimal rotation angles. This is to construct the partially diagonalizing transformation matrix U_1 that brings H into the blocked form shown in Equation 6.11, signaling that the fitness function has been maximized. As stressed already, the optimal rotation angles are to be searched out iteratively starting from a set of guessed $\left\{\theta^i_{1k}\right\}_{k=2,n}$ making up the potential solution strings $S_1, S_2, \ldots, S_{np}$. Initially, the search space therefore has information about (n − 1) $\times n_p$ angles, which must be exploited by the GA to generate (n − 1) optimal angles $\left\{\theta^0_{1k}\right\}_{k=2,n}$ for constructing the partially diagonalizing transformation U_1. The task is accomplished by allowing the n_p number of strings in the population to evolve genetically on the defined fitness landscape so as to maximize the fitness value. The following steps are to be executed for achieving the objective.

   a. Generate the cumulative trial partially diagonalizing transformation matrix U_1^k from the angles contained in the string $S_k \left(\theta^k_{12} \; \theta^k_{13} \ldots \theta^k_{1n}\right)$ and transform the matrix H into $\tilde{H}_1 = \left(U_1^k\right)^\dagger H \left(U_1^k\right)$. Evaluate the sum of the squares of the off-diagonal elements on the first row and column of the matrix $\tilde{H}_1$ produced by the transformation on H and call it $\sigma_k = \dfrac{1}{2} \sum_{j=2}^{n} \left(\left|H_{1j}\right|^2 + \left|H_{j1}\right|^2\right)$, and calculate the fitness value of the string S_k as $f_k = e^{-\lambda \sigma_k}$ ($\lambda > 0$). Execute the process for all the n_p number of strings $k = 1, 2, \ldots, n_\mathrm{p}$ in the population and compute the average fitness of the population $f_{av}(t)$.

   b. Perform a fitness proportional selection on the population by the traditional roulette wheel method or a tournament selection procedure with predefined tournament size, and create a breeding pool of n_p number of strings (some of them expectedly with higher than average fitness of the population) for generating new solution strings by crossover breeding. The crossover is carried out with a probability p_c (preset) and new rotation angles (the genes) are generated in the following fashion.

   Let S_i and S_j be the strings selected for undergoing a single-point crossover at the site k with a probability p_c. Before crossover the compositions of the strings are

$$S_i = \left(\theta^i_1 \theta^i_2 \cdots \theta^i_k \cdots \theta^i_m\right)$$

$$S_j = \left(\theta^j_1 \theta^j_2 \cdots \theta^j_k \cdots \theta^j_m\right). \qquad (6.12)$$

The post crossover strings are S'_i and S'_j where the genetic compositions of S'_i and S'_j are

$$S'_i = \left(\theta^i_1 \theta^i_2 \cdots \theta^{i'}_k \cdots \theta^i_m \right)$$

$$S'_j = \left(\theta^j_1 \theta^j_2 \cdots \theta^{j'}_k \cdots \theta^j_m \right)$$

(6.13)

where

$$\theta^{i'}_k = f\theta^i_k + \left(1 - f\right)\theta^j_k$$

$$\theta^{j'}_k = f\theta^j_k + \left(1 - f\right)\theta^i_k.$$

(6.14)

The mixing parameter f in the arithmetic crossover defined in Equation 6.14 is drawn randomly from a uniform distribution with f in the range $0 < f < 1$. The arithmetic crossover process defined in Equation 6.14 may be called a "minimal arithmetic crossover" as it enforces minimal changes in the genetic makeup of the strings. In other implementations, all the genes beyond the kth gene θ_k are transformed as given in Equation 6.14. Needless to mention perhaps is that any other crossover mechanism (see Chapter 2) could be employed. Exactly n_p number of offspring are generated.

c. Once the crossover operation on the entire population is complete, the post-crossover population is allowed to undergo mutation with a probability p_m ($p_m \ll p_c$). The mutation is carried out in the following manner. Let us assume that the lth entry (gene) on the ith post crossover string S'_i (i.e., the angle θ^i_l has been chosen to undergo mutation randomly with the probability p_m). The mutated angle (call it $\bar{\theta}^i_l$) is obtained by the following transformation:

$$\bar{\theta}^i_l = \theta^i_l + \left(-1\right)^M r \cdot \Delta_m$$

(6.15)

(Postmutation)

where M is a randomly chosen integer, γ is a random number ($0 \le \gamma \le 1$), and Δm is the mutation strength or intensity (a small user-defined number in the range 0.1–0.2) which may either be held fixed (static mutation intensity) or can be dynamically adjusted as a function of the number of generations n_g elapsed.

The set of operations (a) to (c) is repeated until no further improvement in the fitness of the best (the fitness maximal) string takes place. At this stage, the $\tilde{E}_1$ value generated by the partially diagonalizing transformation U_1 encoded by the angles in the best string in the population is accepted as the ground or the lowest eigenvalue of the real symmetric H. A flowchart for the GA-driven stochastic partial diagonalization of the H matrix to obtain the lowest eigenvalue E_1 and the corresponding eigenvector $\bar{U}_1$ is shown below. Readers are encouraged to convert it into an executable code which can be done quite easily.

Steps for Jacobi GA route to the lowest eigenvalue and vector:

i. Read in the matrix ($H(i,j)$, $i = 1, N$; $j = I,N$).
ii. Symmetrize $H(j,i) = H(i,j)$.

iii.   Rearrange rows and columns so as to ensure $H_{11} < H_{jj}$ $(j = 2, N)$.

iv.   Create a population of n_p strings or chromosomes $S_1, S_2, \ldots, Sn_p$ by randomly choosing a set $(n - 1)$ rotation angle $\theta_1, \ldots, \theta_{n-1}$ with $\theta \le \dfrac{\pi}{4}$, for each string.

v.   For each string $S_1, S_2, \ldots, Sn_p$ construct $(n - 1) N \times N$ orthogonal matrices, taking θ_is from S_i.

$$u^1(\theta_1), u^2(\theta_2), \ldots, u^{n-1}(\theta_{n-1}): \ \theta_1 \equiv \theta_{12}; \ \theta_2 \equiv \theta_{13} \ldots \theta_{n-1} = \theta_{1,n}$$

vi.   Cumulatively transform $H \rightarrow \tilde{H} = \ldots u^i(\theta_i) H u^i(\theta_i) \cdots$ and accumulate the product of $(n - 1)$ u^i s matrices in U_1.

vii.   After $(n-1)$ transformations are completed evaluate $\sigma_i = \dfrac{1}{2} \sum_{j=2}^{N} \left(|H_{1j}|^2 + |H_{j1}|^2 \right)$ and then the fitness $f_k = e^{-\lambda \sigma} i$ of the string S_i.

viii.   Find f_{best} (maximum among $f_1, f_2, \ldots, f_{np}$) and the corresponding $(U_L)_{best}$.

ix.   Check if $|f_{best} - 1 \le \epsilon$ (small number $\sim 10^{-12}$).

If yes, Print $\tilde{H}_{11} = \tilde{E}_1, (U_1)$, exit.

Otherwise, go to step v and iterate.

6.2.3 Computing the Largest Eigenvalue of H

The algorithm presented in Section 6.2.2 can be easily tweaked a little to enable it to compute the largest eigenvalue and eigenvector of a real symmetric matrix. The following changes are to be made.

   The first change to be introduced is in the objective of rearrangement of the rows and columns of H. The rows and columns of H are to be so rearranged that the following conditions are satisfied.

$$H'_{nn} \ge H'_{n-1\,n-1} \ge \cdots \ge H'_{22} \ge H'_{11}. \tag{6.16}$$

The second change to be made concerns the objective of $(n - 1)$ Jacobi-like transformation to be applied to the rearranged matrix H'. The purpose now is to annihilate all the off-diagonals on the nth row and by Hermitian symmetry, on the nth column of H'. The objective function is now defined as

$$\sigma_l = \frac{1}{2} \sum_{j=n-1}^{1} \left(|H_{nj}|^2 + |H_{jn}|^2 \right) \tag{6.17}$$

while the fitness function retains the same form

$$f = e^{-\lambda \sigma} \ (0 < \lambda). \tag{6.18}$$

When all the designated off-diagonals have been annihilated by the optimally conditioned U_1 matrix generated by the GA the transformed matrix H' assumes the following blocked form:

$$\bar{H}' = u_n^\dagger H' u_n \equiv \begin{pmatrix} \bar{H}'_{n-1\,n-1} & 0 \\ 0 & \tilde{E}_n \end{pmatrix} \tag{6.19}$$

with

$$\bar{H}'_{nn} = \tilde{E}_n.$$

It is easy to establish that $\tilde{E}_n$ is indeed the largest eigenvalue of the original H matrix, and the nth column of the transformation matrix U_n that has brought the H matrix into the form given in Equation 6.19 represents the corresponding eigenvector.

6.2.4 Algorithms for Extracting Multiple Eigenvalues by GA

1. **The sequential route:** The GA-driven Jacobi-like partial diagonalization method described in the preceding section can be utilized for extracting more than one eigenvalue including the lowest one, along with the relevant eigenvectors. The simplest way to accomplish this objective is to adopt a sequential route—first obtain the lowest eigenvalue and eigenvector (E_1, ψ_1), then go for the first excited eigenvalue and vector (E_2, ψ_2), and so on. The strategy is elaborated as follows.

   Let us assume for the sake of simplicity that all the eigenvalues of the H matrix are negative (i.e., $E_i < 0$, $i = 1, 2, \ldots, N$) and that lowest eigenvalue E_1 and the associated eigenvector ψ_1 have been already determined by the algorithm described in the previous section. Then the projector P_1 onto the ground state is $P_1 = |\psi_1\rangle\langle\psi_1|$ and $Q_1 = 1 - P_1$ is the projector for the complementary subspace of the $(N - 1)$ dimension. To compute the first excited state eigenvalue E_2 and the associated eigenvector $|\psi_2\rangle$, we may proceed as follows:

   a. Form the projected H matrix (call it $\tilde{H}_2$)

      where $\tilde{H}_2 = Q_1^\dagger H Q_1$

      Note that the ground state eigenvector $|\psi_1\rangle$ of H is also an eigenvector of with $\tilde{H}_2$ the eigenvalue $\tilde{E}_1 = 0$ by construction. The remaining $(n - 1)$ eigenvalues of $\tilde{H}_2$ are all negative and have one-to-one correspondence with the $(n - 1)$ excited eigenvalues of H.

   b. Rearrange rows and columns of the projected matrix $\tilde{H}_2$ so that

      $$\left(\tilde{H}_2\right)_{11} \leq \left(\tilde{H}_2\right)_{kk}, \quad k = 2, 3, \ldots, N.$$

   c. Look for the lowest eigenvalue and the corresponding eigenvector of the projected matrix $\tilde{H}_2$ by the GA-Jacobi algorithm already described. It is easy to see that the lowest eigenvalue of $\tilde{H}_2$ is also the energy eigenvalue of the first excited state of H. Let the corresponding eigenvector be $|\psi_2\rangle$ so that $P_2 = |\psi_2\rangle\langle\psi_2|$ is the projector for the first excited state.

   d. Form the projector $Q_2 = 1 - P_2$ and construct the doubly projected matrix $\tilde{H}_3 = Q_2^\dagger H Q_2$. Note that $|\psi_1\rangle$ and $|\psi_2\rangle$ are both eigenvectors of $\tilde{H}_3$ with the eigenvalues equal to zero by construction, while the remaining $(N - 2)$ eigenvectors of H are also the eigenvectors of $\tilde{H}_3$ with unaltered eigenvalues (all < 0). We can therefore follow the same GA-based procedure and go for the lowest eigenvalue and eigenvector of $\tilde{H}_3$ which will also be the second excited state eigenvalue and eigenvector of H. The procedure may be repeated until the small subset of say m eigenvalues and eigenvectors of H (including the ground one) have been computed. The sequential route works well if the projectors and the projected matrices are computed with the desirable level of accuracy.

2. **Determination of all the n-eigenvalues and vectors simultaneously**: Instead of proceeding sequentially, the GA-Jacobi method can be exploited to compute all the n-eigenvalues and eigenvectors of H simultaneously in one go. Only two changes are to be introduced in the computing strategy.

   a. The objective function σ now involves all the $n(n-1)$ off-diagonal elements on the n rows and columns of the H-matrix and reads

   $$\sigma = \frac{1}{2} \sum_{i=1}^{n-1} \sum_{k=i+1}^{n} \left(|H_{ik}|^2 + |H_{ki}|^2 \right). \tag{6.20}$$

   b. The corresponding scaled fitness function is accordingly set as

   $$f = e^{-\lambda\sigma}, \lambda > 0. \tag{6.21}$$

   The strings $S_1, S_2, \ldots, S_{n_p}$ are now composed of the $N = n(n-1)/2$ rotation angles $\theta_{12}, \theta_{13}, \ldots, \theta_{(n-1)n}$ with $\theta_{ij} \leq \frac{\pi}{4}$. Each θ_{ij} defines an orthogonal transformation matrix $U^k(\theta_{ij})$ that acts on H to annihilate the off-diagonals H_{ij} and by Hermitian symmetry H_{ji}. The method remains otherwise the same. When the optimally conditioned rotation matrices have been found by the GA, the n columns of the cumulative transformation U_0 matrix represent the n eigenvectors of H and the n-diagonal elements of the matrix $U_0^\dagger H U_0$ represent the n-eigenvalue $(E_1, E_2, \ldots, E_n)$. The method avoids the need to calculate projectors and form the successively projected matrices $Q_1^\dagger H Q_1$, $Q_2^\dagger H Q_2$, ... and rearrange rows and columns of the projected matrices, before applying the diagonalizing transformations.

3. **GA-Jacobi Diagonalization with simultaneous optimization of the basis function:** It is in this context that the GA-Jacobi method can display its full potential, and have an edge over the deterministic means of handling the problem. Suppose that the $(n \times n)$ H matrix has been formed in a basis set (ϕ) which has M nonlinear parameters embedded (call them $\alpha_1, \alpha_2, \ldots, \alpha_M$). The task now is to diagonalize the H matrix partially and simultaneously optimize the basis set so that the ground-state energy eigenvalue attains a minimum. The objective function (O) for the task here can be written as a sum of two square terms, O_1 and O_2, each weighted by a weighting factor (call them β_1 and β_2). Thus, we have

   $$O = \beta_1 O_1 + \beta_2 O_2$$
   $$= \beta_1 (E - E_L)^2 + \beta_2 \sum_{k=1}^{n-1} \frac{1}{2} \left(|H_{ik}|^2 + |H_{ki}|^2 \right) \tag{6.22}$$

E_L is an estimated lower bound to the ground state energy. We may set $\beta_1 = \beta_2 = 1$ to start with. The objective O_1 ensures that energy is minimized $(E \rightarrow E_L)$ while O_2 takes care of the fact that the diagonalizing transformation annihilates the relevant off-diagonal elements. The only other change that must be brought in now concerns the chromosome or strings to be used to represent the potential solutions. The genetic information encoded by the strings must contain the $(n-1)$ rotation angles $\theta_{12}, \theta_{13}, \ldots, \theta_{1(n-1)}, \theta_{1n}$ and the M nonlinear parameters present on the basis function. The optimally conditioned rotation angles ensure that the matrix H has been brought into the blocked diagonal form (Equation 6.5), while simultaneous optimization of the M basis set parameters guarantees that $\tilde{E}_1$ of Equation 6.5 has attained the minimum value. As the optimization proceeds $\tilde{E}_1 \to E_L$ and $O_2 \to O$ and the fitness function $f = e^{-\lambda O} \to 1$. Nandy et al. [8] demonstrated the viability of the GA-Jacobi method in the present as well as in the previously described context in a number of model calculations, e.g., ground and excited state energy eigenvalues of Morse oscillators, the ground state of H⁻ ion, the ground and first excited states of the He atom. It appears that the GA-Jacobi route with parallelization of the fitness evaluation step can be a viable soft computing option for computing the ground or ground plus a few of the low-lying excited states in a fixed basis set representation of the Hamiltonian matrix. If the basis set is also simultaneously optimized, the GA-based search proceeds quite smoothly. As the basis functions evolve, the relevant elements of the H-matrix are to be recalculated, consuming additional computer time. The overall convergence rate of the global minimum, however, remains impressive [8]. Much more work is needed to make the methodology more attractive computationally.

Although the methodologies detailed here involve one of the most well-known soft computing options, namely genetic algorithms, the same problem could also have been solved by invoking for example the simulated annealing method (SAM). It would be a useful exercise for the motivated reader write a code using the SAM for stochastic diagonalization of the H matrix to obtain the ground eigenvalue $\tilde{E}_1$, and the eigenvector $|\psi_1\rangle$ in a model problem. The point to be noted here is that the SAM can be considered as a single parent method in which a "mutational-like" event is thermally activated causing the parent string to evolve. Since the fitness of only one string is to be evaluated, the method may appear to be computationally attractive— but the need to carry out many reconfiguring moves at every temperature step (so the system has enough time to equilibrate) can offset that advantage. It could be more advantageous to carry out the GA-based diagonalization in the initial phase of the search for the targeted eigenvalue (and the nonlinear basis parameters) and switch over to the SAM for fine-tuning the optimal rotation angles and the basis parameters. Such hybrid techniques may prove more viable in the present context. We anticipate much more activity in this direction in the forthcoming years.

6.3 GA in Solving the Partitioned Matrix Eigenvalue Problem

The stationary states of atomic or molecular electronic Hamiltonian ($\hat{H}$) are solutions of the energy eigenvalue equation.

$$\hat{H}\psi_n = E_n\psi_n, \ n = 1, 2, \ldots. \tag{6.23}$$

Eigen solutions of Equation 6.23 can be obtained only for rather simple systems. For more complex systems, we must fall back on some means of finding approximate solutions of Equation 6.23—solutions which can be systematically improved by refining the level of approximation. As we have already described (Section 6.2) one standard way to obtain approximate solutions is to expand ψ (the unknown wave function) in a fixed finite (N) orthonormal basis (in the limit of a complete orthonormal basis the expansion is exact) $\{\phi_i\}$: $\psi = \Sigma_i C_i \phi_i$ and invoke the linear variational ansatz that converts the operator eigenvalue equation into a matrix eigenvalue equation.

$$HC = E\,\mathbb{1}_N\,C. \tag{6.24}$$

We repeat that for an atomic/molecular problem, H is real and Hermitian. The basis functions can be chosen to be real making H an $N \times N$ real symmetric matrix eigenvalue problem in the N-dimensional space spanned by the N-basis functions. If we are interested in only a small subset of solutions of Equation 6.24 including the lowest or the ground eigenstate of $\hat{H}$, it would be sensible to partition the eigenvalue problem so that we can work in a subspace of much smaller dimension (n_a say) and yet obtain the exact estimates of the designated n_a eigenvalues [9–11]. One of the schemes of partitioning leads to an energy-dependent effective Hamiltonian $H_{eff}(E)$ that operates in n_a dimensional subspace $n_a \ll N$ and generates exact eigenvalues of n_a number of states and the a-subspace projection (C_a) of the corresponding total eigenvector (C). The b-subspace projection (C_b) of C can then be obtained from another relation (see Sections 6.3.1 and 6.3.3). The first part of the problem therefore boils down to forming and diagonalizing $H_{eff}(E)$ in a subspace of much smaller dimension.

$$H_{eff}(E)C_a = E\,\mathbb{1}_a\,C_a. \tag{6.25}$$

The prospect looks attractive, but is beset with difficulties arising from the built-in non-linearity of Equation 6.25 as $H_{eff}(E)$ depends on the unknown eigen energy (E). Soft computing methods like GA are inherently capable of handling strongly nonlinear problems like the one posed in Equation 6.25 with great efficacy. The soft computing option has indeed been explored by Sharma et al. [12] and we will examine in detail the strategy advocated by the authors in handling partitioned matrix eigenvalue problems.

6.3.1 Energy-Dependent Partitioning: Search for the Ground State

Let us suppose the total eigenvector space of N-dimensions has been partitioned into two subspaces of dimension n_a and n_b. Hence, $N = n_a + n_b$. We will impose the additional restriction that $n_a \ll N$, being the number of eigenvalues and eigenvectors of H to be extracted, the number of eigenvalues and eigenvectors left undetermined being n_b. The partitioned eigenvectors (C) and the H matrix of our interest have the following structures:

$$C \equiv \begin{pmatrix} C_a \\ \cdots \\ C_b \end{pmatrix}, \quad H = \begin{pmatrix} H_{aa} & \vdots & H_{ab} \\ \cdots & \cdots & \cdots \\ H_{ba} & \vdots & H_{bb} \end{pmatrix} \tag{6.26}$$

with partitioned vector C and H matrix defined in Equation 6.26, the original matrix eigenvalue equation (Equation 6.25) splits into the following pair of equations:

$$H_{aa}C_a + H_{ab}C_b = E\,\mathbb{1}_a\,C_a \quad (I)$$

$$H_{ba}C_a + H_{bb}C_b = E\,\mathbb{1}_b\,C_b \quad (II). \tag{6.27}$$

From Equation 6.27 (II), we have

$$C_b = \left(E\mathbb{1}_b - H_{bb}\right)^{-1} H_{ba} C_a \tag{6.28}$$

which can be used to eliminate C_b from Equation 6.27 (I), resulting in an effective eigenvalue equation in the n_a dimensional subspace which reads

$$\left\{ H_{aa} + H_{ab}\left(E\mathbb{1}_b - H_{bb}\right)^{-1} H_{ba} \right\} = E\mathbb{1}_a C_a. \tag{6.29}$$

Calling the E-dependent $n_a \times n_a$ matrix within the curly brackets on the left-hand side of Equation 6.29 an energy-dependent effective Hamiltonian $H_{eff}(E)$, we have reduced the problem of finding the n_a eigenvalues of the $(N \times N)$ H matrix to an equivalent problem of solving an $n_a \times n_a$ effective matrix eigenvalue problem for the desired eigenvalue:

$$H_{eff}^a(E) C_a = E\mathbb{1}_a C_a. \tag{6.30}$$

The solution of Equation 6.30 provides the exact (full space) eigenvalues (E_1, E_2, ..., E_{na}), but only the a-subspace projection of C_a of the total eigenvector (C). The b-subspace projection of C can, however, be easily obtained by using Equation 6.28 which asserts that $C_b = \left(E\mathbb{1}_b - H_{bb}\right)^{-1} H_{ba} C_a$, thereby providing the full space eigenvector (C). Turning to Equation 6.30, we can see that the equation is manifestly nonlinear as the effective Hamiltonian depends on the unknown energy (E). The question now is to find a soft computing technique like the GA, for example, to handle the nonlinear (pseudolinear) problem posed in Equation 6.30.

The key to solving the present nonlinear problem lies in guessing an energy E (say, $E = E(0)$), construct $H_{eff}^a\left(E^{(0)}\right)$ after obtaining the inverse $\left(E^{(0)}\mathbb{1}_b - H_{bb}\right)^{-1}$ (note here that if the energy E matches with one of the b-space eigenvalues, the inverse cannot be computed as the concerned matrix becomes singular) and then diagonalizing $H_{eff}^a\left(E^{(0)}\right)$ to obtain an improved estimate of E ($E = E^1$, say), reconstruct $H_{eff}^a(E)$ matrix with $E = E^1$ and continue the iterative process until E fails to improve any further. Authors of reference [13] discarded the repeated direct diagonalizing route and proposed a GA-driven route to search for the full space eigenvalues and eigenvectors within the a-subspace. It would be useful to analyze the GA-driven route threadbare to understand how it operates and the potential pitfalls, if any. The GA-driven strategy involves the following steps:

1. Create a population of n_p number of trial a-subspace eigenvectors ($\left|\psi_i^a\right\rangle$) where

$$\left|\psi_i^a\right\rangle = \sum_{p=1}^{na} C_{pi}\left|\phi_p\right\rangle, \, i = 1, 2, \ldots, n_p, \, C_{pi} = \left\langle\phi_p\,\middle|\,\psi_i^a\right\rangle. \tag{6.31}$$

Since $\left|\phi_p\right\rangle$ represents a fixed set of orthonormal basis vectors spanning the a-subspace, the vectors $\left\{\left|\psi_i^a\right\rangle\right\}$ can be represented as chromosomes or strings $\left\{S_i^a\right\}$ each string containing n_a number of coefficients $\left\{C_{pi}\right\}_{p=1,na}$ (the genes) appearing in the expansion of $\left|\psi_i^a\right\rangle$ in the given orthonormal basis $\left\{\left|\phi_p\right\rangle\right\}$. Thus, the population of n_p strings can be represented as

$$
\begin{aligned}
S_i^a &\equiv \left(C_{1i}, C_{2i}, \ \ldots, \ C_{ki}, \ldots, C_{n_a i}\right) \\
&\vdots \\
S_l^a &\equiv \left(C_{1l}, C_{2l}, \ \ldots, \ C_{kl}, \ldots, C_{n_a l}\right) \\
&\vdots \\
S_{n_p}^a &\equiv \left(C_{1n_p}, C_{2n_p}, \ \ldots, \ C_{kn_p}, \ldots, C_{n_a n_p}\right).
\end{aligned}
\tag{6.32}
$$

The amplitudes are chosen randomly in the range (0,1) but their phases are to be chosen carefully so that the expansion (Equation 6.31) can represent the ground state correctly.

2. Guess a trial value of energy E for the ground state of H, and find the inverse of $(E_b^1 - H_{bb})$, and hence $H_{bb}^a(E) = \left\{H_{aa} + (E\mathbb{1}_b - H_{bb})^{-1} H_{ba}\right\}$.

3. Calculate the Rayleigh quotient (P_i) for the ith wave function string S_i^a and the corresponding gradient of the Rayleigh quotient (P_i) as follows:

$$
\left.
\begin{aligned}
P_i &= \frac{\left(S_i^a\right)^\dagger H_{eff}^a(E) S_i^a}{S_i^a S_i} \\
\nabla P_i &\quad \frac{2\left(H_{eff}^a(E) - \rho_i \mathbb{1}_a\right)}{S_i^a S_i}
\end{aligned}
\right\}.
\tag{6.33}
$$

If S_i happens to be an eigenvector of $H_{eff}^a(E)$, $\Delta P_i = 0$ otherwise, $\Delta P_i \neq 0$. The norm N_i of ∇P_i $\left(\left(N_i = \sqrt{(\nabla P_i)^\dagger (\nabla P_i)}\right)\right)$ or N_i^2 itself can be used to generate a fitness landscape (f_i) by setting

$$
f_i = e^{-\lambda N_i} \text{ (being a user-defined scalar).}
$$

Clearly $f_i \to 1$ as $\nabla P_i \to 1$. Thus, maximization of f_i would lead the search to an eigenvector of $H_{eff}^a(E)$. If the relative phases of expansion coefficients have been chosen carefully (ensuring minimum number of nodes) the maximization of f_i would lead us to the ground state of $H_{eff}^a(E)$. Nandy et al. [14] evaluated $\left\{(\nabla P_i)^\dagger (\nabla P_i)\right\}$ as well as P_i for each string in the population in parallel by distributing the n_p strings over n-processors. In their implementation each processor receives one string at a time and sends back the values of P_i and ∇P_i to the master node. We note here that P_i itself can be exploited to generate another measure of fitness (f_i') by taking

$$
f_i' = e^{-\lambda(P_i - E_L)^2} \ (0 < \lambda < 1)
\tag{6.34}
$$

where E_L is an estimated updatable lower bound to the ground state energy of the system. $f_i' \to 1$ as $P_i \to E_L$. If both f and $f_i' \to 1$, one would be confident that the search is approaching a stationary point which is also the ground state. In other words, by ensuring that both P and $\left\{ (\nabla P_i)^\dagger (\nabla P_i) \right\}$ are minimized, we can home in on the ground state of $H_{eff}^a (E)$ with absolute certainty.

4. Having defined the wave function strings, the fitness functions, and the basic search philosophy, it is time now to explicate the systematism to be adopted so that the wave function strings genetically evolve toward the desired goal. The evolution is brought about by allowing well-defined genetic operators to act on the population wave function strings $\left\{ S_i^a \right\}$. These operators are (a) selection operator implemented through a fitness proportional roulette wheel-based selection, (b) arithmetic crossover with a probability p_c, (c) arithmetic mutation with a probability p_m, and (d) diversification with probability p_d.

In the crossover process, two strings S_i and S_k selected from the postselection population (the so-called breeding pool) and two children strings are produced (S_i', S_k') by mixing the genetic information encoded by the strings concerned. The mixing of genetic information in the present context means mixing the expansion coefficients carried by the strings S_i' and S_k' as follows: assuming that the lth site has been chosen for crossover by following the standard procedure, the post crossover amplitudes become

$$C_{pi}' = f_c C_{pi} + (1 - f_c) C_{pk}$$
$$C_{pk}' = (1 - f_c) C_{pi} + f_c C_{pk}. \tag{6.35}$$

For $p = l+1, l+2, \ldots, n_a$. Other genes located at $p \le l$ are not affected. The mixing factor f_c is chosen randomly in the range $0 < f_c < 1$. The mutation operation is carried out on all the postcrossover strings with a mutation probability p_m and intensity Δm. The amplitude at the mutation site (say the pth) of the string k is transformed by using either

$$C_{pk}'' = C_{pk} + (1 - 2r) r_1 \Delta m \tag{6.36}$$

when r and r_1 are two random numbers in the range $(0 \le r \le 1)$, or else we can use

$$C_{pk}'' = C_{pn} + (1)^l r_1 \Delta m \tag{6.37}$$

l being a random integer, r_1 a random number $(0 \le r_1 \le 1)$.

Every 10 generations an additional operation—the so-called diversification operation was carried out by adopting a mutation-like procedure in which the diversification probability $p_d \ll p_m$. It was a perfectly democratic process carried out on all the strings after they underwent the scheduled crossover and mutation processes. The operation is designed to keep enough diversity or variability in the population with a view to forestalling premature convergence of the search to a suboptimal vector.

5. After crossover and mutation (and diversification, if scheduled) are over, the best set of n_p number of strings (from among the parents and offspring) are chosen to construct the population for the next generation. The Rayleigh quotient for the best of the n_p strings (P_{best}) is used as the new input for the energy E needed to construct $H_{eff}^a(E)$ for the new generation.

6. The iterations are terminated when P_{best} fails to improve any further.

6.3.2 Applications

The algorithm proposed in the preceding section was tried on two sets of real symmetric matrices:

1. The Coope matrix (H_1) [15]
2. A random matrix (H_2). The elements of (H_1) matrix are defined as follows:

$$H_1(i,i) = 2.i - 1, i = 1, 2, \ldots, N$$

$$H_1(i,j) = 1 \text{ for } i \neq j \tag{6.38}$$

$$H_1(i,j) = H_1(j,i).$$

The elements of H_2 were chosen by taking

$$H_2(i,i) = i, i = 1, 2, \ldots, N \tag{6.39}$$

for $i \neq j$, $H_2(i,j) = H_2(j,i) = r * \beta$ ($\beta > 1$), r is random number in $0 \leq r \leq 1$.

The authors of reference [15] implemented the GA-driven algorithm to compute the ground state eigenvalues vectors of both H_1 and H_2 using different choices of partition sizes (values of n_a and n_b), and compared the performances with those of a direct, repeated diagonalization method and Löwdin's method. It turned out that the GA-driven method has significant computational advantage over the two other methods especially when the GA-driven algorithm was parallelized to carry out the fitness evaluation, in the a-subspace as well as the matrix inversion in the b-subspace. In the GA-based calculation, it was found unnecessary to carry out the b-subspace inversion at every generation thereby reducing the computational labor quite significantly.

The algorithm was used to extract all of the n_a number of eigenvalues of $H_{eff}^a(E)$ (hence n_a eigenvalues of H) sequentially. First, the lowest eigenvalue E_0 was obtained along with the a-subspace projection (C_0^a) of the full space eigenvector (C_0). The b-subspace projection (C_0^b) was then obtained from the relation $C_0^b = \left\{ (E_0 H_b - H_{bb})^{-1} H_{ba} \right\} C^a$. The full $C_0 = \begin{pmatrix} C_0^a \\ C_0^b \end{pmatrix}$ so obtained was used to form the ground state projector $P_0 = C_0 C_0^\dagger$. The projected matrices $H_p (1 - P_0)^\dagger H_{1 \text{ or } 2} (1 - P_0)$ were formed. H_p was partitioned following the same procedure as adopted for partitioning H_1 or H_2 while searching for their ground states. The ground state of $H_p^a(eff)$ was then obtained by invoking the GA-driven method which is just the first excited state of either H_1 or H_2 as the case may be. By successively projecting the eigenvectors already obtained and seeking out the ground states of the projected matrices, the method could be (and was) extended to obtain the n_a number of eigenvalues and eigenvectors of H_1 or H_2. A pertinent question that arises at this point concerns whether direct diagonalization of $H_{eff}^a(E)$ by the Jacobi method using the current best estimate of E_0 to obtain the inverse $(E_1 H_b - H_{bb})^{-1}$ and iteration to self-consistency would be computationally completely equivalent to the GA-driven procedure advocated by the authors. It was stressed and demonstrated by the authors that the anticipated equivalence was not there [12]. It would be useful to look at the question a little more closely. It turns out that while diagonalizing the $H_{eff}^a(E)$ by the Jacobi method the standard practice is to start the iterations with a unit matrix of appropriate dimension thereby losing important information obtained in the previous step. In the GA-driven method the accumulated information is not lost, thereby accelerating the convergence of the iterations to the targeted eigenvalues and vector. Thus, the apparent theoretical equivalence is operationally lost. A second source of difference stems from the possibility of swapping of the eigenvalues in the direct diagonalization-based recipe. If such swapping takes place, it becomes difficult to home in on the targeted eigenvalue creating a convergence problem, which the GA-driven method does not suffer from.

The authors carried out a fairly detailed evaluation of the effects of parallelization of the fitness evaluation and inversion steps. They concluded that the computation time (T) grows as $T = C_1 N^{2.5}$ in the parallel implementation as opposed to $T = C_2 N^3$ in the serial execution where N is the dimension of the full matrix ($N \leq 2500$). The authors suggested that the performance of the parallelized GA-driven method could be enhanced further by using gigabit ethernet cards on all the slave nodes and replacing the gigabit switch by an InfiniBand switch for communication between the master and the slave nodes. It would be worthwhile to investigate the possibilities in more depth.

6.3.3 Evolutionary Computing and Energy-Independent Partitioned Matrix Eigenvalue Problem

Let us go back to the partitioned matrix eigenvalue equations for a real symmetric $N \times N$ matrix H. For partition sizes n_a and n_b, the equations are (cf. Equation 6.27; check the number from the scan)

$$H_{aa}C_a + H_{ab}C_b = E.\mathbb{1}_a C_a \tag{6.40}$$

$$H_{ba}C_a + H_{bb}C_b = E.\mathbb{1}_b C_b. \tag{6.41}$$

We have already seen how C_a can be eliminated between the two equations by exploiting the second equation to obtain $C_b = \left\{(E.\mathbb{1}_b - H_{bb})^{-1} H_{ba}\right\} C_a$. The expression within the curly brackets can be regarded an energy-dependent wave operator $W(E)$ which acts on the a-subspace eigenvector to generate the b-subspace projection of the total eigenvector C of H with the eigenvalue $E\left(= \dfrac{C_a}{C_b}\right)$. We can by pass the energy dependence of W by noting that the expression $(E.\mathbb{1}_b - H_{bb})^{-1} H_{ba}$ can be regarded as an $n_b \times n_a$ dimensional matrix, the elements of which are unknown to start with. Let us call the matrix z_{ba} and try to find out a recipe for obtaining the elements without any explicit reference to the energy dependence that is implicitly present. With the introduction of the z_{ba} matrix, the first Equation 6.40 becomes

$$\{H_{aa} + H_{ab}z_{ba}\} = EC_a \tag{6.42}$$

while the Equation 6.41 yields

$$C_b = z_{ba}C_a. \tag{6.43}$$

The expression within {} in Equation 6.42 can be viewed as an energy-independent effective Hamiltonian $H_{eff}^a = H_{aa} + H_{ab}z_{ba}$, which operates within the a-dimensional subspace and generates n_a number of exact eigenvalues of H and the a-subspace projection of the corresponding total eigenvectors (C). A workable recipe for obtaining z_{ba} therefore provides the key to the effective Hamiltonian eigenvalues. If we multiply Equation 6.40 by z_{ba} from the left and use Equation 6.43 for eliminating C_b from Equation 6.41, we have two equations:

$$z_{ba}H_{aa}C_a + z_{ba}H_{ab}C_b = E.z_{ba}C_a \tag{6.44}$$

$$H_{ba}C_a + H_{bb}z_{ba}C_b = E.z_{ba}C_a. \tag{6.45}$$

Subtracting 6.44 from 6.45 leads to the condition that

$$\{H_{ba} + H_{bb}z_{ba} - z_{ba}(H_{aa} + H_{ab})z_{ba}\}C_a = 0 \tag{6.46}$$

for any vector C_a in the a-subspace. It demands that z_{ba} must be such that the matrix expression within the curly brackets in Equation 6.46 vanish [11], i.e.,

$$X_{ba} = \{H_{ba} + H_{bb}z_{ba} - z_{ba}(H_{aa} + H_{ab})z_{ba}\} = 0. \tag{6.47}$$

Note here that H_{ba}, H_{bb}, H_{aa}, and H_{ab}, are known matrices to start with. Equation 6.46 therefore represents a nonlinear matrix equation that z_{ba} must satisfy. The nonlinear matrix equation is somewhat difficult to solve by the traditional deterministic procedure but can hopefully be solved by invoking a soft computing method like the simulated annealing technique, the GA or any other evolutionary computing technique without much difficulty. For example, we can think of invoking GA for determining the $n_b \times n_a$ elements of the unknown z_{ba} matrix by maximizing a suitably defined fitness function $f(z_{ba})$ the maximum of which corresponds to the satisfaction of the condition 6.47. Once such a z_{ba} matrix is obtained, we can form the a-subspace effective Hamiltonian $H_{eff}^a = H_{aa} + H_{ab}z_{ba}$ and diagonalize it (deterministically) to obtain the n_a energy eigenvalues $\{E_k\}$ of H and the a-subspace projections $\{C_a\}$ of the corresponding total eigenvectors $\{C_E\} = \begin{pmatrix} C_a \\ C_b \end{pmatrix}$. The b-subspace projection of each such vector can be obtained by using the relation $C_b = WC_a = z_{ba}C_a$. It must be stressed that the GA is proposed to be exploited to search out the wave operator ($W = z_{ba}$) directly and not to extract the eigenvalues of H^a_{eff}. However, even that can be easily accomplished by making use of the technique of stochastic diagonalization by GA detailed in Section 6.5. The primary emphasis of Sharma et al. was on finding the wave operator by GA. They suggested the following GA-based scheme and implemented it on several model problems bringing out some of the strong features of their GA-based approach to construct the wave operator directly. The algorithm operates with the following steps.

1. **Representation:** The n_p number of strings (chromosomes) $\{S_i\}$ making up the population of potential solutions each contain $n_b \times n_a$ floating-point numbers (genes) each representing an element of the unknown wave operator, i.e., the z_{ba} matrix values were chosen randomly. Thus, the z_{ba} matrix was essentially mapped onto a one-dimensional array—the strings $\{S_i\}$.

2. **Fitness function:** The fitness f_i of a string S_i was measured by taking

$$f(S_i) = f_i e^{-\lambda \sigma_i} \quad (0 < \lambda \le 1, \text{user-defined}) \tag{6.48}$$

with $\sigma_i = \sum_{p=1}^{n_a} \sum_{q=1}^{n_b} X^2_{pq}$,

where $X = \{H_{ba} + H_{bb}z_{ba} - z_{ba}(H_{aa} + H_{ab})z_{ba}\}$.

3. **Selection:** Fitness proportional selection by the roulette wheel procedure was used.

4. **Genetic operator:** Arithmetic single-point crossover with a fixed crossover probability $p_c = 0.8$ was used for information exchange or mixing. The crossover site for strings $S_i \equiv z^i_{ab}$ and $S_j \equiv z^j_{ab}$ was chosen randomly with probability p_c by generating a random integer pair (m, n) with probability p_c. The postcrossover strings S'_i and S'_j representing new $z^{i'}_{ba}$ and $z^{j'}_{ba}$ matrices were produced as follows:

$$\begin{aligned}
\left(z^{i'}_{ba}\right)_{pq} &= f_c \left(z^i_{ba}\right)_{pq} + \left(1 - f_c\right)\left(z^j_{ba}\right)_{pq} \\
\left(z^{j'}_{ba}\right)_{pq} &= \left(1 - f_c\right)\left(z^i_{ba}\right)_{pq} + f_c \left(z^j_{ba}\right)_{pq}
\end{aligned} \tag{6.49}$$

$p = m+1, m+2, \dots, n_b$ where $q = n+1, n+2, \dots, n_b$ f_c was drawn from a random distribution in the range $0 \le f_c \le 1$.

5. Arithmetic mutation with dynamically adjusted mutation intensity $\Delta^M(t)$ and a fixed mutation probability $p_m = 4/N$, N being the length of the strings was implemented. Thus the mutated (z^i_{ba}) matrix elements were produced by taking

$$\left(z^i_{ba}\right)''_{pq} = \left(z^i_{ba}\right)_{pq} (-1)^L r \, \Delta^M_{pq}(t) \tag{6.50}$$

(L being a random integer).

The dynamic adjustment of $\Delta^M(t)$ was achieved through analysis of the performance of different strings in the selection process in the preceding generation.

A diversification operation (a mutation-like process) was carried at with diversification probability $p_d = 4/(N \times n_p)$ on all postcrossover and postmutation strings with a dynamic diversification intensity $\Delta^d(t)$ chosen one order of magnitude lower than the corresponding value of $\Delta^M(t)$.

6.3.4 Applications

The algorithm proposed for the direct search for the wave operator was tested on an interesting model polyene Hamiltonian $H(N \times N)$ which supports a bound state in the continuum [16,17]. The elements of the H-matrix were defined as follows.

$\langle 1| H|1 \rangle = H_{11} = 0, |1\rangle$ representing the perturbed impurity state. $\langle i| H| i \rangle = H_{ii} = 1$ for $i = 2, 3, ..., N$. The off-diagonal elements of the H matrix were defined by taking $\langle i| H| j \rangle = H_{ij} = \lambda$ for $i = 1$, $j = 2, 3, ..., N$. For $i \neq 1$, $\langle i| H| j \rangle = H_{ij} = \mu$, for, $j = i - 1, i + 1, H_{ij} = 0$, otherwise Sharma and Bhattacharyya [16] carried out a fairly detailed performance analysis of the GA-based search for the wave operator (z_{ba} matrix) with that of a deterministic (iterative Gauss–Seidel-like method) means of solving the nonlinear matrix equation that z_{ba} was required to satisfy. It turned out that the GA-driven method converged successfully to the targeted impurity state irrespective of the choices for λ and μ. The deterministic search failed to converge to the target state except for the choice $\mu = 0.5$. It appears that GA can provide a viable strategy for solving similar nonlinear matrix equations in general and in particular for handling the problem directly constructing the wave operator in the context of energy-independent partitioning of the matrix eigenvalue problem. It is feasible to handle the direct search for the z_{ba} matrix (the wave operator) also by adopting the SAM or the method of random mutation hill climbing (RMHC) described in Chapter 4. The advantage of using the RMHC or the SAM lies in that there is only a single string—not a population of strings to deal with, and hence there is only one fitness evaluation at each step. However, at the time of writing no concrete results are available for comparison with the GA-based procedure described here [16].

6.4 Computing Eigenvalues and Eigenvectors of a Hamiltonian Matrix by GA Rayleigh Quotient Method

We have so far described GA-driven strategies for computing eigenvalues and eigenvectors of a Hamiltonian: either by a Jacobi-like diagonalization scheme that partially diagonalizes the H matrix, or by converting the full space eigenvalue problem into the problem of computing eigenvalues and vectors of a much lower dimensional effective Hamiltonian—be it an energy-independent effective Hamiltonian or energy-dependent one. The full space eigenvalue problem itself can be tackled much more directly by interfacing a genetic algorithm with the problem of minimizing the Rayleigh quotient of the Hamiltonian matrix. The interfacing immediately throws several open possibilities which will be considered now.

6.4.1 Sequential Search for Eigenvalues and Vectors

Let us consider the problem of finding m lowest eigenvalues of an $N \times N$ Hermitian Hamiltonian matrix (H) which has real elements thereby ensuring the symmetry $H_{ij} = H_{ji}$. We can either reduce the problem to a sequential search for the m eigenvalues and vectors, or proceed to obtain all the m-eigenvalues in one go. Let us examine the sequential option first.

Let us consider an evolving population of n_p eigenvectors $\left\{ |\psi_1\rangle, |\psi_2\rangle, \ldots, |\psi_i\rangle, \ldots, |\psi_{n_p}\rangle \right\}$ which are expanded on the basis $\left\{ |\phi_1\rangle, \ldots, |\phi_n\rangle, \ldots, |\phi_N\rangle \right\}$ in which H has been constructed so that we have

$$|\psi_i\rangle = \sum_{p=1}^{N} C_{pi} |\phi_p\rangle$$

$$C_{pi} = \langle \phi_p | \psi_i \rangle, \quad i = 1, 2, \ldots, N.$$

(6.51)

The string $\{S_i\}$ representing the eigenvectors are each collection of the N real expansion coefficients into a row vector C_i for each $|\psi_i\rangle$ so that we have

$$S_i \equiv \left(C_{1i}, C_{2i}, \ldots, C_{ki}, \ldots, C_{Ni} \right)$$

$$= C_i^t.$$

(6.52)

The string $S_i = C_i^t$ for the eigenvector $|\psi_i\rangle$ can be used to define the Rayleigh quotient ρ_i, where $\rho_i = C_i^t H C_i / (C_i^t C_i)$. If we are interested in finding the lowest value of the Rayleigh quotient—the ground state of H—we have at hand the problem of minimizing ρ in the absolute sense, for that is what defines the ground state of our problem. The problem thus reduces to an equivalent global search problem—how do we find the wave function string C_i^t that globally minimizes ρ? We propose to do the search by invoking GA. Before we describe the search details, we must define an appropriate fitness function for the wave function strings. If ρ_i is the Rayleigh quotient of the wave function string S_i, the fitness function f_i can be taken simply as

$$f_i = e^{-\beta(\rho_i - E_L)^2}$$

(6.53)

where E_L is an estimated lower bound of the lowest eigenvalue of H, β is a user-defined scalar that takes care of exponential overflow/underflow. E_L can be updated at any stage of the search whenever required. As the search homes in on the true ground eigenvector string (call it $S_0 = C_0^t$), $\rho_i \to E_L$ and $f_i \to 1$. For the particular wave function string the Rayleigh quotient ρ_0 is the ground state energy eigenvalue defined by

$$\rho_0 = \frac{C_0^t H C_0}{\left(C_0^t C_0 \right)}$$

(6.54)

and in addition

$$\nabla\rho_0 = \frac{2[H-\rho_0\mathbb{1}]C_0}{C_0^t C_0}.$$ (6.55)

One can check the stability of the lowest energy found by the search by reducing the value of $E_L\left(E_L' = E_L - shift\right)$ further, redoing the search to see if any further lowering of ρ takes place. We stress here that $\nabla\rho$ needs to be computed only toward the end of the search, that is when f_{ibest} is not improving any further to check that the search has indeed hit a stationary point that is also the global minimum of ρ.

The evolution toward the global minimum of ρ or the maximum of f_i is brought about through the application of the genetic operators like selection, crossover, or mutation, as we have described in the previous section. Such a scheme was proposed and implemented by Nandy et al. [14]. The stepwise details of their algorithm are as follows:

Step 1: Build a population of n_p wave function strings $\left\{C_i^t\right\}$ $i = 1, 2, \ldots, n_p$. The amplitudes $\{C_{ki}\}$ are chosen randomly.

Step 2: Roulette wheel selection is enforced to construct the crossover breeding pool.

Step 3: Perform crossover with probability p_c to produce offspring from a pair of parent strings.

Step 4: Mutate the offspring with probability p_m.

Step 5: Diversify the population if needed.

Step 6: Generate new population of n_p strings by elitist replacement.

Step 7: Exit if stopping criteria are satisfied; otherwise, go back to step 2 and iterate.

6.4.1.1 Notes on Different Steps

In step 2, fitness proportional selection was used to choose 80% of the strings, the remaining 20% were selected randomly from the entire population. The specific crossover scheme used in step 3 was a two-point arithmetic crossover scheme in which two sites m_1 and m_2 were chosen randomly with a probability p_c and all the amplitudes between the sites m_1 and m_2 including m_1 and m_2 were subjected to arithmetic crossover. The mixing parameter $f_c (0 < f_c < 1)$ for crossover was chosen by setting $f_c = 0.75 + 0.25r$, r being a random number in the range $(0,1)$. In step 4 the mutation scheme employed was a variant of arithmetic mutation in which the mutation intensity or mutation step length Δ_m was dynamically adjusted by setting (t is the generation number)

$$\Delta_m^t = 1 - f_t$$ (6.56)

where f_t is the highest value of fitness obtained until the tth generation. The mutation probability was kept fixed at $p_m = 4/N$, N being the length of the strings. In the replacement step a partially elitist strategy was employed in which out of the combined pool of n_p parents and n_p offspring only the two strings with the highest fitness values were chosen.

The remaining $n_p - 2$ strings were randomly selected from the entire parent and offspring population of the current generation. The diversification in step 5 was carried out with probability $p_d = \dfrac{4}{(N \times n_p)}$, and intensity Δ_d was set equal to Δ_m^t of the same generation.

6.4.2 Sequential Search for Higher Eigenvalues

After obtaining the ground state wave function string $S_0 = |\psi_i\rangle$, the search for higher eigenvalues and eigenvectors was carried out sequentially by forming the projectors $P_0, P_1, ..., P_k$ with $P_0 = |\psi_0\rangle\langle\psi_0| \equiv C_0 C_0^t$, $P_1 = |\psi_1\rangle\langle\psi_1| \equiv C_1 C_1^t, ..., P_k = |\psi_k\rangle\langle\psi_k| \equiv C_k C_k^t$, where $C_0^t \equiv |\psi_0\rangle$ is the ground eigenvector, $C_1^t \equiv |\psi_1\rangle$ is the first excited state eigenvector, etc., assumed to have been found by the GA. P_0 is therefore the ground state projector, P_1 the first excited state projector, and P_k being the projector for the kth excited state. Assuming that $E_i < 0$ and the ordering $E_0 < E_1 < \cdots < E_{k-1} < E_k$ is maintained, the search for the $(K+1)$th eigen state of the Hamiltonian H can be construed as a search for the ground state of the projected H matrix (H_p^k) where

$$H_p^k = \left(1 - \sum_{i=0}^{k-1} P_i\right)^{\dagger} H \left(1 - \sum_{i=0}^{k-1} P_i\right). \tag{6.57}$$

It is straightforward to see that the k eigenvalues ($k = 0, 1, ...$) of H_p^k are zero while $N - k$ are still <0 so that the search for the $(k + 1)$th eigenvalue has been reduced to the problem of finding the ground state of H_p^k. If all the eigenvalues of H happen to be >0, the procedure outlined above can still be applied, but the H matrix must be first turned into a shifted matrix H_s^k where

$$H_s^k = H - \lambda \sum_{i=0}^{k-1} P_i \tag{6.58}$$

so that all but the k lowest eigenvalues of H are shifted ($E_i^s = E_i - \lambda$) and become <0 if λ is properly chosen. The GA can therefore be applied to find the ground state of H_s^k which must be the kth excited state of the original matrix. The sequential search therefore remains viable irrespective of the eigenvalue structure once a proper shift value has been chosen.

Nandy et al. [14] implemented the proposed algorithm on the Coope matrix [15] already defined in Section 6.3.2. The smallest matrix was with $N = 500$. It turns out that the fitness of the best string in the population of 10 strings (a rather modest population size) improved very quickly from a rather low value of 0.1 up to a value of 0.9 during a GA run. The 10 individual wave function strings were created by randomly perturbing the vector $(100...0)$ when searching for the ground state, while the population for the search for the first excited state was created from the vector $(010...0)$, and so on. The search progressed qualitatively in the same manner whether it was for the ground states or the excited states. It is clear from the results (fitness evolution profiles) displayed in Nandy's paper that the most time-consuming part of the search was the fine-tuning of the amplitudes that takes place once the fitness has grown to values around 0.9. More research is needed to improve the performance of the search as it moves into the plateau region of the fitness evolution profile.

6.4.3 Simultaneous Search for Multiple Eigenvalues

The GA-based method was extended to address the problem of simultaneous search for more than one eigenvalue by Nandy et al. [14]. Two crucial changes in the strategy outlined in the preceding section are required for the extension—one representing the potential solutions and the other constructing a suitable fitness function for the problem. Suppose that we are looking for simultaneous computation of the m lowest eigenvalues of an $N \times N$ Hamiltonian matrix. The wave function strings $\{S_k\}$ for the problem are conveniently represented as $N \times m$ matrices, each of the m columns representing one eigenvector of H. The first column represents the amplitudes of N-basis vectors in the trial ground state vector, the second column represents the amplitudes of the basis vectors in the trial vector for the first excited state, and so on. A typical wave function string would therefore look like S_1 where S_1 is a N row m column matrix.

$$S_1 \equiv \begin{array}{c} \\ 1 \\ 2 \\ 3 \\ \vdots \\ N \end{array} \begin{array}{cccc} 1 & 2 & \cdots & m \\ \left[\begin{array}{cccc} 1 & 0 & \cdots & 0 \\ 0 & 1 & \cdots & 0 \\ 0 & 0 & \cdots & 0 \\ \vdots & \vdots & \cdots & \vdots \\ 0 & 0 & \cdots & 1 \end{array}\right]. \end{array} \tag{6.59}$$

The other members of the population can be generated by randomly perturbing the $N \times m$ elements of the S_1 matrix by adding to each a term $\pm\beta r$; $(\beta > 0, r \sim U(0,1))$. The second change to be introduced concerns the construction of the objective and the fitness functions of the problem. A good choice for the objective function could be, for example a "state-averaged" objective function O_{av}:

$$O_{av} = \frac{1}{m} \sum_{i=0}^{m-1} \left(\rho_i - E_L\right)^2 \tag{6.60}$$

ρ_i is the Rayleigh quotient of the trial vector for the ith eigenstate, E_L is an adjustable lower bound to the lowest among the m eigenvalues (here the ground eigenvalue E_0) being sought to be calculated. E_L must be less than E_0 and needs to be adjusted ($E_L^{new} = E_L - shift$) whenever $\rho_0 \to E_L$ too closely. However, Equation 6.60 defines only a part of the objective. The second important objective is that the m-column vectors in S_{ks} (the wave function strings) must stay orthonormal. Recalling that $S_k \equiv C_i^l (N \times m)$ we may define a composite objective function O_{av}^k for the kth string in the population by adding a penalty term to the state-averaged objective function O_{av}, the added term taking care of the required orthnormality constraint. The new objective function then reads

$$O_{av}^{k'} = O_{av}^k + \lambda \left[Tr\left\{ \left(C_i^l C_k\right) - \mathbb{1}_{m \times m} \right\}^2 \right]$$

$$= \frac{1}{m} \sum_{i=0}^{m-1} \left(\rho_i - E_L\right)^2 + \lambda \left[Tr\left\{ \left(C_i^l C_k\right) - \mathbb{1}_m \right\}^2 \right] \tag{6.61}$$

λ is the penalty weight factor. Minimization of $O_{av}^{k'}$ satisfies two goals simultaneously: it minimizes the state-averaged error in the computed Rayleigh quotients for the string S_k (first term in Equation 6.61) and it also minimizes the orthonormality error in the kth string as represented by the second term in the same equation, even as $\lambda \to \infty$. Unless we are dealing with a case in which the m-eigenvalues are degenerate, the first term in $O_{av}^{k'}$ can never be reduced to zero as the lower bound E_L is common for all the m-eigenstates. The second term, however, can be reduced to zero as closely as possible, i.e., we can make $Tr\left\{\left(C_i^t C_k\right) - \mathbb{1}_m\right\}^2 < \in$ with $\in$ chosen to be as small as practicable. The objective function $O_{av}^{k'}$ can be easily mapped into the corresponding fitness function f_k by taking

$$f_k = e^{-\beta O_{av}^{k'}}; \beta > 0 \tag{6.62}$$

in which case we have a scaled fitness landscape ($0 < f_k < 1$). Alternatively, we can also choose to work with unscaled fitness landscape, say

$$\overline{f_k} = \frac{1}{O_{av}^{k'} + \sigma} \tag{6.63}$$

where σ is a small positive constant. Except in the case of search for m-degenerate eigenvalues when $O_{av}^{k'}$ may he reduced to zero, we can safely set $\sigma = 0$. The crossover, mutation or diversification operations already described in the preceding section can be applied with a slight change in strategy. Thus, for crossover, for example, a pair of matrix strings $S_k(n \times m)$, $S_l(n \times m)$ are first chosen and then for each column of S_k and S_l two crossover sites are randomly chosen with the probability p_c (the current value of crossover probability). Two-point arithmetic crossover operation is then carried out for each of the m columns of S_k and S_l. Nandy et al. [14] implemented the scheme on a set of Coope matrices of varying dimensions. Figure 6.1a displays how the maximization of the composite fitness

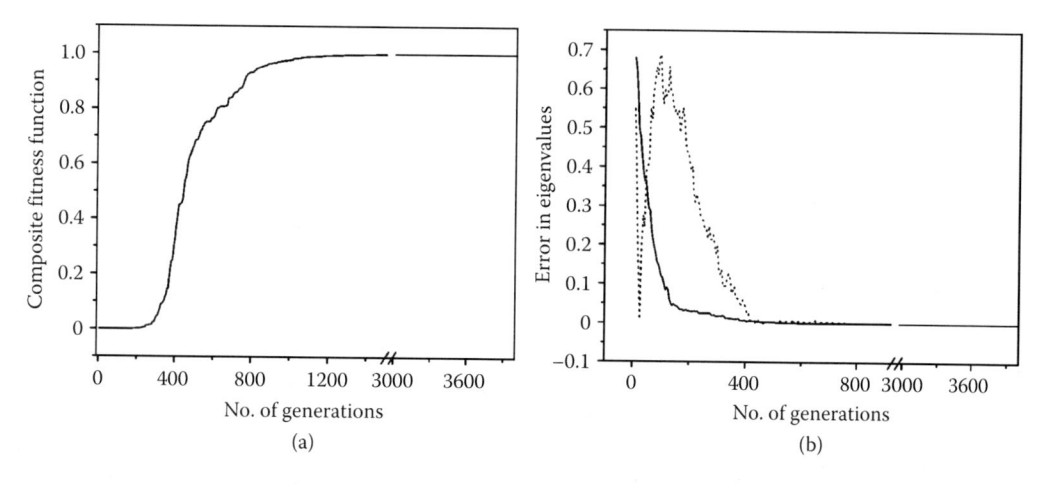

FIGURE 6.1
Evolution of the composite fitness (a) and the error in eigenvalues (b) while calculating the ground and second eigenvector, simultaneously, for the same 30×30 Coope Matrix. (From Nandy, S., et al., *Appl. Soft Comput.*, 11(5), 3946–3961, 2011. Figure 7. With permission.)

function of Equation 6.61 proceeds while Figure 6.1b shows how the error-reduction profiles for two of the lowest eigenvalues of a 30×30 Coope matrix proceeds [14].

The point to note is that the composite fitness grows smoothly and steeply after a few generations have elapsed (Figure 6.1a). The error reductions of individual eigenvalues, however, do not coincide. The error in the ground eigenvalue is steeply reduced from the beginning of the search and runs into a plateau phase where finer tuning in the wave function strings takes place. The second eigenvalue behaves very differently. Along the search path, the error in the computed second eigenvalue rises, passes through a maximum and then starts reducing. The behavior indicates that a "divide and conquer" strategy is perhaps at work when the GA searches for multiple eigenvalues simultaneously. However, the available results are not sufficient to draw a definite conclusion. The authors of reference [14] note, however, that the computation labor grows linearly as the number of eigenvalues (m) being simultaneously searched for, increases. The linear increase of computational labor as a function of m means that the simultaneous search for multiple eigenvalues would have an edge over the sequential search strategy. The simultaneous search does not require the construction of the projectors and projected matrices, which is time-consuming and susceptible to error propagation.

We note here that the search could in principle be based on the simple objective function $O_j^k = \frac{1}{m} \sum_{i=0}^{m-1} \left(\rho_i^k - E_L \right)^2$. The orthonormality constraint then must be imposed on the population of potential solution strings at every generation. That means, after crossover, mutation and diversification steps have been executed, the n_p number of offspring have to be subjected to Gram–Schmidt orthonormalization. It ensures that the search does not stray away from the surface of the constrained objective function. Nandy et al. [14] explored the route involving a population of explicitly orthonormalized vectors along with the use of the simple unconstrained objective function. A detailed performance analysis indicates that an optimal choice of population size can significantly cut down the average computational labor involved in the orthogonalized population-based search for m-eigenvalues.

The authors also analyzed the performance of a parallelized (master–slave) version of their state-averaged GA Rayleigh quotient method (constrained objective function-based search). A typical profile of how the elapsed time required to compute m eigenvalue responds to the increase in the number of processors switched on is shown below (Figure 6.2) [14].

The maximum at two processor appears anomalous. It could have been caused by network latency. The optimal number of processors turns out to be 10 in the instant case as expected (see Chapter 3) on the basis of a Cantu-Paz analysis [18,19]. The parallel efficiency is approximately 50% only. The authors have observed that as the dimension of the matrix (N) and the number of eigenvalues and vectors (m) being searched simultaneously increase, the gain from parallelization becomes more evident. Our feeling is that an asynchronous implementation of the parallel GA could bring in more advantages and improve efficiency.

6.5 Direct Solution of Schrödinger Equation by GA

So far we have explored a number of ways in which a GA can be exploited to obtain information about the bound stationary states of the Schrödinger Hamiltonian ($\hat{H}$) for an atomic or molecular system under the clamped nucleii or frozen nucleii approximation. The central theme of the exploration has been to introduce a linearly independent or an orthonormal

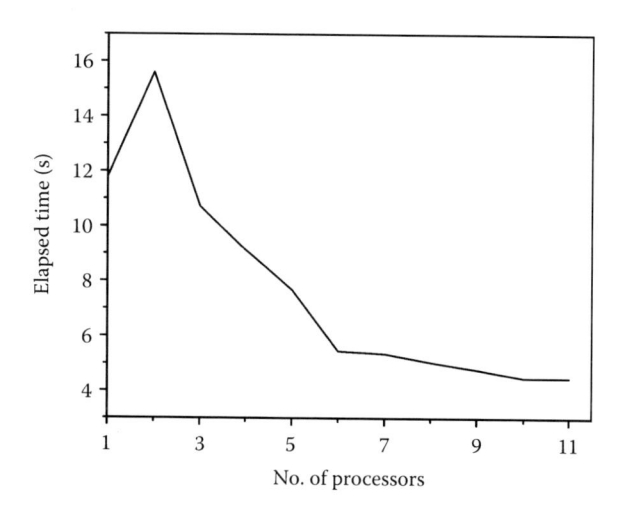

FIGURE 6.2

Elapsed time in computing '10' lowest eigenvalues 100×100 Coope Matrix ($n_p = 10$) by GA-Rayleigh quotient method. (From Nandy, S., et al., *Appl. Soft Comput.*, 11(5), 3946–3961, 2011. Figure 12d. With permission.)

basis and convert the operator eigenvalue equation $\hat{H}\psi = E\psi$ into a matrix eigenvalue problem $HC = CE$ and invoke GA to search for the eigenvectors and eigenvalues of the H matrix. We would now like to turn to the other possibility—the possibility of solving the time-independent Schrödinger equation directly by a GA, without introducing a basis set. That means, we look for or devise techniques of integrating the Schrödinger equation numerically for its stationary states with the help of GAs. Such numerical solutions, if available, may provide important information about the detailed local properties of the stationary states in the most important regions of the space which in turn can be exploited to design analytical basis sets for atoms or perhaps small molecular fragments depending on the chemical environment in which they are placed. The task is not easy; but even a partial breakthrough may provide a rich dividend. There are standard methods of numerical integration of differential equations as such. An alternative can be to form the finite difference version of the equation and solve it iteratively. These approaches have been experimented with, but the success so far has been limited. Invoking GAs or other soft computing options in this context can be a game changer, for the soft computing methods are, in general, tolerant to imprecise information so that we can work with rather poor quality of initial information about the solutions (wave function) being sought. We will describe in the following section some of the possibilities that have been already experimented with, analyze the potential strength of the methods suggested, and future directions of further development.

6.5.1 The Basic Scheme for Ground States

Let us consider Schrödinger equation for the stationary states supported by the one-dimensional Hamiltonian $H(x)$

$$H(x)\psi_n(x) = E_n\psi_n(x). \tag{6.64}$$

We start by uniformly discretizing the coordinate space $-\infty \le x \le \infty$ into a set of N points, $(x_1, x_2, ..., x_k, ..., x_{N-1}, x_N)$. Suppose that we focus on the ground state of the problem which means we wish to find out the lowest possible value of a scalar E_0 (energy) and the corresponding wave function $\psi_0(x)$ such that $H(x)\psi_0(x) = E_0\psi_0(x)$ by invoking GA. We may represent the unknown ground state wave function in the discretized coordinate space as strings $S_1, S_2, \ldots, S_k, \ldots, S_{m-1}, S_m$ where each string is a collection of N-probability amplitudes $b_1, b_2, \ldots, b_k, \ldots, b_{N-1}, b_N$ at the designated grid points. Thus, we have the correspondence

$$b_1 = \phi(x_1), b_2 = \phi(x_2), \ldots, b_N = \phi(x_N). \tag{6.65}$$

For the ground state problem in one dimension ϕ may be any normalizable nodeless function of x. Suppose that we have made a careful choice of n such candidate functions $\phi_1(x), \phi_2(x), \ldots, \phi_k(x), \ldots, \phi_n(x)$ for generating the trial amplitude distributions along the designated grid points. A population of n wave function strings $S_1, S_2, \ldots, S_k, \ldots, S_{n-1}, S_n$ is therefore easily constructed where

$$\left.\begin{array}{l} S_1 \equiv S\big(\phi_1(x_1), \phi_1(x_2), \ldots, \phi_1(x_N)\big) \\ \cdots\cdots\cdots\cdots\cdots\cdots\cdots\cdots\cdots\cdots\cdots\cdots \\ S_n \equiv S\big(\phi_n(x_1), \phi_n(x_2), \ldots, \phi_n(x_N)\big) \end{array}\right\}. \tag{6.66}$$

The collection of n-strings constitutes a population of n-distinct probability amplitude distributions in the discretized coordinate space. Two properties are common to each string: (1) each represents a nodeless distribution of probability amplitudes, and (2) $S_k(x_l) = \phi_k(x_l) \to 0$ as $x_1 \to \alpha$ for $k = 1, 2, \ldots, n$. The amplitude distributions are normalized to unitary for every string.

6.5.1.1 Objective and Fitness Functions

We have to construct an appropriate fitness function f, the maximization of which would amount to solving the problem at hand. We do it in two steps: (a) construct an objective function $O(S_k)$ which takes one of the strings as input and returns a value between 0 and α, and (b) map $O(S_k)$ onto a fitness function f_k for the string S_k such that $0 < f_k \le 1$. The goal of the search can then be defined as finding a probability amplitude distribution (say S_0) in the coordinate space that maximizes the fitness value absolutely. A convenient and easy to implement choice for the objective function could be one of the following possibilities. Let $\phi_k(x)$ analytically represent the probability amplitude distribution that the string S_k encodes as a collection of N probability amplitudes in the discretized coordinate space (read $S_k \equiv S\big(\phi_k(x_1), \phi_k(x_2), \ldots, \phi_k(x_{N-1}), \ldots, \phi_k(x_N)\big)$). Then we can define the objective function $O_k\big(\phi_k(x)\big)$ as

$$O_k\big(\phi_k(x)\big) \equiv O_k(S_k) = \left\{\frac{\langle \phi_k | H | \phi_k \rangle}{\langle \phi_k | \phi_k \rangle} - E_L\right\}^2$$

$$= \left\{\frac{\langle S_k | H | S_k \rangle_q}{\langle S_k | S_k \rangle_q} - E_L\right\}^2 \tag{6.67}$$

$$= \big(\in_k - E_L\big)^2$$

where $\langle S_k|H|S\rangle_g$ simply denotes that the relevant inner products have been evaluated by quadrature, E_L is our estimated lower bound to the ground state energy E_0 of the system. The corresponding fitness function f_k can be defined as

$$f_k(S_k) = exp(-\lambda O_k(S_k)), (\lambda > 0).\tag{6.68}$$

The probability amplitudes encoded by S_k are the search variables. As these amplitudes evolve toward the exact ground state probability amplitude distribution $\in_k$—the energy expectation value for the kth wave function string—approaches E_L and the minimum value of $O_k(S_k)$ is attained if and only if $\in_k = E_L$ when $O_k(S_k)$ becomes exactly zero. This is possible if E_L has been chosen exactly equal to the ground state energy (E_0). Since E_0 would not be known to start with, one can choose a low enough value for E_L and gradually update $E_L\left(E_L^{new} = E_L \pm \sigma\right)$ so that $E_L \to E_0$ as the search to find the true ground state probability amplitude distribution proceeds more and more to completion and $\in_{kbest} \to E_0$. As the objective function becomes minimized and approaches the value zero, the fitness function becomes maximized and approaches the value one ($f_{kbest} \to 1$, as $O_{kbest} \to 0$). An alternative choice of f_k could be

$$f_k = \frac{1}{\left(\in_k - E_L\right)^2 + \beta}; \quad \beta > 0.\tag{6.69}$$

The parameter β shifts the denominator away from 0 in case of $\in_k \sim E_L$ and keeps f_k finite.

6.5.1.2 Selection, Crossover, and Mutation

1. The population of n-wave function strings is allowed to evolve under the action of genetic operators listed above. The selection can be based on the traditional roulette wheel method or the more contemporary tournament selection with carefully chosen tournament size (see Chapter 3).
2. Since the strings are made up of floating-point numbers, there are two simple and easily implementable options for crossover and mutation.

6.5.1.3 Uniform Crossover

Two strings, S_k and S_l, are chosen randomly with the probability p_c (fixed crossover probability) and are crossed to generate two offspring S_k' and S_l' with modified genes (i.e., grid-point amplitudes) on all the N locations. Thus,

$$\begin{aligned}S_k'(x_i) &= a_1 S_k(x_i) + a_2 S_l(x_i) \\ S_l'(x_i) &= a_2 S_k(x_i) + a_1 S_l(x_i)\end{aligned}; \quad i = 1, 2, \ldots, N\tag{6.70}$$

a_1 is chosen randomly from a predefined range, a_2 is set equal to $(1-a_1)$. $S_k(x_i)$ is the value of the probability amplitudes at the ith grid point (x_i) in string S_k and $S_l(x_i)$ has a similar definition.

6.5.1.4 Single-Point Arithmetic Crossover

In this scheme of crossover two strings S_k and S_l are chosen randomly. A crossover site (say, the lth site) is chosen randomly with a preset crossover probability p_c). The two offspring S_k' and S_l' are then generated by adopting the following rule:

$$S_k'(x_i) = a_1 S_k(x_i) + a_2 S_l(x_i)$$
$$S_l'(x_i) = a_2 S_k(x_i) + a_1 S_l(x_i) \quad ; i = l+1, l+2, \ldots, N. \tag{6.71}$$

For all other grid points the probability amplitudes remain unaltered f_c—the information mixing parameter is drawn from a Gaussian distribution in the range (0,1).

6.5.1.5 Mutation

After the requisite number of offspring have been generated by crossover, the newly produced strings are subjected to mutation with a preset fixed mutation probability ($p_m \ll p_c$). Again being floating-point strings, an arithmetic mutation operation can be carried out conveniently either with a fixed (preset) mutation intensity Δ_m or a dynamically adjusted mutation intensity $\Delta_m(t)$, being the generation number that tells us about the current stage of the evolutionary process. The fixed-intensity mutation is carried out on the postcrossover strings (S_k', S_l', …) in the following manner.

The strings $S_1', S_2', \ldots, S_k', \ldots, S_n'$ are chosen serially. Suppose that the kth string has been chosen to mutate. The mutation site on S_k' is chosen randomly with probability p_m. Let the site (gene) chosen to undergo mutation be the lth site. The mutated probability amplitude at the lth site of S_k' is generated by setting

$$S_k''(x_l) = S_k'(x_l) + (1 - 2r)\Delta_m r_1 \tag{6.72}$$

where r is a random number lying in the range (0,1) and r_1 is another random number in a preset range drawn from a normal distribution. The mutation process is important as it fine-tunes the evolving probability amplitudes. Since the mutation process can either increase or decrease the probability amplitudes at the chosen site, it can produce unwanted "kinks" in the wave function strings, specially in the tails. A good practice would be to keep Δ_m proportional to the amplitude being mutated.

A dynamic mutation scheme which automatically adjusts the Δ_m values as a function of the current generation number (t) may be implemented as follows:

$$\Delta_m(t) = y\left[1 - r^{\left(1 - \frac{t}{T}\right)^l}\right] \tag{6.73}$$

where y is chosen from a carefully selected range ($b_{min} \le y \le b_{max}$), t is the generation number, T denotes the maximum number of generations the evolution is designed to cover, and r is a random number lying in the range $0 \le r \le 1$. The parameter l controls the nonlinearity of the evolution of $\Delta_m(t)$. A good choice is to use $l = 2$, but there is scope for optimizing l for a specific problem.

Once the mutation process is over, we have at our disposal n parent wave function strings ($S_1, S_2, \ldots, S_k, \ldots, S_n$) and n offspring ($S_1'', S_2'', \ldots, S_k'', \ldots, S_n''$). The offspring are normalized before evaluating the energy integrals $\epsilon_k = \dfrac{\langle S_k | H | S_k \rangle_q}{\langle S_k | S_k \rangle_q}$ by quadrature, and then the fitness value f_k. The total population of n parents $+n$ offspring strings are then subjected to fitness proportional selection schemes. The string (say, S_i) with the current best fitness value is set aside (call it S_{best}, with fitness f_{best}). As evolution proceeds, the fitness distribution approaches a delta function peaked around f_{best} which approaches 1 as $t \rightarrow T$.

6.5.2 Ground State Problem Explored

Saha et al. [20] implemented variants of the strategy described in this section for computing the ground state energy and wave function of a number of one- and two-dimensional systems, such as the H-atom, symmetric double-well (SDW) potential, or the two-dimensional coupled harmonic oscillator. For the SDW problem the Hamiltonian is of the form

$$H = \frac{-\hbar^2}{2m} \nabla^2 + Ax^4 - Bx^2 \tag{6.74}$$

with $A = 0.01$, $B = 0.02$, and $m = 1836$.

The authors used a population size of $n = 20$ strings. Each string carried 123 genes (i.e., 123 probability amplitudes at the 123 coordinate grid points distributed over the range $-\infty < x < \infty$). Single-point crossover with fixed crossover probability ($p_c = 0.8$) and constant mutation rate ($p_m = 0.05$) and intensity ($\Delta_m = 0.015$) were used along with roulette wheel selection. The quadratures were performed by the trapezoid rule. To improve the accuracy of the quadrature the author [20] employed the cubic spine interpolation scheme to generate eight additional grid-point probability amplitudes between any two adjacent pair of grid points that are used to represent the wave function strings. The genetic evolution process (crossover, mutation) was carried out on only the 123 grid-point probability amplitudes on each string while the quadratures were performed with the additional amplitudes generated by interpolation points added to the grid. This "densification" of the amplitude data on each string improved the quality of quadrature and a smooth evolution of the wave function strings. The results compared well with those obtained by diagonalizing the Hamiltonian in a Fourier grid Hamiltonian (FGH) basis [21]. GA was also exploited to obtain the ground state wave function (radial) and energy (E_0) of the hydrogen atom. The equation to be solved reads

$$-\frac{\hbar^2}{2m}\left[\frac{\partial^2 \phi(r)}{\partial r^2} + \frac{2}{r}\frac{\partial \phi(r)}{\partial r}\right] - \left(\frac{e^2}{r} - E\right)\phi(r) = 0. \tag{6.75}$$

Saha et al. [20] used a population of 20 wave function strings, each representing the probability amplitude distribution in the discretized radial space ($0 < r_i \leq 25$ a.u.). Each string contained 200 genes (probability amplitudes $\{\phi(r_i)\}$ at 200 designated grid points ($r_1 r_2 r_3 \ldots r_{200}$). The amplitudes were generated from a pool of nodeless normalizable radial functions or randomly generated and normalized.

In each case, care was taken to ensure that $S(r_i) \rightarrow 0$ as $r_i \rightarrow \infty$. For smooth quadrature, additional amplitudes were generated between a pair of adjacent grid points (r_i, r_{i+1}) by cubic spline interpolation. The authors experimented with several modes of crossover, and recommended the use of a modified single-point crossover. Suppose that the strings S_k and S_l, each containing 200 genes have been selected for crossover; let the crossover site randomly chosen with probability $p_c(0.8)$ be the pth site. The postcrossover strings generated by the modified crossover mechanism are $\left(f_c \sim U(0,1) \right)$.

$$S_i'(r_i) = f_c S_k(r_i) + (1 - f_c) S_l(r_i)$$

$$\text{and } S_i'(r_i) = (1 - f_c) S_k(r_i) + f_c S_l(r_i), \text{ for } i = 1, 2, \dots, p. \tag{6.76}$$

The modification resulted in a much faster convergence compared to the standard one point or multisite arithmetic crossover schemes. The improved performance can be traced to the fact that the near nuclear region of the radial space (γ_i lying between 0 and 5 a.u.) is the most important region. The modified crossover mechanism ensures that these regions are sampled more extensively under crossover, leading to the rapid reorganization of the probability amplitudes in the near-nuclear regions. It is imperative therefore to examine the nature of the potential carefully before designing a suitable crossover mechanism for the problem at hand. The authors further recommended the use of a dynamic mutation scheme for enhancing the performance of GA in solving the radial Schrödinger equation for the H-atom problem. The quality of $\psi(\gamma)$ predicted by GA tallied well with the exact $\psi(\gamma)$ when compared point-wise. The maximum difference between $|\psi(r_i)|^2$ and $|\phi(r_i)|^2$ was $\sim 10^{-8}$.

Saha et al. also explored their GA-based recipe for obtaining the ground state wave function and energy of a two-dimensional coupled oscillator for which analytical solutions are unavailable. The system is defined by the Hamiltonian $H(x, y)$, where

$$H(x, y) = -\frac{-\hbar^2}{2m} \left(\nabla_x^2 + \nabla_y^2 \right) + \frac{1}{2} k \left(x^2 + y^2 \right) + \lambda x^4 y^4 \tag{6.77}$$

k is set equal to 1 and λ is chosen to have a value equal to 0.1.

The wave function strings in the population ($n = 10$ was used) were represented as probability amplitude distributions in the discretized coordinate space $\left\{ (x_i, y_i) \right\}_{i=1,120; j=1,120}$. On the 2-dimensional grid a string S_k is just an array of probability amplitudes at the designated grid points. Thus $S_k(i,j)$ represents the value of the wave function $\psi(x,y)$ at the grid point (x_i, x_j). The number of grid-point amplitudes that would evolve under the action of genetic operators was therefore restricted to 120×120 per string. Additional amplitudes were produced at every generation by adopting a two-dimensional bicubic interpolation scheme and used for quadrature. The computed wave function and energy for the ground state of the coupled oscillator compared well with those predicted by the standard FGH recipe.

6.5.3 Excited State Calculation

The extension of the method to compute excited states of one- and two-dimensional problems requires a few changes in the strategy. Having determined the ground state wave function (ψ_0) and energy (E_0 in the first step, we can go for the calculation of the first excited state with a redefined objective function $\tilde{\mathcal{F}}(\phi)$ where

$$\tilde{\mathcal{F}}(\phi) = \left\{ \frac{\langle \phi | H | \phi \rangle_q}{\langle \phi | \phi_q \rangle} - E_L \right\}^2 + \gamma \left| \langle \phi | \psi_0 \rangle_q \right|^2. \tag{6.78}$$

The fitness $f(\phi)$ is, as usual, set equal to $e^{-\lambda \tilde{\mathcal{F}}(\phi)}$. ϕs are the appropriate evolving wave function strings and the inner products $\langle\rangle_q$ are estimated by quadrature. γ is a penalty weight factor that ensures a sharp rise in the value of the objective function if ϕ does not remain orthogonal to the exact ground state wave function (ψ_0) already determined by GA. The initial population is generated by creating strings of probability amplitudes that have one node each. The lower bound estimate E_L can be set equal to E_0 (in this case) and updated at every generation (shifted up or down, depending on the current estimate of the excited state energy and its closeness to E_L). The role of the second term in the objective function is of paramount importance in the calculation of the excited state energy and wave function—it ensures that ϕ remains orthogonal to ψ_0 while maximizing the fitness value. While the presence of the orthogonality constraint in the objective function ensures that optimization of ϕ by GA does not move away from the constrained fitness surface, it would be advisable to orthogonalize the evolving wave function strings explicitly to the already determined ground state at regular intervals. Saha et al. [20] implemented the strategy outlined here for the determination of the first excited state of the H-atom, symmetric double-well potential, and the coupled oscillator Hamiltonian. In general, the search for the first excited state wave function appears to proceed smoothly as reflected in the quick, monotonic rise in the fitness value f_{best} of the "best string" in the population. If we analyze how the orthogonality integral $\langle\phi|\psi_0\rangle$ of the same string evolves during the search it becomes clear that $\langle\phi|\psi_0\rangle$ of the best string does not evolve monotonically. During the initial part of the search there is rise and fall in the overlap across the targeted value zero; but as the search progresses the oscillations die out and $\langle\phi|\psi_0\rangle$ smoothly attains the desired zero value. (see Figure 6c of reference [20]).

Earlier, Chaudhury and Bhattacharyya [22] experimented with a similar GA-based scheme for solving the one-dimensional Schrödinger equation for its stationary states directly as well as perturbatively. The fitness function used by Chaudhury et al. was different. The idea explored was to minimize the norm of the vector $|\theta_n\rangle = (H - \epsilon_n)|\psi_n\rangle$ where $|\psi_n\rangle$ is a randomly generated wave function string (probability amplitude distribution) in the uniformly discretized coordinate space and ϵ_n is the corresponding scalar measuring the energy of the string $\left(\epsilon_n = \dfrac{\langle\psi_n|H|\psi_n\rangle}{\langle\psi_n|\psi_n\rangle}\right)$. The norm of the vector θ_n is a measure of error σ_n in the solution where

$$\sigma_n = \left\{|\phi_n|\phi_n|^2\right\}^{1/2}. \tag{6.79}$$

The fitness f_n of the wave function was generated by taking

$$f_n = e^{-\sigma_n} \tag{6.80}$$

f_n values are clearly distributed in the range $0 < f_n \le 1$. The scheme was successfully tested on finding the ground state of screened coulomb and one-dimensional anharmonic oscillator problems. These authors also experimented with the possibility of solving the inhomogeneous differential equation of the Rayleigh Schrödinger perturbation theory by a simple-minded extension of their GA-based algorithm for solving the Schrödinger equation. Suppose that the Hamiltonian H has been partitioned into a zeroth order Hamiltonian

H_0 and a perturbation V $(H = H_0 + V)$ and that the zeroth order solution $H_0 \left| \psi_n^0 \right\rangle = E_n^0 \left| \psi_n^0 \right\rangle$ has been determined. The first order correction $\left| \psi_n^1 \right\rangle$ to the wave function satisfies the equation

$$\left(H_0 - E_n^0 \right) \left| \psi_n^1 \right\rangle = \left(E_n^1 - V \right) \left| \psi_n^0 \right\rangle \tag{6.81}$$

with $E_n^1 = \left\langle \psi_n^0 \right| V \left| \psi_n^0 \right\rangle$.

The right-hand side therefore contains quantities that are already known or can be calculated in terms of quantities already known. Let us call $\left(E_n^1 - V \right) \psi_n^0 = \chi_n$ a function of the coordinates. Chaudhury and Bhattacharyya [22] proposed to solve the first order perturbation equation by GA by tweaking it into an optimization problem. The equation, in terms $\left| \chi_n \right\rangle$ reads

$$\left(H_0 - E_n^0 \right) \left| \psi_n^1 \right\rangle - \left| \chi_n \right\rangle = 0, \tag{6.82}$$

or in function form $\left(H_0 - E_n^0 \right) \psi_n^1 (x) = \chi_n (x)$. So the problem is to find $\psi_n^1 (x)$ that minimizes the norm of the function $\xi_n (x) = \left(H_0 - E_0^0 \right) \psi_n^1 (x) - \chi_n (x)$ which may be taken as $\sigma_n = \left[\left| \left\{ \left(H_0 - E_0^0 \right) \psi_n^1 (x) \right\} - \chi_n (x) \right|^2 \right]^{1/2}$. ψ_n^1s are just like the wave function strings in the discretized coordinate space. The fitness of the string ψ_n^1 is given by $f_n = e^{-\sigma_n}$. The rest of the GA search remains unaltered. The higher-order equations can be solved similarly. Chaudhury and Bhattacharyya [22] showed that it was possible to estimate ψ_n^1 quite accurately for the quartic anharmonic oscillator problem. It is unfortunate that no further explorations are available in the literature so far.

Nakanishi and Sugawara [23] proposed and implemented a novel technique for solving a Schrödinger equation based on a microgenetic algorithm called $\mu - GA$. The solutions were represented on a feed-forward artificial neural network (see Chapter 8). Each solution was assigned a fitness score by a random point evaluation method. $\mu - GA$ was used to maximize the fitness score by breeding new solutions genetically. These authors used a deterministic optimizer to fine-tune the solutions toward the end of the search thereby accelerating the convergence of the search to the optima. The workability of the method was demonstrated by solving the one-dimensional harmonic oscillator problem and the problem of finding the quantum states in double-well potentials.

6.5.4 Solving a Two-Electron Schrödinger Equation by GA

Let us consider the radial Schrödinger equation for two-electron atoms or ions in the ground state:

$$H\phi_0 (r_1, r_2) = E_0 \phi_0 (r_1, r_2) \tag{6.83}$$

where

$$H = -\frac{\hbar^2}{2m} \sum_{i=1}^{2} \left[\frac{\partial^2}{\partial r_i^2} + \frac{2}{r_i} \frac{\partial}{\partial r_i} \right] - \sum_{i=1}^{2} \frac{ze^2}{r_i} + \left(\frac{e^2}{r_{12}} \right). \tag{6.84}$$

We may represent the unknown $\psi_0(r_1, r_2)$ as a distribution of probability amplitudes $S(i, j)$ in the uniformly discretized two-dimensional radial space. $\{S(i, j)\}^2$ in this representation stands for the probability of finding electron 1 at the point r_1^i when electron 2 is at the point r_2^j irrespective of their spin. $S(i, j)$s must reflect the permutation symmetry of the total wave function $\psi_0 = \phi_0^{space}(1, 2) \times \chi_0^{spin}(1, 2)$. The ground state, as we know, is a spin singlet ($s = 0$) so that $\chi_0^{spin}(1, 2)$ must be antisymmetric with respect to the interchange of spin coordinates of electrons 1 and 2. That in turn demands that $\phi_0^{space}(1, 2)$ must be symmetric with respect to the interchange of the spatial coordinates of the two-electron so that the product $\phi_0^{space}(1, 2) \times \chi_0^{spin}(1, 2)$ remains antisymmetric under the interchange of the space-spin coordinates of the two electrons. The amplitude distributions $S_k(i, j); (k = 1, 2, \ldots, n)$ are therefore constrained to be symmetric. One may start by creating a population of n-individual strings $S_1, S_2, \ldots, S_k, \ldots, S_n$, each of which is a distinct two-dimensional array representing the probability amplitudes on a two-dimensional radial grid of $n_1 \times n_2$ points, which remain symmetric under $1 \leftrightarrow 2$ interchange. The amplitudes are generated from an assorted set of two-particle symmetric functions $\psi_k(r_1, r_2)$. For example, $\psi_k(r_1, r_2) = A_k e^{-\beta_k(r_1 + r_2)}$ ($\beta > 0$, chosen randomly), or, $\psi_k(r_1, r_2) = B_k e^{-\alpha_k(r_1 + r_2)}\{1 + \gamma_k r_{12} + \cdots\}$ or else, $\psi_k(r_1, r_2) = D_k\{e^{-(\alpha_l r_1 + \beta_l r_2)} + e^{-(\beta_l r_1 + \alpha_l r_2)}\}$, with α_ks, γ_ks, α_ls, β_ls chosen randomly.

The incorporation of information coming from both correlated and uncorrelated two-electron wave functions selected randomly in the probability amplitudes represented by the initial population of strings $S_k(i, j)$ may allow the genetic operators to exploit the information and generate better, possibility correlated probability amplitude distributions for the ground state. The genetic operators like crossover and mutation must be designed to work on the two-dimensional amplitude distributions. The crossover mechanism defined earlier for the H-atom problem can be used along with the simple arithmetic mutation operation.

Saha et al. [24,25] proposed and implemented a GA-driven strategy for solving the radial SE for He, H$^-$, Li$^+$, etc., perhaps for the first time with considerable success. Salient features of their strategy are as follows:

1. Generate a population of n number of two-dimensional strings $S_k(i, j)$; $(i = 1, 2, \ldots, n_1; j = 1, 2, \ldots, n_2)$. $S_k(i, j)$s containing floating-point numbers, the squares of which represent the probability of finding electron 1 at the radial grid point r_1^i when electron 2 is at the radial grid point r_2^j. $S_k(i, j)$s are constrained to be symmetric under the interchange of the coordinates of electrons 1 and 2, i.e., $S_k(i, j) = S_k(j, i)$. A square grid is used, i.e., $n_1 = n_2 = N$, and the probability amplitude distributions for each string are normalized.

2. The fitness landscape is generated by defining the fitness function f_k of the string S_k as

$$f_k = e^{-O_k} \tag{6.85}$$

where

$$O_k = \left\{ \frac{\langle S_k | H | S_k \rangle_s}{\langle S_k | S_k \rangle_s} - E_L \right\}^2 \tag{6.86}$$

E_L being an estimated lower bound to the ground state energy of the two-electron atom or ion being investigated and $\langle\ \rangle_s$ indicates that the concerned integrations have been done by quadrature. For performing accurate quadratures, an additional number of M grid-point probability amplitudes were generated by bicubic interpolation and the quadratures were carried out over the $(N+M)\times(N+M)$ grid while genetic operators such as crossover and mutation were restricted to act only on the $N\times N$ grid-point amplitudes coded by a string S_k.

3. Roulette wheel selection was made for creating the pool for crossover breeding, followed by mutation (a tournament selection could be used).

4. A modified arithmetic single-point crossover, followed by arithmetic mutation was used to generate new strings of probability amplitude distribution. Suppose that the strings S_k and S_l have been chosen to undergo crossover at the crossover site marked by the row index i and the column index j with the probability P_c. Then the pair of new strings produced under the adopted crossover process are

$$S_k'(p,q) = fS_k(p,q) + (1-f)S_l(p,q)$$

$$S_l'(p,q) = (1-f)S_k(p,q) + fS_l(p,q) \tag{6.87}$$

where $p = 1, 2, \ldots, i; q = 1, 2, \ldots, j$ while for all other grid points $(p = i+1, i+2, \ldots, N; q = j+1, j+2, \ldots, N)$

$$S_k'(p,q) = S_k(p,q)$$

$$S_l'(p,q) = S_l(p,q). \tag{6.88}$$

A simple arithmetic mutation with probability p_m was carried out on each of the postcrossover strings generating new mutated strings $S_k''(p,q)$, where

$$S_k''(p,q) = S_k'(p,q) + (-1)^l . r . \Delta_m \tag{6.89}$$

assuming that the amplitude at the Pth row and Q^{th} column of S_k' was selected to undergo mutation. After all the mutations had been carried out, all the strings were symmetrized to bring the strings in conformity with the permutational symmetry requirement of the two-electrons ground single wave function, i.e., for each string S_k'' $(k = 1, 2, \ldots, n)$

$$S_k''(p,q) = S_k''(q,p) \tag{6.90}$$

where

$$p = 1,2, \ldots, N; q = 1,2, \ldots, N.$$

TABLE 6.1

Comparison of the Ground State Energies of Two-Electron Atom/Ion Predicted by the GA-Driven Method with the Corresponding HF Energies

Atom/Ion	State	Energy Predicted by GA (a.u.) [24]	Hartree–Fock Energy (a.u.) [26]
He	$^1S_0$	−2.87505	−2.86168
H−	$^1S_0$	−0.51198	−0.48793

5. From the postcrossover mutated strings only 80% were screened on the basis of their fitness values for the next generation. The remaining 20% were chosen randomly from the same pool.

6. The selection crossover, mutation, symmetrization, and screening process were continued until the fitness distribution in the evolving population approached δ distribution as closely as possible. Once this stopping criterion is satisfied, the string with the highest fitness value represents the best radial wave function for the ground state. The following set of unoptimized GA parameters were used:

$$p_c = 0.75 \ (\text{fixed})$$

$$p_m = 0.05 \ (\text{fixed})$$

$$n(\text{population size}) = 10$$

$$f(0 < f < 1) \text{chosen randomly.}$$

The radial grid lengths used were $0 \leq r_1 \leq 20$ a.u., which were uniformly discretized into 500 grid points. An additional 1000 grid points were generated along each of the two radial coordinates by two-dimensional bicubic interpolations. Genetic evolution was carried out on the probability amplitudes defined on the 500×500) grid while quadrature (Simpson's rule) was performed over the (1500×1500) grid. Table 6.1 compares the ground state energies predicted by the GA-derived method (see Saha et al. [24,25]) with other available results.

It appears from Table 6.1 that the GA-driven scheme has been able to incorporate at least a part of the radial correlation which is absent in the standard Hartree–Fock calculations. A significant effect of this correlation is reflected in the computed ground state energy of the H ion which is unbound at the Hartree–Fock level, but is predicted to be bound by the GA-based direct solution of the two-electron radial Schrödinger equation. The predicted binding energy is still underestimated, presumably due to the neglect of angular correlations. We must mention here that the two-electron wave functions are inherently factorizable into their space and spin components. The overall antisymmetry of the total wave function and the singlet two-electron spin wave function automatically fix the symmetry of the space part of the two-electron wave function to be symmetric under the interchange of the coordinates of two electrons. The spatial symmetry of the amplitude distribution or its nodal structure is therefore uniquely defined. This simplicity disappears when we try to handle system with more than two electrons. No completely satisfactory resolution of the problem is available at the time of writing.

The nodal structure of the corresponding Hartree Fork wave function may be used to fix the nodal structures of the evolving amplitude distributions.

6.5.5 GA-Driven Hartree Method

Şahin et al. [27] developed and implemented a GA-driven method to predict the energy levels of N-electron quantum dots. Simultaneous solution of the coupled Schrödinger and Poisson equations has been achieved by invoking GA at the level of Hartree approximation. These authors considered a spherically symmetric quantum dot of GaAs embedded in bulk matter of AlGaAs. The difference between the band gaps of the two semiconducting materials (GaAs and AlGaAs) forms a confining potential $V_0(r)$ felt by an electron inside the quantum dot. For a spherically symmetric confining potential $V_0(r)$ the single-particle Schrödinger equation at the level of Hartree approximation turns out to be

$$\left[\frac{-\hbar^2}{2m} \vec{\nabla} \left(\frac{L}{m^*(r)} \right) \vec{\nabla} + \frac{l(l+1)\hbar^2}{2m^*(r)r^2} + V_0(r) - eV_H(r) \right] R_i(r) = \epsilon_{il} R_i(r). \tag{6.91}$$

The Hartree potential $V_H(r)$ is determined from the Poisson equation that it satisfies

$$\left(\frac{d^2}{dr^2} + \frac{2}{r}\frac{d}{dr} \right) V_H(r) = \frac{4\pi e}{\in(r)} \rho(r) \tag{6.92}$$

where

$$\rho(r) = \sum_{i=1}^{N} |\psi_i(r)|^2, \quad \psi_i(r) = R_i(r) y_{lm}(\theta, \phi_i). \tag{6.93}$$

The one-electron SE and the Poisson equation are solved self-consistently in tandem. The radial wave functions $R_i(r)$ are represented as probability amplitudes on a discretized radial grid. Each individual is represented by a four-dimensional array and contain an s-shell and a p-shell radial function value at different grid points. The following steps are implemented:

1. Generate an initial population of 100 individuals, each individual represented by a probability amplitude distributed on the chosen radial grid. Note: The probability amplitudes are generated from a set of normalized candidate functions.

2. Set $V_H(r)$, for the first iteration. For each individual, determine the single-particle energies ϵ_{il} in the confining potential $V_0(r)$ by quadrature (Simpson's rule) with the derivatives determined by five-point differences.

$$\epsilon_{il} = R_{il} | \hat{H} | R_{il}. \tag{6.94}$$

3. Determine the total energy E_i of each individual (i) as

$$E_i = \sum_l \epsilon_{il} - \frac{1}{2} \int R_{il}^*(r) V_H(r) R_{il}(r) r^2 dr; \quad (l = 0,1). \tag{6.95}$$

These energies are used to determine the fitness (f_i) value of an individual by defining the fitness f_i of the ith individual as $f_i = e^{-(E_i - E_{av})}$ where E_{av} is the average energy of the population.

4. Sort out the best individual in terms of fitness.

5. New individuals are produced by arithmetic crossover and mutation.

6. The wave function of the best individual is used to determine the charge density inside the dot and the Poisson equation is solved to obtain $V_H(r)$, which is added to $V_0(r)$. Return to step 2. For the second iteration onwards, check if the stopping criterion has been reached. If yes, exit; iterate otherwise. The best individual the exit is the solution of the many-electron problem at hand. The following points may be noted.

   a. The initial population of radial wave functions was generated from analytical functions $R_{kl}(r)$ where

$$R_{kl}(r) = A_k r^l e^{-\alpha_k r^{M_k}} \tag{6.96}$$

   A_k is the normalization constant fixed by the requirement

$$\int_0^\infty R_{kl}^2(r) r^2 dr = 1 \tag{6.97}$$

   $l = 0, 1$ for the s- and p-shell functions, respectively. M_k is an integer between 0 and 4, α_k is a random number drawn from the range $(0, R_{max})$.

   b. The crossover process produces two new radial functions $R_1'(r)$ and $R_2'(r)$ from the individuals $R_1(r)$ and $R_2(r)$ in the following manner [28]:

$$R_1'(r) = R_1(r) f(r) + R_2(r)(1 - f(r)) \tag{6.98}$$

$$R_2'(r) = R_2(r) f(r) + R_1(r)(1 - f(r)) \tag{6.99}$$

   For $f(r)$ a step function was chosen with

$$f(r) = \frac{1}{2}\left[1 + \tanh \frac{(r - r_0)}{c^2}\right] \tag{6.100}$$

   r_0 being a random number $(0, \infty)$, c is a parameter introduced for controlling the sharpness of the step function.

   c. The mutation used was a modified arithmetic mutation with $R_{kl}'(r) = R_{kl}(r) + \Delta_m x R_m(r)$, where Δ_m is the chosen mutation intensity, x is a random number in $(0,1)$ and $R_m(r)$ is a mutation function which was chosen to have the same form as $R_{kl}(r)$.

6.5.6 Molecular Schrödinger Equation

Saha et al. [24] extended their GA-based strategy to solve the molecular electronic Schrödinger equation for H_2^+ and obtain the ground state properties of the molecular ion. These authors experimented with two alternative approaches. In approach I, they solved the equation for different internuclear distances (R), constructed the potential energy curve from which equilibrium ground state energy, dissociation energy (D_e), and the values of the equilibrium bond length (R_e) were obtained. In their approach II, the internuclear distance was allowed to evolve along with the probability amplitudes of the electronic wave function in the three-dimensional Cartesian space and GA was used to obtain all the equilibrium properties of the ground state simultaneously—that is, the electronic probability amplitudes and the internuclear separation were optimized by GA simultaneously. To the best of our knowledge, Saha et al.'s communication is the only piece of work devoted to solve the molecular electronic Schrödinger equation directly by GA. In what follows, the salient features of the work are presented. The coordinate system used to describe the H_2^+ ion is presented in Figure 6.3 [24].

In the depicted coordinate system, the Schrödinger equation for H_2^+ under Born–Oppenheimer (BO) approximation becomes

$$H\left(x,y,z,\bar{R}\right)\phi_n\left(x,y,z,\bar{R}\right) = E_n\left(\bar{R}\right)\phi_n\left(x,y,z,\bar{R}\right) \tag{6.101}$$

where the molecular electronic Hamiltonian H is parametrically dependent on the internuclear separation $\bar{R}$ and reads

$$H\left(x,y,z,\bar{R}\right) \equiv H\left(\bar{R}\right) = \frac{-\hbar^2}{2m}\left(\frac{d^2}{dx^2} + \frac{d^2}{dy^2} + \frac{d^2}{dz^2}\right) - e^2\left(\frac{1}{r_A} - \frac{1}{r_B}\right) + \frac{e^2}{\bar{R}} \tag{6.102}$$

$$r_A = \sqrt{x^2 + y^2 + \left(z - z_A\right)^2}, \quad r_B = \sqrt{x^2 + y^2 + \left(z - z_B\right)^2}.$$

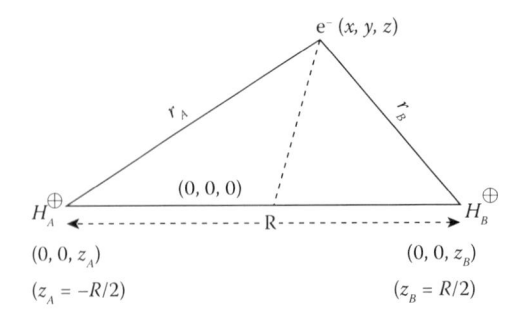

FIGURE 6.3

The coordinate system used to represent the hydrogen molecule ion (H_2^+). (Redrawn from Saha, R., and Bhattacharyya, S.P., *Int. J. Modern Phys. C*, 18(2), 163–175, 2007. With permission.)

In the BO approximation nuclei are fixed, so $\dfrac{e^2}{R}$ is just an additive constant to the electronic energy $E_n^{el}\left(\bar{R}\right)$ which is an eigenvalue of the molecular electronic Hamiltonian at the internuclear distance $\bar{R}$ as given in

$$H\left(\bar{R}\right)\phi_n\left(x,y,z,\bar{R}\right)=E_n^{el}\left(\bar{R}\right)\phi_n\left(x,y,z,\bar{R}\right). \tag{6.103}$$

The total energy $E_n^{total}\left(\bar{R}\right)$ is obtained by adding the internuclear repulsion energy $\dfrac{e^2}{R}$ to $E_n^{el}(\bar{R})$:

$$E_n^{total}\left(\bar{R}\right)=\frac{e^2}{R}+E_n^{el}\left(\bar{R}\right). \tag{6.104}$$

Approach 1: In this approach, Saha et al. [24] used GA to obtain the optimal electronic probability amplitude distribution in discretized Cartesian coordinate space in three dimensions for a fixed value of $\bar{R}$ and estimate the corresponding value of electronic energy $E_n^{el}\left(\bar{R}\right)$ to which $\dfrac{e^2}{R}$ was added to obtain the total energy $E_n^{total}\left(\bar{R}\right)$. The calculations were carried out for different values of $\bar{R}$ and the potential energy curve was constructed from $E_n^{total}\left(\bar{R}\right)$ values by plotting $E_n^{total}\left(\bar{R}\right)$ against $\bar{R}$.

Approach II: In this approach, GA was invoked to obtain the optimal electronic probability amplitude distribution in the three-dimensional discretized Cartesian coordinate space, not at a fixed internuclear separation R, but at genetically evolving values of R which was also optimized by GA simultaneously.

In their approach I, the wave function strings $\{S_1\}$ encoded the amplitudes of the ground state wave function at different points $\left\{x_i,y_j,z_k\right\}$ of the discretized Cartesian space for a given (fixed) values of $\bar{R}$. The S_Is were one-dimensional arrays containing $n_x\times n_y\times n_z=N$ number of floating-point numbers (n_x,n_y,n_z are the number of grid points along the x-axis, y-axis, and z-axis), each such number representing a value of the probability amplitude for the ground state wave function of the H_2^+ ion at the internuclear separation $\bar{R}$ at the point (x_i,y_j,z_k) of the coordinate space. Thus,

$$S_I^{\bar{R}}\left(x_i,y_j,z_k\right)\equiv S_I^{\bar{R}}\left(i,j,k\right)=\phi_I\left(x_i,y_j,z_k,\bar{R}\right) \tag{6.105}$$

for $I=1,2,3,\ldots,n_p$, n_p the number of wave function strings in the population.

In approach II, Saha et al. [24] defined and used a more general type of wave function strings. Here each wave function string encoded the ground state probability amplitudes at different points $\left(x_i,y_j,z_k\right)$ of the electronic coordinate space at randomly chosen value R_l of the internuclear separation (the internuclear space was discretized into a set of n_l number of points while the electronic space was discretized into a grid of $n_x\times n_y\times n_z$ points. The wave function strings S_I $\left(I=1,2,\ldots,n_p\right)$ were mapped onto one-dimensional arrays

$$S_I\left(i,j,k,l\right)\equiv S_I\left(x_i,y_j,z_k,R_l\right)\equiv S_I\left(ijkl\right)=\phi_I\left(x_i,y_j,z_k,R_l\right) \tag{6.106}$$

$\left|S_I\left(ijkl\right)\right|^2$ represents the probability of finding the electron at the point (x_i,wy_j,z_k) when the internuclear separation is R_l. In approach II, genetic operators were allowed to act both on the amplitude and the internuclear distance variable causing both to evolve. In both the approaches a starting population of wave function strings was generated from a set of normalized candidate wave functions such as $\phi_I = N_I \left\{ e^{-(\alpha r_A + \beta r_B)} + e^{-(\beta r_A + \alpha r_B)} \right\}$. Once the population was created, the fitness of each individual string was evaluated following a three-step procedure.

1. In the first step the energy E_I of the string S_I was calculated by defining

$$E_I = \frac{\langle S_I | H | S_{Iq} \rangle}{\langle S_I | S_{Iq} \rangle} \tag{6.107}$$

where q indicates that the integrals were evaluated by quadratures.
In the second step an objective function O_1 was defined by taking

$$O_I = (E_I - E_L)^2 \tag{6.108}$$

where E_L represented an estimated (updatable) lower bound to the ground state energy (E_0) of H_2^+ ion.
O_I was finally mapped onto a scaled fitness function f_I where

$$f_I = e^{-\lambda O_I} \tag{6.109}$$

λ being a user-defined scalar in the range $0 \le \lambda \le 1$.

The kinetic energy part of E_I was computed by using three-dimensional fast Fourier transform (FFT) while the potential part of E_I was computed by using trapezoidal or Simpson's rule. The use of FFT requires that the three-dimensional grid has number of points along each of the coordinate axes expressible as 2^n (n an integer). Accordingly, the authors used grids of $64 \times 64 \times 64$ or $128 \times 128 \times 128$ or $256 \times 256 \times 256$ points. The problem here has cylindrical symmetry about the z-axis which is chosen to coincide with the internuclear axis (R). While the initial choice of amplitude distribution was made to conform to the requirement of cylindrical symmetry of the problem, there was no guarantee that post-crossover and postmutation strings would also automatically conform to the symmetry requirement. Two symmetrization operations were therefore carried out on the postcrossover and postmutation wave function strings in the following manner:

1. The left–right symmetrization along the internuclear axis, i.e., the z-axis, was effected by making

$$S_I \left(x_i, y_j, -z_k \right) = S_I \left(x_i, y_j, z_k \right) \tag{6.110}$$

for $I = 1, 2, \ldots, n_p$; $i = 1, 2, \ldots, n_x$; $j = 1, 2, \ldots, n_y$; $k = 1, 2, \ldots, n_z$ when the internuclear distance was not large, or the nucleii not far away from the equilibrium internuclear separation. This symmetrization ensures that the evolving strings produce symmetric charge distributions about the origin.

2. The second symmetry requirement concerns the equivalence of the x and y axes which demands that the amplitude distributions obey the condition

$$S_I\left(\underbrace{x_i, y_j}, z_k\right) = S_I\left(\underbrace{y_j, x_i}, z_k\right). \tag{6.111}$$

That is, the amplitude retains the same value when the x_i and y_i values are interchanged.

The predicted values of equilibrium internuclear separation (R_e) and the dissociation energy (D_e) are displayed in Table 6.2. The calculations were done on a three-dimensional grid of $128 \times 128 \times 128$ points.

The PEC predicted by the GA (approach I) is displayed in Figure 6.4 [24].

In approach II the equilibrium internuclear separation was optimized simultaneously with the probability amplitude distribution and therefore no PEC could be constructed.

The authors noted that the left–right symmetry constraint was withdrawn for calculations carried out under approach I for large values of internuclear separation R. This was

TABLE 6.2

The Values of R_e and D_e for the Ground State of H_2^+ Estimated from the Potential Energy Curve (PEC) Obtained from the Two Sets of Calculations (64 Grid Points (Set I) and 128 Grid Points (Set II)) using the GA-Based Type I Method

Method	R_e (a.u.)	D_e (a.u.)
Set I	2.098	0.106
Set II	2.031	0.107
Experimental [29,30]	2.00	0.103

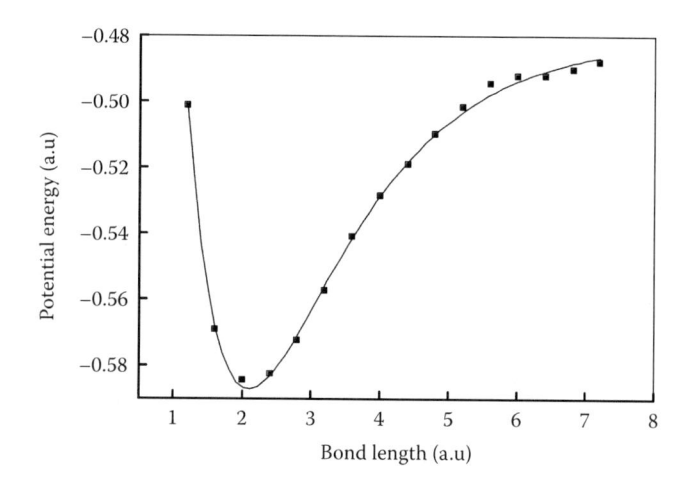

FIGURE 6.4

Energy–internuclear distance profile (computed by GA-based search (approach I, set I) for the ground state of H_2^+ at different internuclear distances) is fitted to a Morse curve. (From figure 3 of Saha, R., and Bhattacharyya, S.P., *Int. J. Modern Phys. C*, 18(2), 163–175, 2007. With permission.)

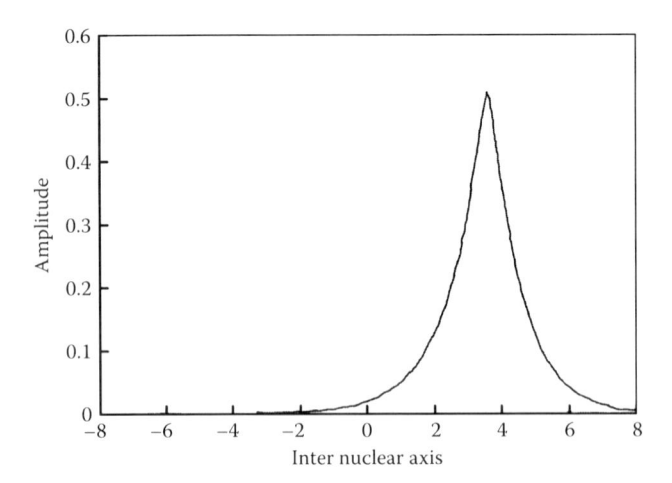

FIGURE 6.5
The predicted probability amplitude distribution (GA-based search (approach I, set 2)) at R=7.2 a.u. Note the lack of symmetry about z=0. (From figure 5 of Saha, R., and Bhattacharyya, S.P., *Int. J. Modern Phys. C*, 18(2), 163–175, 2007. With permission.)

needed to allow the electronic charge distribution the flexibility to distort itself so that as $R \to \infty$, the ion could dissociate into H^+ and H (and not into $H^{+0.5} + H^{+0.5}$ because of the left–right symmetry constraint). That the GA-driven method without the left–right symmetry constraint enforces tended to predict electronic probability amplitude distribution correctly as $R \to \infty$ is displayed in Figure 6.5 [24]. Clearly, the electronic charge distribution has lost symmetry and is asymptotically approaching what is expected when H_2^+ dissociates into H^+ and H as $R \to \infty$.

The coupled simultaneous evolution of electronic probability amplitudes and the internuclear distance (R) to their equilibrium values was found to take place smoothly [24] and led to the equilibrium R_e value, the energy (E_e) and charge density of equilibrium in one go. We will see later that it is possible to achieve coupled (simultaneous) optimization of electron density distribution and the nuclear position variables in much bigger systems within the framework of soft computing methods (Chapter 7).

6.6 Application of Genetic Computing

6.6.1 Fitting Potential Energy Surfaces

Accurate molecular electronic structure calculations under the BO approximations at many nuclear configurations $\{R_i\}_{i=1,2,...,N}$ provide us with molecular energy values (numerical) as functions of $\{R_i\}$. The collection of these energy values $E\{R_i\}$ defines a multidimensional potential energy surface on which the nucleii move. The analytical form of the function $E\{R_i\}$ hidden in the energy data is, however, unknown in the general case. One way to construct such a function can be through an elaborate fitting procedure which starts by guessing an analytical form of the unknown function, say $f(R_i, a_1, a_2, ..., a_n) \equiv f(R_i, a)$ where

a_ks are parameters embedded in the function chosen. The fitting procedure works by minimizing the overall error in the energy predicted by the analytical function, with respect to the parameter set $(a_1, a_2, \ldots, a_n) = \{a_\beta\}$ say, which is allowed to vary. The error $\epsilon(a)$ is the sum of errors for all the points $\{R_i\}$ on the potential energy surface (PES) and reads

$$\epsilon_\beta(a) = \sum_{i=1}^{N} \left(E(R_i) - f(R_i, a)\right)^2. \tag{6.112}$$

Alternatively, the error function $\epsilon_\beta(a)$ can be mapped onto that fitness function $F\beta(a)$ where

$$F_\beta(a) = exp\left\{\frac{-\lambda\epsilon_\beta(a)}{\sum_i E^2(R_i) + \delta^2}\right\} \text{or} \; exp\left[-\lambda\epsilon_\beta(a)\right]; \;\; (\lambda, \delta > 0). \tag{6.113}$$

The goal then becomes searching out a parameter set $a^0 \equiv \left(a_1^0, a_2^0, \ldots, a_n^0\right)$ such that the fitness value is maximized. This traditional approach is limited by the fact that a choice of the function $f(R_i, a)$ has to be made by the user. Is it possible to devise a much more flexible fitting regimen in which the search is not restricted to finding an optimal set of parameters for the given choice of the analytical function, but extends also to the search for an optimal analytical form of the function itself along with the relevant embedded parameters? The answer to the question is "yes." Makarov and Metiu [31,32] explored genetic programming (see Chapter 3) as a possible tool for the expanded search and came up with what they called directed genetic programming (DGP)—"directed," because the user specifies the forms of functions to the sampled during the search, based on analysis and prior knowledge without losing the flexibility of the search. It turns out that the directed GP performs better than the free GP when it comes to fitting the PES to analytical forms. A crucial element of this procedure is the random function generator that throws in different analytic functions with embedded parameters and creates a population of n_p individual functions $f(x, a), g(x, b), h(x, c), \ldots$. These functions are given parse tree representation (Chapter 3), the structural elements of the tree being nodes, branches, and leaves. Nodes are functions such as $+, (-), /, *, sin, cos$. The leaves are literals—the nuclear coordinates or numerical values of parameters appearing in the function being represented as a tree. The fitness values of all the n_p member functions are evaluated and a selection procedure is applied to the individual population. New functions are generated by a careful application of grafting (replacing a short branch by a tree generated by the restricted function generator (RFG), parameter alteration (replacing a leaf (say b) in a tree by $b+R$, R being a random number in a specified range $b_{min} \le b \le b_{max}$), scaling (a variable x is replaced with $x(1+r)$, r being a random number in a specified range), mutation, and crossover (see Chapter 3). Once new functions are generated by genetic operations, they are checked for their "badness values" (reciprocal fitness) and functions with high badness values are discarded. The authors considered the PES of a water molecule (triatomic) modeled by three parameters x_1, x_2, and ψ where

$$x_1 = \frac{r_1 - r_1^0}{r_1^0}; \quad x_2 = \frac{r_2 - r_2^0}{r_2^0} \quad \text{and} \quad \psi = (\pi - \theta) \tag{6.114}$$

θ being the H-O-H angle, x_1 and x_2 are the lengths of the O-H bonds, and r_1^0 and r_2^0 are their values at the ground state equilibrium geometry (= 0.95AÅ). One of the best fit functions predicted by the DGP for the MP_2 energy values predicted by Gaussian package was

$$
\frac{f(x_1, x_2, \psi)}{V_0} = 3.785 \left\{ \left(1 - e^{-2.07 x_1}\right)^2 + \left(1 - e^{-2.07 x_2}\right)^2 \right\}
$$

$$
+ \left(\left(\cos\psi - 0.307 + 0.755(x_1 + x_2)\cos\psi\right) - \left(0.289 + x_1 + x_2 + 0.115(x_1 - x_2)^2\right. \right.
$$

$$
+ (x_1 - x_4)^4 + 0.481(x_1 + x_2)^2 + 1.592\cos\psi
$$

$$
\left. \left. + 0.193\left(-0.28 + 0.00015(x_1 - x_4)^4 + \cos^3\psi\right)\right)\right) e^{-(x_1 + x_2)}
$$

(6.115)

V_0 is the energy of H_2O molecule in the linear configuration. Clearly, such a function could probably not have been guessed by any amount of prior analysis of the energy data points. The directed GP method has great potential in the context of suggesting a possible analytical representation of the PES in multiatom systems, which can subsequently be further refined, if needed, by adopting deterministic procedures.

6.6.2 Finding Analytical Solutions of SE

The directed GP-driven method for obtaining good analytic fit to PES was further tweaked to find good analytic solutions of time-independent Schrödinger equation (SE). The trick lies in inverting the time-independent SE

$$
[T + V]\psi = E\psi \quad \text{to} \quad V = -\frac{\{T \psi\}}{\psi} + E
$$

(6.116)

and posing $V(x)$, the known potential function of coordinates now as a functional of the unknown wave function ψ (a function of coordinates) and energy E (a scalar). The energy eigenvalue problem then becomes a problem of fitting, i.e., the problem boils down to finding a function ψ and the corresponding energy (E) such that the right-hand side of Equation 6.116 laid out on a grid fits accurately to the known potential $V(x)$ at every point. The fitting is done by invoking the DGP tool of Makarov and Metiu [32].

Let us consider, for the sake of simplicity, a one-dimensional time-independent Schrödinger equation ($\hbar = 1$, $m = 1$):

$$
-\frac{1}{2}\frac{d^2\psi(x)}{dx^2} + V(x)\psi(x) = E\psi(x).
$$

(6.117)

Inverting, we have

$$
V(x) = \frac{1}{2}\frac{1}{\psi(x)}\frac{d^2\psi(x)}{dx^2} + E.
$$

(6.118)

Let $V(x)$ be represented on a discrete grid of points $\{x_i\}_{i=1,N}$, and $f(x, a)$ analytically (approximately though) represents the unknown wave function on the same grid, a representing the collection of parameters embedded in $f(x, a)$, and E being the energy associated with $f(x, a)$ for a given choice of the parameters a. The total error in the fit is then obtained by summing up the error ϵ_i over all the points on the grid:

$$\epsilon\left[f,a\right] = \sum_{i=1}^{N}\left[V(x_i) - \left\{\frac{1}{2}\frac{1}{f(x_i,a)}\frac{d^2 f(x,a)}{dx^2}\bigg|_{x=x_i} + E\right\}\right]^2$$

$$= \sum_{i=1}^{N}\epsilon_i. \tag{6.119}$$

The directed GP (DGP) simultaneously searches through the space of a list of candidate functions $f(x)$ and the parameters (a) embedded in them, until the fitness $F(f, a)$ is maximized. The fitness may be generated by

$$F\left(f,a\right) = e^{-\lambda\epsilon\left[f,a\right]}; \quad (\lambda > 0, \text{user-defined}). \tag{6.120}$$

The energy $E(f, a)$ was not calculated directly by evaluating the expectation value $E = \langle H \rangle = \dfrac{\langle f|H|f \rangle}{\langle f|f \rangle}$. Instead, $f(x, a)$ created by the DGP was used in the expression for $\epsilon[f, a]$ and ϵ (a quadraticism E) was minimized with respect to E leading to the estimate of energy associated with the wave function of $f(x, a)$.

As an illustration, we may consider the problem of finding bound states of a particle of unit mass moving in a one-dimensional Eckert potential well $V(x)$ [32]:

$$V(x) = \frac{1}{\cosh^2\left(\dfrac{x}{2}\right)}. \tag{6.121}$$

The potential supports found bound states which can be qualytically represented by appropriate hypergeometric functions. As noted by the authors, a search based on the traditional GP was not fruitful. The search space—the space spanned by all the normalizable functions—was too large and the GP failed to generate good quality solutions. The authors of reference [32] therefore toyed with the idea of providing a direction to the GP search so that it operates in a much more restricted, but relevant portion of the search space.

The ground state wave function of the system, the authors argue, is expected to be a bell-shaped function and could be represented as

$$f(x,a) = \frac{1}{\cosh\left(g(x,a)\right)} \tag{6.122}$$

where $g(x, a)$ is an unknown function that the GP searches for using only four arithmetic functions—the so-called nodes of parse trees—such as $+, -, /, *$ to construct trial forms for $g(x, a)$. The search, as reported by the authors, produced a function

$$f(x) = \frac{1}{\cosh\left[1.086x + 0.023x(-1.78 + 1.086x)(1.658 + 1.086x)\right]}. \tag{6.123}$$

The corresponding energy $E_0 = -1.5576$ eV compared well with the exact ground state energy $E_0^{exact} = -1.5586$ eV. A pertinent question to be asked at this point concerns the role played by the guiding function. The authors claim that the quality of the final solution (in terms of energy) is independent of the choice of the guiding function itself. That means, any other guiding function $\phi(g(x,a))$ capable of restricting the search to normalizable bell-shaped functions would be equally valid and the GP will produce $f(x, a)$ with a good estimate of energy for the ground state.

The question that needs to be addressed now concerns the generation of excited states of the potential. The authors did not go for orthogonality constrained calculations (see Chapter 4) for the excited states directly. Indirectly, however, they constrained the search to look for functions that would satisfy the orthogonality requirements. The constraint on the search came in the form of an appropriate choice for the guiding function capable of generating the excited state in question. Let us suppose that the ground state of the potential $V(x)$ is $f_0(x)$. The first excited state can then be generated as $f_1(x) = a^{\dagger} f_0(x)$, the second excited state as $f_2(x) = \left(a^{\dagger}\right)^2 f_0(x)$, and so on to the higher excited states, where a $a^{\dagger}$ is the step-up ladder operator for the problem. Unfortunately, $a^{\dagger}$ for the problem at hand is not known. So they approximated $a^{\dagger}$ by the ladder for a harmonic oscillator of frequency ω where ω is determined by the value of the second derivative of $V(x)$ at the minimum (x_0) of the potential. Written explicitly, we have, for a particle of mass μ, moving in the potential $V(x)$

$$\omega = \sqrt{\frac{V''(x_0)}{\mu}} \tag{6.124}$$

and

$$a^{\dagger} = \sqrt{\frac{\omega}{2}}\left(x - \frac{1}{\omega}\frac{d}{dx}\right). \tag{6.125}$$

The guided search for the first excited state wave function is then carried out by maximizing the fitness over all functions $f_1(x)$ where

$$f_1(x) = a^{\dagger}\Phi(g(x,a)) \tag{6.126}$$

Φ being a bell-shaped function already introduced for generating the ground state, the argument $g(x, a)$ which is optimized by the GP. The GP search effectively looks for the ground state, $a^{\dagger}$ excites it into the corresponding first excited state which is used to estimate fitness. The strategy is interesting but needs deeper exploration, especially in the context of the search for the higher excited states.

The authors [32] also provided additional examples of the viability of their DGP. These examples include finding the ground and excited states of symmetric double-well potential (SDWP) in one- and two-dimensional anharmonic oscillators. For the SDWP $V(-x) = V(x)$, the authors exploited the inversion symmetry about $x = 0$ which suggests that wave functions $\psi_L(x)$ and ψ_R localized within the left and right potential wells bear the symmetry relation $\psi_L(x) = \psi_R(-x)$. This is assuming that the barrier height separating the wells is high (∞) $\psi_L(x)$ and ψ_R are degenerate eigenfunctions of the energy operator that can be superposed to produce parity eigenstates $\psi_+(x)$ and $\psi_-(x)$ of the same energy where $\psi_\pm(x) = \frac{1}{\sqrt{2}}\{\psi_L(x) \pm \psi_R(x)\}$. If the barrier height is finite $\psi_L(x)$ and ψ_R will have slightly different energies and the parity eigenstates $\psi_+(x)$ and $\psi_-(x)$ will not be equal super positions of $\psi_L(x)$ and $\psi_R(x)$, and there will be a small energy splitting $\Delta E = (E_- - E_+)$ called tunnel splitting. That means, the directed GP search would have to be carried out separately for $\psi_+(x)$ and $\psi_-(x)$. They chose the template function for guiding the search for $\psi_+(x)$ in the form

$$\psi_+(x) = \frac{1}{\cosh(\phi(x))} + \frac{1}{\cosh(\phi(-x))} \tag{6.127}$$

so that the lobes in the left and right wells are each bell-shaped and invoke GP to construct the best $\phi(x)$. One of the solutions that produced a very good estimate of the ground state energy E_+ in the given potential was $\phi(x)$, where

$$\phi(x) = 1.63 + 0.85(-4.07 + x) + 0.19x + 0.05(-0.98 + 0.945x)x^2(-2.94 + x^2) \tag{6.128}$$

for $\psi_-(x)$, a separate search for $\phi(x)$ had to be carried out.

Chaudhury and Bhattacharyya [33] deftly exploited genetic algorithm for calculating the ground state wave functions and energies of screened-coulomb two-electron atoms or ions within the framework of a discretized generator coordinate (DGC) method already introduced in Chapter 4 in the context of one- or two-dimensional anharmonic oscillators and used the information provided by the optimal DGC wave functions to construct model two-electron noninteracting Hamiltonians that reproduced the energies of the systems accurately. It would be instructive to look into their strategy in detail and examine the results.

The Hamiltonian of a screened Coulomb hydrogenic atom (screening length λ) is $(\hbar = 1, m = 1, a_0 = 1)$

$$H_\lambda = -\frac{1}{2}\nabla^2 - \frac{ze^{-\lambda r}}{r}. \tag{6.129}$$

Chaudhury and Bhattacharyya [33] represented the ground state wave function of $H\lambda$ by a linear superposition of five three-dimensional harmonic oscillator eigenfunctions with $n = 0,2,4,6,8,10$, $l = 0$ optimized the generator coordinates (GCs) by GA/SAM and computed the ground state energy E_λ for many values of λ in the range $0 \leq \lambda \leq 1.1$. The information so obtained was used to estimate the critical screening length (λ_c, say) for which bound S-states ($l = 0$) of the screened coulomb hydrogenic atom disappear. The estimated λ_cs turned out to be weakly z-dependent.

The method was extended to the ground states of many other two-electron atoms/ions (He, Li$^+$, H$^-$, etc.). The space parts of the wave functions for the two-electron atoms/ions were represented as

$$\psi(r_1, r_2) = \sum_{ij} C_{ij} \phi_{ij}(r_1, r_2; n_i, n_j; l_i = 0, l_j = 0) \qquad (6.130)$$

where the GC configuration functions were chosen as

$$\phi_{ij}(r_1, r_2; n_i, n_j; l_i = 0, l_j = 0) = N\left[\chi(r_1, \alpha_{n}) \chi(r_2, \alpha_{n}) + \chi(r_1, \alpha_{n}) \chi(r_2, \alpha_{n})\right] \qquad (6.131)$$

$\chi(r, \alpha_n)$ are three-dimensional harmonic oscillator wave functions with $n = 0, 2, 4, 6, 8, 10$, $l = 0$, N being the normalization constant. Since the GC basis contained only $l = 0$ functions, the target was to predict S-limit energies. All the discrete GCs were fully optimized by GA/SAM. The ground state energies predicted by the calculations tallied well with exact S-limit energies of all the two-electron species.

The authors of reference [33] succeeded in condensing their results to construct model two-electron atomic or ionic Hamiltonians that were noninteracting (electron–electron repulsion-free) but contained an optimally screened electron–nucleus interaction. Interestingly, they found that the optimal screening constant (λ_0) was virtually z-independent (at least over the range of z studied) so a model (generic) two-electron noninteracting optimally screened coulomb Hamiltonian could be written as

$$H_0^z(r_1, r_2) = -\frac{1}{2}\nabla_1^2 - \frac{1}{2}\nabla_2^2 - z\left\{\frac{e^{-0.3077 r_1}}{r_1} + \frac{e^{-0.3077 r_2}}{r_2}\right\}. \qquad (6.132)$$

H_0^z almost exactly reproduced the ground state S-limit energies of the two-electron atoms or ions. The authors demonstrated the usefulness of their H_0^z by using H_0^z to calculate the ground state polarizabilities of two-electron atoms/ions in very good agreement with known accurate values of polarizabilities of the same species. These calculations demonstrate the power of the soft computing method in the elucidation of electronic structure and properties of atoms in a nonconventional manner. Needless to mention, perhaps, is that the optimization of GCs and the search for the optimal screening constant (λ_0) were made easy by invoking soft computing options. It would be very interesting to explore similar models for two-electron molecules or ions, or for many electron atoms or ions.

6.6.3 GA in Search of Pattern-Forming Conditions

Instability and pattern formation in reaction diffusion systems has been a subject of vigorous research ever since Turing [34] proposed that a homogeneous steady state in such systems may experience instability. This is due to asymmetry in diffusion coefficients of the two species participating in the chemical reaction, leading to the appearance of symmetry breaking spatial structures (the Turing patterns). Turing's stability analysis is, however, moored in the linearization of the dynamics around the steady state and is therefore somewhat limited. Riaz et al. [35] in a recent communication probed the role of nonlinearity, if any, in the emergence of instability and pattern formation by taking into account higher-order spatiotemporal perturbation terms in the relevant concentration variables. In other words, the authors of reference [35] investigated (theoretically) if pattern formation could exclusively originate from nonlinearity in the system even when linear stability is guaranteed under the same condition. They considered the prototypical fish pigmentation model [36] described by the equation

$$\frac{\partial u(x,y,t)}{\partial t} = \alpha u\left(1 - r_1 v^2\right) + v\left(1 - r_2 u\right) + \delta d\left(\frac{\partial^2 u}{\partial x^2} + \frac{\partial^2 v}{\partial y^2}\right) \tag{6.133}$$

$$\frac{\partial v(x,y,t)}{\partial t} = \beta v\left(1 + \alpha r_1 uv / \beta\right) + u\left(\gamma + r, v\right) + \delta d\left(\frac{\partial^2 u}{\partial x^2} + \frac{\partial^2 v}{\partial y^2}\right) \tag{6.134}$$

where $u \equiv u(x,y,t)$ and v represent the dimensionless concentrations of the reactants, the activator, and the initiator, and $\alpha, \beta, \gamma, r_1, r_2$ are parameters of the dynamics. δ is simply the length scale while d represents the ratio of the diffusion coefficient (the diffusion asymmetry parameter). Riaz et al. attempted to identify a parameter regime (if it existed) where the system is linearly stable but nonlinearly unstable and ascertain if the nonlinear instability would drive a transition to pattern formation. The authors adopted a stochastic global search method (GA) to navigate through the entire parameter space and marshaling the power of a real coded genetic algorithm with selection (roulette wheel), crossover (arithmetic), and mutation (dynamic), they succeeded in discovering parameter regimes where pattern formation indeed takes place due to nonlinearity.

While representing the potential solution strings $\{S_k\}$ with $S_k \equiv S(\alpha^k, \beta^k, \gamma^k, r_1^k, r_2^k)$ as arrays containing floating-point numbers each representing a possible value of one of the search variables (one of the five parameters) poses no problem, defining a fitness function for the strings is not so straightforward. The authors noted that the variance of a concentration variable u $\left(\sigma = \left\{\left\langle u^2 \right\rangle - \left\langle u^2 \right\rangle^2 \right\}^{1/2}\right)$ is much larger in an inhomogeneous state than in a homogeneous one, and set their search objective as the maximization of the variance of the relevant concentration variables as functions of the parameters of the problem (the floating-point variables in the strings). Since analytic expression for the variance was unavailable, the authors solved numerically the coupled Equations 6.133 and 6.134 for u and v for the evolving parameter sets in different strings as GA produced these by invoking the Euler algorithm under zero flux boundary condition and computed the required variances numerically. The genetic search through the parameter space discovered several regimes in the parameter space where instability had a purely nonlinear origin and pattern formation driven by the nonlinear instability was noticed. For example, a parameter set $\alpha = 0.217, \beta = -0.256, \gamma = -2.82, r_1 = 0.11, r_2 = 0.99$ discovered by the GA produces a spatial pattern in the form of a spiral although all the eigenvalues of the linear stability matrix for the same parameter set yielded one positive eigenvalue (λ) for a number of values of u^2.

The study clearly indicates the enormous ability of genetic algorithms as a pure search device irrespective of the nature of the problem being studied. The requirement is to be able to give a chromosome or string representation to the potential solutions and define a physically meaningful fitness landscape on which the vector of search variables evolves under the action of genetic operators.

6.6.4 GA in Optimal or Coherent Control of Quantum Dynamics

If we are given a quantum system in a specific initial state $\psi(0)$ and are required to shape the dynamics of quantum evolution so that a specific target state, say $\phi(T)$ is reached at $t = T$ with high probability, we have to design a control Hamiltonian $[H'(t)]$ that nudges the evolution of $\psi(0)$ into $\phi(T)$. It is convenient to choose the external perturbation $H'(t)$ in the form of an optical field $(\in(t))$, the field parameters serving as control variables or design variables which are optimized so as to maximize the probability of transition of $\psi(t)$ into $\phi(T)$ after the lapse of time $\Delta t = (T - t)$. A lot of theoretical and experimental research has been undertaken into the problems associated with the practical realization of optimal or coherent control and its possible solutions. Soft computing techniques, if applied correctly, can be viable tools for generating the appropriate control Hamiltonian. One of the soft computing methods most frequently used for the purpose is a genetic algorithm of one variety or another. The idea here has been to use feedback-modified genetic exploration and adaptive learning to construct the appropriate control fields.

Ghosh et al. [37], for example, investigated whether a suitably designed time varying field (say, the electric field of light) could be generated to excite a single carrier two-dimensional quantum dot to a predecided target level. To achieve the objective, one must be able to exercise optimal quantum control of the dynamics of the system either by designing an appropriately shaped pulse of coherent light or by generating a suitably chirped laser pulse. They chose the chirped-pulse route to coherent control. The Hamiltonian of the two-dimensional dot, in their model, reads

$$H = -\frac{\hbar^2}{2m^*}\left(\frac{\partial^2}{\partial x^2} + \frac{\partial^2}{\partial y^2}\right) + \frac{1}{2}m^*\omega_0^2 x^2 + \frac{1}{2}m^*\left(\omega_0^2 + \omega_C^2\right)y^2 - i\hbar\omega_C y\frac{\partial}{\partial x}$$

$$= H_x(\omega_0) + H_y(\Omega) + V(x,y)$$

(6.135)

where $H_x(\omega_0)$ represents a harmonic oscillator Hamiltonian with frequency of oscillation ω_0 along the x-axis while $H_y(\Omega)$ represents a similar harmonic oscillator of frequency $\Omega = \sqrt{\omega_0^2 + \omega_C^2}$ (oscillation along the y-axis). The Hamiltonian H, a complex Hermitian operator, was diagonalized in the product basis set $\left\{\phi_n(\alpha,x)\phi_m(\beta,y)\right\}_{n,m}$ where $\phi_n(\alpha,x)$ and $\phi_m(\beta,y)$ are eigenstates of $H_x(\omega_0)$ and $H_y(\Omega)$, respectively, with α determined by ω_0 and β by Ω. The pth eigenstate of H is then expressed as a linear combination of the product basis set:

$$\psi_p(x,y,t=0) = \sum_{n,m} C_{nm}^{(p)}(0)\phi_n(x,\alpha)\phi_m(y,\beta).$$

(6.136)

The problem now is to start from $\psi_0(x,y,t=0)$—the ground state of the dot, and turn on a suitably designed optical field that takes the system to, say the kth state of the system with high probability $(k > 0)$ in the long time limit. Instead of randomly choosing the field, the authors chose the control field in the form of a chirped pulse with Gaussian chirping frequency $[\gamma(t)]$ and peak field intensities $(\in_x^0, \in_y^0)$ together with the pulse duration (T_p) dynamically adjusted by feedback-modified genetic learning so as to maximize the population P_k of the state k at time T ($>T_p$). A convenient measure of $P_k(T)$ is provided by the squared overlap amplitude $\langle \psi(x,y,t=0)|\psi(x,y,t)\rangle^2$ where $\psi(x,y,t)$ is obtained by integrating the time-dependent Schrödinger equation of the quantum dot in the presence of the control field as described below:

$$i\hbar \frac{\partial \psi(t)}{\partial t} = \left(H(x,y,t=0) + H'(t)\right)\psi(t). \tag{6.137}$$

The control field $H'(t)$ was chosen as

$$H'(x,y,t) = \in_0^x S_x(t,T_p) x \sin(\gamma_x(t)t) + \in_0^y S_y(t,T_p) y \sin(\gamma_y(t)t) \tag{6.138}$$

$S_x(t) = S_y(t) = S(t)$ are pulse-shape functions of triangular, Gaussian or rectified sine varieties, T_p being the pulse duration. The frequency $\gamma(t)$ was determined by a Gaussian chirping formula

$$\gamma(t) = \omega + \Delta\omega \, exp \left(\frac{t}{\tau}\right)^2 \tag{6.139}$$

where $\gamma(0) = \omega + \Delta\omega = \omega_{0k}$, $\hbar\omega_{0k}$ being the $0 \to k$ transition energy. ω was kept fixed and $\Delta\omega$ allowed to take any value in the range $0 \leftrightarrow \omega_{0k}$. The GA worked with a population of 10 strings $R_1, R_2, ..., R_{10}$, each string being a one-dimensional array of five floating-point numbers representing the five parameters of the control field, namely $\in_0^x, \in_0^y, T_p, \tau, \Delta\omega$. For calculating the fitness values of the genetically evolving strings under roulette wheel selection, arithmetic crossover, and mutation, the time-dependent SE was integrated from $t = 0$ to $t = T(T = 1.5T_p)$ with the control field $H'(x,y,t)$ set up by the five parameters of an individual strings (i) (10 such integrations were performed in every generation) and $P_k^i(T) = \left|\langle \psi_k(x,y,t=0)|\psi^i(x,y,T)\rangle\right|^2$ was evaluated in each case. The fitness value f_i of the string S_i was then measured by taking

$$f_i = e^{-\left(P_k^i(T)-1\right)^2} \tag{6.140}$$

so that the maximum fitness value of 1 is possible only when the control field is able to populate the target level (k) maximally ($P_k(T) = 1$). The study reveals the possibility of achieving a good level of target excitation in quantum dots by designing chirped laser pulse (the control field) through feedback-guided genetic learning. It is interesting that the authors found that the optimal control field along the x and y axes to be complementary in nature. Their surmise was that the genetic algorithm was, as if sequentially, maximizing the transition rate from $\phi_0(x) \to \phi_m(x)$ and $\phi_0(y) \to \phi_n(y)$ when the initial state was $\phi_0(x)\phi_0(y)$ and the target state was $\phi_m(x)\phi_n(y)$.

6.6.5 Annealing Route to Optimal Control Field

Pulse shaping to achieve targeted cleavage of a bond has been an exciting proposition investigated by theoreticians and experimentalists [38]. Shandilya et al. [39] explored the possibility of selectively cleaving the O-H or the O-D bond in the H-O-D molecule by shaping infrared (IR) and UV pulses theoretically through feedback control within the framework of the method of simulated annealing. The authors addressed the problem in two steps: first they invoked SAM to design an infrared laser pulse that would maximize the population in low-lying vibrational excited states on the PES of the ground electronic state (ψ_g). These one-quantum ($|01\rangle$ or $|10\rangle$) or two-quanta ($|02\rangle$ or $|20\rangle$) states are known to be localized predominantly along O-H or O-D directions. After the IR pulse maximally populated the targeted state, a suitably designed UV pulse would transfer the population to the repulsive electronically excited state (ψ_g). The net result would be cleavage of the targeted O-H or O-D bond. For the dynamic feedback-modified pulse shaping, the authors had to solve the time-dependent Schrödinger equation for a coupled two-state problem as described below:

$$i\hbar \begin{pmatrix} \dot{\psi}_g(t) \\ \dot{\psi}_e(t) \end{pmatrix} = \begin{pmatrix} H_g + H'_{IR}(t) & H'_{uv}(t) \\ H'_{uv}(t) & H_e \end{pmatrix} \begin{pmatrix} \psi_g \\ \psi_e \end{pmatrix}. \tag{6.141}$$

The Hamiltonians for nuclear motion on the ground and excited PES (V_g or V_e) are

$$H_g = T + V_g(r_1, r_2)$$

$$H_e = T + V_e(r_1, r_2) \tag{6.142}$$

$$T = \frac{p_1^2}{2\mu_{O-H}} + \frac{p_2^2}{2\mu_{O-D}} + \frac{p_1 p_2}{m_0} \cos\theta$$

r_1, r_2 stand for O-H and O-D bond stretching coordinates, μ_{O-H}, μ_{O-D} are the reduced masses of the O-H and O-D pairs. V_g and V_e represent the ground and excited states of the PES respectively.

There are two control fields to be optimized, ie., $H'_{IR}(t)$ and $H'_{uv}(t)$. The $H'_{IR}(t)$ is the IR control field for achieving the targeted vibrational excitation on the ground PES while $H'_{uv}(t)$ stands for the UV control field for transferring the population to the excited PES (V_e) which, being repulsive, would lead to O-D or O-H bond breaking depending on which vibrational level on V_g was maximally populated. One can write the IR control field as

$$H'_{IR}(t) = -\mu_g(r_1, r_2) S_{IR}(t) \tag{6.143}$$

where $S_{IR}(t)$ stands for the optimizable and externally controlled pulsed IR laser field, and $\mu_g(r_1, r_2)$ denotes the electric dipole moment of the H-O-D molecule in the ground state. $S_{IR}(t)$ was chosen as a superposition of eight individual Gaussian laser pulses as

$$S_{IR}(t) = \sum_{i=1}^{8} C_i \left\{ S_i \left(\epsilon_{0i}, \omega_i, \gamma_i, t \right) \right\} \tag{6.144}$$

with

$$S_i = \epsilon_{Oi} \left(\frac{8\gamma_i t_I^2}{\pi} \right)^{1/4} \exp\left(-\gamma_i (t - t_I)^2\right) \cos\omega_i t. \tag{6.145}$$

Each IR pulse component (i) has four optimizable parameters $\epsilon_{Oi}, \omega_i, \gamma_i$, and the combining weight C_i so designing the control field $H'_{IR}(t)$ involves a 32-parameter optimization problem. The objective function O_{IR}^k for optimizing the control IR field was linked to the population $P_k(t = T)$ in the targeted vibrational level, say ϕ_k after the lapse of time T where $P_k(t) = \left|\langle \phi_k | \psi_g(t) \rangle\right|$ and was set as

$$O_{IR}^k = \left(P_k(t = T) - 1\right)^2 \tag{6.146}$$

SAM was invoked to minimize O_{IR}^k which amounted to the maximum buildup of population in the targeted state. The optimization of UV control field $H'_{uv}(t)$ was much simpler. The authors [39] used a single UV pulse the parameters of which were optimized by SAM using a similar objective function $O_{uv}(t)$ where

$$O_{uv}(t) = \left(f(t) - 1\right)^2 \tag{6.147}$$

$f(t)$ being the integrated flux in the O-H or O-D mode. Thus, the optimized IR control field populates one or two quantum excited vibrational levels selectively while the designed UV control field promotes the H-O-D molecule into the excited state where it suffers O-H or O-D bond breaking depending on whether the selectively populated excited vibration level on the ground surface dominantly lie along the O-D or the O-H stretching modes. The method proposed produced good selectivity in the targeted bond breaking and apparently offers a practicable, yet simple solution to the long-standing problem of optimal control of quantum dynamics.

Talukder et al. [40] extended the method just described to the technically more difficult problem of selectively cleaving one O-O bond or the other in $O^{16}O^{16}O^{18}$ molecules. The annealing optimized IR and UV control fields produced good selectivity and the product channel was dominated by either the $O^{16} + O^{16}\text{-}O^{18}$ or $O^{18} + O^{16}\text{-}O^{16}$ dissociation process. The integrated fluxes along the targeted dissociation channel with optimal control fields was found to be as high as 70% or more.

It appears that a genetic learning algorithm or simulated annealing can be used to control optimal (UV+IR) pulse shapes for quantum control of photo induced production of atoms and molecules in a dissociation event. However, an unresolved issue concerns how to use the solutions discovered by GA or SA or for that matter, by any other soft computing route, to learn about the quantum dynamics of the system. The information hidden in a 30- or 50-parameter pulse may make little physical sense unless the dominant degree of freedom in the control Hamiltonian can be unearthed. White et al. [41] proposed a simple method based on covariance analysis of the control space to achieve this objective and successfully applied the technique to study the stimulated Raman scattering in liquid methyl alcohol. A similar analysis would have provided very useful information about the most important degrees of freedom in the control fields designated by the GA or SA responsible for achieving quantum control of the O-H/O-D or $O^{16}\text{-}O^{16}/O^{16}\text{-}O^{18}$ dissociation channels.

6.6.6 Designing Control Fields by SAM for Tunneling Suppression

The SDWP appears everywhere in modeling physical and chemical events. A symmetric DWP with a very large barrier (infinite) separating the two wells reveals exactly degenerate eigenstates appearing in pairs, one component of the degenerate pair being localized in the left well (ψ_L) while the other is confined to the right well. When the barrier is finite, tunneling takes place across the barrier and eigenstates of the SDW Hamiltonian appear as linear combination of the localized infinite barrier eigenstates, which are also eigenstates of the parity operator. The lowest pair of opposite parity eigenstates designated as 0_+ and 0_-, are nondegenerate and the small energy splitting $\Delta\in = |\in_{0+} - \in_{0-}|$ is called tunneling splitting. In general, lower the barrier height, the lower the magnitude of tunnel splitting. The splitting becomes smaller as higher eigenstates like $1_\pm, 2_\pm$ are considered. Tunneling is a manifestation of the ability of a wave function to delocalize across an energy barrier even if the energy (E) associated with it is less than the barrier energy. It gives rise to many interesting quantum phenomena. The question then arises if this natural quantum process can be controlled, especially suppressed. A signature of the extent of control achieved can be found in the extent of reduction in the tunnel splitting of the $0_\pm$ eigenstates by manipulating the quantum dynamics of the system with the help of an externally imposed control field (the so-called control Hamiltonian). In the case of complete control, the splitting would reduce to zero and the delocalized parity eigenstates would be converted into symmetry broken wave functions localized completely in one well or the other.

Ghosh et al. [42] reported in 2013 how a SAM could be exploited to design a polychromatic low-intensity, low-frequency dipolar field that could drive the system externally and completely halt the tunneling. The full Hamiltonian of the system reads

$$H(x,t) = H_0(x) + \sum_{i=1}^{n} x \in_i \cos(\omega_i t)$$

$$= H_0(x) + H_c\left[\{\in_i\}\{\omega_i\}\right]$$

(6.148)

where $H_0(x)$ is the unperturbed SDW Hamiltonian. $H(x,t)$ is transformed into the time-independent Floquet Hamiltonian [43] $H_F(x,t)$ using system-field product eigenstates as a basis. Diagonalization of H_F produces eigenstates and eigenvalues which are called quasi energies, labeled as $\in_1, \in_2, \dots$. The objective function f_o used by the authors [42] is the absolute difference $\Delta\in = |\in_1 - \in_2|$ between the two lowest quasi-energy eigenvalues and the target of the optimization by SAM was to design H_c, i.e., to find the sets of $\{\in_i\}$ and $\{\omega_i\}$ values that would minimize $\Delta\in$ globally. Of course, the global minimum is zero and achieving global minimum signals that tunneling has halted completely. Thus, one starts from an initial choice of control field parameters $\{\in_i\}$, $\{\omega_i\}$, chooses and reconfigures one of the parameters randomly, constructs the Floquet Hamiltonian matrix, diagonalizes it fully and estimates $\Delta\in_{new} = \left|\in_1^{new} - \in_2^{new}\right|$. If $\Delta\in_{new} \leq \Delta\in_{old}$, one checks if such a reconfiguration is thermally accessible at the prevailing temperature (T) at which samplings are made. If it is allowed (i.e., if it passes the Metropolis test), the new field configuration is accepted even though $\Delta\in_{new} > \Delta\in_{old}$. As the samplings are carried out the temperature is slowly reduced at fixed intervals. Slow enough cooling ensures that $\Delta\in$ is globally minimized as $T \to 0$. The corresponding control field configuration is the optimal one. The authors [42] carried out a quantum dynamical calculation with the optimized control field and showed that tunneling has indeed completely ceased and a state completely localized in one well has been created.

In a sequel to the paper just discussed Ghosh et al. [44] investigated if the tunneling isomerization of malonaldehyde in the ground state could be quenched by allowing the molecule to interact with an optimal control field. The isomerization process in malonaldehyde in described by a model Hamiltonian $H(x,y)$ in two dimensions of which x described the in-plane cis and transisomerization coordinate. The motion along x takes place in a symmetric double-well potential $V(x)$ while that along y takes place in an asymmetric double-well potential. The modes are coupled by an interaction potential $V(x,y) = \lambda x^2 y^2$; $(\lambda > 0)$. Polychromatic dipolar time-dependent fields are allowed to interact with the molecule and the Floquet method described already is used to construct the objective function $\Delta\epsilon = |\epsilon_1 - \epsilon_2|$ for a particular field configuration (ϵ_1, ϵ_2 are the lowest quasi-energy levels at the current control field configuration) and the method of simulated annealing is invoked to globally minimize the objective function. The authors [44] claim that the tunnel splitting could be virtually reduced to zero at the optimal control field designed by SAM which means cis–cis isomerization by tunneling could be completely frozen.

The application demonstrates that optimal control of tunneling is indeed possible by adopting soft computing techniques to optimize the control fields through global minimization of appropriate objective function. Although the authors make use of the method of simulated annealing, RMHC (simple as well as adaptive) and genetic algorithms could provide equally viable (if not better) techniques of addressing the same problem. We anticipate many more investigations of a similar nature would be forthcoming and establish the power of soft computing in the area of control of quantum dynamics.

One lacuna of the work discussed in the preceding paragraphs concerns the physical understanding of the control process that a complex polychromatic field activates. Stated simply, having achieved control of tunneling through an optimally designed control field, it is equally important to decipher the modus operandi of the control action—that is, what is the physical mechanism that succeeds in freezing the tunneling process? Once again we may refer the reader to the work of White et al. [41] cited earlier. A similar analysis could provide a better understanding of the dynamics of control action.

References

1. Epstein, S.T., A.C. Hurley, R.E. Wyatt, and R.G. Parr. 1967. Integrated and Integral Hellmann–Feynman Formulas. *The Journal of Chemical Physics* 47, no. 4: 1275–1286.
2. Hylleraas, E.A., and B. Undheim. 1930. Numerische Berechnung Der 2S-Terme von Ortho- Und Par-Helium. Zeitschrift für Physik 65, no. 11–12: 759–772.
3. MacDonald, J.K.L. 1933. Successive Approximations by the Rayleigh-Ritz Variation Method. *Physical Review* 43, no. 10: 830–833.
4. Wilkinson, J.H., and J.H. Wilkinson. 1965. *The Algebraic Eigenvalue Problem*. Vol. 87. Oxford: Clarendon Press.
5. Parlett, B.N., 1998. The Symmetric Eigenvalue Problem. *Journal of Applied Mechanics*. Philadelphia, PA: Society for Industrial and Applied Mathematics, 48.
6. Davidson, E.R. 1975. The Iterative Calculation of a Few of the Lowest Eigenvalues and Corresponding Eigenvectors of Large Real-Symmetric Matrices. *Journal of Computational Physics* 17, no. 1: 87–94.

7. Cullum, J.K., and R.A. Willoughby. 1985. *Lanczos Algorithms for Large Symmetric Eigenvalue Computations*. Vol. 1 Theory. Philadelphia PA: Society for Industrial and Applied Mathematics.

8. Nandy, S., P. Chaudhury, and S.P. Bhattacharyya. 2002. Stochastic Diagonalization of Hamiltonian: A Genetic Algorithm-Based Approach. *International Journal of Quantum Chemistry* 90, no. 1: 188–194.

9. Löwdin, P. 1951. A Note on the Quantum-Mechanical Perturbation Theory. *The Journal of Chemical Physics* 19, no. 11: 1396–1401.

10. Löwdin, P.-O. 1968. Studies in Perturbation Theory XIII. Treatment of Constants of Motion in Resolvent Method, Partitioning Technique, and Perturbation Theory. *International Journal of Quantum Chemistry* 2, no. 6: 867–931.

11. Killingbeck, J.P., and G. Jolicard. 2003. The Bloch Wave Operator: Generalizations and Applications: Part I. The Time-Independent Case. *Journal of Physics A: Mathematical and General* 36, no. 20: R105–R180.

12. Sharma, R., S. Nandy, and S.P. Bhattacharyya. 2008. On Solving Energy-Dependent Partitioned Real Symmetric Matrix Eigenvalue Problem by a Parallel Genetic Algorithm. *Journal of Theoretical and Computational Chemistry* 7, no. 6: 1103–1120.

13. Sharma, R., S. Nandy, P. Chaudhury, and S.P. Bhattacharyya. 2011. A Density-Genetic Algorithm Method for Computing Electronic Structures of Doped and Undoped Polythiophene Oligomers: A Modified Su–Schrieffer–Heeger Hamiltonian-Based Study. *Materials and Manufacturing Processes* 26, no. 3: 354–362.

14. Nandy, S., R. Sharma, and S.P. Bhattacharyya. 2011. Solving Symmetric Eigenvalue Problem via Genetic Algorithms: Serial versus Parallel Implementation. *Applied Soft Computing* 11, no. 5: 3946–3961.

15. Coope, J.A., and D. Sabo. 1977. A New Approach to the Determination of Several Eigenvectors of a Large Hermitian Matrix. *Journal of Computational Physics* 23, no. 4: 404–424.

16. Sharma, R., and S.P. Bhattacharyya. 2007. Direct Search for Wave Operator by a Genetic Algorithm (GA): Route to Few Eigenvalues of a Hamiltonian. In *IEEE Congress on Evolutionary Computation (CEC)*, September 25–28, pp. 3812–3817. Singapore: IEEE.

17. Durand, P., J. Savrda, and I. Paidarova. 1994. Convergence Studies in Quantum Perturbation Theory. *Theoretica Chimica Acta* 88, no. 3: 243–256.

18. Cantú-Paz, E., and D.E. Goldberg. 1999. On the Scalability of Parallel Genetic Algorithms. *Evolutionary Computation* 7, no. 4: 429–449.

19. Cantú-Paz, E. 2000. *Efficient and Accurate Parallel Genetic Algorithms*. Genetic Algorithms and Evolutionary Computation. New York: Springer.

20. Saha, R., P. Chaudhury, and S.P. Bhattacharyya. 2001. Direct Solution of Schrödinger Equation by Genetic Algorithm: Test Cases. *Physics Letters A* 291, no. 6: 397–406.

21. Marston, C.C., and G.G. Balint-Kurti. 1989. The Fourier Grid Hamiltonian Method for Bound State Eigenvalues and Eigenfunctions. *The Journal of Chemical Physics* 91, no. 6: 3571–3576.

22. Chaudhury, P., and S.P. Bhattacharyya. 1998. Numerical Solutions of the Schrödinger Equation Directly or Perturbatively by a Genetic Algorithm: Test Cases. *Chemical Physics Letters* 296, no. 1–2: 51–60.

23. Nakanishi, H., and M. Sugawara. 2000. Numerical Solution of the Schrödinger Equation by a Microgenetic Algorithm. *Chemical Physics Letters* 327, no. 5–6: 429–438.

24. Saha, R., and S.P. Bhattacharyya. 2007. On Directly Solving Schrödinger Equation for ${\rm H}^+_2$ Ion by Genetic Algorithm. *International Journal of Modern Physics C* 18, no. 2: 163–175.

25. Saha, R., and S. Bhattacharyya. 2004. On Solving Schrodinger Equation for the Ground State of a Two-Electron Atom Using Genetic Algorithm. *Current Science* 86: 960–962.

26. Roothaan, C.C.J., and A.W. Weiss. 1960. Correlated Orbitals for the Ground State of Heliumlike Systems. *Reviews of Modern Physics* 32, no. 2: 194–205.

27. Şahin, M., Ü. Atav, and M. Tomak. 2006. Applications of Genetic Algorithm to Quantum Mechanical Systems. *Turkish Journal of Physics* 30, no. 4: 253–275.

28. Şahin, M., and M. Tomak. 2005. The Self-Consistent Calculation of a Spherical Quantum Dot: A Quantum Genetic Algorithm Study. *Physica E: Low-Dimensional Systems and Nanostructures* 28, no. 3: 247–256.

29. Herzberg, G., and J.W.T. Spinks. 1950. *Molecular Spectra and Molecular Structure: Diatomic Molecules.* Prentice-Hall Physics Series, Condon, U. (ed.). Princeton NJ: Van Nostrand.

30. Richardson, O.W. 1935. On the Ground State of (HFormula), the Molecular Ion (HFormula +) and Wave Mechanics. *Proceedings of the Royal Society A: Mathematical, Physical and Engineering Sciences* 152, no. 877: 503–514.

31. Makarov, D.E., and H. Metiu. 1998. Fitting Potential-Energy Surfaces: A Search in the Function Space by Directed Genetic Programming. *The Journal of Chemical Physics* 108, no. 2: 590–598.

32. Makarov, D.E., and H. Metiu. 2000. Using Genetic Programming To Solve the Schr{ö}dinger Equation. *The Journal of Physical Chemistry A* 104, no. 37: 8540–8545.

33. Chaudhury, P., and S.P. Bhattacharyya. 1999. Bound States in Screened and Bare Coulomb Potentials: A Nonorthogonal CI-Based Route to Effective Hamiltonians for Two-Electron Systems. *International Journal of Quantum Chemistry* 74, no. 2: 153–161.

34. Turing, A.M. 1952. The Chemical Basis of Morphogenesis. *Philosophical Transactions of the Royal Society B: Biological Sciences* 237, no. 641: 37–72.

35. Riaz, S.S., R. Sharma, S.P. Bhattacharyya, and D.S. Ray. 2007. Instability and Pattern Formation in Reaction-Diffusion Systems: A Higher Order Analysis. *The Journal of Chemical Physics* 127, no. 6: 64503.

36. Barrio, R. A. and C. Varea, J.L. Aragón, and P.K. Maini. 1999. A Two-Dimensional Numerical Study of Spatial Pattern Formation in Interacting Turing Systems. *Bulletin of Mathematical Biology* 61, no. 3: 483–505.

37. Ghosh, M., R. Sharma, and S.P. Bhattacharyya. 2007. Target Excitation in 2-D Quantum Dots by Optimized Chirped Pulses. *Chemical Physics Letters* 449, no. 1–3: 165–170.

38. Warren, W.S., H. Rabitz, and M. Dahleh. 1993. Coherent Control of Quantum Dynamics: The Dream Is Alive. *Science* 259, no. 5101: 1581–1589.

39. Shandilya, B.K., S. Sen, T. Sahoo, S. Talukder, P. Chaudhury, and S. Adhikari. 2013. Selective Bond Breaking Mediated by State Specific Vibrational Excitation in Model HOD Molecule through Optimized Femtosecond IR Pulse: A Simulated Annealing Based Approach. *The Journal of Chemical Physics* 139, no. 3: 34310.

40. Talukder, S., S. Sen, B.K. Shandilya, R. Sharma, P. Chaudhury, and S. Adhikari. 2015. Enhancing the Branching Ratios in the Dissociation Channels for O 16 O 16 O 18 Molecule by Designing Optimum Laser Pulses: A Study Using Stochastic Optimization. *The Journal of Chemical Physics* 143, no. 14: 144109.

41. White, J.L., B.J. Pearson, and P.H. Bucksbaum. 2004. Extracting Quantum Dynamics from Genetic Learning Algorithms through Principal Control Analysis. *Journal of Physics B: Atomic, Molecular and Optical Physics* 37, no. 24: L399–L405.

42. Ghosh, S., S. Talukder, S. Sen, and P. Chaudhury. 2013. Coherent Destruction of Tunneling with Optimally Designed Polychromatic External Field. *Chemical Physics* 425: 73–79.

43. Floquet, G. 1883. Sur Les {é}quations Diff{é}rentielles Lin{é}aires {à} Coefficients P{é}riodiques. *Annales Scientifiques de l'{É}cole Normale Sup{é}rieure* 12: 47–88.

44. Ghosh, S., S. Talukder, S. Sen, and P. Chaudhury. 2015. Optimised Polychromatic Field-Mediated Suppression of H-Atom Tunnelling in a Coupled Symmetric Double Well: Two-Dimensional Malonaldehyde Model. *Molecular Physics* 113, no. 23: 3826–3838.

7

Soft Computing in Chemistry

7.1 Introduction

Our focus in the previous chapter has been primarily on exploring various soft computing options for solving the time-independent Schrödinger equation (SE) for the ground and a few excited (bound) states of simple systems. Both basis-set and basis-free approaches have been investigated within the framework of evolutionary computing methods such as genetic algorithms (GA), natural methods like the Metropolis simulated annealing method (SAM), and the genetic programming technique. In the current chapter our emphasis will be on larger systems—molecules and clusters, i.e., systems that evoke strong chemical interest. The cornerstone of molecular quantum mechanics has been the Born–Oppenheimer (BO) approximation [1] which exploits the two widely different timescales of motion in molecules, namely timescales of electronic motion and those of nuclear motion. The electronic movement is fast (electronic degrees of freedom are treated as fast variables). The nuclei are much heavier and motion is sluggish allowing the nuclear degrees of freedom to be treated as slow variables. The slow movements of nuclei are therefore immediately sensed by the molecular electrons which respond almost instantaneously by readjusting their position or distribution in the molecule. BO approximation starts by freezing the nuclear motion to start with (the nuclear kinetic energy operator $\hat{T}_N = 0$) and then solving the electronic problem in the fixed or frozen nuclear framework. That means one looks for the eigenstates of the molecular electronic Hamiltonian $H_e(\underline{R})$ with $H_e(r,\underline{R}) = \hat{T}_e(r) + \hat{V}_{ee}(r) + V_{eN}(r,\underline{R}) + V_{NN}(\underline{R})$. Here $\underline{R}$ stands for the collection of frozen nuclear coordinates (they are treated as parameters), and r is the collection of coordinates of all the electrons in the molecule. $\hat{V}_{ee}(r)$ represents the electron–electron interaction potential operator, $V_{eN}(r,\underline{R})$ stands for the operator for electron–nucleus interaction while $V_{NN}(\underline{R})$ is the internuclear potential energy operator which for frozen nuclear coordinates ($\underline{R}$) becomes just an additive constant in the molecular electronic Hamiltonian $H(r,\underline{R})$. The molecular energy eigenvalue equation under the BO approximation reads

$$H(r,\underline{R})\psi_n^e(r,\underline{R}) = E_n^e(\underline{R})\psi_n^e(r,\underline{R}). \tag{7.1}$$

The molecular electronic energy $E_n^e(\underline{R})$ (including the internuclear repulsion energy $V_{NN}(\underline{R})$ for different frozen nuclear coordinates $\{\underline{R}_i\}$) provides the multidimensional potential energy surface (the adiabatic surfaces in the nth electronic states) $U_n(\underline{R})$ on which the nuclei move. The global minimum of $E_n(\underline{R})$ defines the equilibrium configuration of the nuclei in the nth eigenstate of the molecular electronic Hamiltonian—the so-called nth adiabatic electronic state. If we take the lowest value of n ($n = 0$), we have the ground adiabatic potential energy surface $U_n(R)$ and the nuclear motion on this surface is dictated by the nuclear SE:

$$\left\{ \hat{T}_N + U_0(R) \right\} \chi_v^0(R) = \in_v^0 \chi_v^0(R) \tag{7.2}$$

where $\in_v^0$ is the total energy in the vth eigenstate of the nuclear Hamiltonian for the ground electronic state ($n = 0$). The total molecular wave function $\psi_{vu}(r, R)$ is represented as a product of the adiabatic molecular wave function $\psi_n^c(r, \underline{R})$ and the corresponding nuclear wave function $\chi_v^n(R)$. Thus, the molecular energy eigenvalue problem is solved in two steps in tandem: first, freeze the nuclei and solve the molecular electronic SE for the energy eigenvalue $E_n(\underline{R_i})$ for many nuclear configurations generating the adiabatic PES ($U_n(R)$) for the nth electronic state; next, solve the nuclear SE on the nth adiabatic potential energy surface for total energy in the nth electronic state and the nuclear wave function $\chi_v^n(R)$. Our primary focus is now on the PES of an N-atom molecule. Let us suppose that such a surface has been constructed and analytically represented as a function of $3N$ Cartesian coordinates $U_n(R_1, R_2, ..., R_N)$ with $R_i \equiv (x_i, y_i, z_i)$, $i = 1, 2, ..., N$. Considering all possible types of nuclear displacements on the surface $U_n(R)$, we can see that there are 3 degrees of freedom associated with the translational motion of the center of mass of the molecule (supposedly non-linear) and 3 degrees of freedom associated with rotation about the center of mass, that do not experience any restoring force. The 3N-6 remaining degrees of freedom are associated with relative displacements of nuclei with respect to each other bringing restoring forces into play. These modes represent how the nuclei oscillate about their equilibrium positions on the nth adiabatic surface.

The task of constructing the multidimension PES for molecules and clusters of many atoms is a formidable one, requiring us to solve the molecular electronic Schrödinger equation (Equation 7.1) for many configurations of the nuclei or the nuclear framework. We have already seen (in Chapter 6) how soft computing can be exploited to handle the problem of computing energy eigenvalues for small systems. Once the eigenvalues are available for a wide range of configurations $\{R_i\}$, the problem of fitting the data to an appropriate analytical form comes up. The fitting (done by a deterministic or nondeterministic procedure) produces an analytical representation of the adiabatic PES ($U_n(R)$).

Now a multidimensional PES usually supports multiple minima, the deepest one defining the global minimum energy molecular structure on the adiabatic surface. The other minimum energy structures represent higher energy configurations of the molecule. The different minima are separated by barriers having points that are maxima or saddle points of different orders. A full analysis of a PES in any electronic state (and notably in the ground state) involves identification of all the stationary points of the PES. These are points $\{R_0\}$ at which

$$\left. \frac{\partial U_n(R)}{\partial x_i} \right|_{R_0} = 0, \quad \left. \frac{\partial U_n(R)}{\partial y_i} \right|_{R_0} = 0, \quad \left. \frac{\partial U_n(R)}{\partial z_i} \right|_{R_0} = 0 \tag{7.3}$$

for $i = 1, 2, ..., N$; that is, the first derivative of the potential with respect to nuclear displacements along the three orthogonal directions about the point $\{R_0\}$ is zero. In other words, the norm of the 3N-dimensional gradient vector g is zero (with $g_{ki} = \dfrac{\partial U_k}{\partial x_i}$ etc.). To characterize the stationary point completely, we need to compute the second derivatives of $U_n(R)$ with respect to nuclear displacements about the point $\{R_0\}$. The second derivative, i.e., $\left.\dfrac{\partial^2 U_n(R)}{\partial R_i\,\partial R_j}\right|_{R_0}$ $\left(R_i \equiv x_i, y_i, z_i;\ R_j \equiv x_j, y_j, z_j;\ i = 1, 2, \ldots, N;\ j = 1, 2, \ldots, N\right)$ can be collected in a $3N \times 3N$ matrix H, which we call the Hessian matrix. If all the eigenvalues of H are ≥ 0, i.e., the H matrix is positive semidefinite, the point R_0 is characterized as a true minimum. If only one of the eigenvalues is <0, while all others are ≥ 0 the configuration R_0 of the molecule is a saddle point of first order on the PES (if two eigenvalues are ≤ 0, and rest are ≥ 0, R_0 this defines a saddle point of order 2, and so on for still higher-order saddles). If all the eigenvalues are ≤ 0, the configuration defines a true maximum on the PES. We note here that "all eigenvalues" we have mentioned excludes the eigenvalues associated with the three translational modes of the center of mass of the system along with its three rotational modes (those eigenvalues, as we have already noted, are of zero frequency as no restoring forces are operative). The analysis of the many-dimensional PES in chemistry focuses on ascertaining whether a stationary point located is a stable and true minimum, or whether it can be characterized as a first-order saddle point. The minima located may be local minima and the deepest of them is the global minimum, the associated structure defining the true equilibrium molecular structure on the PES at 0K. If the local minima lie close to the global one and we are interested in the thermally averaged value of some property of the system, the local as well as the global minima must be taken into account. The first-order saddle points are important for identifying the structure associated with the transition state of a chemical process in which the nuclei move from one minimum (R_1 = reactant) to another (P_1 = product) through a first-order saddle point (S^1) the curve joining $R_1 \rightarrow S^1 \rightarrow P_1$ defining what may be called the reaction path (RP). Our focus in the following section will be on a soft computing-based search for locating the local as well as the global minima on a given PES, identifying first-order saddle points and tracing out the so-called minimum energy RPs where they exist. The potential energy surfaces, unless stated explicitly, refer to the ground electronic state only.

7.2 Soft Computing in Search for Global and Local Minima

We will focus on the several soft computing alternatives for the search, the basic template of which is best illustrated with the Metropolis SAM already introduced in Chapter 4.

7.2.1 Simulated Annealing-Based Search

Suppose that the electronic problem has been solved and the potential energy surface $U(q_1, q_2, \ldots, q_{3N})$ has been constructed for the ground electronic state of the molecule under consideration. Our objective is to find out the global minimum energy nuclear configuration $\left(q_1^0, q_2^0, \ldots, q_{3N}^0\right) \equiv R^0$ and the associated molecular electronic energy E^0. The SAM requires us to define an objective function for the problem, the global minimum of which is the target of this search: one such objective function could be found in reference [2]:

$$F_{obj}\left(\{q_i\}, U_L\right) = \left[U\left(q_1, q_2, \ldots, q_{3N}\right) - U_L\right]^2 \tag{7.4}$$

where U_L is an estimated lower bound to the ground state energy E_G of the system. Let the current configuration of the nucleii be $R^i \equiv \left(q_1^0, q_2^0, \ldots, q_{3N}^0 \right)$ and the corresponding energy be $U(R_i) = E^i$. Let the current temperature of the simulated thermodynamic system be T_s. Let one of the nuclear coordinates, say q_k^i, be picked up randomly and a small random change be made in it producing a new value of the coordinate q_k^{i+1} where

$$q_k^{i+1} = q_k^i + (1 - 2r)\Delta_s r_1 \tag{7.5}$$

where r is a random number in $(0,1)$, r_1 is another random number in the same range drawn from a Gaussian distribution. Δ_s is a user-defined step length. Let the energy of the system in the new nuclear configuration $R^{i+1} \equiv \left(q_1^i, q_2^i, \ldots, q_k^{i+1}, \ldots, q_{3N}^i \right)$ be E^{i+1}. If $\Delta F = F^{i+1} - F^i$ is ≤ 0, $[F^{i+1} = (U^{i+1} - U_L)^2, F^i = (U^i - U_L)^2]$ the new configuration is accepted, i.e., R^{i+1} replaces R^i ($R^i \leftarrow R^{i+1}$), and the probability of the reconfiguring move being accepted is $p_i = 1$. If $F^{i+1} > F^i$, the probability of such a reconfiguring move taking place in the system at the prevailing temperature T_i is $p_i = e^{-\Delta F / k_B T_S}$ [3,4]. Let r be a random number in the range $(0,1)$. If $r < p_i$, the uphill reconfiguring move is accepted as a thermally feasible move and R^{i+1} replaces as before. If $r > p_i$, the move is rejected and the current configuration R^i is retained for further sampling. These reconfiguring moves are carried out M times ($M \approx 100 \times 3N$) so that thermal equilibrium sets in at the current temperature of the simulated thermodynamic system and the temperature is lowered by a predetermined amount—say $T_{S+1} = \chi T_S$, χ being a fraction (say 0.9). The whole process of M samplings is carried out on the last accepted configuration in the previous temperature step (T_S), following which the temperature is lowered further and the process is repeated until the temperature T becomes low enough ($T \to 0$) or no further lowering of the objective function takes place. A good way to monitor the progress of the search is to compute the average value of the objective function F_{av} at every temperature step T_S where

$$F_{av}(T_S) = \frac{\sum_i p_i \left(U^i - U_L \right)^2}{\sum_i p_i} \tag{7.6}$$

p_i being the probability of realizing the ith accepted configuration at the temperature T_S. If T_S is low and F_{av} is already close to zero (i.e., $E^i \approx E_L$) it is time to lower the value of E_L by a preassigned amount say ΔU_L, reset the temperature to a higher value and carry out the search as described already. If lowering U_L does not lead to the discovery of a lower minimum energy than what was found during the first cycle it is probable that the search has already hit the global minimum on the PES and the nuclear configuration reached is the global minimum energy structure the molecule is expected to have on the given adiabatic PES. The following points may be noted by readers:

1. If at some stage of the search, F_{av} remains fixed or shows only small oscillations over a number of temperature steps, it is always advisable to reset the temperature to a higher value and cool the system as prescribed. It helps the system to escape from being trapped in a local minimum. The heating and cooling schedule must be preprogrammed so that no external intervention is required.

2. If we find that T_S is already low, F_{av} is not decreasing and $(U^i - U_L)$ is not sufficiently small it would be prudent to reset U_L to a higher value in the next cycle of

search ($U_L^{new} = U_L^{old} + \lambda\sqrt{U^i - U_L}$ is a possible choice where U^i is the energy at the end of the current cycle).

3. The step length Δ_S could be reduced as the system is cooled ($\Delta_S(T+1) = \Delta_S(T) \times f$, ($0 < f < 1$); being a possible choice).

Dutta et al. [5] proposed and implemented the simulated annealing-based molecular geometry optimization scheme described here that was successful in locating both local and global minima on the PES, which was available either in an analytical form or generated on the fly by a semiempirical method (MNDO). Some of the points made by the authors of reference [5] are worth noting:

1. If the equilibrium structure of a molecule has several geometrical parameters related by a symmetry operation of the molecular point group, a SAM-based search does not destroy the symmetry when it hits the minimum, although in the intermediate steps of the search many symmetry broken structures are generated. In other words, symmetry constraints need not be imposed to guide the search.

2. The structure discovered is always the same irrespective of the initial structure chosen provided cooling has been slow and a low enough value of U_L has been chosen.

3. By choosing a suitable U_L, it is possible to nudge the search to a stationary point with energy closest to U_L.

Once the search has converged to a particular structure, i.e., a particular point on the PES, it is necessary to check if the absolute value of all the components of the gradient vector (call them g_k, $k = 1, 2, \ldots, 3N$) have been reduced below a threshold ($\in$). If the condition is satisfied signaling that the point located on the PES is a stationary point, the eigenvalues of the Hessian matrix at that point are to be computed to ascertain if the stationary point reached is indeed a true minimum on the PES.

7.2.2 Alternative Choices of the Objective Function for Minima Search

Instead of locating the minimum of the objective function of Equation 7.4 and then checking if the random walk has indeed terminated at a stationary point, it is possible to incorporate the local gradient information in the objective function in such a way that the minimization of the modified objective function ensures that a stationary point is reached. One choice is to modify the objective function of Equation 7.4 by adding the gradient norm to the energy term. That is, the new objective of the search is now set to seek the global minimum of a modified objective function $F_1(q, U_L, \lambda)$

$$F_1(q, U_L, \lambda) = F_{obj}(q, U_L) + \lambda \sum_{i=1}^{N} \left[\left(\frac{\partial V}{\partial x_i} \right)^2 + \left(\frac{\partial V}{\partial y_i} \right)^2 + \left(\frac{\partial V}{\partial z_i} \right)^2 \right]$$

$$= \left[U(q) - U_L \right]^2 + \lambda \sum_{k=1}^{3N} \left(\frac{\partial V}{\partial q_k} \right)^2$$

(7.7)

In Equation 7.7, λ is a penalty weight factor (>0), U_L is the estimated lower bound to the global minimum being sought. Noting that $\dfrac{\partial V}{\partial q_k}$ is the kth component (g_k) of the gradient vector g, we can rewrite the objective function as

$$F_1\left(q^i, U_L^i, \lambda^i, \beta\right) = \beta\left[U\left(q^i\right) - U_L^i\right]^2 + \lambda^i Tr\left(g^i g^{i\dagger}\right). \tag{7.8}$$

So, the global minimization of F_1 would not only ensure that $U\left(q^i\right) \to U_L^i$ as closely as possible, but also guarantee that the norm of the gradient vector also vanishes when such a nuclear configuration $\{q^0\}$ has been identified by the search. ($\{q^i\}$ denotes the nuclear configuration (i.e., molecular geometry) at the ith stage of the search). It is generally advisable to begin with $\beta = 1$ and use a small value of λ^i, and gradually increase its value as $U\left(q^i\right) \to U_L^i$, although a constant moderate value of λ works quite well in many applications.

It is possible to base the search on minimizing the gradient norm alone—i.e., we may define a new objective function $F_2(q)$ where

$$F_2\left(q\right) = Tr(gg^{\dagger}). \tag{7.9}$$

Clearly, this is just the previously defined objective function $F_1\left(q^i, U_L^i, \lambda^i, \beta\right)$ with $\beta = 1$, and $\lambda^i = 1$. The minimization of $F_2(q)$ will locate a stationary point on the PES $U(q)$ without any guarantee that it would be the global minimum or even a local minimum on the PES. The search can terminate even at a saddle point. That is why it is safe and convenient to base the search for the global as well as local minima on the objective function $F_1\left(q^i, U_L^i, \lambda^i, \beta\right)$. Choudhury et al. [6] experimented successfully with the search for local as well as global minima on a PES based on minimization of the objective function F_1 by the method of Metropolis simulated annealing with $\beta = 1$ and a small λ to start with. As $U(q) \to U_L$, λ can be increased to 1 gradually. The initial stage of the search is dominated by energy minimization while the final stage is dictated by the requirement of reducing the gradient norm to zero. It turns out that the gradient need not be calculated after every reconfiguring move, as it is unlikely that a large change would take place in each step. It may be evaluated at intervals, depending on the progress of the search.

7.2.3 Objective Function for Locating First-Order Saddle Points on the PES

We have already mentioned that a first-order saddle point on the PES has one negative eigenvalue of the Hessian matrix, excluding those associated with the translation and rotation of the center of mass of the molecule. What modifications in the objective function F_1 could ensure that the annealing-based search homes in on to a saddle point of order one (S_1)? The answer appears to be that we must include information about the local curvature in the objective function in a way that the global minimization of the curvature-constrained objective function so defined ensures that the stationary point reached is indeed a saddle point of order one and also energetically the lowest one in the vicinity. Choudhury and Bhattacharyya [7] suggested the following form of the objective function F_S for the search for a first-order saddle point on the PES $U(q_1, q_2, \ldots, q_{3N})$:

$$F_S\left(q, U_L^S, \alpha, \beta, \gamma\right) = \alpha\left[U\left(q\right) - U_L^S\right]^2 + \beta\, Tr\left(gg^{\dagger}\right) + \sum_{i=1}^{n} \gamma_i e^{p_i \lambda_i} \tag{7.10}$$

where α, β, and γ_i s are penalty weight factors. These authors uniformly set $\alpha = 1$, $\beta > 0$. γ_is and p_is are set >0 or <0 depending on what the sign of the ith eigenvalue of the Hessian matrix is. Thus, if the kth eigenvalue $(\lambda_k) < 0$, they make the choice $p_k < 0$, and $\gamma_k < 0$. On the other hand, if $\lambda_k > 0$, they set $p_k > 0$, and $\gamma_k > 0$. It essentially means that a negative eigenvalue of the Hessian will bring in a sharp reduction in the value of the objective function F_S for the configuration that has generated one negative eigenvalue (exclusive of the six eigenvalues for the translation and rotation of the center of mass). Since the search is for a first-order saddle point, any reconfiguring move that generates more than one negative eigenvalue of the Hessians is rejected as minimization of F_S along that direction will possibly end up in a second- or higher-order saddle point on the given PES. The suggested SAM-based recipe involves the following steps:

1. Start from a local or global minimum already discovered on the PES, the coordinates of the minimum energy configuration being $\left(q_1^0, q_2^0, ..., q_n^0\right)$ and the energy being U^0.

2. Set a lower bound for the saddle point search with $U_L^S = U^0 + \sigma$ (σ is user specified; set the initial temperature T^0 at a high value).

3. Reconfigure the system by randomly perturbing the coordinates $\left(q_1^0, q_2^0, ..., q_n^0\right)$ defining the minimum from where the search begins: That is set for $i = 1, 2, ..., n$, $q_i = q_i^0 + \delta \Delta q_i$ ($\delta > 0$, Δq_i is randomly chosen from the range $(0,1)$).

4. Minimize the objective function $F_S\left(q, U_L^S, \alpha, \beta, \gamma\right)$ by invoking SAM with $\lambda = 1$, $\beta > 0$, $\gamma_k > 0$ for all values of k. β may range between 0 and 1, while γ_k s have small positive values to start with. If the search discovers a configuration in which the Hessian has one and only one negative eigenvalue (say the kth one); γ_k is reset to have a negative value ($\gamma_k = -1$, say). All configurations with more than one negative eigenvalue of the Hessian are rejected.

5. If $\left|U(q) - U_L^S\right|$ has become small, but the norm of the gradient vector is still not small enough while one eigenvalue of the Hessian is still negative, it is an indication that the nearest first-order saddle point must be higher in energy. U_L^S is therefore reset by shifting it up as follows:

$$U_L^S(new) = U_L^S(old) + \sigma. \tag{7.11}$$

The minimization of F_S is continued; if necessary, the temperature T (if it has become low) is increased (T (*new*) = T(*current*) + ΔT).

6. Exit when F_S has become small and is not improving, $Tr(gg^\dagger) < \epsilon_1$ (user defined), $\left|U(q) - U_L^S\right| < \epsilon_2$ and one of the γ_k s < 0. The $q_1, q_2, ..., q_n$ values at this stage define the molecular geometry associated with the saddle point of order one that the search just discovered. If these conditions have not been met but the other termination criterion, namely maximum number of heating–cooling cycles, have been already executed, the search is discontinued.

It is likely that the search has led to a first-order saddle point as close as possible to U_L^S. It could be that a lower energy saddle point of order one exists; it is always advisable to lower the terminal value of U_L^S, rest the temperature to a higher value of and minimize F_S by SAM. If no other saddle point of order one is discovered, one may confidently claim that the saddle point discovered is indeed the lowest energy first-order saddle in the neighborhood of the reference minimum energy configuration from which the search was initiated. Chaudhury and Bhattacharyya [7] applied the technique to locate global and local minima and first-order saddle points on a number of model potential energy surfaces like the Cerjen–Miller potential, Culot–Dive–Nguyen–Ghuysen potential, Halgren–Lipscomb potential, and Hoffman–Nord–Ruedenberg potential surfaces. They analyzed details of the annealing path leading to the first-order saddle points, starting from either the global minimum or a local minimum. Although not tried so far in this context, an alternative objective function for locating a first-order saddle point could be

$$F_S\left(q, U_L^s, \beta, \gamma\right) = \left[U(q) - U_L^s\right]^2 + \beta\, Tr\left(gg^\dagger\right) + \gamma(n_- - 1)^2 \tag{7.12}$$

where is the number of negative eigenvalues of the Hessian matrix.

7.2.4 Construction of RPs (Annealing Based)

Suppose that the annealing-based search has already discovered all the minima on the PES and has also located a first-order saddle point S_1 (it is mandatory). Let L_1 and L_2 be two minima ($E(L_1) < E(L_2)$). What could be the path of transformation from L_1 to L_2? Suppose S_1 is the energetically lowest first-order saddle point (S_{1m}) between L_1 and L_2. Then, we are required to trace the path leading from L_1 to L_2 through S_{1m}. Such a path may or may not exist as the path of transformation from L_1 to L_2 may involve passage through another first-order saddle point, if it exists on the PES. Each point on the path represents particular arrangements of the nuclei and analysis of these arrangements provide a visual picture of the different stages of the transformation process—a chemical reaction—and help us understand microscopic details of the transformation taking place. The entire enterprise may be called annealing-based RP analysis. The term reaction path (RP) is a somewhat vaguely defined concept. On a multidimensional PES, the RP can be viewed as a curving potential energy channel linking two minima on the PES through the first-order saddle point of minimum energy on the same surface (call it S_{1m}). One of the minima is called the reactant (R) configuration while the other is defined as the product configuration (P). The saddle point S_{1m} is a point with the highest energy along the RP and plays a very important role in the transition state theory. The point S_{1m} defines what has become known as the transition state and the associated structure is called the transition structure. We may immediately note the usefulness of the concept of RP in so much as it reduces the exercise of RP analysis to computing one-dimensional curves on a many-dimensional PES. There is, however, a degree of arbitrariness associated with the concept—different mathematical definitions leading to different curves emanating from the stationary (minimum) points (R, P) and passing through S_{1m}. In practice this arbitrariness does not matter much in RP analysis in chemistry as the important parameters of the reaction are the energies $E(R)$, $E(P)$, and $E(S_{1m})$, and the vibration frequencies at the configuration R, P, and S_{1m} which are uniquely defined for the reaction concerned. Nevertheless, we may ask the question: is it possible to construct a meaningful RP within the framework of the SAM-based search for stationary points on the PES? The answer turns out to be in the affirmative.

Suppose that the reactant configuration (R), the product configuration (P), and the lowest energy first-order saddle point (S_{1m}) have been already located. To construct the reaction path $R \rightarrow S_{1m} \rightarrow P$, we start from the reactant minimum on the PES and try to minimize the objective function F_S by SAM after resetting the lower bound U_L ($U_L' = U_L + \Delta U$, where ΔU is a fraction of the energy difference $U(S_{1m}) - U(R)$), and setting high values for penalty weight factors (γ_is) (PWF) in the curvature constraint terms. The initial configuration is generated by randomly perturbing the equilibrium reaction configuration. The high values of the PWFs (γ_is) should ensure that the minimization does not stray away from the direction leading to the first-order saddle point S_{1m}. Let us suppose that the minimization leads to an intermediate configuration $\{q_i\}$ of energy $E\{q_i\}$ with $E\{q_i\}$ as close as possible to U_L'.

A new value of U_L ($U_L'' = U_L' + \Delta U$) is set, the coordinates of the intermediate structure $\{q_i\}$ discovered in the first step are perturbed to generate the initial structure for the search for the next intermediate structure $\{q_i'\}$ lying on the reactions path, and the procedure is repeated until the sequence of intermediate structures turns over into the transition state structure at S_{1m}. This completes construction of half of the reaction path ($R \rightarrow S_{1m}$). The other half of the path, namely $S_{1m} \rightarrow P$ is constructed similarly with the initial structure determined by perturbing the product structure (P) randomly and resetting U_L to $U_L' = U_L + \Delta U$, where ΔU is a fraction of the energy difference ($U_{S_{1m}} - U_R$). Here also the minimization of the objective function is carried out by the method of simulated annealing. Let d_{RI} and d_{PI} be the Euclidean distances of the intermediate structure (I) from the reactant or the product structures held in maximum coincidence. We can assign an RP coordinate (P_I) for the intermediate structure by taking

$$P_I = d_{RI}\left(d_{RI} + d_{PI}\right)^{-1} \tag{7.13}$$

where $0 \le P_I \le 1$ as the intermediate structure evolves from the reactant to the product configuration, laying out energies of different intermediate minimum energy structures as a function of the path coordinate P_I. We can thus trace out a representation of the reaction path—in fact a minimum energy reaction path constructed from stochastic minimization of an appropriately constrained objective function. The path coordinates where the transition structure is located give us some ideas as to the relative proximity of the transition state (TS) to the reactant, the product configurations, or structures. Thus, p or $P_I = \frac{1}{2}$ would signify a TS that is equidistant from both reactant or product configurations. $p < \frac{1}{2}$ indicates an early TS which is reactant-like, while $p > \frac{1}{2}$ implies a product more like TS. Chaudhury et al. [6] demonstrated the workability of the annealing-based algorithm for constructing RPs in an Ar_6 cluster described by the Lennard–Jones potential. The RP for transformation of the cluster from its global minimum energy configuration to one local minimum which is energetically only slightly above the global minimum. The saddle point is located at a path coordinate $p = 0.49$ (Figure 7.1).

The authors claim that their method could be expected to perform better than either the hill climbing or the coordinate driving methods for locating the saddle point even when the PES has a complicated structure, since no explicit choice has to be made in advance about the reaction coordinate in the proposed annealing-based method.

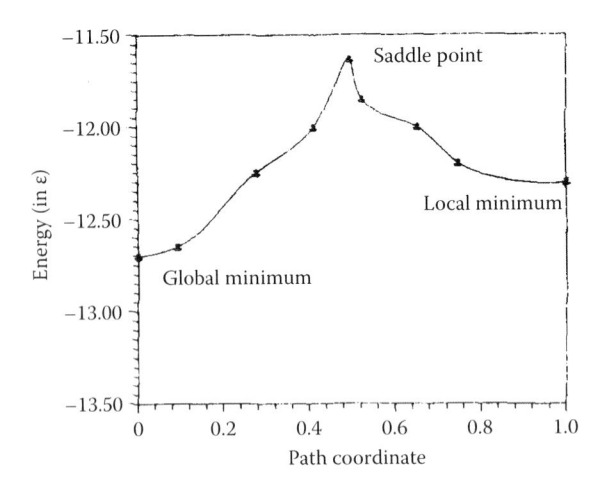

FIGURE 7.1
Energy of various conformations against path coordinates during construction of an RP in an Ar_6 cluster while going from global minimum to local minimum through a saddle point. (From Chaudhury, P., and Bhattacharyya, S. P., J. Mol. Struct. THEOCHEM, 429, 175–186, 1998. Figure 11. With permission.)

7.3 Genetic Algorithms for Locating Critical Points on PES

Chaudhury and Bhattacharyya [8] also experimented with a GA-driven search for critical points. The basic strategy described in the previous section for an annealing-guided search remains the same. The objective function for the PES of an n-atom clusters has three components:

1. Energy cost: $\left[V\left(x_1, y_1, z_1, \ldots, x_n, y_n, z_n\right) - V_L\right]^2$, V_L being an estimated updatable lower bound to the energy at the critical point being searched

2. Gradient cost: $Tr\left(gg^\dagger\right) = \lambda \sum_{i=1}^{n}\left[\left(\frac{\partial V}{\partial x_i}\right)^2 + \left(\frac{\partial V}{\partial y_i}\right)^2 + \left(\frac{\partial V}{\partial z_i}\right)^2\right]$; g is the gradient vector with $3n$ components ($\lambda > 0$)

3. Curvature cost: $\sum_{i=1}^{3n}\eta_i e^{p_i\lambda_i}$, λ_is represents the Hessian eigenvalues, η_is are penalty weight factors, p_i has the same sign as λ_i

The objective function is a sum of the three cost terms weighted by appropriate penalty weight factors:

$$F_{obj}\left(x, y, z\right) = \left[V\left(x, y, z\right) - V_L\right]^2 + \beta Tr\left(gg^\dagger\right) + \sum_{i=1}^{3n}\eta_i e^{p_i\lambda_i}$$

$$x \equiv x_1, x_2, \ldots, x_n$$
$$y \equiv y_1, y_2, \ldots, y_n \quad\quad (7.14)$$
$$z \equiv z_1, z_2, \ldots, z_n.$$

Unlike the annealing method which can be viewed as a single parent or string-based search heuristic, the real coded GA works with a population (n_p) of evolving potential solutions, each solution being an array of $3n$ floating point numbers which represent the Cartesian coordinates of the n-atoms. The fitness value of each string (k) is computed by taking

$$f_k = e^{-\alpha F_{obj}^k \left(x^k, y^k, z^k \right)}, \ \lambda > 0 \ \left(\text{user specified} \right).$$ (7.15)

F_{obj}^k being the value of the objective function evaluated at the configuration encoded by the kth string. The strings undergo the normal genetic processes such as

1. Selection (70% of strings selected by roulette wheel, rest chosen randomly)
2. Crossover (arithmetic, single point)
3. Mutation (arithmetic)
4. Screening (elitist)
5. Diversification (carried out every five generations, only the best string remains immune to diversification)

If the search is meant for locating the minima, global as well as local, the curvature constraints are switched off (i.e., all η_is are set equal to zero during the search). They are switched on when the search is carried out to discover saddle points on the PES with $\eta_i < 0$ if $\lambda_i < 0$.

The method was tested on Ar_n clusters with $n = 4 - 30$ argon atoms interacting via appropriate Lennard–Jones potential as well as on $Ar_m X$ clusters where X is a neutral atom or a cation. They also exploited the GA-based recipe for mapping out an RP. The strategy adopted was a rehearsal of the RP construction method by SAM described already, with the change that here a population of strings evolves so as to achieve fitness maximization. The viability of RP analysis under GA was demonstrated with reference to the structural reorganization in the Ar_{13} cluster in which the cluster at the global minimum energy ($E_G = -44.3268\epsilon$) geometry passes over into a local minimum energy geometry ($E_{LM} = -41.4719\epsilon$) through a first-order saddle point S_1 (energy $= -40.403\epsilon$).

We note here that the intermediate structures used in the RP construction in a GA-based approach refers to the structure encoded by the string of the highest fitness in the current population. The search for the intermediate structures on the RP is carried out with a fixed value of the lower bound U_L in the energy cost term in the objective function. If n – points on a particular segment of the RP (say, $R - S_1$) are to be generated, U_L^q for the qth point on the path is determined as

$$U_L^q = U_L + q\Delta U$$ (7.16)

where U_L is the energy at the reactant configuration, and

$$\Delta U = \left(U_{S_1} - U_R \right) / n.$$ (7.17)

A similar approach is adopted for generating points on the product segment $P - S_1$ of the RP. The following points may be noted:

1. The RP is constructed by a nondeterministic search performed through a population-based evolutionary computing technique (GA).

2. The individual members (geometry strings) of the population define different paths. Out of the ensemble of n_p number of RPs traced out by the evolving population, the one traced out by the string of the highest fitness at that stage of the search is accepted as the representation of the reaction path. One may view the procedure as a stochastic technique of RP construction and analysis.

3. Note that the search initiated from the minimum energy structures (R and P) may not each be able to reach the lowest first-order saddle point (S_1). It would mean that there is possibly a higher energy path through a higher energy first-order saddle point (S_1' with $U(S_1') > U(S_1)$) or the transformation is not feasible.

4. The eigenvalues of the full Hessian matrix at the geometry encoded by string of the highest fitness at the kth stage can be used for evaluating the curvature constraint terms and fitness evaluation of all other strings in the $(k + 1)$th step. These eigenvalues need not be calculated in every generation of the search—one evaluation in every five generations would usually be sufficient.

5. The algorithm can be interfaced with any standard electronic structure calculation routine so that $U(q)$ is computed on the fly.

6. We have assumed that the first-order saddle points have already been identified. This restriction is not necessary as it can be located while constructing the path.

7.3.1 An Alternative GA-Based Strategy

Bungay et al. [9] proposed and implemented an interesting method for optimization of TS structure with a genetic algorithm as a tool. The salient features of the method are as follows:

1. Binary strings are used for encoding structures. The geometry variables such as bond lengths, bond angles, dihedral angles, or torsional angles defining a unique geometry of the species are first converted into integers by multiplicative encoding or by range encoding. The integers, one for each geometry variable, are then converted into their binary equivalents and concatenated to produce bit strings (chromosomes) encoding the individual structures ($\{S_i\}$).

2. Fitness function $f(S)$: The fitness function for evaluating the suitability of the individual structures (S_i) making up the population for the saddle point search is

$$f = \frac{1}{Tr\left(gg^+\right) + \epsilon_1} \times \frac{1}{|n_- - 1| + \epsilon_2} \tag{7.18}$$

where g represents the local energy gradient vector for structure S, n_- is the number of negative eigenvalues of the Hessian for the same structure, and ϵ_1, ϵ_2 are small numbers added to the denominators to avoid division by zero. The saddle point search objective is to find structures with zero gradient norm (critical point) and exactly one negative eigenvalue of the Hessian matrix (first-order saddle point identifier). (Note: There is no energy constraint and thus the first-order saddle being sought may not necessarily be the lowest energy first-order saddle point).

3. Roulette wheel or tournament selection was used along with single point crossover (binary) and standard binary ($0 \rightarrow 1$ or $1 \rightarrow 0$ flip) mutation protocol with preset values of crossover and mutation probabilities (p_c and p_m).

4. Calculations were carried out at the Hartree–Fock level with an STO-3G basis set. The reactions studied were

    a. HCN↔HNC (rearrangement)

    b. HCCH↔CCH$_2$ (rearrangement)

    c. HOCl↔O+HCl (reaction)

    d. HCN+H$_2$↔H$_2$CNH (reaction)

The authors note that the search time required is rather long, but the algorithm correctly identifies the first-order saddle points and the TS structure irrespective of the initially generated structures in the population. This robustness is a positive feature when viewed against the deterministic methods of locating TS structures.

7.3.2 A Method Bypassing Hessian Diagonalization

Saddle point searches are in general much more difficult than searches for minima on a given PES. It is because little can be guessed beforehand about their location and disposition on the PES. The problem may be aggravated when we are dealing with weakly bound systems like Lennard–Jones clusters as the topography of PES around the saddle points may be rather flat. Wales [10] located saddle points of LJ clusters of different sizes by the standard deterministic eigenvector-following method. Quapp [11] traced out the saddle points in Ar$_4$ cluster by a quasi gradient-only recipe. Jensen [12] extensively investigated the saddle points on the PES of Ar$_8$ clusters identifying 42 saddle points and the associated structures by a deterministic method. We have already described several stochastic or nondeterministic methods (based on simulated annealing and GA) for locating saddle points on PES, optimizing TS structures, and constructing RPs that represent structural transformations of LJ clusters from one minimum to another. The nondeterministic or stochastic searches proposed by Chaudhury and Bhattacharyya. [7,8] are based on objective functions that incorporate in them an energy cost term together with cost components dependent on local energy gradient and curvature with respect to changes in nuclear coordinates. The incorporation of curvature constraints demands calculation of eigenvalues of the Hessian matrix at the current geometry. For large clusters, the cost of computing Hessian eigenvalues can be high. A method that avoids repeated computation of Hessian eigenvalues, but still directs the nondeterministic search toward a first-order saddle point on the PES could be a welcome progress in saddle points search by GA or SAM. As we may recall, an objective function for the search of first-order saddles on the PES of an *n*-atom cluster closest in energy to U_L (user specified) can be written as

$$F_{obj}\left(U_L, \beta, \{\eta_i\}\right) = \left[U(q) - U_L\right]^2 + \beta Tr\left(gg^\dagger\right) + \sum_{i=1}^{3n-6} \eta_i e^{p_i \lambda_i} \tag{7.19}$$

where λ_is are the $3n - 6$ eigenvalues (excluding the translational and rotational modes) of the Hessian matrix at the current configuration (q). The scaled fitness function for a GA-based search is generated by the mapping

$$f = e^{-\gamma F_{obj}}. \tag{7.20}$$

Note that $\eta_i, p_i < 0$ if $\lambda_i < 0$. One can argue that the term corresponding to the lowest eigenvalue in the curvature constraint part of the objective function is expected to have the largest contribution in the increase of fitness value of a string. The higher eigenvalues of the Hessian will have negligible impact in the maximization of the fitness value in a GA-based search for the first-order saddle point. Neglecting all but the lowest eigenvalue of the Hessian in the curvature constraint term, we have a much simpler looking objective function:

$$F'(q, U_L, \beta, \eta_1, \lambda_1) = \left[U(q) - U_L \right]^2 + \beta Tr(gg^\dagger) + \eta_1 e^{p_1 \lambda_1}. \tag{7.21}$$

Chaudhury et al. [13] argued that along the minimum energy path leading to a first-order saddle point, the eigenvector for the negative eigenvalue (λ_1) is expected to be oriented nearly parallel to the local gradient vector, and can be approximated by λ_1' where λ_1' is the difference of gradients $|grad'| - |grad|$. $|grad'|$ is the local gradient at a test point generated by taking a small step toward the nearest first-order saddle point along the direction of the local gradient, $|grad|$ being the gradient at the current point reached by the search. Assuming that the search is already in the neighborhood of a first-order saddle point, λ_1' should be negative. The curvature constraint term in the objective function with the proposed gradient-based recipe for computing λ_1 should be

$$C_1 = \eta_1 Sign\left(|grad'| - |grad|\right) e^{p_1\left(|grad'| - |grad|\right)}. \tag{7.22}$$

The authors claim that the modified GA-based first-order saddle point search algorithm has two advantages over the previously suggested methods.

1. It does not require diagonalization of the full Hessian matrix.
2. Once a direction leading to the first-order saddle point is found, it is not lost easily. The workability of the proposed algorithm was demonstrated on Ar_n clusters with $n = 7 - 30$. It would be interesting and useful to interface the algorithm with an electronic structure computing code and use it in the RP analysis.

Biring and Chaudhury [14] made extensive use of the technique of search for the global minimum, local minima, and saddle points of order one on the PES of clusters like Ar_n perturbed by an alkali metal ion (M^+). The PES of resulting clusters $Ar_n M^+$ have been described by interatomic potentials that are purely Lennard–Jones type. In addition to locating various minima and saddle points of order one, the author also mapped out representations of minimum energy paths describing the structural transformation of the system from one minimum to another through an appropriate first-order saddle point using the technique proposed earlier by Chaudhury and Bhattacharyya [8]. The small perturbed clusters turn out to be rich in diversity and provide a very good testing ground for new techniques of minima and saddle point search.

Talukder et al. [15] recently proposed and demonstrated the viability of a simplified general recipe for constructing RPs for elementary or even multistep reactions. The strategy hinges on the method of simulated annealing as the search technique. Although the demonstrations are limited to describing structural transformation in Ar_n or mixed $Ar_{n-1}Xe$ ($n = 7 - 25$), the method appears to be capable of handling far more complex chemical reactions with or without the formation of reaction intermediates.

The proposed method works in two steps. First, the global and local minima on the PES (V) are located by constrained minimization of the objective function F_{obj} by SAM where

$$F_{obj} = [V - V_L]^2 + \beta Tr(gg^\dagger) \tag{7.23}$$

where V_L is an estimated updatable lower bound to V, g is the gradient vector with components $\dfrac{\partial V}{\partial x_i}, \dfrac{\partial V}{\partial y_i}, \dfrac{\partial V}{\partial z_i}, i = 1, 2, \ldots, n$. $\beta > 0$ is a user-specified penalty weight factor for the constraint that at the critical point being sought, the norm of g must be zero. A pair of minima (L_1, L_2) on the PES is then selected, one (say L_1) serving as the reactant well, the other (L_2, say) serving as the product well. The RP connecting the two minima is then sought to be constructed by constrained minimization of a new objective function (F'_{obj}) where

$$F'_{obj} = \alpha V + \beta \left| P_c\left(\text{target}\right) - P \right| \tag{7.24}$$

where P_c (target) is the targeted path coordinate, while P is the general path coordinate, i.e., the path coordinate defined by the current structure.

7.4 Completely Adaptive Random Mutation Hill Climbing Method: Further Applications

The completely adaptive random mutation hill climbing (CARMHC) method was introduced and its elementary applications were discussed in Chapter 4. We will focus on a few more interesting applications of CARMHC in this section that demonstrates its viability in solving more complex cluster optimization problems, or in probing potential energy surfaces of clusters. The strongest advantage of RMHC as already discussed lies in its simplicity—there is only a single string, a single parent as such if we choose to retain its link to the standard GAs. The single string containing full information about the coordinates of particles making up the cluster undergoes mutations with probability and intensity determined by the past experience of success or failure rates. In nature, mutations occur randomly—only beneficial mutations, i.e., mutations that enhance the survival probability of the species are retained by nature in the long run. Nonbeneficial mutations die out somewhere along the evolutionary path. Adaptive control of mutation probability and intensity in CARMHC simulates this long-time behavior of the mutation process in evolutionary biology, in the computer-based execution of an evolutionary model on a much shorter timescale in which improving quality of the emerging solutions replaces the goal of enhancing reproductive viability in the real life genetic of evolution. Unlike the single parent evolutionary algorithm (EA) of Rata et al. [16] CARMHC [17] does not make use of any specialized mutation operator like "piece reflection" or "piece rotation," but uses experience in deciding how frequently and to what extent the coordinates must change so that the best solution slowly and finally emerges. The structural evolution in a harmonically confined system of N unit positive charges has been monitored and analyzed by invoking the CARMHC when anisotropy is introduced in the confining potential. The potential in a three-dimensional harmonic trap with $k_x = k_y = k \neq k_z$ experienced by N unit positive charges reads

$$V_C\left(x, y, z\right) = \frac{1}{2} k \sum_{i=1}^{N} \left(x_i^2 + y_i^2\right) + \frac{1}{2} k_z \sum_{i=1}^{N} z_i^2 \tag{7.25}$$

while the total potential energy $V(x, y, z)$ is a sum of Coulomb repulsion energy (V_q) among the charges and the confinement energy V_c already defined. Thus,

$$V_N\left(x, y, z\right) = V_C\left(x, y, z\right) + V_q\left(x, y, z\right)$$
$$= \frac{1}{2} k \sum_{i=1}^{N} \left(x_i^2 + y_i^2\right) + \frac{1}{2} k_z \sum_{i=1}^{N} z_i^2 + \frac{1}{2} \sum_{i=1}^{n} \sum_{j(\neq i)=1}^{n} \frac{1}{|\vec{r}_i - \vec{r}_j|}. \quad (7.26)$$

At $T = 0$, the total energy $E_N\left(x, y, z\right) = V_N\left(x, y, z\right)$. We have already seen in Chapter 4 that CARMHC predicts charges arranging themselves in concentric shells about the origin of the confinement potential when it is harmonic as well as isotropic. The anisotropy in confinement brings in subtle effects of competition between confinement and Coulomb forces, and a rich variety of response is revealed. Some of the responses are as follows:

$N = 50; k_x = k_y = 1, k_z$ **varying** $(0.5 \geq k_z \leq 5.0)$: The anisotropy destroys the symmetry of the structures predicted for the isotropic system with $k_x = k_y = k_z = 1.0$. As k_z is varied the lowest energy structures that emerge are far less symmetric as can be seen in Figure 7.2 [18] where the predicted structures along with the point symmetry of each are displayed. The computed energy per particle (E_N/N) varies smoothly and continuously as a function of k_z (Figure 7.3a). The first and second difference in energy per particle $\Delta(E_N/N)$ and $\Delta^2(E_N/N)$ also very continuously (Figure 7.3b and c).

The behavior of E_N/N, $\Delta(E_N/N)$, and $\Delta^2(E_N/N)$ as a function of k_z seem to indicate that the structural evolution displayed in Figure 7.2 is continuous, at least in the 50 unit charge system in the confinement range $0.5 \leq k_z\ 5.0$ ($k_x = k_y = 1.0$).

If $k_z \rightarrow 0$ while k_x and k_y are nonzero, the charged particle would be progressively free to move along z-axis as the restoring force acting on the charges along the z-direction tend to vanish. We can anticipate that charges would be organized into a linear arrangement along the z-axis as k_z becomes zero. The question then arises if the signature of the

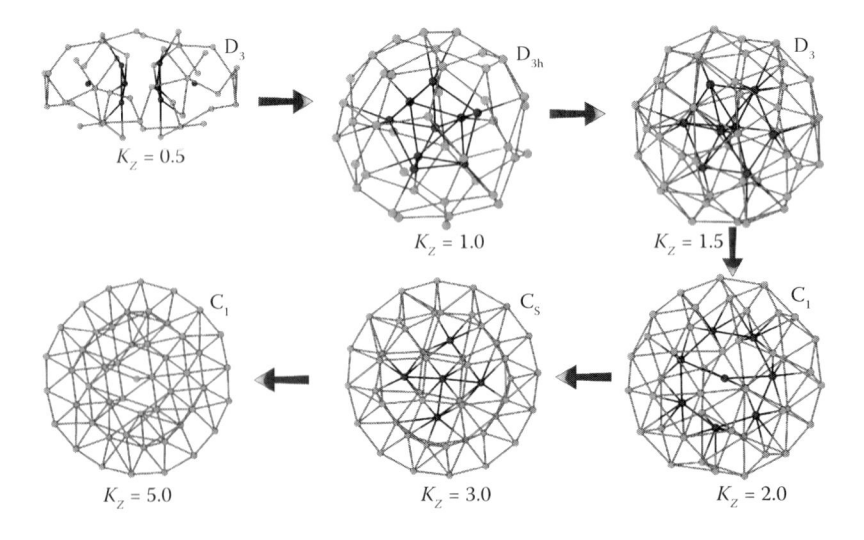

FIGURE 7.2
Structural evolution for a system consisting of 50 charged particles with varying confinement strength along the z-axis (K_z). Confinement strength along the x- and y-axes ($K_x = K_y = 1.0$) are kept constant. The point symmetry of each structure are also displayed. (Courtesy of Sarkar, K., *Soft-Computing algorithms for solving some problems of Quantum Chemistry*, Indian Association for the Cultivation of Science, Jadavpur University Kolkata, India, 2014, 95.)

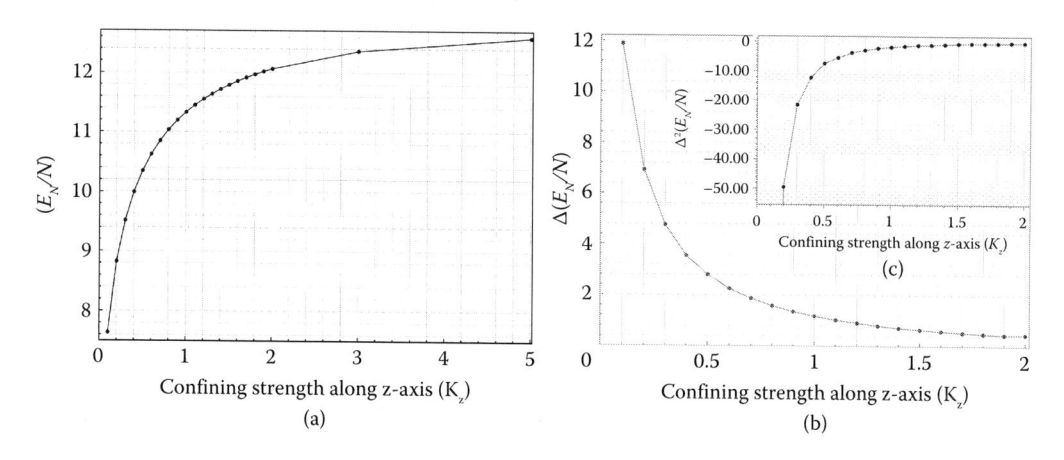

FIGURE 7.3

(a) Corresponding energy evolution profile and (b) first and (c) second difference in energy of Figure 7.2. (Courtesy of Sarkar, K., *Soft-Computing algorithms for solving some problems of Quantum Chemistry*, Indian Association for the Cultivation of Science, Jadavpur University Kolkata, India, 2014, 96.)

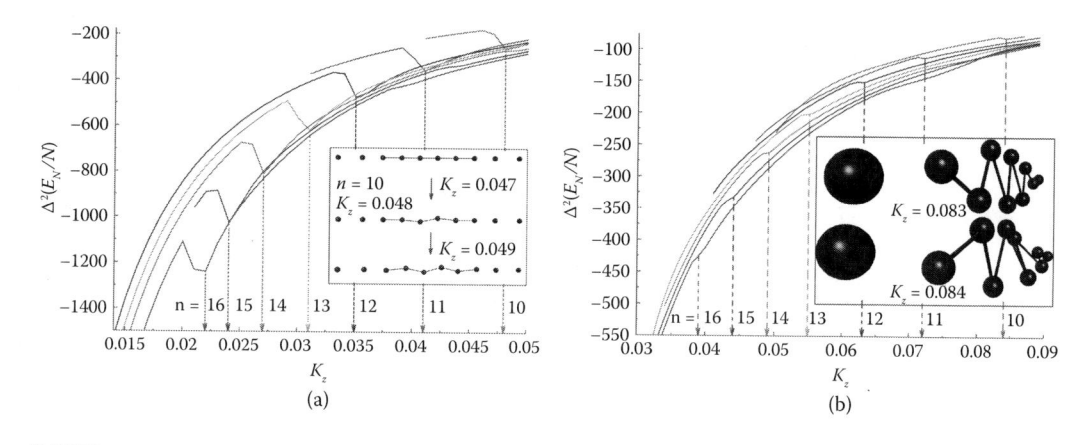

FIGURE 7.4

(a) Signature of one- to two-dimensional transition in a harmonically but not isotropically confined system of N interacting unit positive charges as k_z ($\neq k_x, k_y$) increases from zero value. Note the dependence of k_z critical on the number of charges. (b) Signature of two- to three-dimensional transition in the same systems as a function of k_z. (Courtesy of Sarkar, K., *Soft-Computing algorithms for solving some problems of Quantum Chemistry*, Indian Association for the Cultivation of Science, Jadavpur University Kolkata, India, 2014, 92.)

anticipated one- to two-dimensional transition could be detected by the CARMHC procedure as k_z becomes nonzero. Figure 7.4a and b displays the computed second difference of energy per charged particle $\Delta^2(E_N/N)$ for $N = 10 - 16$. It appears that an N-dependent critical value of k_z exists at which a transition from the one- to two-dimensional arrangement of charges takes place (for $N = 10$, $k_z \sim 0.048$, $k_x = k_y = 1$). On further increasing the value of k_z a second critical value of k_z is reached where the two-dimensional arrangement of charges passes over into a three-dimensional arrangement. Here also the critical k_z value for the onset of one- to two-dimensional transition depends on the number of charges in the harmonically confined system as displayed in the $\Delta^2(E_N/N)$ versus k_z plots.

In a harmonically confined system of 50 unit positive charges with $k_x = k_y = 1.0$, $k_z \neq k_x, k_y$, the computed CARMHC energies per particle at the global minima for different values of

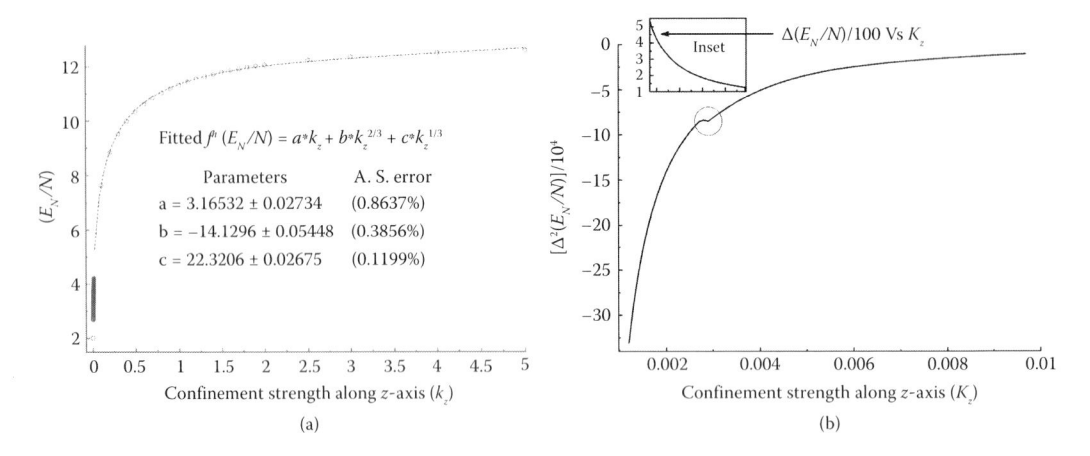

(a)   (b)

FIGURE 7.5

(a) Energy per charge particle for coulomb clusters of 50 unit positive charges under asymmetric confinement (the confining strength has been varied along the z-axis) (b) Discontinuous one- to two-dimensional structural phase transition with respect to the second difference in energy for the same system. Two- to three-dimensional structural transition seems to be continuous. (Inset: First difference in energy vs K_z.) (Courtesy of Sarkar, K., *Soft-Computing algorithms for solving some problems of Quantum Chemistry*, Indian Association for the Cultivation of Science, Jadavpur University Kolkata, India, 2014, 93.)

k_z fit nicely into the form $\dfrac{E_N}{N} = ak_z + bk_z^{2/3} + ck_z^{1/3}$ (Figure 7.5a) which does not reveal any breaks identifying onsets of possible structural phase transition. Similar plots for the first and second differences in the computed energies per particle (E_N/N) (Figure 7.5b) however reveal a clear break in the $\Delta^2(E_N/N)$ versus k_z plot while no such break appears in the plot of $\Delta(E_N/N)$ against k_z (inset). The break in $\Delta^2(E_N/N)$ against k_z plot can be identified with a one- to two-dimensional structural phase transition point in the system. The two- to three-dimensional transition point in this system remained elusive, perhaps implying a continuous transition from two- to three-dimensional structures as the anisotropic strength is increased. It could also be that a finer search for the global minimum is necessary to settle the issue unambiguously.

The foregoing results of applications of the CARMHC procedure to a system of point charges confined in a three-dimensional harmonic potential with axial anisotropy reveal that a simple soft computing algorithm like CARMHC can be so useful in understanding the emergence of structural order in the system. It would certainly be interesting to extend the procedure to polydisperse confined charges with a view to understanding segregation and phase transformation in such systems as a function of various system parameters. We anticipate significant activity in this direction in the near future.

7.5 Global Search on PES by Artificial Bee Colony Algorithm

We may recall (Chapter 5) that the artificial bee colony (ABC) algorithm inspired by the foraging behavior of honey bees works by employing three types of random fliers to sample the search space, i.e., move across the objective function hypersurface. The first type of random fliers are called scouts—they move on the hypersurface in a perfectly random manner without any string of conditionality attached. The second type of fliers called "employed bees" also engage in random movements accepting new positions only if those positions correspond to lower objective function values than those in their current positions. The third type of bees are called onlookers. They take their cue from the employed bees and randomly explore around the regions of the search space already marked by the employees as promising. The onlookers thus increase the frequency with which promising areas are sampled. The ABC algorithm has been remarkably effective in handling global optimization problems in computer science as well as engineering. The impressive success of ABC in these fields has stimulated interest in exploring ABC in global optimization problems in chemistry and physics. The problem of optimization of molecular structure or cluster geometry has been at the focus of such explorations. The initial experience apparently suggested that some modification in the original ABC could enhance its performance in the search for the global minimum on the PES of many interacting particles. Wehmeyer et al. and others [19–21] recently proposed and demonstrated the viability of a modified ABC algorithm in global optimization of structures in systems of up to 57 particles interacting via different interatomic potentials. This modified ABC algorithm is described in some detail in what follows:

Let the single vector X represent the coordinate of N interacting particles,

$$X \equiv (R_1, R_2, \ldots, R_N). \tag{7.27}$$

X thus defines configurations of the system of particles. The potential energy surface $E(X)$ is then obtained by evaluating the interaction energy among the N-particles for different configurations. $E(X)$ can be transformed into an unscaled fitness function $f(E(X))$ by the mapping

$$f\left(E\left(X\right)\right) = \begin{cases} 1 - E(X), & \text{if } E(X) < 0 \\ \dfrac{1}{1 + E(X)} & \text{if } E(X) \geq 0 \end{cases}. \tag{7.28}$$

Suppose that n_B number of random but interacting fliers (the swarm of artificial bees) are foraging on the fitness landscape $f(E(X))$ with the objective of locating the configuration X_G of the N particles where the global maximum on the fitness landscape is attained. By definition it is also the configuration or geometry of the N interacting particles defining the global minimum on the PES.

At a given instant, the search agents—the n_B number of bees (B_i) are each tagged by a structure identifier or a configuration vector (X_i) and a scalar representing its fitness value (f_i) with respect to the search objective. Thus, we have

$$B_i \equiv \left(X_i, f_i\right); \; i = 1, 2, \ldots, n_B$$

$$X_i \equiv \left(R_1^i, R_2^i, \ldots, R_N^i\right) \tag{7.29}$$

$$f_i = f\left(E(X_i)\right) \equiv f(E_i).$$

Wehmeyer et al. [19] introduced a fundamental change in the original scheme of ABC optimization in the sense that the bees in the swarm are not partitioned into the three behavioral classes. Instead, a more democratic scheme is adopted in which there is only a single class of foraging agents which can explore the search space by adopting different strategies for moving from one part of the PES to another. Two types of movement strategies are defined [19]:

1. A free move strategy assigns the calling bee (agent B_i) to a randomly chosen location on the PES enabling the algorithm to explore the entire breadth of the search space. All the agents are initially distributed on the PES following the free move strategy.

2. A local move strategy is used by default and allows the calling agent (B_i) to explore locally by modifying its position vector

$$X_i \rightarrow X_i' = X_i + r \cdot d_S \tag{7.30}$$

where r is a random number (0,1) drawn from a Gaussian distribution and d_S is a step length parameter. The agent (B_i) accepts the move only if $f_i' = f\left(E\left(X_i'\right)\right) > f_i$. The local move strategy is used by all agents subsequent to their initial placement on the PES by the free move operator. If the failure rate of the free moves increases beyond a limit for an agent B_i, the step length d_s is reduced successively ($d_S' = y d_S$; $y < 1$) until d_S' becomes less than a preassigned threshold value when agent B_i switches over to adopt a free move strategy again. In summary, steps (i) and (ii) amount to n_B number of simultaneous but independent Markovian downhill searches. Unless agents exchange or share information about their positions and fitness and act on that basis, the promising low-energy areas of the PES may not receive the attention they deserve. In the democratic setup of the modified ABC algorithm of Wehmeyer et al. [19] the search agents are made to carry out explorations of the surface in two phases, namely an employee phase and an onlooker phase which ensure that all the agents make some progress and the agents who have already reached promising areas of the search space are allocated additional moves based on their relative fitness with respect to the entire swarm. Thus, in the employer phase all the agents {B_i} are allowed to adopt either a free move or a local move strategy of exploration depending on their current status. In the onlooker phase, the relative fitness of the individual agents with respect to the entire swarm are evaluated to assign the probability of the ith agent to qualify for getting additional local or free moves to be allocated to them. The onlooker phase apparently seeks to mimic the feature of GAs wherein schemas with higher fitness values receive exponentially increasing numbers of trials.

The modified ABC algorithm was put to test on a global optimization of molecular clusters of up to 57 atoms. Two types of clusters were studied: (i) Morse clusters and (ii) water clusters at the level of five point transferable intermolecular potential (TIP5P)-based representation of the PES. The applications confirm the unbiased global search capability of the modified ABC algorithm albeit it still cannot compete with approaches like minima hopping. Many avenues, however, exist for further exploration and experimentation which can perhaps make it competitive with the currently available techniques of global optimization.

We may note here that the ABC may be viewed as a dynamical system of many interacting agents, which has the ability to self-organize. The emergence of self-organization requires fluctuations, multiple interactions, and positive as well as negative feedback to be present in the dynamical system. Zhang and Dolg [22] proposed a modified ABC algorithm in 2015 for discovering the global minimum on the PES of clusters represented by a variety of interatomic or many-body potentials. They incorporated a number of novel features in their algorithm and used deterministic local search techniques wherever needed without compromising the global search capability of the swarm, or the emergence of self-organization. In view of the rapidly increasing interest in the ABC algorithm, especially among the cluster community, we will discuss the ABC recipe advocated by Zhang and Dolg [22] in detail. The following steps are to be executed:

1. Initialization: A population of SN number honey bees is foraging on the PES. Each bee (B_i) is represented by an M component vector X_i^1 (superscript 1 denotes the first cycle), the components representing values of all the positional coordinates of the particles in the cluster that are required to define a candidate structure uniquely. The M components are either chosen randomly from a user-specified range or by "seeding"; the population with structures is generated from the already available coordinates of the particles of $(N-1)$ particle cluster and placing the Nth particle randomly on the $(N-1)$ particle seed structure. The initial swarm is thus represented by the collection of vectors $\{X_i^1\}; i = 1, 2, …, SN$.

2. Modeling Foraging by Employee Bees: In cycle g a new trial vector Z_i is generated for each X_i^g (the cluster configuration vector for the ith bee) by exploiting information from all other bees (except the ith one) by adopting a triangular mutation operation which generates the new configuration vector as follows:

$$Z_i = \frac{1}{3}\left(X_k^g + X_l^g + X_m^g\right) + \left(p_l - p_k\right)\left(X_k^g - X_l^g\right) + \left(p_m - p_l\right)\left(X_l^g - X_m^g\right) + \left(p_k - p_m\right)\left(X_m^g - X_k^g\right) \quad (7.31)$$

where k, l, m are random integers drawn from $1, 2, …, SN$ with the restriction that $k \neq l \neq m \neq i$, p_k, p_l and p_m are chosen as follows (the philosophy adopted in constructing Z_i is borrowed from differential evolution (Chapter 3)). The parameter p_k is defined as follows:

$$p_k = \frac{\left|\bar{U}(X_k)\right|}{\left|\bar{U}(X_k)\right| + \left|\bar{U}(X_l)\right| + \left|\bar{U}(X_m)\right|}, \quad (7.32)$$

p_l and p_m being defined similarly. Note that

$$\bar{U}(X) = \min\{U(X)\} \quad (7.33)$$

which means $\bar{U}(X)$ has been obtained by locally minimizing $U(X)$ by a deterministic gradient search recipe starting from X. Such a transformation reduces the ruggedness of the actual PES by removing barriers on a downhill path to a funnel. X_i^g is replaced with Z_i only if $U(Z_i) < U(X_i^g)$.

Otherwise X_i^g is not updated at all.

3. Modeling search by onlooker bees: Each onlooker bee chooses a good solution, say X_k^g and generates a new solution (trial) from other solutions X_j^g; $(j = 1, 2, …, SN \neq k)$. A tournament of size N_t is played and the best one is chosen to be updated into Z_k where $(N_t = 5)$

$$Z_k = \begin{cases} X_k^g + R\left(X_{k_1}^g + X_{k_2}^g - X_{k_3} - X_{k_4}\right) & \text{if } \eta < 0.5 \\ X_{best}^g + R\left(X_{k_1}^g + X_{k_2}^g - X_{k_3} - X_{k_4}\right) & \text{if } \eta \nless 0.5 \end{cases}. \quad (7.34)$$

In the above updating procedure, k_1, k_2, k_3, k_4 are random integers under the constraint that $k_1 \neq k_2 \neq k_3 \neq k_4 \neq k$, R and are η random numbers in [0,1].

4. Modeling exploration by scout bees: After the onlooker bee search phase, each X_i^g is examined to ascertain whether its fitness (i.e., energy) improved over the last g_{lim} number of cycles (g_{lim} is user specified). If not, it is replaced by a random configuration vector X_i^{g+1} without any consideration for the change in energy associated with such a replacement. The scout bee search is an avenue for maintaining diversity in the population. It also helps the search get out of local traps.

5. Termination: If $g \geq g_{max}$, exit; $X_{best}^{g_{max}}$ is most likely to be the global minimum energy configuration or geometry of the cluster.

Zhang and Dolg claim that their ABC method is an unbiased swarm intelligence-based recipe for locating global minimum energy structures of LJ and metal clusters which requires only three parameters, namely SN (swarm size), g_{lim} (decides, if scout bee search is needed), and g_{max} (decides if termination conditions have been reached) to be specified from outside. The algorithm displays the self-organization ability of the swarm and performs extremely well for clusters dominated by long-range interactions. Its performance deteriorates substantially for metal clusters in which short-range and many-body interactions are dominant. In such systems, the search may get trapped in local minimum energy funnels. The authors felt that a hybrid of ABC and the basin-hopping method could be successful in avoiding such traps and improving the global search capability.

We note here that the methods described so far in this chapter work with empirical or semiempirical potential energy surfaces. The parameters in these potentials are often derived from bulk properties of matter and may not be appropriate for describing a variety of structures including the GM that the actual, microscopic potential energy surface supports. It is therefore necessary to interface the soft computing algorithms with methods of electronic structure calculation so that the PES is generated on the fly from first principles and the soft-search for optima is conducted simultaneously.

There are several possibilities to be explored.

1. One way to accomplish the task may be to integrate a soft-global-search algorithm (SGSA) with a standard electronic structure computing code (ESCC) which computes electronic structure and total energy (sum of electronic and nuclear repulsion energies) for any nuclear configuration generated by the SGSA. The SGSA then uses the information about structures and energies received from the electronic structures computing routines for generating better structures. The two, namely SGSA and ESCC, can work in tandem until the global minimum is located. The primary role of SGSA in this scheme is to generate progressively better molecular structures using feedback from the ESCC.

2. A second option could be to use the ESCC to generate energies for many nuclear configurations and construct an artificial neural network representation of the PES which is later exploited by the SGSA for locating global minimum energy configuration or structure.

3. A third option may be to exploit the structures generated by the SGSA chosen to guide a direct and optimal evolution of the wave function or electronic charge distribution in the atom, molecule, or cluster of any kind.

Of these options, (1) and (2) have already been used successfully; we will review such applications of option (1) later in this chapter (see Section 7.7). An account of the usage of option (2) will be presented in Chapter 8 where neural networks are discussed. Option (3) has found some use and is still under development. We propose to give an account of the exploration relating to option (3) in the next section.

7.6 Simultaneous Soft-Search for the Global Minimum and Associated Electronic Structure of a Large Molecule

The problem of locating the global minimum of a multimodal PES is a formidable one which defeats commonly used techniques of minimization. The traditional methods generally converge to a local minimum close to the starting point where the norm of energy gradient vector becomes zero just as it would have been at the global minimum of the PES. If the PES is not known but must be generated on the fly, the problem becomes twofold:

1. We must have a method capable of globally searching through the space spanned by nuclear coordinates (R) avoiding or escaping from local traps.

2. Any movement through the nuclear space must simultaneously be able to cause corresponding changes in the electronic degrees of freedom (change distribution) so that the electronic charge density evolves optimally toward the global minimum of the PES.

Problem (i) can be addressed by invoking an appropriate computing method with global search capacity, while problem (ii) is an open issue requiring attention. Our focus now would be look for algorithms that provide a facile solution to the problem posed in (ii).

7.6.1 Coupled Search through the Spaces of Nuclear and Electronic Variables

Let $H(R)$ be the Born–Oppenheimer Hamiltonian of a molecule, where R is a vector of all the N number of nuclear coordinates or geometrical parameters that are required to define the nuclear framework uniquely. Let the initial choice of the geometry be coded by R_0

$$R_0 \equiv \left(R_1^0, R_2^0, ..., R_N^0 \right). \tag{7.35}$$

Let $H_{eff}(R_0)$ be an effective single particle (electronic Hamiltonian (mean-field model)) and $\left\{ \phi_i^0 \right\}_{i=1,n}$ be the set of n lowest energy single particle functions (the molecular orbitals) which are double occupied by electrons of opposite spins. That means we are dealing with a closed shell single determinant ground state. These orbitals are provided as a starting guess and depend on the initial geometry (R_0), and can be generated in many ways. In the LCAO MO approximation, the MOs are expressed as linear combinations of a basis set of M atom-centered orbitals (χ_k) (assumed orthonormal, for the sake of simplicity) so that

$$\phi_i = \sum_{p=1}^{M} \chi_p C_{pi}^0 ; i = 1, 2, ..., n \tag{7.36}$$

where C_{pi}^0s depend on the nuclear geometry variables. In matrix notation we can collect all C_{pi}^0s for the n-MOs and write

$$(\phi_1, \phi_2, \ldots, \phi_n) = (\chi_1, \chi_2, \ldots, \chi_M) C_{M \times n}^0 (R_0). \tag{7.37}$$

The starting single particle density matrix in the AO(χ) basis is given by

$$P(R_0) = C^0(R_0) \, C^0(R_0)^\dagger. \tag{7.38}$$

The corresponding charge density bond order matrix $Q(R_0)$ is obtained by multiplying $P(R_0)$ by the occupation number for each MO (which is two for the closed shell single determinant representable ground state) so that

$$Q(R_0) = 2P(R_0). \tag{7.39}$$

As defined, $P(R_0)$ has the following properties:

$$P(R_0)^\dagger = P(R_0) \text{ (hermiticity)}$$

$$\left(P(R_0) \right)^2 = P(R_0) \text{ (idempotency)} \tag{7.40}$$

$$2Tr\left(P(R_0) \right) = 2n \text{ (constancy of the total number of electrons).}$$

Let the nuclear repulsion energy for the configuration R_0 be $E_{NU}(R_0)$ and the electronic energy be $E_{el}(R_0)$. At the present level of description $P(R_0)$ alone is sufficient to calculate the electronic energy (a two-electron density matrix is factorizable in terms of a one-electron density matrix) and for an appropriately defined $H_{eff}(R_0)$ we can write

$$E_{el}(R_0) = 2 \, Tr\left[P(R_0) H_{eff}(R_0) \right]. \tag{7.41}$$

The total energy $E_T(R_0)$ then becomes a sum of $E_{el}(R_0)$ and the nuclear framework energy $E_{NU}(R_0)$ (at the BO level $E_{NU}(R_0)$ is just an additive constant that changes with changes in R_0). Thus,

$$E_T(R_0) = Tr\left(R_0 H_{eff}(R_0) \right) + E_{NU}(R_0). \tag{7.42}$$

The problem is to locate the global minimum of $E_T(R)$ and identify the nuclear configuration R^* at which the global minimum is located. A general recipe can be as follows.

Suppose that a single string-based soft computing method (e.g., RMHC or SA) is employed to generate a new configuration R_1 by randomly changing some of the components of the initial nuclear configuration vector R_0. For the new nuclear configuration, the nuclear repulsion energy $E_{NU}(R_1)$ can be easily calculated. Let the approximate effective Hamiltonian at R_1 be $H'_{eff}(R_1)$. Using the hermiticity of $H'_{eff}(R_1)$, we can define a unitary transformation $U(R_1)$ where

$$U(R_1) = e^{i\lambda H'_{eff}(R_1)} \text{ (λ is a real scalar).} \tag{7.43}$$

$U(R_1)$ can be used to generate the new single-electron density matrix for the configurations in the same AO basis (in which H'_{eff} is formed) by the following transformation:

$$
P(R_1) = \begin{cases} U(R_1)P(R_0)U^{\dagger}(R_1) \\ P(R_0) + i\lambda\left[H'_{eff}(R_1), P(R_0)\right] + \dfrac{i^2\lambda^2}{2}\left[H'_{eff}(R_1),\left[H'_{eff}(R_1), P(R_0)\right]\right] \end{cases}. \quad (7.44)
$$

The new density matrix $P(R_1)$ is then used to construct the actual effective Hamiltonian $H_{eff}(R_1)$ at R_1 assuming that the density-dependent forces are completely determined by $P(R)$. The total energy at the configuration R_1 is

$$
E_T(R_1) = 2Tr\left(P(R_1)H_{eff}(R_1)\right) + E_{NU}(R_1). \quad (7.45)
$$

The next step depends on the soft computing method being used. The soft computing method may either be a single-string or a multiple-string method. We consider two single parent strategies.

7.6.2 RMHC-Based Approach

If $E_T(R_1) \le E_T(R_0)$, the new nuclear configuration is accepted, and R_1 becomes the new R_0 ($R_0 \leftarrow R_1$), while $P(R_0)$ is replaced with $P(R_1)$. If the test fails, i.e., $E_T(R_1) > E_T(R_0)$, the old configuration is retained. In either case, the next move consists of randomly reconfiguring the current nuclear geometry string, and the entire cycle is repeated until the total energy remains unchanged over a preset number of reconfigurations or the limit of the maximum number of cycles to be repeated is reached. A good way to test the convergence is to check whether $\left[P(R_0), H_{eff}(R_0)\right] = 0$ (a null matrix).

7.6.3 SAM-Based Approach

Just as in the RMHC-based approach, here too the initial geometry is randomly perturbed into a new nuclear configuration R_1 and the total energy $E_T(R_1)$ is calculated. Two scenarios emerge. Assuming that the reconfigurations are carried out currently at temperature T_0 there are two possibilities: (a) If $E_T(R_1) \le E_T(R_0)$, the new nuclear configuration is accepted ($R_0 \leftarrow R_1$). If the test fails, i.e., $E_T(R_1) > E_T(R_0)$, a Metropolis test is conducted to ascertain if such an uphill movement is allowed by thermal fluctuations of the geometry parameters at the prevailing annealing temperature (T_0). If it passes the Metropolis state, R_0, is replaced with R_1 and $P(R_1)$ replaces $P(R_0)$. If the limiting number of sampling have been carried out at temperature T_0, the temperature is lowered in accordance with a preset schedule (say $T_{new} = \chi T_0$, $\chi < 1$) and the entire process is repeated until T_{new} becomes too small or no new configuration has been accepted over a huge number (N_{lim}) of trials at the current temperature, signifying that the search has reached a dead end. It is a safe practice to heat up the system and make new efforts for discovering a better nuclear configuration as the system is cooled again following a predesigned annealing schedule.

In both RMHC- and SAM-based implementations of the direct density method, the scale parameter λ has so far been treated as a disposable scalar ($\lambda < 1$). It is possible to provide an analytical near optimal estimate of λ and use it at every density matrix updating step (Equation 7.44). It increases the computational overhead, but leads to better and smoother convergence of the search to the global minimum. The λ–estimation method proceeds as follows.

Suppose that Equation 7.44 has been truncated such that terms containing λ^3 and higher order terms in λ have been neglected. An approximate expression of electronic energy up to second order in λ is then obtained from

$$E_{el}(R_1,\lambda) = 2\,Tr\big[P(R_1)H_{eff}(R_1)\big]$$

$$\approx 2\left\{\begin{array}{l} Tr\big(P(R_1)H_{eff}(R_1)\big) + i\lambda\,Tr\big(\big[H_{eff}(R_1),\big[H_{eff}(R_1),P(R_1)\big]\big]\big) \\[2ex] + \dfrac{i^2\lambda^2}{2!}Tr\big(\big[H_{eff}(R_1),\big[H_{eff}(R_1),P(R_1)\big]\big]H_{eff}(R_1)\big) \end{array}\right\}. \qquad (7.46)$$

Making $E_{el}(R_1,\lambda)$ stationary with respect to the variation in the scale parameter λ, we arrive at an analytical near optimum value of λ (λ_{opt}, say) where

$$\lambda_{opt} = -i\,\frac{Tr\big(\big[H_{eff}(R_1),P(R_1)\big]H_{eff}(R_1)\big)}{Tr\big(\big[H_{eff}(R_1),\big[H_{eff}(R_1),P(R_1)\big]\big]H_{eff}(R_1)\big)}. \qquad (7.47)$$

Sarkar et al. [23] demonstrated that—at a modified Su–Schrieffer–Heeger (SSH) level [24]—the RMHC-based direct density method proposed for a simultaneous search for the global minimum on a molecular PES, and the corresponding one-electron density, provided a facile route to optimal bond length and charge distribution in undoped as well as doped polythiophene oligomers (described by a modified SSH Hamiltonian). The calculations identified the principle charge storage configurations in hole-doped polythiophene oligomers (the bipolarons) and predicted that the bandgap is expected to decrease as a function of dopant concentration and their distribution along the chain. A simulated annealing-based variant [25] in which the one-electron density matrix was calculated at each reconfiguring step by diagonalizing the $H_{eff}(R)$, and exploiting eigenvectors for computing the electronic density, proved to be costlier in comparison with the λ–optimized RMHC direct density method. We note here that the λ–optimized SA-based direct density method appears quite feasible and it would be worthwhile to explore the avenue carefully since the global nature of the SA-based search is guaranteed to locate the GM if cooling remains slow.

7.6.4 Population-Based Direct Density Methods

The search through the space of the nuclear variables (R) can also be carried out by suitably invoking a population-based soft computing method, while the one-electron density matrix may be nudged to evolve quantum mechanically under an appropriate unitary transformation. In what follows, we describe a population-based soft computing implementation of the direct density method. The particular population-based method of our choice is a swarm intelligence-guided recipe, such as the PSO method [26]. A second change to be introduced concerns the nature of the unitary transformation $U(R)$ acting on the one-electron density matrix. In the RMHC-based direct density method the unitary transformation was complex; the generator was the real Hermitian BO Hamiltonian at the nuclear configuration R. It could be advantageous to avoid complex transformations and arithmetic, and work with unitary transformations that are real. That means we have to work with a generator $A(R)$ that is antihermitian $(A^\dagger(R) = -A(R))$ and define $U(R)$ [27] as

$$U(R) = e^{\lambda A(R)}, \ (\lambda > 0, \text{ real}). \tag{7.48}$$

We will defer an explicit choice of $A(R)$ and only note that $U(R)$ so defined is manifestly real.

Let us assume that the geometry of our molecule is completely specified by a collection of N number of structural parameters $(R_1, R_2, ..., R_N)$ where R_is can be regarded as the components of a real-valued N-dimensional vector R^N, the corresponding molecular electronic Hamiltonian being $H(R)$. To set the search in motion, we start by creating a swarm of n_p-particles flying through the N-dimensional search space spanned by the N-geometry variables or parameters. The particles are dynamical objects, the position of the ith particle in the search space at time t (t, a continuous parameter describing the progress of the search) is defined by an N-dimensional vector $R^{(i)}(t)$, the N-components of which are $R_1^{(i)}(t), R_2^{(i)}(t), ..., R_N^{(i)}(t)$. Together they define the structure of the molecule at that time as encoded by $R^{(i)}(N)$. The particles are treated as classical entities and are therefore also assigned velocity vectors $\{v^{(i)}(t)\}; i = 1, 2, ..., n_p$. The N-components of the velocity vector of the ith particle are $v_1^{(i)}(t), v_2^{(i)}(t), ..., v_N^{(i)}(t)$. They indicate the propensity of changes of the N-structural parameters at the instant t.

Let the energies of the n_p particles (i.e., of n_p different structures that the particles represent) of the swarm be $E^{(1)}(t), E^{(2)}(t), ..., E^{(n_p)}(t)$ (we assume that the structure to energy mapping can be easily done once the corresponding Hamiltonian and density are available). The lowest among these energies defines the best geometry so far discovered by the swarm as a whole. Let us call this structure $z_{best}(t)$. Let the velocity vector of the ith particle at the next instant be $V^{(i)}(t+1)$ while the geometry or the position vector be $R^{(i)}(t+1)$. These quantities can be estimated following the Kennedy–Eberhart recipe [28] of cognitive and social modes of learning of individuals of a swarm. Following Kennedy and Eberhart, we write (in vector form)

$$V^{(i)}(t+1) = \eta V^{(i)}(t) + C_1 r_1 \left(R_{best}^{(i)} - R^{(i)}(t) \right) + C_2 r_1 \left(Z_{best} - R^{(i)}(t) \right)$$
$$R^{(i)}(t+1) = R^{(i)}(t) + \chi v^{(i)}(t+1). \tag{7.49}$$

In the velocity-updating formula (Equation 7.49) $R_{best}^{(i)}$ is the best position (i.e., geometry, in the present problem) that the ith particle has been able to find for itself so far (i.e., up to the instant t), r_1 and r_2 are uniformly distributed random numbers between 0 and 1 ($0 \le r_{1,2} \le 1$), C_1 and C_2 are constants called acceleration coefficients and η is the inertial weight. In the position (geometry) updating formula, χ is a damping constant that moderates the actual extent of change in the geometry represented by the position vector of the ith particle. The first term in Equation 7.49 expresses the tendency of the particle i to remain in its current dynamical state, while the second term represents the cognitive component of learning by which it tries to remain close to the best position it has found for itself so far. The third term of Equation 7.49 represents the social component of learning. The combined effects of the second and third terms in the velocity-updating formula drive the individuals of the swarm relentlessly toward the energetically better regions of the PES and eventually to the global minimum.

As the nuclei fly through the search space spanned by the nuclear position variables $(R_1, R_2, \ldots, R_N)$, the electronic charge distribution too must undergo appropriate change immediately. That means the one-electron density matrix $P^{(i)}R(t)$ must quantum-mechanically evolve into $P^{(i)}R(t+1)$ as soon as $R^i(t)$ evolves classically into $R^i(t+1)$. The evolution can be brought about by a unitary matrix $U\left(R^i(t+1)\right)$ generated by a nuclear position dependent antihermitian matrix $AR^i(t+1)$. To make things simpler, we introduce two changes: drop the superscript i indexing individuals of the swarm, and replace $R(t)$ with R and $R(t+1)$ with R'. With these changes we can describe the evolution of the density matrix as

$$P(R') = U(R')P(R)U^{\dagger}(R') \tag{7.50}$$

where $U(R') = e^{\lambda A(R')}$, and $A(R')^{\dagger} = -A(R')$. Expanding the exponential and collecting terms containing identical powers in λ, we have

$$P(R') = P(R) + \lambda\left[A(R'), P(R)\right] + \frac{\lambda^2}{2}\left[A(R'), \left[A(R'), P(R)\right]\right]. \tag{7.51}$$

Two questions need to be addressed at this point:

1. How do we choose $A(R)$?
2. How do we fix λ, the scale parameter of the transformation?

The optimal choice of $A(R)$ is an open issue. A simple but effective choice is to make use of $H(R)$, the Hermitian Hamiltonian matrix in the AO basis. Thus, we can define [27]

$$A_{ii}(R) = 0; i = 1, 2, \ldots, n$$
$$A_{ij}(r) = H_{ij}(R); i = 1, 2, \ldots, n-1; j = i+1, \ldots, n \tag{7.52}$$
$$A_{ji}(R) = -H_{ij}(R).$$

A near optimal value of λ can be estimated by evaluating the electronic energy $E(R')$ at new geometry (R')

$$\in_{el}(R', \lambda) = 2\, Tr\{P(R')H(R')\}, \tag{7.53}$$

to second order in λ and making it stationary with respect to changes in the scale parameter, which yields

$$\lambda_{opt} = -\frac{Tr\left(\left[A(R'), P(R)\right]H(R')\right)}{Tr\left(\left[A(R'), \left[A(R'), P(R)\right]\right]H(R')\right)}. \tag{7.54}$$

The density-swarm algorithm is now completely defined. A few clarifications on the operational aspect of the density-swarm algorithm would be in order:

1. The geometry strings are generated randomly. However, the population may be seeded with specific types of strings on the basis of prior knowledge about the system.

2. Although a population of n_p density matrices $P^i(R'); (i = 1, 2, \ldots, n_p)$ are evolving, we need only one trial density $P(R_0)$ at a guessed geometry (R_0) to start with. $P(R_0)$ can be constructed directly by diagonalizing $H(R_0)$ and using the eigenvectors $\{C_i(R_0)\}$ to form $P(R_0)$.

3. Since the exponential defining $U(R')$ is truncated, we may suspect that the transformed density $P(R')$ may not be idempotent, or the trace of $P(R')$ may not be conserved. With λ optimization at every step, expansion up to second order in λ is usually adequate. It is advisable, however, to check idempotency and trace of P occasionally during the evolution and purify the density matrix by the standard procedures if significant violation is detected.

The proposed density-swarm algorithm has been tested [26] on hole-doped polythiophene oligomers to understand how the defect states evolve upon doping—how the electron density reorganizes, how the bond lengths are modified, or the bandgap changes. Since the electron density reorganizes extensively and so do the bond lengths along the chains, a good search method is necessary to navigate through the search space and discover the optimum structure at all levels of doping. Using a modified SSH Hamiltonian for modeling the oligomers, and a 20-particle swarm to represent the population, it was always possible to arrive at the doped structures that correctly reproduced the main charge storage configurations (bipolaron) and predicted correct distribution of purely aromatic and quinoid regions along the chain for different levels of bipolaron doping. Figure 7.6a through c displays the evolution of the global best energies discovered by the swarm for a 30-ring polythiophene chain at different levels of bipolaron doping. Figure 7.7a through c displays how the computed density of π-electrons for the best particle in the swarm varies for different rings in the polythiophene chain at various doping levels. The predicted π-density distribution is uniform in the neutral oligomer, all rings being electroneutral. When one bipolaron is doped, there is a pronounced dip in the π-electron distribution along the chain—the broad dip indicating π-electron deficient region created in the middle of the chain by bipolaron doping. On the introduction of a second bipolaron, the π-electron density distribution reveals two symmetrically disposed dips in the middle of the chain of polythiophene rings indicating two π-deficient regions are equal and these regions define the primary charge storage structure in hole-doped oligomers. The molecular structure responds to the π-deficiency created by doping by producing structural distortion in the π-deficient regions—in these regions the rings assume a quinoid character. The emergence of the quinoid regions can be noticed in Figure 7.7d where the optimized C–C (single) bond lengths in PT rings are plotted as a function of the ring number for a neutral polythiophene chain (a), a single bipolaron doped chain (b), two bipolaron doped chains, and (c) 30 polythiophene rings. The bond-length distribution pattern clearly identifies the emergence of quinoid structures in the π-electron deficient region. It is thus demonstrated that a deft combination of classical learning algorithm like the PSO method to evolve structures and exploiting it to induce the electronic density to evolve quantum mechanically produces a viable soft computing method of electronic structure calculation at the simplest mean-field level. We note, however, that Equation 7.49 represents a

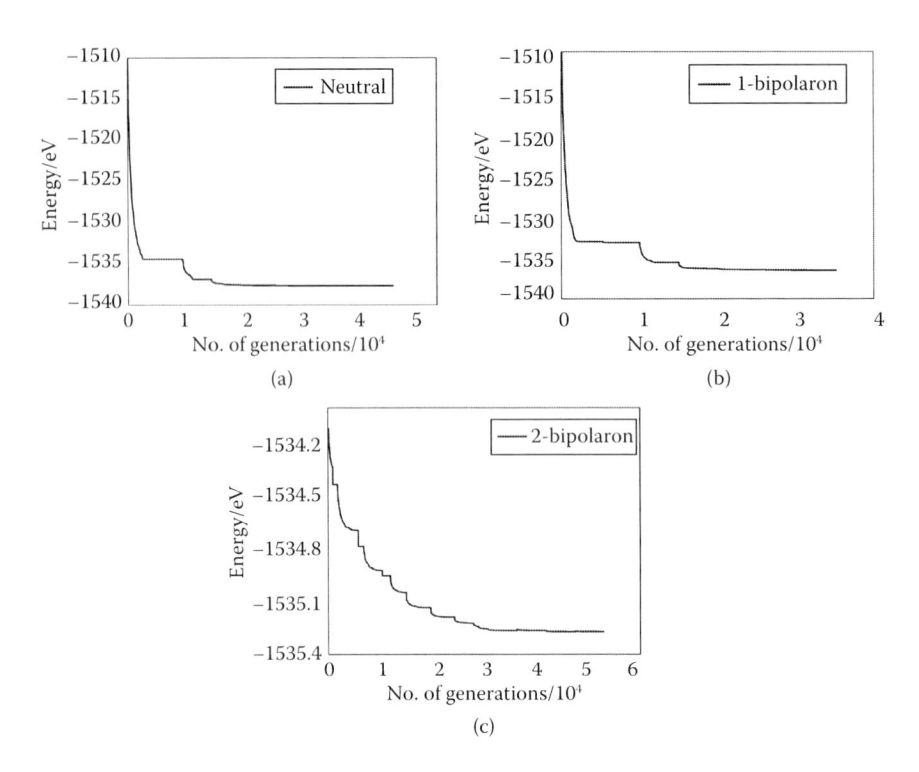

FIGURE 7.6
Profiles of evolution of the computed energy of the best particle in the swarm with time (generation) in (a) neutral, (b) 1-bipolaron doped, (c) 2-bipolaron doped 30-ring PT chain. (From Shukla, R., et al., *Int. J. Quantum Chem.*, 117(5), e25328, 2017. With permission.)

static swarm-learning model. A more sophisticated scheme in which the inertia weight (η) and the acceleration coefficients C_1, C_2 are dynamically adjusted based on analysis of the history of performance can improve the overall performance of density-swarm algorithm. An additional dimension of the problem is worth taking note of. In the velocity-updating Equation 7.49, we have used the globally best particle to influence all other individuals in the swarm to adjust their velocities (and hence positions) accordingly. As argued by Kennedy and Eberhart [28], it accelerates convergence but kills diversity and therefore curbs the ability of the swarm to explore the search space. An alternative velocity-updating equation, which maintains the swarm diversity better and also explores the search space better replaces z_{best} by $z^i_{lbest}(t)$, where $z^i_{lbest}(t)$ is the best solution so far discovered by the ith particle and particles in its immediate neighborhood ($i-1$ and $i+1$). One may consider using the two updating equations keeping in view whether better exploration is the need of the hour or better exploitation is. If better exploration seems to describe at the tth step, we may use the $z^i_{lbest}(t)$ in Equation 7.49 while if we feel that it is time to exploit the information already gained by the swarm the z_{best} updating may be used. The analysis of the past performance history is crucial for successful switching between the two schemes. More careful work is needed for further progress in the direction of developing a completely adaptive density-swarm algorithm for electronic structure calculation in which multilevel adaptation take place. It is quite possible to invoke other population-based techniques of global search in the place of the PSO technique discussed here. Thus, there could be density-GA or density-firefly algorithms for simultaneous search for the nuclear configurations and the corresponding electron density distributions.

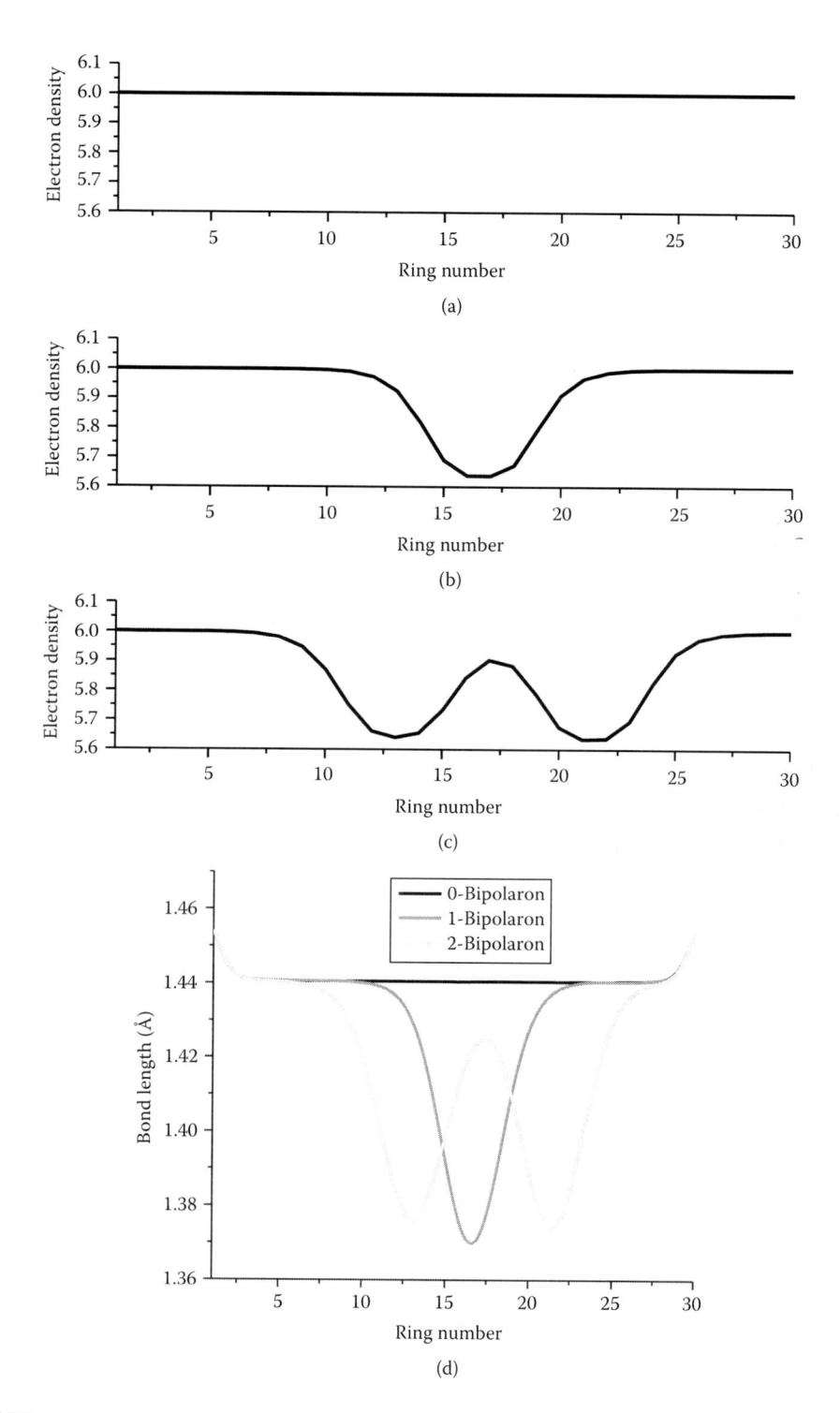

FIGURE 7.7

Density of π-electrons in different rings of a 30-ring PT chain (a) undoped, (b) 1-bipolaron doped, and (c) 2-bipolaron doped PT chain. (d) Optimized C-C ring (single) bond length vs ring number plots with 0-bipolaron, 1-bipolaron, 2-bipolarons. (From Shukla, R., et al., *Int. J. Quantum Chem.*, 117(5), e25328, 2017. With permission.)

In the density-GA method, there would be no velocity-updating equation—the strings representing the N-structural parameters ($R_1^{(i)}, R_2^{(i)}, \ldots, R_N^{(i)}$) evolve genetically (see Chapters 2–3) under the action of crossover, mutation, and fitness-guided selection operators. The geometrical parameters in a string (R, say) can be exploited to construct the molecular electronic Hamiltonian $H(R)$ which would act as the generator of the unitary transformation $U(R)$ for producing $P(R)$ either directly ($P(R) = e^{i\lambda H(R)} P(R_0) e^{-i\lambda H(R)}$) or through an antihermitian matrix $A(R) \left(P(R) = e^{\lambda A(R)} P(R_0) e^{\lambda A\left(R^\dagger\right)} \text{with } A(R)^\dagger = -A(R) \right)$ as elaborated in the density-swarm method. One of the two realizations—the one that makes use of the complex unitary transformation generated by $H(R)$—was tested by Sharma et al. [25] without optimization of the scale parameter λ. There is thus scope for further refinement of the strategy advocated by them.

An alternative would be to make use of the density-firefly algorithm. Here too, there is no velocity vector, but the position vectors in the search space, i.e., the vectors representing the nuclear configurations or molecular or cluster geometry variables, are directly updated by invoking, for example, Equation 5.22 (for the normal firefly algorithm) or Equation 5.24 (for the firefly algorithm with Levy flights). The fitness of the strings computed by evaluation of energy must be mapped onto the intensity of flashes emitted by the fireflies' interactions among which are controlled by flashing intensities and mutual distance of two fireflies (see Chapter 5).

In 2016, Avendano-Franco and Romero [29] leveraged that the FF algorithm for carrying out a structural search on fixed potential energy surfaces of van der Waals clusters and looked for the minimum energy crystal structures including the global one. These authors have demonstrated that the problem of predicting structures of materials can be successfully handled by the firefly algorithm. Starting from the chemical composition and optimally exploiting prior knowledge of similar structures, their firefly algorithm not only successfully predicted the known stable structures, but also discovered a variety of novel energetically metastable but competitive structures. The authors have carefully analyzed the strength and weakness of the firefly algorithm as a multimodal, global search strategy. The method has been implemented in the Pychemia package which serves as an open-source python library for analysis of materials. We anticipate many more explorations leveraging the power and simplicity of the firefly algorithm in computational molecular science and materials chemistry. The firefly algorithm can be easily interfaced with any ESCC or can be used more directly in a density-firefly code by replacing the PSO part in the density-PSO code developed in this chapter. Going beyond the simplest mean-field level is still an open challenge in density soft computing methods as two-electron density matrices are difficult to deal with directly.

7.6.5 Property Constrained Search for Structures

So far we have considered problems in which the molecular structure (R) defines the Hamiltonian $H(R)$ which in turn has been exploited to evolve the starting electron density into the density at the global minimum of the PES. We may now venture to ask the inverse question and seek its solution. The question, broadly speaking, is the following.

Suppose that we want to design a molecule with a targeted property. The question that is to be answered theoretically is: what kind of structures in a class of molecules will generate the desired property or properties? The property-to-structure mapping is difficult to execute. One possibility to carry out the task involves variationally minimizing (globally) energy subject to one or more property constraints. Such constrained variational calculations can be carried out conveniently in a soft computing framework.

Let us consider a conjugated π-electron system like thiophene or selenophene which polymerizes under appropriate conditions to polythiophenes or polyselenophenes with well-defined π-bands. The neutral polythiophenes or polyselenophenes have bandgaps of the order of 2eV. There is extended π-conjugation in these neutral oligomers which are therefore fully aromatic. The question now is to predict theoretically what kind of modifications of the π-electron distribution through structural tuning of the oligomers will lead to considerable reduction in the bandgaps so that we can think of organic conducting materials based on such "structurally tuned" molecular chromophores. Posed theoretically we have a multiobjective optimization problem at hand. We are to minimize the energy of the oligomer subject to the constraint that the bandgap is as low as possible (optimally zero). Sarkar et al. [30] suggested a method of solving the bimodal problem within the framework of the random mutation hill climbing method introduced in Chapter 4. Needless to say, SA or GA could have been used with equal facility.

Sarkar et al. start by constructing a weighted functional $L(R, \alpha, \beta)$—the objective function of the problem where

$$L(R,\alpha,\beta) = \alpha\left(E_T(R) - E_L\right)^2 + \beta\left(\epsilon_{LUMO}(R) - \epsilon_{HOMO}(R)\right)$$
$$= \alpha f_1(R) + \beta f_2(R). \tag{7.55}$$

$E_T(R)\left(=E_\pi(R) + E_\sigma(R)\right)$ represents the total energy (π electronic plus σ compression) of the oligomer at the geometry coded by the string R; E_L is an estimated lower bound to the constrained total energy. $\epsilon_{LUMO}(R)$ and $\epsilon_{HOMO}(R)$ are the binding energies of the lowest unoccupied and the highest occupied molecular orbitals of the oligomer at the geometry R. α and β are penalty weight factors to chosen by the user—but the authors made a particular choice of $\beta = 1 - \alpha$ (i.e., $\alpha + \beta = 1$), and used a modified Su–Schrieffer–Heeger Hamiltonian $H(R)$ to model the oligomers [24]. The π-molecular orbitals (ϕ_is) are formed by linear combinations of carbon $2p_z$ and sulfur $3p_z$($4p_z$ for selenium) atomic orbitals (x_ps). Therefore, with $\{C_{pi}\}$s as optimizable parameters, $\phi_i = \sum_{p=1}^{N} C_{pi}x_p$. The number of occupied (doubly) π-molecular orbitals (noc) are determined by the numbers of π-electrons (n_π) in the system with $noc = n_\pi / 2$. The π-electron density matrix $P(R)$ is obtained from the occupied molecular orbitals by the definition

$$P(R) = \sum_{i=1}^{noc} C_i(R)C_i^\dagger(R) \tag{7.56}$$

where $\{C_i\}$s are eigenvectors of the $H(R)$ matrix (real, symmetric Hamiltonian matrix) with eigenvalues $\{\epsilon_i\}$. $\{C_i\}$s define the π-molecular orbitals and ϵ_is are the π-molecular orbital energies. Total energy $E_T(R)$ of the oligomer, in the SSH model is given by

$$E_T(R) = 2\, Tr\left(P(R)H(R)\right) + \frac{1}{2}\sum_{ij} f(R_{ij})$$
$$= E_\pi(R) + E_\sigma(R) \tag{7.57}$$

where $f(R_{ij})$ is the carefully parametrized function of interatomic distances (only bonded atom pairs (i, j) are considered) and the sum defines the σ-framework energy $E\sigma(R)$. The variational task at hand is to look for structures that minimize the total energy (therefore minimizing the objective function $L(R)$ in Equation 7.55) for a particular choice of weighting factors α and β. This is clearly a multiobjective optimization problem that Sarkar et al. addressed by invoking their CARMHC method. They proceed by converting the bi-objective functional of Equation 7.55 into the corresponding fitness function $F(R, \alpha, \beta)$ by the standard mapping

$$F\left(R, \alpha, \beta\right) = e^{-\lambda L(R,\alpha,\beta)} \tag{7.58}$$

and then looked for maxima of the fitness function for different choices of α (hence, β) within the framework of their CARMHC method. A random modification of the structure (R') is generated from R, $H(R')$ is constructed and diagonalized to obtain the π-molecular orbitals $(C_i(R'))$ and hence the π-electron density matrix $P(R')$, the total energy and bandgap are computed. If $F(R', \alpha, \beta) \geq F(R, \alpha, \beta)$, the modified structure is accepted and the procedure is repeated until no further improvement of fitness value is possible.

If we plot the values of the bandgap constraint $f_2(R)$ and energy constraint $f_1(R)$ at the optimized structures (R) for different choices of α, we get the curve displayed in Figure 7.8a in which the points on the curve correspond to the Pareto-optimal solution of the bi-objective problem. By plotting $f_2(R)$ and $f_1(R)$ against α and β in the same figure (Figure 7.8b), we can graphically determine the α value (call it α^*) where $f_1(R) = f_2(R)$. α^* represents the weight factor for which both the constraints are equally satisfied and the corresponding structure could hold the key to designing thiophene- or selenophene-based low bandgap material. Polythiophene turns out to have $\alpha^* = 0.4837$ ($\beta^* = 0.5163$). For this optimal choice of α^*, β^* values, the zero bandgap constraint is found to force the polythiophene oligomer to adopt a structure in which approximately 5% of the polythiophene rings in the middle of the chain assume a perfect quinoid character.

These quinoid rings in the middle are electron deficient and carry a net positive charge (Δq), while the remaining rings are aromatic and carry negative charge;

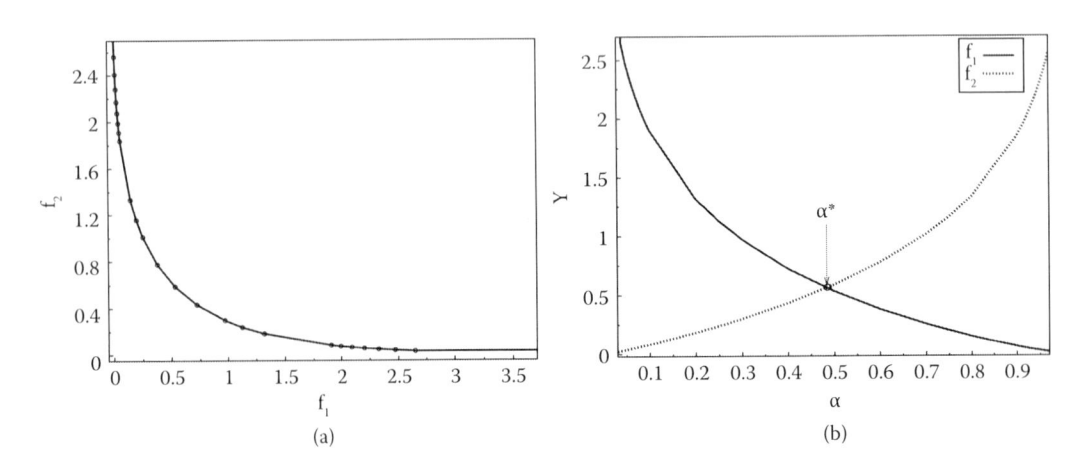

FIGURE 7.8
(a) Pareto-front for zero bandgap constrained 100 ring PT chain; the points in the graph represent the Pareto-front solutions, (b) plot of energy constraint (f_1) and HOMO–LUMO gap (f_2) with penalty factor (α) for energy constraint; α^* is the optimum penalty factor. (From Sarkar, K., *Int. J. Quantum Chem.*, 112(6), 1547–1558. With permission.)

the polythiophene chain as a whole remains electroneutral. For the optimal choice of α^*, the electron deficiency in the quinoid rings in the middle of the bandgap constrained chain is ~0.015, the energy sacrifice relative to the unconstrained polythiophene chain is ~0.743 eV—the bandgap being lowered to 0.748 eV. The designing clue coming out from the bandgap constrained energy minimization is crystal clear—we must introduce quinoid rings in the chain, i.e., create defects—either by hole doping or by creating an alternative AB type of polymer, where A carries an electron donating and B an electron withdrawing group.

There are other soft computing ways of solving the bi-objective problem discussed here. A nondeterministic sorting GA (NSGA-II) developed by Deb et al. [31] could be an interesting way of handling the present or similar multiobjective optimization problems in which minimization of energy has to be carried out under p-number of property constraints.

$$Tr\left(PA_k\right) = a_k^0;\ k = 1, 2, \ldots, p. \tag{7.59}$$

The objective function may be written as

$$F\left(E_L, \{\lambda_k\}\right) = \left(E - E_L\right)^2 + \sum_k \lambda_k\ Tr\left[PA_k - a_0^k\right]^2 \tag{7.60}$$

where A_k is the matrix representing the kth one-particle property in the atomic orbital basis, P is the one-electron density matrix in the same basis and a_0^k s represent the desired property values the optimized structures are expected to produce. The minimization can be done by invoking any one of the soft computing methods. Very little work in this area seems to have been reported. Once fully developed these soft computing techniques of search may replace the trial and error method of finding suitable chromophores for materials with targeted properties.

We have given a fairly comprehensive account of different types of methodological developments that have taken place in the interface of soft computing and molecular electronic structure calculations. It is time now that we briefly review numerous applications of soft computing in the search and design of new structures that have already appeared in literature.

7.7 Reviewing the Rapidly Growing Soft Computing Structural Search Interface

Clusters may have properties which are different from those of discrete molecules or bulk matter; for example, some metals (e.g., palladium), which are nonmagnetic in the solid state, are magnetic, with relatively high local magnetic moments in both neutral and anionic Pd_N clusters [32]. The coulomb explosion of large rare-gas clusters releases huge energy (approaches the energy of nuclear processes). At very low temperatures ($< 2K$ for ^{4}He), He clusters may form superfluid droplets. Metal nanoparticles play a crucial role as heterogeneous catalysts in petrochemical, pharmaceutical, and clean energy sectors. Novel and interesting nanostructures, having useful chemical and physical properties, could be obtained by controlling the nanoparticle size, composition, surface site preferences, and degree of segregation between the metal constituents [33]. Therefore, computing ground state configurations of atomic and molecular clusters is a challenging task from the theoretical perspective. The number of local minima of the clusters rises exponentially with the growth in cluster size. It increases further with composition heterogeneity because structures more complex than their homogeneous counterparts are possible. It is due to the existence of isomers with the same geometry and composition but showing different distributions of the constituent particles that the optimization task becomes notoriously difficult [34–42]. For a cluster made of n_A and n_B atoms, with $n_A + n_B = N$, one finds that a single geometrical isomer can have $\dfrac{N!}{n_A!\,n_B!}$ different homotopes. Various attempts have been presented in the literature to find the global minimum energy structure of these systems, for example, the dynamic lattice searching method [43], the basin-hopping (BH) approach [44–47], the adaptive immune optimization algorithm (AIOA) [48–51], and EAs [52–69]. Hybrid approaches combining an EA and a local search procedure have been increasingly used to handle these problems. We review some of the recent work in the area in what follows.

7.7.1 LJ and Morse Clusters

The ab initio quantum mechanical calculation for finding the global minimum geometry of a cluster of atoms or molecules is often extremely expensive [70,71]. Computational cost increases with the accuracy of the calculation and size of the system. Generally, empirical or semiempirical pair potentials are employed for large systems. LJ and Morse functions are the two most widely used and relatively simple pairwise additive potential models describing interactions between atoms which have been regularly used as benchmarks to assess the performance of global search methods for cluster geometry optimization. The LJ and Morse potentials are respectively given by

$$V_{LJ}\left(r_{ij}\right) = 4\,\epsilon \sum_{i=1}^{N-1}\sum_{j>i}^{N}\left[\left(\frac{\sigma}{r_{ij}}\right)^{12} - \left(\frac{\sigma}{r_{ij}}\right)^{6}\right] \tag{7.61}$$

$$V_{M}\left(r_{ij}\right) = \epsilon \sum_{i=1}^{N-1}\sum_{j>i}^{N}\left[e^{-2\beta\left(r_{ij}-r_0\right)} - 2e^{-\beta\left(r_{ij}-r_0\right)}\right] \tag{7.62}$$

where r_{ij} stands for the Euclidean distance between atoms i and j, r_0 is the equilibrium bond length, ϵ is the well depth of the pair potential, σ is the separation at which the pair potential between the atoms goes through zero, and β is the interaction range scaling parameter. For the prediction of the ground state structure of crystals, the USPEX method [72–81] has turned out to be extremely powerful and already guided materials scientists in finding interesting and unexpected crystal structures. Zhu et al. [82] enhanced the efficiency of USPEX method by designing additional variation operators and constraints for partially or completely fixed molecules. Goedecker et al. [83–85] offered improved minima hopping with a softening method and a stronger feedback mechanism to predict structures of homoatomic and binary clusters with LJ interaction as well as structures of silicon and gold clusters described by force fields. Cheng et al. [38] employed a funnel hopping algorithm coupled with GA to locate the putative global minima of the LJ clusters and the Morse clusters up to $N = 160$. Wales et al. showed that the use of approximate symmetry provides a more productive way to explore the configuration space and substantially improves the efficacy of global optimization for the atomic clusters [86–89]. Froltsov et al. [90] developed the "cut and splice" GA augmented with twinning mutation moves for the structural optimization of both a series of single-funnel LJ clusters up to 70 atoms and the double-funnel LJ_{38} cluster. Binary Lennard–Jones (BLJ) cluster optimization is an even more challenging problem from the point of view of combinatorial complexity. This is also interesting because the catalytic properties of such clusters depend on the composition and structure. Comparatively little work has been done on the mixed clusters [91–93] as the search space is dramatically enlarged with the inclusion of more atom types. Mixed LJ clusters of widely varying compositions offer highly interesting additional perspectives on how variations in preferred structures emerge. Hybrid approaches such as combining a global optimizer with a local search procedure have proved to be useful for mixed clusters [94–100], and 17 new putative global minima for BLJ clusters in the size range of 90–100 particles have been predicted by coupling hidden-force algorithms and non-Markovian parallel Monte Carlo search [96]. El Dor et al. proposed a multiswarm-based algorithm called PSO-2S [97] and tested it successfully on a number of benchmark functions. It uses charged particles in a partitioned search space and applies a repulsion heuristic on particles to increase the diversity for solving both unimodal and multimodal problems. Using a GA-based global structure optimization framework and new set of fitted parameters for LJ potential from high-end ab initio calculations, Dieterich et al. optimized strongly mixed binary to quinary rare-gas clusters [98,99]. An average offspring method [83] designed for cluster structure prediction was shown to perform better in systems with compact optimal structures. Two individuals are randomly selected and for each atom of the first cluster, the closest lying atom of the other cluster is identified. The corresponding atom of the child is placed randomly on the connecting line between the two parent atoms. The randomness of this operator is necessary to prevent the algorithm from producing a lot of identical offspring. Leitao et al. [101] claimed that they applied for the first time an island model to the optimization of Morse clusters ranging from 41 to 80 atoms, combined with a hybrid steady-state EA and a local optimization method. The performance was slightly more robust than that of the sequential approach. Dieterich et al. [99,102,103] designed an EA program suite called OGOLEM for structure optimization of mixed clusters. The OGOLEM framework provides both an MPI frontend framework for MPP parallelization and a threading fronted framework for SMP parallelization, omitting unnecessary MPI overhead. They demonstrated the possibility to design molecules with targeted properties in the area of photochemistry using the OGOLEM framework [102]. There are prescriptions for the inclusion of tabu search features [104,105] into the EAs which might help reduce the amount of time spent in local optimizations rediscovering already known minima. Daskin et al. [106] presented the group leaders optimization algorithm in which the influence of the leaders in social groups is the inspiration for the evolutionary progress. The method is applied to locate the geometric structures of LJ clusters as well as to the quantum circuit design problems. Deep et al. [107] made an attempt to solve LJ problem by incorporating a multiorbit dynamic neighborhood topology in PSO. In multiorbit topology, the swarm has heterogeneous connectivity with some subsets of the swarm strongly connected while the others are relatively isolated. This heterogeneity of connections balances the exploration–exploitation trade-off in the swarm. The dynamic neighborhood topology helps avoid entrapment in local optima, as well.

7.7.2 Ionic Clusters

Oleksy et al. [108] calculated the equilibrium geometries and dissociation energies of the electronic ground state of He^+_N clusters ($N = 3 - 35$) via an extended GA method and employing a semiempirical valence-bond model of intracluster interactions. They found that for the cluster sizes of $N = 3, 4, 7–9$, and $14–35$, the positive charge of He_N^+ delocalizes over a trimer ionic core, which is more or less linear and centrosymmetric. For $N = 2, 10–13$, a dimer ionic core develops, while for $N = 5$ the positive charge is delocalized over the whole cluster; for $N = 6$, a tetramer ionic core is formed. Atoms outside the ionic core are neutral and are distributed among several solvation rings. The electronic and geometrical structures of sodium cluster anions [109] [Na_n^-, $n = 20 – 57$] and cationic Na_n^+ clusters [110] were determined by applying GA-density functional theory (DFT). Structures of indium oxide nanoclusters and zirconia nanoclusters have been predicted by a method combining a robust EA with classical interatomic potential and quantum chemical models [111,112]. Kim et al. [113] used GA to design nanoporous TiO_2 for low-temperature processable photoanodes of dye-sensitized solar cells. Darwinian and Lamarckian schemes within EAs have been compared in the context of structure prediction of the titania (TiO_2) polymorphs. Lamarckism in natural evolution regards the effects of "inheritance of acquired characters" as the motive force of evolution, while Darwinism claims that evolution is nothing but the cumulative processes of natural selection with random mutation and denies the possibility of inheritance of acquired characters. Although the mainstream of today's evolutionary theory follows Darwinism, it has been found that the Lamarckian scheme is more successful and efficient at generating the target structures [114]. A combination of Buckingham and LJ potential functions describe the interactions between different atomic species in the system:

$$V(r_{ij}) = A_{ij}e^{-\frac{r_{ij}}{\rho_{ij}}} + \frac{B_{ij}}{r_{ij}^{12}} - \frac{C_{ij}}{r_{ij}^{6}}. \tag{7.63}$$

The parameters A_{ij}, B_{ij}, C_{ij}, and ρ_{ij} are dependent on the species involved in the interaction. To perform a global geometry optimization of clusters resulting from the microsolvation of alkali metal ions (i.e., Na^+, K^+, and Cs^+) with benzene molecules, Marques et al. [115] used modified LJ function for nonelectrostatic contributions. Zhu et al. [116] explored all the possible stoichiometries for Mg-O systems at pressures up to 850 GPa. They found that two extraordinary compounds MgO_2 and Mg_3O_2 became thermodynamically stable at 116 GPa and 500 GPa, respectively. They predicted the existence of thermodynamically stable Xe-O compounds [117] at high pressures (XeO, XeO_2, and XeO_3 became stable at pressures above 83, 102, and 114 GPa, respectively). Al-Sunaidi et al. [118] predicted a greater stability of tetrahedral and trigonal coordinations compared to the tetragonal one for zinc oxide clusters [$(ZnO)_n$, $n = 1 – 32$] using an EA with polarizable shell interatomic potentials. Wang et al. [119] successfully applied PSO for the prediction of elemental (Li, C, Mg, Si), binary (SiO_2, SiC, ZnO, TiH_2, TiB_2, MoB_2), and ternary ($MgSiO_3$ and $CaCO_3$) compounds in various chemical-bonding environments (metallic, ionic, and covalent bonding). Li et al. [120] obtained the geometrical structures of $(Al_2O_3)_n$ ($n = 1 – 7$) clusters via GA coupled with the DFT. Avaltroni et al. [121] identified peculiar low-energy isomers of small clusters (C_2Al_4 and CB_6^{2-}) using both standard random search and GA procedures. Pal et al. [122,123] determined the lowest energy structure of ZnS quantum dots of different sizes by using a search based on GA coupled with the density-functional tight-binding method (DFTB) and found new ring-like configurations of ZnS quantum dots which had higher HOMO–LUMO gaps compared to other ZnS quantum dot structures. Pereira et al. developed an EA for the global minimum energy search for water clusters up to $(H_2O)_{20}$, benzene clusters up to $(C_6H_6)_{30}$ and $(C_6H_6)_n^+$ with $n = 2–20$ [124].

Do et al. [125] identified the energy minima of water, methanol, water + methanol, protonated water, and protonated water + methanol clusters with DFT combined with basin hopping. Addicoat et al. [126] optimized a GA in order to identify the minimum energy structure of arbitrarily hydrogenated and hydroxylated fullerenes, and implemented the method for exhaustive calculations on all possible isomers. They suggested that the crossover operator did not have a significant effect on the search efficiency of GA and hence the most efficient version of their GA does not employ crossover, thereby reducing it to an EA. Using the GA, Zhang et al. [127] predicted the most stable structures and a number of low-energy metastable structures for Si[001] symmetric tilted grain boundaries with various tilt angles, which are found to be in very good agreement with the results of first-principles calculations. Yao et al. [128] investigated the high-pressure structures of solid nitrogen through GA combined with first-principles electronic structure calculations. Hooper et al. [129] studied generic defect association complexes in metal oxide materials at experimentally relevant dopant concentrations by a specialized GA-inspired search procedure for doped metal oxides. The search algorithms had been tested on lanthanide-doped ceria (L = Sm, Gd, Lu) with various dopant concentrations.

7.7.3 Metal Clusters and Nanoalloys

Chen et al. [130] proposed a parallel differential evolution for cluster optimization (PDECO) with triangle mutation and migration operators, and applied it to Pt clusters with great efficiency. Rogan et al. [131] implemented a GA-based method on small Pd clusters. The lowest-lying isomers of the copper cluster, Cu_9, have been obtained by combining a GA-driven approach with DFT [132]. Assadollahzadeh et al. [133] employed a seeded GA technique using DFT together with a relativistic pseudopotential to search for global and energetically low-lying minimum energy structures of neutral gold clusters Au_n (n = 2–20). GA coupled with a tight-binding potential was employed by Li et al. [134] to optimize neutral lead clusters Pb_n (n = 2–20). Pereira et al. [135] made a study on the effectiveness of different crossover operators in the global optimization of atomic clusters. They came up with a cut and splice crossover operator which proved to be extremely useful. The first parent string is "cut" along a random horizontal cutting plane into two complementary parts and the second parent string is cut in such a way that the number of the atoms beneath and above the cutting plane are equal to the number of atoms of the two complementary parts of the first parent. "Splice" joins the head of one string with the tail of the other. Pereira et al. [135] modified this operator by the way it determines the subclusters to be exchanged. Here a random atom from the parent string is selected and placed on the offspring. A random number ($M \in [1, N - 2]$) is generated (N is the number of atoms in the cluster). M atoms from the first parent string closer to the first selected atom are chosen and placed on the offspring. The remaining ($N - M - 1$) atoms are taken from another parent string closer to the first selected atom. Those atoms which are already occupied are skipped. Figure 7.9 schematically depicts the cut and splice crossover. The energetic ground states of gold clusters, for some magic numbers, with up to 318 atoms were obtained by the minima hopping method [136]. Meng et al. [137] found that $NbCl_5$ and ZnMg intercalation in graphite results in p- and n-type doping, respectively. To model clusters they used Gupta semiempirical potential and obtained globally optimal structures of those metal clusters employing GA with cut and splice crossover and triangle mutation. The optimized clusters were then placed between two layers of graphite, each of which contained primitive 12 × 12 cells and 288 atoms. The whole system was optimized with a universal potential and the electronic structures at the optimized doped cluster geometries

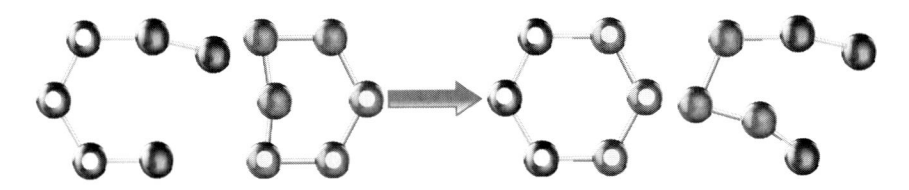

FIGURE 7.9
Cut and splice crossover: To make the operation consistent and produce descendants with same number of atoms, the center of mass of the parent structures have to be translated to the origin of the coordinate system and have to rotate the plane of the parent clusters until the generated offspring contain the correct number of atoms.

were finally calculated using DFT. The atomic arrangement makes an immense contribution to the catalytic activities of a nanoalloy. So, the development of a reliable method for predicting atomic arrangements has become increasingly important. Hong et al. [138] combined GA with first-principles electronic structure calculations to find the most stable configurations of elementary Au_m and Ag_n clusters, as well as binary clusters Au_mAg_n ($5 \leq m + n \leq 12$). The search led to the critical Au:Ag ratios for which the two- to three-dimensional transition takes place. The study showed that Ag atoms prefer peripheral positions with a lower coordination number while Au atoms tend to occupy central sites with higher coordination numbers. Johnston et al. [139] investigated small Sn-Bi clusters employing a DFT/GA method. Oh et al. [140] developed a combination of GA and classical molecular dynamics (MD) simulations to evaluate the optimal arrangement of the multi-layered core-shell structure for the Pt-Cu nanoalloy with 1,654 atoms, regardless of the composition ratio. Using a tabu search in the descriptor space and DFT, Orel et al. [141] found the global minima of the neutral binary Sn_mPb_n atomic clusters, $7 \leq m + n \leq 12$, for all the possible stoichiometric ratios. Fournier et al. [142] found the minimum-energy structures of Ag_nRb_n ($n = 2 - 10$) clusters by a combination of DFT and tabu search. Chu et al. [143] described a new fragment-based EA for de novo optimization, specifically developed to handle organometallic and transition metal compounds. Au atoms preferentially segregate on the surface of Pd-Au nanoalloys. DFT calculations on FCC-type cubo-octahedral Pd-Au nanoparticles have indicated that the surface Pd atoms occupy (111) rather than (100) faces, thereby maximizing the number of relatively strong surface Pd-Au bonds [144]. Larger cohesive energy, lower electronegativity, and smaller atomic radius (minimize bulk elastic strain) of Pd favors a Pd core and an Au surface shell by lowering the surface energy of the cluster. Besides electron transfer from Pd to more electronegative Au, greater order of the Pd-Au than the Pd-Pd and Au-Au bonds favors Pd-Au mixing [37]. Ferrando et al. had considered a methodology based on extensive global optimization within empirical potential models and subsequent DFT based on local relaxation of low-energy structures pertaining to different structural motifs (or basins on the energy landscape) of gas-phase alloy nanoclusters [35,36,58,145–148] modeled by the many-body Gupta potential. Pittaway et al. [37] performed global optimization of Pd-Au bimetallic clusters using a GA, coupled with the Gupta many-body empirical potential (EP) to model interatomic interactions. Johnston and coworkers [35,36,41,42,58,149–155] worked on a series of developments on theoretical simulation of alloys and bimetallic clusters (e.g., Cu-Ag, Ni-Ag, Cu-Au, Ag-Pd, Ag-Au, Au-Pd) with GA coupled with the Gupta many-body potential. $Ag_{core}Pd_{shell}$ nano-catalysts could be important in the development of micropolymer electrolyte membrane fuel cells for portable devices, and could also be applied to the promotion of other catalytic reactions [156]. An Ag_3Pd_{10} cluster was identified as a wedge-shaped nanoshell with

C_S point group symmetry by a combination of GA global optimization and DFT calculation. Wu et al. [49] performed global optimization of Ag_mPd_n ($m + n = 15$) and $Ag_{3m}Pd_{38-3m}$
($m = 1 - 12$) clusters and came to the conclusion that silver atoms had a strong tendency
toward segregating at the surface. Combined experimental and theoretical studies have
indicated the preference for $Pd_{core}Au_{shell}$ (~5 nm) and an "onion-like" Pd-Au-Pd configuration for larger Pd-Au nanoalloys (~12 nm). Johnston et al. [36] have studied mixed coinage
metal clusters, using the Gupta potential.

7.7.4 Crystal Structure Prediction

EAs [73,75,76,77,80,81,82,83,108,111,112,114,119,157–163] have provided a reliable and
powerful means of exploring the PES of crystals leading to the identification of the most
stable, interesting, and sometimes unexpected crystal structures. There are several other
methods such as basin hopping [83,84,164–166], metadynamics [167,168], PSO [119,169],
and simulated annealing [170] for performing similar searches. For large-dimensional
problems, due to the complex nature of PES and the presence of high-energy disordered
structures (random initial population) these algorithms often encounter serious challenges. Lyakhov et al. introduced an additional variation operator—the coordinate mutation or soft mutation [157] for the purpose of crystal structure prediction. Instead of using
complete randomness in the mutation operator, concerted mutation which directs the
system to choices that have a higher probability to improve the fitness of the solution
string is introduced. The idea of soft-mode mutation or softmutation [77,157] is to move
the atoms along the softest modes, i.e., the lowest frequency eigenmodes that correspond
to directions of the lowest curvatures of the energy surface. The eigenvector corresponding to lowest nonzero eigenvalue determines the direction of softmutation. To calculate
the softest modes one has to construct the dynamical matrix. It is often enough to have
an approximate direction and sufficiently large mutation amplitude to arrive at a new
low-energy structure. There is no need therefore to perform computationally expensive ab initio dynamical matrix calculation. Cheaper methods for example a dynamical
matrix computed from bond hardness coefficients can be used. The soft mutation favors
atoms with higher local order; Lyakhov et al. [77] constructed the first generation using
pseudosubcells with fractional atomic occupancies to improve the order and diversity
of the structures. They used a fingerprint function that improves the selection process
through removing clones. Superconducting high pressure phase of germane [73], structural characterization of compressed silane [171], rhombohedral structure of superhard
BC_2N [172], superconducting structures of BC_5 [173], superhard monoclinic polymorph of
carbon [174], high-pressure orthorhombic polymorph of MgB_2 stable above 190 GPa [175],
high-pressure phases of $CaLi_2$ [176], metallic structures of oxygen at pressures in the range
of 100–250 GPa [75], two new hexagonal ultra-hard phases of WN_2 [177], two unique high-
pressure metallic phases of Stannane (SnH_4) at superconducting temperatures of 15–22 K
for the Ama2 phase at 120 GPa and 52–62 K for the $P6_3/mmc$ phase at 200 GPa [178],
and $SiH_4(H_2)_2$ at high superconducting temperatures of 98–107 K at 250 GPa [178,179] are
some of the systems explored through ab initio evolutionary methodology.

By combining PSO with first-principles calculations, Xiang et al. [180] discovered a new
metastable Si phase (Si_{20}-T has a quasidirect gap of 1.55 eV) which could be a promising
solar energy absorber. Oganov et al. [76] found a new boron phase (comprising icosahedral B_{12} clusters and B_2 pairs in an NaCl-type arrangement), stable between 19 and 89 GPa,
and exhibiting evidence for charge transfer. In the context of reliably predicting ground
state geometry, the Universal Structure Predictor: Evolutionary Xtallography (USPEX)

[72,80–82,181] system has turned out to be extremely successful. Oganov et al. using USPEX methodology first demonstrated the possibility of computing hardness (at least for insulators and semiconductors) just from the crystal structure, which opens up the possibility of global optimization of hardness, aimed at the computational discovery of new superhard materials. They had introduced local measures of the quality of structure to locate defective regions in the crystal and used fingerprint niching, soft-mode mutation, symmetry, and pseudosymmetry-enabled generation of structures to greatly speed up the search for the global minimum [182,183]. In an article published in 2009 [184] they summarized the different applications of the USPEX method as a tool for crystal structure prediction and showed that this method could have enormous applications in both computational materials design and studies of matter under extreme conditions. In a recent application Oganov et al. showed that in conjunction with first-principles calculations, EAs will lead to the discovery of novel dielectrics [185]. In another application, they successfully predicted an energetically lower novel two-dimensional boron structure than the α-sheet structure [186], stable hafnium carbides [187]. Woodley et al. predicted ground state geometry for a series of transparent conducting indium oxide [111] and zirconia nanoclusters [112] by combining an EA with classical interatomic potential and quantum chemical models. Wu et al. showed that structure exploration by classical potentials with the accuracy of DFT is an efficient scheme for complex crystal structure prediction [188]. In another work, they favorably compared Lamarckian schemes within EAs with Darwinian schemes in the context of minimum energy structure prediction of titania phases [114]. O'Keeffe [189] described several difficulties encountered in the crystal structure prediction. Study of GA and minima hopping method-based crystal structure prediction reveals that for relatively smaller sizes, both methods show comparable efficiency while for larger systems GA becomes advantageous over minima hopping [164]. Oganov et al. explained how and why evolutionary crystal structure prediction works the way it does [190]. Using adaptive GA Zhao et al. [191] studied the structures and stabilities of the alkaline earth metal peroxide XO_2. They also predicted complex crystal structures of the orthorhombic, rhombohedral, and hexagonal polymorphs close to the Zr_2Co_{11} intermetallic compound [192]. Li et al. [193] explored the high-pressure crystal structures of Mg by the PSO algorithm. Wang et al. have developed a software package titled Crystal Structure Analysis by Particle Swarm Optimization (CALYPSO), enforcing symmetry constraints on structural generation, bond characterization matrix for elimination of similar structures, partial random structures per generation for enhancing structural diversity, and penalty function, etc. [119,169] for predicting crystal structure from random initial starting geometries. They have applied CALYPSO to crystal structure prediction, earth, and planetary materials [194–201]. Luo et al. [202] predicted new stable structures of two-dimensional boron-carbon compounds for a wide range of boron concentrations by the PSO algorithm implemented in the CALYPSO code. XtalOpt [203] and EVO [160] are the two other packages that have proved to be useful in this context. Zhu et al. proposed a method based on metadynamics and EAs [167] and found stable and metastable states for Al_2SiO_5, SiO_2, $MgSiO_3$, and carbon clusters [168] from reasonable initial structures, providing insight into the mechanisms of phase transitions. They designed an evolutionary algorithm-based method [181] to automatically explore low-energy surface reconstructions with variable surface atoms and reconstruction cells, and illustrated it by the identification of N_3 trimeric reconstruction of GaN($10\bar{1}1$) surfaces. Liu [204] constructed a multialgorithm-collaborative universal structure-prediction environment (MUSE) to efficiently find the stable and metastable structures of materials under given conditions. In MUSE the EA was coupled with simulated annealing and basin hopping algorithms. With the inclusion of two new variation operators, slip and twist, the performance of MUSE was greatly enhanced. Fadda et al. [205] described an EA based on symmetry-preserving and symmetry-breaking mutations for the exploration of the space of the conjugacy classes of crystal structures. Johnston et al. have implemented and parameterized both Darwinian and Lamarkian GA searches to identify the structure of clusters from experimental STEM images [206]. Meredig et al. [207] devoted considerable theoretical and computational efforts for the development of first-principles-assisted structure solution (FPASS) that automatically solve crystal structures. Nguyen et al. [208] performed GA to find two metastable Si-IX phases with good experimental agreement in the lattice parameters.

The evolutionary computing techniques and their hybrids have registered impressive success in the elucidation of minimum energy structures of atomic clusters and crystalline solids. These methods have not been explored that extensively for weakly bonded molecular clusters and crystals. We anticipate rapid growth of hybrid EC-based exploration of such structures. Since these surfaces are dominated by shallow minima separated by small barriers, special mutation and crossover operators would have to be designed for exhaustive exploration of the PES and locating global as well as local close-lying minima. The information on such structures and their energies could then lead to the correct prediction of thermally averaged properties of such species. The use of EC for directly solving the molecular Schrödinger equation has been very limited. The density-based method described in reference [23] has the potential to be a viable, even a superior alternative for computing the equilibrium molecular structure and the corresponding one-electron density resolved in a given basis. We anticipate extensive exploration along that line in future years. Intelligent interfacing of the standard electronic structure codes and EC codes could prove instrumental in further growth of EC techniques in the exploration of molecular structures. One area of great promise where EC could decisively prove superior is in designing molecules and materials with targeted properties. Such problems can be cast in the mold of a multimodal optimization problem [30] which can be efficiently handled by the EC techniques using for example, the prey–predator model for multiobjective constrained optimization. Along with artificial neural networks, EA can meet state-of-the-art methods for powerful quantitative structure–property relationships modeling [209,210]. Recently a hybrid Taguchi-genetic algorithm [211] has been developed for global numerical optimization problems with continuous variables. The systematic reasoning ability of the Taguchi method [212] was exploited to select better genes for crossover, thereby enhancing the performance of the GA. The method holds promise in cluster and crystal geometry optimization. In addition, the Taguchi method can be very useful with optimization of GA parameters. We anticipate rapid growth in such studies in the next decade.

The cross-talk between EC and computational chemistry is expected to remain productive. We foresee new hybrid algorithms for exploring electronic structures of atoms, molecules, and clusters evolving from it in the near future.

References

1. McWeeny, R., and B.T. Sutcliffe. 1969. *Methods of Molecular Quantum Mechanics*. Theoretical Chemistry. Berkeley, CA: Academic Press.
2. Fiacco, A.V., and G.P. McCormick. 1968. *Nonlinear Programming: Sequential Unconstrained Minimization Techniques*. Research Analysis Corporation. Research Series. Philadelphia: Wiley.
3. Kirkpatrick, S., C.D. Gelatt, and M.P. Vecchi. 1983. Optimization by Simulated Annealing. *Science* 220, no. 4598: 671–680.

4. Metropolis, N., A.W. Rosenbluth, M.N. Rosenbluth, A.H. Teller, and E. Teller. 1953. Equation of State Calculations by Fast Computing Machines. *The Journal of Chemical Physics* 21, no. 6: 1087–1092.

5. Dutta, P., D. Majumdar, and S.P. Bhattacharyya. 1991. Global Optimization of Molecular Geometry: A New Avenue Involving the Use of Metropolis Simulated Annealing. *Chemical Physics Letters* 181, no. 4: 293–297.

6. Chaudhury, P., P. Dutta, P. Bandyopadhyay, P. Sarkar, and S.P. Bhattacharyya. 1996. A Random Walk to Local Minima and Saddle Points on a Potential Energy Surface. *A Strategy Based on Simulated Annealing. Chemical Physics Letters* 250, no. 2: 238–246.

7. Chaudhury, P., and S.P. Bhattacharyya. 1998. A Simulated Annealing Based Technique for Locating First-Order Saddle Points on Multidimensional Surfaces and Constructing Reaction Paths: Several Model Studies. *Journal of Molecular Structure: THEOCHEM* 429: 175–186.

8. Chaudhury, P., and S.P. Bhattacharyya. 1999. Locating Critical Points on Multi-Dimensional Surfaces by Genetic Algorithm: Test Cases Including Normal and Perturbed Argon Clusters. *Chemical Physics* 241, no. 3: 313–325.

9. Bungay, S.D., R.A. Poirier, and R.J. Charron. 2000. Optimization of Transition State Structures Using Genetic Algorithms. *Journal of Mathematical Chemistry* 28, no. 4: 389–401.

10. Wales, D.J. 1989. Finding Saddle Points for Clusters. *The Journal of Chemical Physics* 91, no. 11: 7002–7010.

11. Quapp, W. 1996. A Gradient-Only Algorithm for Tracing a Reaction Path Uphill to the Saddle of a Potential Energy Surface. *Chemical Physics Letters* 253, no. 3–4: 286–292.

12. Jensen, F. 1995. Locating Transition Structures by Mode Following: A Comparison of Six Methods on the Ar 8 Lennard-Jones Potential. *The Journal of Chemical Physics* 102, no. 17: 6706–6718.

13. Chaudhury, P., S.P. Bhattacharyya, and W. Quapp. 2000. A Genetic Algorithm Based Technique for Locating First-Order Saddle Point Using a Gradient Dominated Recipe. *Chemical Physics* 253, no. 2–3: 295–303.

14. Biring, S.K., and P. Chaudhury. 2010. A Stochastic Optimization Method Based Technique for Finding out Reaction Paths in Noble Gas Clusters Perturbed by Alkali Metal Ions. Article. *Chemical Physics* 377, no. 1–3: 46–53.

15. Talukder, S., S. Sen, R. Sharma, S.K. Banik, and P. Chaudhury. 2014. A Generalized Recipe to Construct Elementary or Multi-Step Reaction Paths via a Stochastic Formulation: Application to the Conformational Change in Noble Gas Clusters. *Chemical Physics* 431–432: 5–14.

16. Rata, I., A.A. Shvartsburg, M. Horoi, T. Frauenheim, K.W.M. Siu, and K.A. Jackson. 2000. Single-Parent Evolution Algorithm and the Optimization of Si Clusters. *Physical Review Letters* 85, no. 3: 546–549.

17. Sarkar, K., and S.P. Bhattacharyya. 2013. Single String Based Global Optimizer for Geometry Optimization in Strongly Coupled Finite Clusters: An Adaptive Mutation-Driven Strategy. *The Journal of Chemical Physics* 139, no. 7: 74106.

18. Sarkar K. 2014. Soft-Computing algorithms for solving some problems of Quantum Chemistry. PhD thesis, Indian Association for the Cultivation of Science, Jadavpur University, Kolkata, India.

19. Wehmeyer, C., G. Falk von Rudorff, S. Wolf, G. Kabbe, D. Schärf, T.D. Kühne, and D. Sebastiani. 2012. Foraging on the Potential Energy Surface: A Swarm Intelligence-Based Optimizer for Molecular Geometry. *The Journal of Chemical Physics* 137, no. 19: 194110.

20. Schiffmann, C., and D. Sebastiani. 2011. Artificial Bee Colony Optimization of Capping Potentials for Hybrid Quantum Mechanical/Molecular Mechanical Calculations. *Journal of Chemical Theory and Computation* 7, no. 5: 1307–1315.

21. Ihrig, A.C., C. Schiffmann, and D. Sebastiani. 2011. Specific Quantum Mechanical/molecular Mechanical Capping-Potentials for Biomolecular Functional Groups. *The Journal of Chemical Physics* 135, no. 21: 214107.

22. Zhang, J., and M. Dolg. 2015. ABCluster: The Artificial Bee Colony Algorithm for Cluster Global Optimization. *Physical Chemistry Chemical Physics* 17, no. 37: 24173–24181.

23. Sarkar, K., R. Sharma, and S.P. Bhattacharyya. 2010. Blending Determinism with Evolutionary Computing: Applications to the Calculation of the Molecular Electronic Structure of Polythiophene. *Journal of Chemical Theory and Computation* 6, no. 3: 718–726.

24. Lavarda, F.C., M.C. dos Santos, D.S. Galvão, and B. Laks. 1994. Insulator-to-Metal Transition in Polythiophene. *Physical Review B* 49, no. 2: 979–983.

25. Sharma, R., S. Nandy, P. Chaudhury, and S.P. Bhattacharyya. 2011. A Density-Genetic Algorithm Method for Computing Electronic Structures of Doped and Undoped Polythiophene Oligomers: A Modified Su–Schrieffer–Heeger Hamiltonian-Based Study. *Materials and Manufacturing Processes* 26, no. 3: 354–362.

26. Kennedy, J., and R. Eberhart. 1995. Particle Swarm Optimization. *Proceedings of ICNN'95— International Conference on Neural Networks* 4: 1942–1948.

27. Shukla, R., D. Ray, K. Sarkar, M. Kumar Dixit, and S. Prasad Bhattacharyya. 2017. Flying onto Global Minima on Potential Energy Surfaces: A Swarm Intelligence Guided Route to Molecular Electronic Structure. *International Journal of Quantum Chemistry* 117, no. 5: e25328.

28. Kennedy, J., and R.C. Eberhart. 2001. *Swarm Intelligence.* San Francisco, CA: Morgan Kaufmann.

29. Avendaño-Franco, G., and A.H. Romero. 2016. Firefly Algorithm for Structural Search. *Journal of Chemical Theory and Computation* 12, no. 7: 3416–3428.

30. Sarkar, K., R. Sharma, and S.P. Bhattacharyya. 2012. A Constrained Variational Approach to the Designing of Low Transport Band Gap Materials: A Multiobjective Random Mutation Hill Climbing Method. *International Journal of Quantum Chemistry* 112, no. 6: 1547–1558.

31. Deb, K., A. Pratap, S. Agarwal, and T. Meyarivan. 2002. A Fast and Elitist Multiobjective Genetic Algorithm: NSGA-II. *IEEE Transactions on Evolutionary Computation* 6, no. 2: 182–197.

32. Moseler, M., H. Häkkinen, R.N. Barnett, and U. Landman. 2001. Structure and Magnetism of Neutral and Anionic Palladium Clusters. *Physical Review Letters* 86, no. 12: 2545–2548.

33. Nørskov, J.K., T. Bligaard, J. Rossmeisl, and C.H. Christensen. 2009. Towards the Computational Design of Solid Catalysts. *Nature Chemistry* 1, no. 1: 37–46.

34. Ferrando, R. 2012. Computational Methods for Predicting the Structures of Nanoalloys. Incollection. In *Nanoalloys*, eds. D. Alloyeau, C. Mottet, and C. Ricolleau, pp. 259–286. Engineering Materials. London: Springer.

35. Paz-Borbón, L.O., R.L. Johnston, G. Barcaro, and A. Fortunelli. 2007. A Mixed Structural Motif in 34-Atom Pd–Pt Clusters. *The Journal of Physical Chemistry C* 111, no. 7: 2936–2941.

36. Paz-Borbon, L.O., R.L. Johnston, G. Barcaro, and A. Fortunelli. 2008. Structural Motifs, Mixing, and Segregation Effects in 38-Atom Binary Clusters. *The Journal of Chemical Physics* 128, no. 13: 134517.

37. Pittaway, F., L.O. Paz-Borbon, R.L. Johnston, H. Arslan, R. Ferrando, C. Mottet, G. Barcaro, and A. Fortunelli. 2009. Theoretical Studies of Palladium–Gold Nanoclusters: Pd–Au Clusters with up to 50 Atoms. *The Journal of Physical Chemistry C* 113, no. 21: 9141–9152.

38. Cheng, L., Y. Feng, J. Yang, and J. Yang. 2009. Funnel Hopping: Searching the Cluster Potential Energy Surface over the Funnels. *The Journal of Chemical Physics* 130, no. 21: 214112.

39. Huang, W., X. Lai, and R. Xu. 2011. Structural Optimization of Silver Clusters from to Using a Modified Dynamic Lattice Searching Method with Constructed Core. *Chemical Physics Letters* 507, no. 1–3: 199–202.

40. Wilcoxon, J.P., and B.L. Abrams. 2006. Synthesis, Structure and Properties of Metal Nanoclusters. *Chemical Society Reviews* 35, no. 11: 1162–1194.

41. Ferrando, R., J. Jellinek, and R.L. Johnston. 2008. Nanoalloys: From Theory to Applications of Alloy Clusters and Nanoparticles. *Chemical Reviews* 108, no. 3: 845–910.

42. Chen, F., and R.L. Johnston. 2008. Energetic, Electronic, and Thermal Effects on Structural Properties of Ag–Au Nanoalloys. *ACS Nano* 2, no. 1: 165–175.

43. Shao, X., L. Cheng, and W. Cai. 2004. A Dynamic Lattice Searching Method for Fast Optimization of Lennard–Jones Clusters. *Journal of Computational Chemistry* 25, no. 14: 1693–1698.

44. Grosso, A., M. Locatelli, and F. Schoen. 2007. A Population-Based Approach for Hard Global Optimization Problems Based on Dissimilarity Measures. *Mathematical Programming* 110, no. 2: 373–404.

45. Rossi, G., and R. Ferrando. 2009. Searching for Low-Energy Structures of Nanoparticles: A Comparison of Different Methods and Algorithms. *Journal of Physics: Condensed Matter* 21, no. 8: 84208.

46. Oakley, M.T., and R.L. Johnston. 2013. Exploring the Energy Landscapes of Cyclic Tetrapeptides with Discrete Path Sampling. *Journal of Chemical Theory and Computation* 9, no. 1: 650–657.

47. Cerbelaud, M., R. Ferrando, G. Barcaro, and A. Fortunelli. 2011. Optimization of Chemical Ordering in AgAu Nanoalloys. *Physical Chemistry Chemical Physics* 13, no. 21: 10232.

48. Wu, X., W. Cai, and X. Shao. 2009. Optimization of Bimetallic Cu-Au and Ag-Au Clusters by Using a Modified Adaptive Immune Optimization Algorithm. *Journal of Computational Chemistry* 30, no. 13: 1992–2000.

49. Wu, X., Y. Wu, X. Kai, G. Wu, and Y. Chen. 2011. Structural Optimization of Ag–Pd Clusters Based on Different Potential Parameterizations. *Chemical Physics* 390, no. 1: 36–41.

50. Wu, X., C. Huang, Y. Sun, and G. Wu. 2013. Structural Distribution in Mixed Ternary Noble Gas and Lennard-Jones Clusters. *Chemical Physics* 415: 69–75.

51. Wu, X., G. Wu, Y. Chen, and Y. Qiao. 2011. Structural Optimization of Cu–Ag–Au Trimetallic Clusters by Adaptive Immune Optimization Algorithm. *The Journal of Physical Chemistry A* 115, no. 46: 13316–13323.

52. Deaven, D.M., and K.M. Ho. 1995. Molecular Geometry Optimization with a Genetic Algorithm. *Physical Review Letters* 75, no. 2: 288–291.

53. Pereira, F.B., J.M.C. Marques, T. Leitao, and J. Tavares. 2006. Analysis of Locality in Hybrid Evolutionary Cluster Optimization. In *IEEE Congress on Evolutionary Computation (CEC)*, July 16–21, pp. 2285–2292. Vancouver, BC: IEEE.

54. Pereira, F.B., J.C. Marques, T. Leitão, and J. Tavares. 2008. Designing Efficient Evolutionary Algorithms for Cluster Optimization: A Study on Locality. In *Advances in Metaheuristics for Hard Optimization*, eds. P. Siarry and Z. Michalewicz, pp. 223–250. Natural Computing Series. Berlin: Springer.

55. Pereira, F.B., and J.M.C. Marques. 2008. A Self-Adaptive Evolutionary Algorithm for Cluster Geometry Optimization. In *2008 Eighth International Conference on Hybrid Intelligent Systems*, September 10–12, pp. 678–683. Barcelona: IEEE.

56. Pereira, F.B., and J.M.C. Marques. 2009. A Study on Diversity for Cluster Geometry Optimization. *Evolutionary Intelligence* 2, no. 3: 121–140.

57. Marques, J.M.C., F.B. Pereira, and T. Leitão. 2008. On the Use of Different Potential Energy Functions in Rare-Gas Cluster Optimization by Genetic Algorithms: Application to Argon Clusters. *The Journal of Physical Chemistry A* 112, no. 27: 6079–6089.

58. Ferrando, R., A. Fortunelli, and R.L. Johnston. 2008. Searching for the Optimum Structures of Alloy Nanoclusters. *Physical Chemistry Chemical Physics* 10, no. 5: 640–649.

59. Schwerdtfeger, P., N. Gaston, R.P. Krawczyk, R. Tonner, and G.E. Moyano. 2006. Extension of the Lennard-Jones Potential: Theoretical Investigations into Rare-Gas Clusters and Crystal Lattices of He, Ne, Ar, and Kr Using Many-Body Interaction Expansions. *Physical Review B* 73, no. 6: 64112.

60. Marques, J.M.C., A.A.C.C. Pais, and P.E. Abreu. 2010. Generation and Characterization of Low-Energy Structures in Atomic Clusters. *Journal of Computational Chemistry* 31, no. 7: 1495–1503.

61. Cassioli, A., M. Locatelli, and F. Schoen. 2009. Global Optimization of Binary Lennard–Jones Clusters. *Optimization Methods and Software* 24, no. 4–5: 819–835.

62. Alberto Fernandez-Lima, F., O.P. Vilela Neto, A. Silva Pimentel, M.A.C. Pacheco, C.R. Ponciano, M.A.C. Nascimento, and E.F. da Silveira. 2009. Theoretical and Experimental Study of Negative LiF Clusters Produced by Fast Ion Impact on a Polycrystalline 7 LiF Target †. *The Journal of Physical Chemistry A* 113, no. 52: 15031–15040.

63. Fernandez-Lima, F.A., O.P. VilelaNeto, A.S. Pimentel, C.R. Ponciano, M.A.C. Pacheco, M.A.C. Nascimento, and E.F. da Silveira. 2009. A Theoretical and Experimental Study of Positive and Neutral LiF Clusters Produced by Fast Ion Impact on a Polycrystalline LiF Target. *The Journal of Physical Chemistry A* 113, no. 9: 1813–1821.

64. Pullan, W. 2010. Unbiased Geometry Optimisation of Morse Atomic Clusters. In *IEEE Congress on Evolutionary Computation (CEC)*, July 18–23, pp. 1–7. Barcelona: IEEE.

65. Oakley, M.T., D.J. Wales, and R.L. Johnston. 2011. Energy Landscape and Global Optimization for a Frustrated Model Protein. *The Journal of Physical Chemistry B* 115, no. 39: 11525–11529.

66. Buck, U., C.C. Pradzynski, T. Zeuch, J.M. Dieterich, and B. Hartke. 2014. A Size Resolved Investigation of Large Water Clusters. *Physical Chemistry Chemical Physics* 16, no. 15: 6859.

67. Larsson, H.R., A.C.T. van Duin, and B. Hartke. 2013. Global Optimization of Parameters in the Reactive Force Field ReaxFF for SiOH. *Journal of Computational Chemistry* 34, no. 25: 2178–2189.

68. Thornton, A.R., T. Weinhart, V. Ogarko, and S. Luding. 2013. Multi-Scale Methods for Multi-Component Granular Materials. *Computer Methods in Materials Science* 13, no. 2: 197–212.

69. Li, Y., and B. Hartke. 2013. Assessing Solvation Effects on Chemical Reactions with Globally Optimized Solvent Clusters. *ChemPhysChem* 14, no. 12: 2678–2686.

70. Alexandrova, A.N. 2010. H·(H 2 O) N Clusters: Microsolvation of the Hydrogen Atom via Molecular Ab Initio Gradient Embedded Genetic Algorithm (GEGA). *The Journal of Physical Chemistry A* 114, no. 48: 12591–12599.

71. Doll, K., J.C. Schön, and M. Jansen. 2010. Ab Initio Energy Landscape of LiF Clusters. *The Journal of Chemical Physics* 133, no. 2: 24107.

72. Glass, C.W., A.R. Oganov, and N. Hansen. 2006. USPEX—Evolutionary Crystal Structure Prediction. *Computer Physics Communications* 175, no. 11–12: 713–720.

73. Gao, G., A.R. Oganov, A. Bergara, M. Martinez-Canales, T. Cui, T. Iitaka, Y. Ma, and G. Zou. 2008. Superconducting High Pressure Phase of Germane. *Physical Review Letters* 101, no. 10: 107002.

74. Oganov, A.R., S. Ono, Y. Ma, C.W. Glass, and A. Garcia. 2008. Novel High-Pressure Structures of $MgCO_3$, $CaCO_3$ and CO_2 and Their Role in Earth's Lower Mantle. *Earth and Planetary Science Letters* 273, no. 1–2: 38–47.

75. Ma, Y., A.R. Oganov, and C.W. Glass. 2007. Structure of the Metallic ζ Phase of Oxygen and isosymmetric nature of the ε-ζ Phase Transition: *Ab Initio* Simulations. *Physical Review B* 76, no. 6: 64101.

76. Oganov, A.R., J. Chen, C. Gatti, Y. Ma, Y. Ma, C.W. Glass, Z. Liu, T. Yu, O.O. Kurakevych, and V.L. Solozhenko. 2009. Ionic High-Pressure Form of Elemental Boron. *Nature* 457, no. 7231: 863–867.

77. Lyakhov, A.O., A.R. Oganov, H.T. Stokes, and Q. Zhu. 2013. New Developments in Evolutionary Structure Prediction Algorithm USPEX. *Computer Physics Communications* 184, no. 4: 1172–1182.

78. Valle, M., and A.R. Oganov. 2010. Crystal Fingerprint Space—A Novel Paradigm for Studying Crystal-Structure Sets. *Acta Crystallographica Section A Foundations of Crystallography* 66, no. 5: 507–517.

79. Ma, Y., M. Eremets, A.R. Oganov, Y. Xie, I. Trojan, S. Medvedev, A.O. Lyakhov, M. Valle, and V. Prakapenka. 2009. Transparent Dense Sodium. *Nature* 458, no. 7235: 182–185.

80. Oganov, A.R., and C.W. Glass. 2008. Evolutionary Crystal Structure Prediction as a Tool in Materials Design. *Journal of Physics: Condensed Matter* 20, no. 6: 64210.

81. Woodley, S.M., and R. Catlow. 2008. Crystal Structure Prediction from First Principles. *Nature Materials* 7, no. 12: 937–946.

82. Zhu, Q., A.R. Oganov, C.W. Glass, and H.T. Stokes. 2012. Constrained Evolutionary Algorithm for Structure Prediction of Molecular Crystals: Methodology and Applications. Article. *Acta Crystallographica Section B* 68, no. 3: 215–226.

83. Schönborn, S.E., S. Goedecker, S. Roy, and A.R. Oganov. 2009. The Performance of Minima Hopping and Evolutionary Algorithms for Cluster Structure Prediction. *The Journal of Chemical Physics* 130, no. 14: 144108.

84. Amsler, M., and S. Goedecker. 2010. Crystal Structure Prediction Using the Minima Hopping Method. *The Journal of Chemical Physics* 133, no. 22: 224104.

85. Sicher, M., S. Mohr, and S. Goedecker. 2011. Efficient Moves for Global Geometry Optimization Methods and Their Application to Binary Systems. *The Journal of Chemical Physics* 134, no. 4: 44106.

86. Oakley, M.T., R.L. Johnston, and D.J. Wales. 2013. Symmetrisation Schemes for Global Optimisation of Atomic Clusters. *Physical Chemistry Chemical Physics* 15, no. 11: 3965–3976.

87. Wales, D.J., and J.M. Carr. 2012. Quasi-Continuous Interpolation Scheme for Pathways between Distant Configurations. *Journal of Chemical Theory and Computation* 8, no. 12: 5020–5034.

88. Wales, D.J., and T. Head-Gordon. 2012. Evolution of the Potential Energy Landscape with Static Pulling Force for Two Model Proteins. *The Journal of Physical Chemistry B* 116, no. 29: 8394–8411.

89. Chakrabarti, D., and D.J. Wales. 2009. Simulations of Rigid Bodies in an Angle-Axis Framework. *Physical Chemistry Chemical Physics* 11, no. 12: 1970.

90. Froltsov, V.A., and K. Reuter. 2009. Robustness of 'cut and Splice' Genetic Algorithms in the Structural Optimization of Atomic Clusters. *Chemical Physics Letters* 473, no. 4–6: 363–366.

91. de Souza, V.K., and D.J. Wales. 2009. Connectivity in the Potential Energy Landscape for Binary Lennard-Jones Systems. *The Journal of Chemical Physics* 130, no. 19: 194508.

92. Parodi, D., and R. Ferrando. 2007. Influence of the Potential Range on the Structure of Binary Nanoclusters. *Physics Letters A* 367, no. 3: 215–219.

93. Marques, J.M.C., and F.B. Pereira. 2013. A Detailed Investigation on the Global Minimum Structures of Mixed Rare-Gas Clusters: Geometry, Energetics, and Site Occupancy. *Journal of Computational Chemistry* 34, no. 6: 505–517.

94. Marques, J.M.C., and F.B. Pereira. 2010. An Evolutionary Algorithm for Global Minimum Search of Binary Atomic Clusters. *Chemical Physics Letters* 485, no. 1–3: 211–216.

95. Pereira, F.B., and J.M.C. Marques. 2010. Towards an Effective Evolutionary Approach for Binary Lennard-Jones Clusters. In *IEEE Congress on Evolutionary Computation (CEC)*, July 18–23, pp. 1–7. Barcelona: IEEE.

96. Kolossváry, I., and K.J. Bowers. 2010. Global Optimization of Additive Potential Energy Functions: Predicting Binary Lennard-Jones Clusters. *Physical Review E* 82, no. 5: 56711.

97. El Dor, A., M. Clerc, and P. Siarry. 2012. A Multi-Swarm PSO Using Charged Particles in a Partitioned Search Space for Continuous Optimization. *Computational Optimization and Applications* 53, no. 1: 271–295.

98. Dieterich, J.M., and B. Hartke. 2011. Composition-Induced Structural Transitions in Mixed Lennard-Jones Clusters: Global Reparametrization and Optimization. *Journal of Computational Chemistry* 32, no. 7: 1377–1385.

99. Dieterich, J.M., and B. Hartke. 2010. OGOLEM: Global Cluster Structure Optimisation for Arbitrary Mixtures of Flexible Molecules. A Multiscaling, Object-Oriented Approach. Article. *Molecular Physics* 108, no. 3–4: 279–291.

100. Dzhurakhalov, A.A., I. Atanasov, and M. Hou. 2008. Calculation of Binary and Ternary Metallic Immiscible Clusters with Icosahedral Structures. *Physical Review B* 77, no. 11: 115415.

101. Leitao, A., F.B. Pereira, and P. Machado. 2012. Enhancing Cluster Geometry Optimization with Island Models. In *IEEE Congress on Evolutionary Computation (CEC)*, June 10–15, pp. 1–8. Brisbane: IEEE.

102. Ole Carstensen, N., J.M. Dieterich, and B. Hartke. 2011. Design of Optimally Switchable Molecules by Genetic Algorithms. *Physical Chemistry Chemical Physics* 13, no. 7: 2903–2910.

103. Dieterich, J.M., and B. Hartke. 2012. Empirical Review of Standard Benchmark Functions Using Evolutionary Global Optimization. *Applied Mathematics* 3, no. 10A: 1552–1564. doi: 10.4236/am.2012.330215.

104. Stepanenko, S., and B. Engels. 2009. Tabu Search Based Strategies for Conformational Search†. *The Journal of Physical Chemistry A* 113, no. 43: 11699–11705.

105. Stepanenko, S., and B. Engels. 2008. New Tabu Search Based Global Optimization Methods Outline of Algorithms and Study of Efficiency. *Journal of Computational Chemistry* 29, no. 5: 768–780.

106. Daskin, A., and S. Kais. 2011. Group Leaders Optimization Algorithm. *Molecular Physics* 109, no. 5: 761–772.

107. Deep, K., and Madhuri. 2013. Liquid-Drop-Like Multi-Orbit Topology versus Ring Topology in PSO for Lennard-Jones Problem. In *Proceedings of Seventh International Conference on Bio-Inspired Computing: Theories and Applications (BIC-TA 2012) SE - 20*, eds. J.C. Bansal, P. Singh, K. Deep, M. Pant, and A. Nagar, vol. 202, pp. 229–243. Advances in Intelligent Systems and Computing. India: Springer.

108. Oleksy, K., F. Karlický, and R. Kalus. 2010. Structures and Energetics of Helium Cluster Cations: Equilibrium Geometries Revisited through the Genetic Algorithm Approach. *The Journal of Chemical Physics* 133, no. 16: 164314.

109. Huber, B., M. Moseler, O. Kostko, and B. V. Issendorff. 2009. Structural Evolution of the Sodium Cluster Anions Na_{20}^--Na_{57}^-. *Physical Review B* 80, no. 23: 235425.

110. Huang, X., L. Sai, X. Jiang, and J. Zhao. 2013. Ground State Structures, Electronic and Optical Properties of Medium-Sized Nan + (N = 9, 15, 21, 26, 31, 36, 41, 50 and 59) Clusters from Ab Initio Genetic Algorithm. *The European Physical Journal D* 67, no. 2: 43.

111. Walsh, A., and S.M. Woodley. 2010. Evolutionary Structure Prediction and Electronic Properties of Indium Oxide Nanoclusters. *Physical Chemistry Chemical Physics* 12, no. 30: 8446.

112. Woodley, S.M., S. Hamad, and C.R.A. Catlow. 2010. Exploration of Multiple Energy Landscapes for Zirconia Nanoclusters. *Physical Chemistry Chemical Physics* 12, no. 30: 8454.

113. Kim, S., K.-S. Sohn, and M. Pyo. 2011. Genetic Algorithm-Assisted Optimization of Nanoporous TiO 2 for Low-Temperature Processable Photoanodes of Dye-Sensitized Solar Cells. Article. *ACS Combinatorial Science* 13, no. 2: 101–106.

114. Woodley, S.M., and C.R.A. Catlow. 2009. Structure Prediction of Titania Phases: Implementation of Darwinian versus Lamarckian Concepts in an Evolutionary Algorithm. Article. *Computational Materials Science* 45, no. 1: 84–95.

115. Marques, J.M.C., J.L. Llanio-Trujillo, M. Albertí, A. Aguilar, and F. Pirani. 2012. Alkali-Ion Microsolvation with Benzene Molecules. *The Journal of Physical Chemistry A* 116, no. 20: 4947–4956.

116. Zhu, Q., A.R. Oganov, and A.O. Lyakhov. 2013. Novel Stable Compounds in the Mg–O System under High Pressure. *Physical Chemistry Chemical Physics* 15, no. 20: 7696.

117. Zhu, Q., D.Y. Jung, A.R. Oganov, C.W. Glass, C. Gatti, and A.O. Lyakhov. 2012. Stability of Xenon Oxides at High Pressures. *Nature Chemistry* 5, no. 1: 61–65.

118. Al-Sunaidi, A.A., A.A. Sokol, C.R.A. Catlow, and S.M. Woodley. 2008. Structures of Zinc Oxide Nanoclusters: As Found by Revolutionary Algorithm Techniques. *The Journal of Physical Chemistry C* 112, no. 48: 18860–18875.

119. Wang, Y., J. Lv, L. Zhu, and Y. Ma. 2010. Crystal Structure Prediction via Particle-Swarm Optimization. *Physical Review B* 82, no. 9: 94116.

120. Li, R., and L. Cheng. 2012. Structural Determination of (Al2O3)n (n=1–7) Clusters Based on Density Functional Calculation. *Computational and Theoretical Chemistry* 996: 125–131.

121. Avaltroni, F., and C. Corminboeuf. 2012. Identifying Clusters as Low-Lying Mimina-Efficiency of Stochastic and Genetic Algorithms Using Inexpensive Electronic Structure Levels. *Journal of Computational Chemistry* 33, no. 5: 502–508.

122. Pal, S., R. Sharma, B. Goswami, and P. Sarkar. 2009. Theoretical Prediction of Ring Structures for ZnS Quantum Dots. *Chemical Physics Letters* 467, no. 4–6: 365–368.

123. Pal, S., R. Sharma, B. Goswami, P. Sarkar, and S.P. Bhattacharyya. 2009. A Search for Lowest Energy Structures of ZnS Quantum Dots: Genetic Algorithm Tight-Binding Study. *The Journal of Chemical Physics* 130, no. 21: 214703.

124. Llanio-Trujillo, J.L., J.M.C. Marques, and F.B. Pereira. 2011. An Evolutionary Algorithm for the Global Optimization of Molecular Clusters: Application to Water, Benzene, and Benzene Cation. *The Journal of Physical Chemistry A* 115, no. 11: 2130–2138.

125. Do, H., and N.A. Besley. 2012. Structural Optimization of Molecular Clusters with Density Functional Theory Combined with Basin Hopping. *The Journal of Chemical Physics* 137, no. 13: 134106.

126. Addicoat, M.A., A.J. Page, Z.E. Brain, L. Flack, K. Morokuma, and S. Irle. 2012. Optimization of a Genetic Algorithm for the Functionalization of Fullerenes. *Journal of Chemical Theory and Computation* 8, no. 5: 1841–1851.

127. Zhang, J., C.-Z. Wang, and K.-M. Ho. 2009. Finding the Low-Energy Structures of Si[001] Symmetric Tilted Grain Boundaries with a Genetic Algorithm. *Physical Review B* 80, no. 17: 174102.

128. Yao, Y., J.S. Tse, and K. Tanaka. 2008. Metastable High-Pressure Single-Bonded Phases of Nitrogen Predicted via Genetic Algorithm. *Physical Review B* 77, no. 5: 52103.

129. Hooper, J., A. Ismail, J.B. Giorgi, and T.K. Woo. 2010. Genetic Algorithm Based Approach to Investigate Doped Metal Oxide Materials: Application to Lanthanide-Doped Ceria. *Physical Review B* 81, no. 22: 224104.

130. Chen, Z., X. Jiang, J. Li, S. Li, and L. Wang. 2013. PDECO: Parallel Differential Evolution for Clusters Optimization. *Journal of Computational Chemistry* 34, no. 12: 1046–1059.

131. Rogan, J., G. García, M. Ramírez, V. Muñoz, J. Alejandro Valdivia, X. Andrade, R. Ramírez, and M. Kiwi. 2008. The Structure and Properties of Small Pd Clusters. *Nanotechnology* 19, no. 20: 205701.

132. Assadollahzadeh, B., P.R. Bunker, and P. Schwerdtfeger. 2008. The Low Lying Isomers of the Copper Nonamer Cluster, Cu9. *Chemical Physics Letters* 451, no. 4–6: 262–269.

133. Assadollahzadeh, B., and P. Schwerdtfeger. 2009. A Systematic Search for Minimum Structures of Small Gold Clusters Au[sub N] (n=2–20) and Their Electronic Properties. *The Journal of Chemical Physics* 131, no. 6: 64306.

134. Li, X.-P., W.-C. Lu, Q.-J. Zang, G.-J. Chen, C.Z. Wang, and K.M. Ho. 2009. Structures and Stabilities of Pb N (N ≤ 20) Clusters. *The Journal of Physical Chemistry A* 113, no. 22: 6217–6221.

135. Pereira, F.B., and J.M.C. Marques. 2011. Analysis of Crossover Operators for Cluster Geometry Optimization. In *Computational Intelligence for Engineering Systems*, eds. A. Madureira, J. Ferreira, and Z. Vale, vol. 46, pp. 77–89. Intelligent Systems, Control and Automation: Science and Engineering. Dordrecht: Springer.

136. Bao, K., S. Goedecker, K. Koga, F. Lançon, and A. Neelov. 2009. Structure of Large Gold Clusters Obtained by Global Optimization Using the Minima Hopping Method. *Physical Review B* 79, no. 4: 41405.

137. Meng, X., S. Tongay, J. Kang, Z. Chen, F. Wu, S.-S. Li, J.-B. Xia, J. Li, and J. Wu. 2013. Stable P- and N-Type Doping of Few-Layer Graphene/Graphite. *Carbon* 57: 507–514.

138. Hong, L., H. Wang, J. Cheng, X. Huang, L. Sai, and J. Zhao. 2012. Atomic Structures and Electronic Properties of Small Au–Ag Binary Clusters: Effects of Size and Composition. Article. *Computational and Theoretical Chemistry* 993: 36–44.

139. Heiles, S., R.L. Johnston, and R. Schäfer. 2012. Bismuth-Doped Tin Clusters: Experimental and Theoretical Studies of Neutral Zintl Analogues. *The Journal of Physical Chemistry A* 116, no. 29: 7756–7764.

140. Oh, J.S., H.-S. Nam, J.-H. Choi, and S.-C. Lee. 2013. Prediction of Atomic Arrangement of Pt-Cu Nanoalloy by Genetic Algorithm. *Journal of Physics: Conference Series* 410, no. 1: 12084.

141. Orel, S., and R. Fournier. 2013. Density Functional Theory and Global Optimization Study of Sn M Pb N Clusters (7 ≤ M + N ≤ 12, 0 ≤ M /(M + N) ≤ 1). *The Journal of Chemical Physics* 138, no. 6: 64306.

142. Fournier, R., S. Zamiruddin, and M. Zhang. 2009. Competition between Mixing and Segregation in Bimetallic Ag N Rb N Clusters (N = 2–10) 1, 2. *Canadian Journal of Chemistry* 87, no. 7: 1013–1021.

143. Chu, Y., W. Heyndrickx, G. Occhipinti, V.R. Jensen, and B.K. Alsberg. 2012. An Evolutionary Algorithm for de Novo Optimization of Functional Transition Metal Compounds. *Journal of the American Chemical Society* 134, no. 21: 8885–8895.

144. Yuan, D., X. Gong, and R. Wu. 2008. Peculiar Distribution of Pd on Au Nanoclusters: First-Principles Studies. *Physical Review B* 78, no. 3: 35441.

145. Paz-Borbón, L.O., T.V. Mortimer-Jones, R.L. Johnston, A. Posada-Amarillas, G. Barcaro, and A. Fortunelli. 2007. Structures and Energetics of 98 Atom Pd–Pt Nanoalloys: Potential Stability of the Leary Tetrahedron for Bimetallic Nanoparticles. *Physical Chemistry Chemical Physics* 9, no. 38: 5202.

146. Barcaro, G., A. Fortunelli, G. Rossi, F. Nita, and R. Ferrando. 2007. Epitaxy, Truncations, and Overhangs in Palladium Nanoclusters Adsorbed on MgO(001). *Physical Review Letters* 98, no. 15: 156101.

147. Barcaro, G., E. Aprà, and A. Fortunelli. 2007. Structure of Ag Clusters Grown on Fs-Defect Sites of an MgO(1 0 0) Surface. *Chemistry—A European Journal* 13, no. 22: 6408–6418.

148. Ferrando, R. 2013. Global Optimization of Free and Supported Clusters. In *Metal Clusters and Nanoalloys SE—6*, pp. 195–214. Nanostructure Science and Technology. New York: Springer.

149. Núñez, S., and R.L. Johnston. 2010. Structures and Chemical Ordering of Small Cu–Ag Clusters. *The Journal of Physical Chemistry C* 114, no. 31: 13255–13266.

150. Tran, D.T., I.P. Jones, J.A. Preece, R.L. Johnston, and C.R. van den Brom. 2011. TEM Characterization of Chemically Synthesized Copper–Gold Nanoparticles. *Journal of Nanoparticle Research* 13, no. 9: 4229–4237.

151. Paz-Borbón, L.O., A. Gupta, and R.L. Johnston. 2008. Dependence of the Structures and Chemical Ordering of Pd–Pt Nanoalloys on Potential Parameters. *Journal of Materials Chemistry* 18, no. 35: 4154.

152. Logsdail, A., L.O. Paz-Borbón, and R.L. Johnston. 2009. Structures and Stabilities of Platinum–Gold Nanoclusters. *Journal of Computational and Theoretical Nanoscience* 6, no. 4: 857–866.

153. Logsdail, A.J., and R.L. Johnston. 2012. Predicting the Optical Properties of Core–Shell and Janus Segregated Au–M Nanoparticles (M = Ag, Pd). *The Journal of Physical Chemistry C* 116, no. 44: 23616–23628.

154. Heiles, S., A.J. Logsdail, R. Schäfer, and R.L. Johnston. 2012. Dopant-Induced 2D–3D Transition in Small Au-Containing Clusters: DFT-Global Optimisation of 8-Atom Au–Ag Nanoalloys. *Nanoscale* 4, no. 4: 1109–1115.

155. Bruma, A., R. Ismail, L. Oliver Paz-Borbón, H. Arslan, G. Barcaro, A. Fortunelli, Z.Y. Li, and R.L. Johnston. 2013. DFT Study of the Structures and Energetics of 98-Atom AuPd Clusters. *Nanoscale* 5, no. 2: 646–652.

156. Tedsree, K., T. Li, S. Jones, C.W.A. Chan, K.M.K. Yu, P.A.J. Bagot, E.A. Marquis, G.D.W. Smith, and S.C.E. Tsang. 2011. Hydrogen Production from Formic Acid Decomposition at Room Temperature Using a Ag–Pd Core–shell Nanocatalyst. *Nature Nanotechnology* 6, no. 5: 302–307.

157. Lyakhov, A.O., A.R. Oganov, and M. Valle. 2010. How to Predict Very Large and Complex Crystal Structures. *Computer Physics Communications* 181, no. 9: 1623–1632.

158. Oganov, A.R., Y. Ma, A.O. Lyakhov, M. Valle, and C. Gatti. 2010. Evolutionary Crystal Structure Prediction as a Method for the Discovery of Minerals and Materials. *Reviews in Mineralogy and Geochemistry* 71, no. 1: 271–298.

159. Lyakhov, A.O., A.R. Oganov, and M. Valle. 2010. Crystal Structure Prediction Using Evolutionary Approach. Article. In *Modern Methods of Crystal Structure Prediction*, Oganov, A.R. (ed.), Weinheim, Germany: Wiley-VCH Verlag GmbH. KGaA. pp. 147–180.

160. Bahmann, S., and J. Kortus. 2013. EVO—Evolutionary Algorithm for Crystal Structure Prediction. *Computer Physics Communications* 184, no. 6: 1618–1625.

161. Hu, M., Z. Zhao, F. Tian, A.R. Oganov, Q. Wang, M. Xiong, C. Fan, et al. 2013. Compressed Carbon Nanotubes: A Family of New Multifunctional Carbon Allotropes. Article. *Scientific Reports* 3: 1331.

162. Ciobanu, C. V, C.-Z. Wang, and K.-M. Ho. 2013. Crystal Structure Prediction. In *Atomic Structure Prediction of Nanostructures, Clusters and Surfaces*, pp. 37–69. Weinheim, Germany: Wiley-VCH Verlag GmbH. KGaA. https://leseprobe.buch.de/images-adb/da/79/da7901b6-c2f9-4208-a7c5-c70ad42256da.pdf

163. Abraham, N.L., and M.I.J. Probert. 2008. Improved Real-Space Genetic Algorithm for Crystal Structure and Polymorph Prediction. *Physical Review B* 77, no. 13: 134117.

164. Ji, M., C.-Z. Wang, and K.-M. Ho. 2010. Comparing Efficiencies of Genetic and Minima Hopping Algorithms for Crystal Structure Prediction. Article. *Physical Chemistry Chemical Physics* 12, no. 37: 11617.

165. Wales, D.J. 2010. Energy Landscapes and Structure Prediction Using Basin-Hopping. Article. In *Modern Methods of Crystal Structure Prediction*, Oganov, A.R. (ed.), Weinheim, Germany: Wiley-VCH Verlag GmbH. KGaA. pp. 29–54.

166. Goedecker, S. 2010. Global Optimization with the Minima Hopping Method. Article. In *Modern Methods of Crystal Structure Prediction*, Oganov, A.R. (ed.), Weinheim, Germany: Wiley-VCH Verlag GmbH. KGaA. pp. 131–145.

167. Zhu, Q., A.R. Oganov, and A.O. Lyakhov. 2012. Evolutionary Metadynamics: A Novel Method to Predict Crystal Structures. *CrystEngComm* 14, no. 10: 3596.

168. Zhu, Q., Q. Zeng, and A.R. Oganov. 2012. Systematic Search for Low-Enthalpy sp(3) Carbon Allotropes Using Evolutionary Metadynamics. *Physical Review B* 85, no. 20: 201407.

169. Wang, Y., J. Lv, L. Zhu, and Y. Ma. 2012. CALYPSO: A Method for Crystal Structure Prediction. *Computer Physics Communications* 183, no. 10: 2063–2070.

170. Doll, K., J.C. Schön, and M. Jansen. 2008. Structure Prediction Based on Ab Initio Simulated Annealing for Boron Nitride. *Physical Review B* 78, no. 14: 144110.

171. Martinez-Canales, M., A.R. Oganov, Y. Ma, Y. Yan, A.O. Lyakhov, and A. Bergara. 2009. Novel Structures and Superconductivity of Silane under Pressure. *Physical Review Letters* 102, no. 8: 87005.

172. Li, Q., M. Wang, A.R. Oganov, T. Cui, Y. Ma, and G. Zou. 2009. Rhombohedral Superhard Structure of BC2N. *Journal of Applied Physics* 105, no. 5: 53514.

173. Li, Q., H. Wang, Y. Tian, Y. Xia, T. Cui, J. He, Y. Ma, and G. Zou. 2010. Superhard and Superconducting Structures of BC5. *Journal of Applied Physics* 108, no. 2: 23507.

174. Li, Q., Y. Ma, A.R. Oganov, H. Wang, H. Wang, Y. Xu, T. Cui, H.-K. Mao, and G. Zou. 2009. Superhard Monoclinic Polymorph of Carbon. *Physical Review Letters* 102, no. 17: 175506.

175. Ma, Y., Y. Wang, and A.R. Oganov. 2009. Absence of Superconductivity in the High-Pressure Polymorph of MgB_2 *Physical Review B* 79, no. 5: 54101.

176. Xie, Y., A.R. Oganov, and Y. Ma. 2010. Novel High Pressure Structures and Superconductivity of $CaLi_2$ *Physical Review Letters* 104, no. 17: 177005.

177. Wang, H., Q. Li, Y. Li, Y. Xu, T. Cui, A.R. Oganov, and Y. Ma. 2009. Ultra-Incompressible Phases of Tungsten Dinitride Predicted from First Principles. *Physical Review B* 79, no. 13: 132109.

178. Gao, G., A.R. Oganov, P. Li, Z. Li, H. Wang, T. Cui, Y. Ma, et al. 2010. High-Pressure Crystal Structures and Superconductivity of Stannane (SnH4). *Proceedings of the National Academy of Sciences* 107, no. 4: 1317–1320.

179. Li, Y., G. Gao, Y. Xie, Y. Ma, T. Cui, and G. Zou. 2010. Superconductivity at 100 K in Dense SiH4(H2)2 Predicted by First Principles. *Proceedings of the National Academy of Sciences* 107, no. 36: 15708–15711.

180. Xiang, H.J., B. Huang, E. Kan, S.-H. Wei, and X.G. Gong. 2013. Towards Direct-Gap Silicon Phases by the Inverse Band Structure Design Approach. *Physical Review Letters* 110, no. 11: 118702.

181. Zhu, Q., L. Li, A.R. Oganov, and P.B. Allen. 2013. Evolutionary Method for Predicting Surface Reconstructions with Variable Stoichiometry. *Physical Review B* 87, no. 19: 195317.

182. Oganov, A.R., and A.O. Lyakhov. 2010. Towards the Theory of Hardness of Materials. *Journal of Superhard Materials* 32, no. 3: 143–147.

183. Oganov, A.R., A.O. Lyakhov, and Q. Zhu. 2014. Theory of Superhard Materials. In *Comprehensive Hard Materials*, Sarin, V.K., and Nebel, C.E. (eds.), Oxford: Elsevier. pp. 59–79.

184. Krevelen, D.W., and K. Te Nijenhuis. 2009. *Properties of Polymers* (4th ed.). Amsterdam: Elsevier.

185. Zeng, Q., A.R. Oganov, A.O. Lyakhov, C. Xie, X. Zhang, J. Zhang, Q. Zhu, et al. 2014. Evolutionary Search for New High-{\it K} Dielectric Materials: Methodology and Applications to Hafnia-Based Oxides. *Acta Crystallographica Section C* 70, no. 2: 76–84.

186. Zhou, X.-F., X. Dong, A.R. Oganov, Q. Zhu, Y. Tian, and H.-T. Wang. 2014. Semimetallic Two-Dimensional Boron Allotrope with Massless Dirac Fermions. *Physical Review Letters* 112, no. 8: 85502.

187. Zeng, Q., J. Peng, A.R. Oganov, Q. Zhu, C. Xie, X. Zhang, D. Dong, L. Zhang, and L. Cheng. 2013. Prediction of Stable Hafnium Carbides: Stoichiometries, Mechanical Properties, and Electronic Structure. *Physical Review B* 88, no. 21: 214107.

188. Wu, S.Q., M. Ji, C.Z. Wang, M.C. Nguyen, X. Zhao, K. Umemoto, R.M. Wentzcovitch, and K.M. Ho. 2014. An Adaptive Genetic Algorithm for Crystal Structure Prediction. *Journal of Physics: Condensed Matter* 26, no. 3: 35402.

189. O'Keeffe, M. 2010. Aspects of Crystal Structure Prediction: Some Successes and Some Difficulties. *Physical Chemistry Chemical Physics* 12, no. 30: 8580.

190. Oganov, A.R., A.O. Lyakhov, and M. Valle. 2011. How Evolutionary Crystal Structure Prediction Works—And Why. *Accounts of Chemical Research* 44, no. 3: 227–237.

191. Zhao, X., M.C. Nguyen, C.-Z. Wang, and K.-M. Ho. 2013. Structures and Stabilities of Alkaline Earth Metal Peroxides XO2 (X = Ca, Be, Mg) Studied by a Genetic Algorithm. *RSC Advances* 3, no. 44: 22135.

192. Zhao, X., M.C. Nguyen, W.Y. Zhang, C.Z. Wang, M.J. Kramer, D.J. Sellmyer, X.Z. Li, et al. 2014. Exploring the Structural Complexity of Intermetallic Compounds by an Adaptive Genetic Algorithm. *Physical Review Letters* 112, no. 4: 45502.

193. Li, P., G. Gao, Y. Wang, and Y. Ma. 2010. Crystal Structures and Exotic Behavior of Magnesium under Pressure. *The Journal of Physical Chemistry C* 114, no. 49: 21745–21749.

194. Zhu, L., H. Liu, C.J. Pickard, G. Zou, and Y. Ma. 2014. Reactions of Xenon with Iron and Nickel Are Predicted in the Earth's Inner Core. Article. *Nature Chemistry* 6, no. 7: 644–648.

195. Wang, Y., and Y. Ma. 2014. Perspective: Crystal Structure Prediction at High Pressures. Article. *The Journal of Chemical Physics* 140, no. 4: 40901.

196. Li, Q., D. Zhou, W. Zheng, Y. Ma, and C. Chen. 2013. Global Structural Optimization of Tungsten Borides. *Physical Review Letters* 110, no. 13: 136403.

197. Lu, C., M. Miao, and Y. Ma. 2013. Structural Evolution of Carbon Dioxide under High Pressure. *Journal of the American Chemical Society* 135, no. 38: 14167–14171.

198. Xu, L.-C., R.-Z. Wang, M.-S. Miao, X.-L. Wei, Y.-P. Chen, H. Yan, W.-M. Lau, L.-M. Liu, and Y.-M. Ma. 2014. Two Dimensional Dirac Carbon Allotropes from Graphene. *Nanoscale* 6, no. 2: 1113–1118.

199. Chen, Y., F. Peng, Y. Yan, Z. Wang, C. Sun, and Y. Ma. 2013. Exploring High-Pressure Lithium Beryllium Hydrides: A New Chemical Perspective. *The Journal of Physical Chemistry C* 117, no. 27: 13879–13886.

200. Scheler, T., F. Peng, C.L. Guillaume, R.T. Howie, Y. Ma, and E. Gregoryanz. 2013. Nanocrystalline Tungsten Hydrides at High Pressures. *Physical Review B* 87, no. 18: 184117.

201. Gao, G., R. Hoffmann, N.W. Ashcroft, H. Liu, A. Bergara, and Y. Ma. 2013. Theoretical Study of the Ground-State Structures and Properties of Niobium Hydrides under Pressure. *Physical Review B* 88, no. 18: 184104.

202. Luo, X., J. Yang, H. Liu, X. Wu, Y. Wang, Y. Ma, S.-H. Wei, X. Gong, and H. Xiang. 2011. Predicting Two-Dimensional Boron–Carbon Compounds by the Global Optimization Method. *Journal of the American Chemical Society* 133, no. 40: 16285–16290.

203. Lonie, D.C., and E. Zurek. 2011. XtalOpt: An Open-Source Evolutionary Algorithm for Crystal Structure Prediction. *Computer Physics Communications* 182, no. 2: 372–387.

204. Liu, Z.-L. 2014. Muse: Multi-Algorithm Collaborative Crystal Structure Prediction. *Computer Physics Communications* 185, no. 7: 1893–1900.

205. Fadda, A., and G. Fadda. 2010. An Evolutionary Algorithm for the Prediction of Crystal Structures. *Physical Review B* 82, no. 10: 104105.

206. Logsdail, A.J., Z.Y. Li, and R.L. Johnston. 2012. Development and Optimization of a Novel Genetic Algorithm for Identifying Nanoclusters from Scanning Transmission Electron Microscopy Images. *Journal of Computational Chemistry* 33, no. 4: 391–400.

207. Meredig, B., and C. Wolverton. 2012. A Hybrid Computational–experimental Approach for Automated Crystal Structure Solution. *Nature Materials* 12, no. 2: 123–127.

208. Nguyen, M.C., X. Zhao, Y. Wang, C.-Z. Wang, and K.-M. Ho. 2014. Genetic Algorithm Prediction of Crystal Structure of Metastable Si-IX Phase. *Solid State Communications* 182: 14–16.

209. Katritzky, A.R., M. Kuanar, S. Slavov, C.D. Hall, M. Karelson, I. Kahn, and D.A. Dobchev. 2010. Quantitative Correlation of Physical and Chemical Properties with Chemical Structure: Utility for Prediction. *Chemical Reviews* 110, no. 10: 5714–5789.

210. Le, T., V.C. Epa, F.R. Burden, and D.A. Winkler. 2012. Quantitative Structure–Property Relationship Modeling of Diverse Materials Properties. *Chemical Reviews* 112, no. 5: 2889–2919.

211. Tsai, J.-T., T.-K. Liu, and J.-H. Chou. 2004. Hybrid Taguchi-Genetic Algorithm for Global Numerical Optimization. *IEEE Transactions on Evolutionary Computation* 8, no. 4: 365–377.

212. Taguchi, G., S. Chowdhury, and S. Taguchi. 2000. *Robust Engineering*. New York: McGraw-Hill.

8

Artificial Neural Networks

8.1 Introduction

The idea of building an abstract computing model of the human brain or for that matter, the brain of any intelligent species is inherently exciting. The seat of human intelligence is believed to be the brain which turns out to be a hugely complex network of interconnected tiny cells called neurons. In 1911, S. R. y Cajal advanced the proposition that neurons are the fundamental constituents of the brain, thereby paving the way to understanding how it functions. The number of neurons in the human brain is stupendously large, being between 10^{10} and 10^{11}, and the total number of interconnections among the neurons is also huge, being between 10^{14} and 10^{15}. The complex interconnected assembly of neurons is believed to be the primary functional source of our intelligence as characterized by our ability to perceive, learn, recognize, recall, etc. Can there be a computing model that is endowed with all these abilities? The artificial neural networks (ANNs) or more simply, the neural networks (NNs) are mathematical constructs or models that attempt to provide an affirmative answer to the question posed.

The ANNs are assemblies of interconnected artificial neurons, which are the fundamental or basic functional units of the network. Structurally, artificial neurons are far too simple compared to the natural or living neurons. The physical characteristics of the ANNs such as their sizes, complexities, and processing speed are markedly different from what the living NNs have. Thus, the ANNs have typically 10–1000 artificial neurons as opposed to 10^{10}–10^{11} neurons in the human brain. Likewise, the number of interconnections, too, is much smaller in the ANNs (100–10000). The processing speed of ANNs is much higher (in the order of a nanosecond) compared to the much slower processing speed (millisecond) of the natural neural nets. Nevertheless, the ANNs, like the living NNs, have the capability to learn when inputs are presented to them in the form of patterns and can later exploit the learning to perform classification, prediction, optimization, and control. The ANNs, therefore, appear to display a kind of artificial intelligence. However, at the present state of development, the ANNs fall far short of displaying the real intelligence of a human brain that they are supposed to mimic. In view of their learning ability, the ANNs are also called adaptive (learning) systems [1,2].

In an ANN, the neurons are the basic processing elements. They can be thought of as occupying vertices (nodes) of a complex graph, the interconnections among the neurons representing the edges of the graph. The complexity of an NN is a reflection of the complexity of the corresponding graph. The NNs have therefore been called the connectionist model [3,4].

A given class of neurons in the network can carry out their tasks in parallel just like the brain, while different groups of neurons can perform different tasks. The NNs have therefore been often classified as parallel distributed processing models of computation. ANNs have found many technological and industrial applications. In the context of physical and chemical sciences, there has been a sustained growth of interest in ANNs over the last two decades or so. They are being increasingly used for analysis of data, pattern identification, structure-activity (biological) mapping, classification of compounds, identification of drug targets, prediction of physicochemical properties, online and offline data analysis in high energy physics, from spectrum to molecule mapping, etc. New areas of applications are emerging. Before we consider a cross section of these applications we will introduce the reader to some of the most commonly used ANNs, their learning mechanisms, and their potential applications.

8.2 Natural and Artificial Neurons

A natural neuron is a tiny cell that has a nucleus called a soma to which many filament-like structures known as dendrites are attached (Figure 8.1). The dendrites appear as irregularly shaped tentacles that serve as input channels to a neuron from other neurons or from external agents or sources. Emanating from the soma there is a special type of link called the axon which essentially works as the output channel for the neuron. The axon terminates in a rather specialized contact known as the synaptic junction or simply "synapse" which connects the neuron to the dendrites of another neuron. If the integrated (summed up) inputs that the soma of the neuron of our concern receives cause the internal electrical potential of the cell to rise above a threshold value, the neuron fires. That means it sends a "voltage pulse" down the relevant axon to other connected neurons. The synaptic junction, a tiny gap, contains a neuron transmitter fluid which is believed to be mainly responsible for enhancing or inhibiting the electrical charge flow to the soma. A large area of the synaptic junction signifies its excitatory nature while a small area or size of the junction makes it inhibitory in character. Learning and memory formation are believed to be controlled by the size of the synapses. Large synapses imply increased neuronal activity which in turn tends to activate learning and memory-forming processes in the brain.

Functionally, the natural neurons may be thought of as simple processing elements performing input signal integration and controlling output via a nonlinear

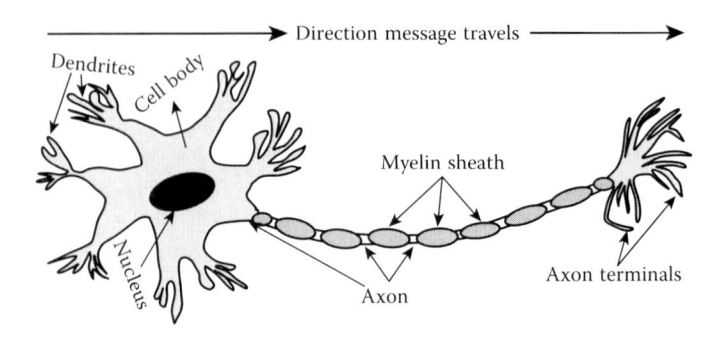

FIGURE 8.1
A natural or living neuron.

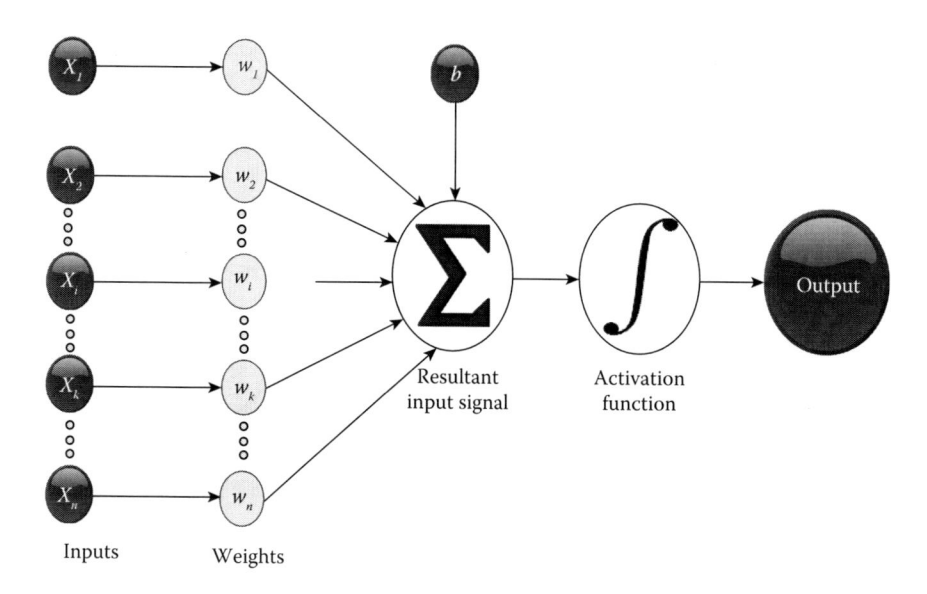

Inputs   Weights

FIGURE 8.2

Artificial neuron. (Courtesy of Sarkar, K., *Soft-Computing algorithms for solving some problems of Quantum Chemistry*, Indian Association for the Cultivation of Science, Jadavpur University Kolkata, India, 2014, 36.)

thresholding device. These processing elements, as we have described, are interconnected into a massively complex network—the brain, which works like a huge parallel distributed computing system.

An ANN is a miniaturized network of much simpler processing elements called artificial neurons which still captures the basic functional features of natural neurons. An artificial neuron is depicted in Figure 8.2 [5].

$x_1, x_2, \ldots, x_k, \ldots, x_n$ in Figure 8.2 denote n input signals arriving at the neuron L through dendritic links. $\omega_1, \omega_2, \ldots, \omega_k, \ldots, \omega_n$ represent weights associated with various synaptic links. Weights determine whether the arriving signal is enhanced or quenched. A large weight factor implies a large synaptic area (high synaptic strength) and enhanced neuronal activity. A small weight similarly implies the opposite. Each input (x_i) is multiplied by the corresponding weight (ω_i) and the product $(\omega_i x_i)$ is fed into the neuron. The neuron sums up all such products and produces what is commonly known as net input, abbreviated simply as *net*. If the neuron carries the identifier or index j, the net input is identified as net_j, b is the biasing term. In Figure 8.2, the net input is $net = \sum_{i=1}^{n} x_i \omega_i$. In vector notation, we may represent X as an n component (row) vector $(1 \times n)$ and W as a column vector $(n \times 1)$. In this representation *net* is just the scalar product of X and W:

$$net = X.W. \tag{8.1}$$

Denoting the total input to neuron i by net_i, we have

$$net_i = \sum_{j=1}^{n} x_j \omega_j^i. \tag{8.2}$$

The neuron not only performs an integrating task over the input signals (Equation 8.2) received, it also undertakes a (generally nonlinear) thresholding job. Thus it compares net_i

against a threshold parameter θ_i (say). If the threshold is exceeded, then it only produces an output (y_i). The natural thresholding action is usually simulated by a nonlinear filter, $\varphi(net_i)$ called the activation function or the transfer function. Thus the output y_i from the ith neuron turns out to be

$$y_{i\,=}\,\varphi(net_i). \tag{8.3}$$

The value of y_i will, of course, depend on the actual functional form of the transfer function used and the value of net_i.

8.3 Activation Functions

The exact form of the nonlinear filter or the activation function used is dictated by the nature of the application being considered. The following forms are the most widely used:

1. Step function: The activation function performs a pure thresholding job. The sum
   $net_i = \sum_{j=1}^{n} x_j \omega_j^i$ is compared against a threshold value θ_i. If $net_i > \theta_i$ the output y_i
   is equal to 1, otherwise $y_i = 0$. Mathematically, the action is simulated perfectly by the Heaviside function or the step function with the argument $(net_i - \theta_i)$. Hence we can write

$$y = \begin{cases} \varphi\left(net_i - \theta_i\right) \\ \varphi(\sum_{j=1}^{n} x_j \omega_j^i - \theta_i) \end{cases} \tag{8.4}$$

with the condition that

$$y = \begin{cases} 1 \text{ if } net_i > \theta_i \\ 0 \text{ if } net_i \le \theta_i \end{cases}. \tag{8.5}$$

Figure 8.3a illustrates the thresholding action.

2. Signum function: For some applications, we may require a thresholding function $\varphi(x)$ which produces the output $y = -1$ if the argument x is ≤ 0, and generates output $y = 1$ if $x > 0$. Recalling that the argument $x_i = (net_i - \theta_i)$ we can set

$$\varphi\left(net_i - \theta_i\right) = \begin{cases} 1 \text{ if } net_i > \theta_i \\ -1 \text{ if } net_i \le \theta_i \end{cases}. \tag{8.6}$$

Figure 8.3b displays the signum activation function graphically.

3. Sigmoidal activation function: We may require, in many applications, an activation function that produces output continuously and asymptotically varying from

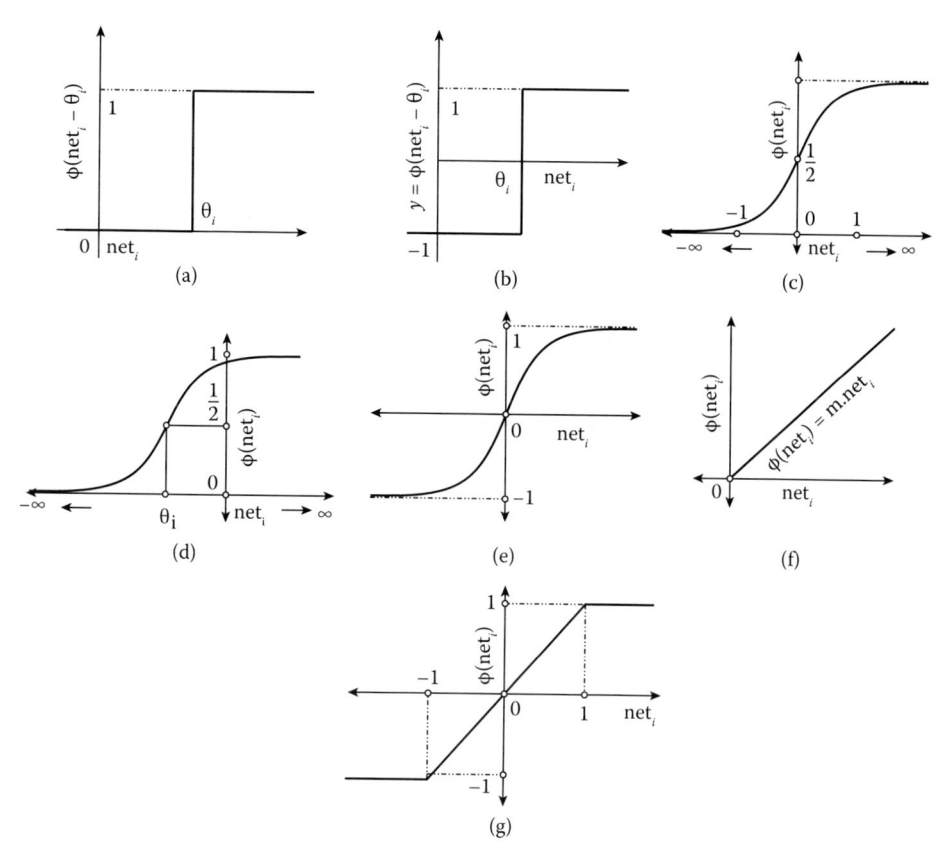

FIGURE 8.3
(a) Step function as activation function, (b) signum activation function, (c) sigmoidal activation function, (d) sigmoidal activation function with a threshold, (e) bipolar sigmoidal activation function, (f) linear activation function, and (g) piecewise linear activation function.

0 to 1 depending on the value of net_i. A continuous, differentiable function that fits the requirements is provided by the sigmoidal activation function $\varphi(net_i)$ where

$$y = \varphi(net_i) = \frac{1}{1+e^{-net_i}}. \tag{8.7}$$

As $net_i \to \infty$, $y \to 1$, and as $net_i \to -\infty$, $y \to 0$ with y becoming $\frac{1}{2}$ as net_i assumes the value 0.

The nature of the sigmoidal activation function without threshold is displayed in Figure 8.3c. The function rises abruptly at $net_i = 0$ The sigmoidal activation function can be modified further by multiplying the exponent (net_i) with a parameter β (the slope parameter) which controls how abruptly $\varphi(net_i)$ rises between the two asymptotic values (0,1) as net_i approaches zero from the left. The modified function is

$$\varphi(net_i) = \frac{1}{1+e^{-\beta net_i}}; \ \beta > 0. \tag{8.8}$$

In many applications, a sigmoidal activation function with a threshold θ_i is required. Such a function is represented as

$$\varphi(net_i) = \frac{1}{1+e^{-(net_i+\theta_i)}}.$$ (8.9)

The growth of the function between its asymptotic values as net_i increases and crosses the threshold at $-\theta_i$ is displayed in Figure 8.3d. In cases where the output of the activation varies continuously between the asymptotic values -1 and $+1$, the activation function is chosen in the so-called bipolar sigmoidal form displayed in Figure 8.3e. Mathematically, it is represented by the hyperbolic tangent function

$$\varphi(net_i) = \tanh(net_i)$$

$$\text{or by } \varphi(net_i) = \tanh(\beta.net_i); \ \beta > 0.$$ (8.10)

The sigmoidal activation functions have the strongest resemblances with the nonlinear thresholding action of biological neurons. The output here varies continuously but nonlinearly between the asymptotic values. Naturally, they are the most commonly used forms of activation function in an ANN. However, in many scientific and technological applications, it is often convenient to use the other forms of activations already described (e.g., step function or signum function). In addition, sometimes linear activation functions (Figure 8.3f) or a piecewise linear activation function (Figure 8.3g) can be useful. The functional form of the graph in Figure 8.3g can be defined as follows:

$$\varphi(net_i) = \begin{cases} 1 \text{ if } m(net_i) > 1 \\ m.(net_i) \text{ if } m(net_i) < 1. \\ -1 \text{ if } m(net_i) \ll 1 \end{cases}$$ (8.11)

Two other forms of activation function have been used in ANNs, namely the unipolar multimodal nonlinear activation function and the radial activation function. Their analytical forms are summarized below.

1. Unipolar multimodal:

$$\varphi(net) = \frac{1}{2}[1 + \frac{1}{N}\sum_{p=1}^{N}\tanh(g^p(netY))]$$ (8.12)

2. Radial basis function:

$$\varphi(net) = e^{-net},$$ (8.13)

$$\text{where, } net = \sum_{i=1}^{N}\frac{(\omega_i(t)-net_i(t))^2}{2\sigma^2}.$$

Many other mathematical forms can be chosen, depending on the type of problem being addressed. Having described how artificial neurons can be endowed with integrating and thresholding actions, it would be natural to investigate the simplest way of assembling artificial neurons into a simple network architecture capable of performing an intelligent task. The simplest architecture was proposed for the first time by Rosenblatt [6] in 1958, in the form of his perceptron model. This is a computational model of the functioning of the retina of the eye—hence the name "perceptron" (visual perception). In the following section, we take up the construction and working of the perceptron model in fair detail and examine its strengths and weaknesses.

8.4 Single-Layer Perceptron

Let us consider n number of input neurons arranged in a layer (input layer) connected with an m number of output neurons (output layer) through synapses, the strengths of which are denoted by a weight matrix with the element ω_{ij} ($i = 1, 2, \ldots, n; j = 1, 2, \ldots, m$). The network architecture is displayed in Figure 8.4a.

$x_1, x_2, \ldots, x_k, \ldots, x_n$ are signals arriving at input layer neurons $i_1, i_2, \ldots, i_n$ (or simply 1, 2, \ldots, n). The net input to the jth output neuron ($o_j, j = 1, 2, \ldots, m$) is obtained by summing up input signals $x_1, x_2, \ldots, x_k, \ldots, x_n$ from the input layer neurons, $i_1, i_2, \ldots, i_k, \ldots, i_n$ multiplying each by the weight of the corresponding synaptic element ω_{kj} ($k = 1, 2, \ldots, n$). Thus, the net input strength perceived by o_j is

$$net_j = \sum_{i=1}^{n} x_i \omega_{ij}; \quad (j = 1, 2, \ldots, m). \tag{8.14}$$

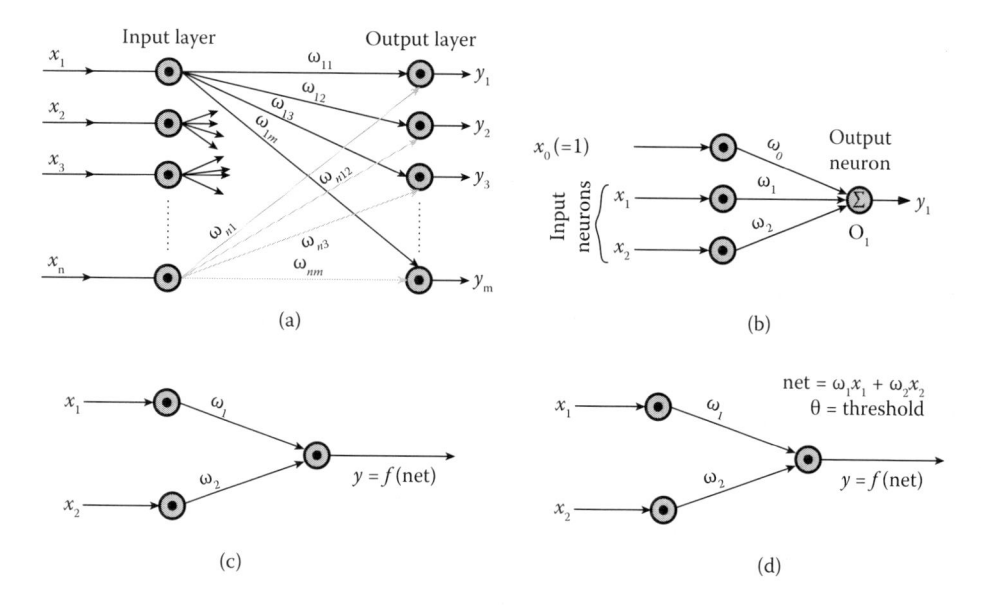

FIGURE 8.4

(a) A simple perceptron; (b) a two-input, one-bias, and one-output perceptron; (c) a two-input, one-output perceptron; and (d) a perceptron model for *XOR* function.

The output signal (y_j) from o_j the jth output layer neuron is determined by the thresholding function $f(z)$ where z represents the argument of the function (here $z = net_j$). Thus, we have

$$y_j = f(net_j). \tag{8.15}$$

Suppose that the thresholding or activation function $f(z)$ has been so chosen that the output from o_j is either 1 or 0. Such a thresholding or activation action can be simulated by choosing $f(z)$ as a step function

$$f(net_j) = \begin{cases} 1 \text{ if } net_j > 0 \\ 0 \text{ otherwise} \end{cases}.$$

Now comes the question about the determination of the values of synaptic weight matrix elements $\omega_{ij}(i = 1, 2, \ldots, n; j = 1, 2, \ldots, m)$. These are the unknowns of the perceptron model and are to be determined through a process called supervised learning.

Let us consider a problem of pattern classification in which only two classes are there (0 and 1) and the input has n features. Suppose that for a given input pattern $p = (x_1, x_2, \ldots, x_n$, say) we know the correct output pattern ($y_1, y_2, \ldots, y_m$) and that we have randomly assigned the values of the weight matrix elements $\{\omega_{ij}^0\}$ initially. Since we know the correct input $\rightarrow$ output mapping for the pattern $p_0 = (x_1^0, x_2^0, \ldots, x_n^0)$, we may try to train the perceptron by adjusting iteratively the weight matrix element $\{\omega_{ij}\}$ such that the predicted outputs $\{y_j\}$ slowly approach $\{y_j^0\}$. The following steps may be executed to train the perceptron:

1. Assign initial weights $\omega_{ij}(0)$ randomly between $(-a, a)$.
2. Read the input signals $(x_1^k, x_2^k, \ldots, x_n^k)$ for patterns p_k ($k = 0, 1, 2, n_p$) and for each pattern do the following.
3. Form $net_j = \sum_i x_i^k \omega_{ij}(t)$; ($t = 0, 1, 2, \ldots$).
   Evaluate outputs from the output neurons $o_j \{y_j^k\}$, where $y_j^k = f(net_j^k)$.
4. If y_j^k matches with the targeted or the correct value y_j^0, then the weight matrix elements are kept unaltered, i.e.,

$$\omega_{ij}(t+1) = \omega_{ij}(t) + 0. \; x_i^k. \tag{8.16}$$

If all the n_p number of patterns have been classified correctly, the current weight matrix elements are printed before exiting. If not, weight elements are updated as follows.

a. If the output is incorrect, say p_k, $y_j^k = 1$ whereas the targeted output is 0, the corresponding weight matrix elements are decreased by taking

$$\omega_{ij}(t+1) = \omega_{ij}(t) - \alpha. \; x_i^k \; (i = 1, 2, \ldots, n) \tag{8.17}$$

(t is the iteration counter)

b. On the other hand, if the output $y_j^k = 0$, where it should have been 1, the weight matrix elements are incremented as

$$\omega_{ij}(t+1) = \omega_{ij}(t) + \alpha. \; x_i^k \;\; (i=1,\,2,\,\ldots,\,n) \tag{8.18}$$

α is a parameter (≥ 0) that determines the learning rate. Obviously, a small value of α means the learning is slow while large α signifies accelerated learning. We can now go back to step 3, replace $\omega_{ij}(t)$ by $\omega_{ij}(t+1)$ and iterate until the perceptron outputs the targeted response correctly for all patterns in the training set. If an unknown with n features is now presented to the perceptron, it should be able to classify it into one of the two output classes correctly. What kind of values should be assigned to α? As we have indicated already, a small value of α leads to slow learning. A larger value of α will accelerate the learning, but runs the risk of creating an oscillation in the values of the weight matrix elements about the desired values, i.e., values that would have enabled correct classification. It is not mandatory to keep the value of α fixed for all the iterations. However, it is simpler to work with a constant value of α in which case it is called a fixed increment learning algorithm. A pertinent question is: does the algorithm converge? Rosenblatt proved the very important perceptron convergence theorem (1958). Let us first focus on the limitations of the perceptron model with a two-input (x_1, x_2) and one output (o_1) neuron model of a simple perceptron as an example. A third input neuron $x_0 = 1$ is introduced as a biasing element (Figure 8.4b). The weighted sum of the inputs including the bias element is

$$net = \omega_0 x_0 + \omega_1 x_1 + \omega_2 x_2 = \omega_0 + \omega_1 x_1 + \omega_2 x_2. \tag{8.19}$$

Equation 8.19 represents a straight line which functions as a decision boundary separating the points on the x_1, x_2 plane into two classes p_1, p_2 if and when the correct set of weights ω_0, ω_1, ω_2 have been found (Figure 8.2) through training. In an example where n input neurons $(x_1, x_2, \ldots, x_n)$ together with a biasing element x_0 is considered, the weighted sum of inputs is

$$net = \sum_{i=0}^{n} \omega_i x_i = \omega_0 + \omega_1 x_1 + \cdots + \omega_n x_n \tag{8.20}$$

which defines the equation of a hyperplane that can, in principle, partition the points on the n-dimensional space into regions when the correct weights have been discovered during the training process. The biasing weight ω_0 tunes the orientation of the decision boundary (straight line or hyperplane as the case may be). It is clear now that the perceptron functions as a linear classifier and the iterative determination of the network weights would ultimately converge to the solution (identification of the correct decision boundary) only if the problem space is linearly separable. The linear separability condition is a serious limitation of the perceptron model, a limitation first articulated by Minsky and Papert [7]. It affected (practically halted) further development in perceptron research and research in ANN for about two decades. It would be useful to look back into Minsky and Papert's demonstration of the limitations of the perceptron model as it would help us to find means of overcoming the limitations.

8.4.1 The Limitations of A Single-Layer Perceptron

The limitations can be best understood by considering the problem of perceptron representation. Representation in the context of ANNs means whether a neural network would be able to generate a specific function at all. If it is capable of representing a function, it would be able to learn (i.e., determine the weights correctly). The learning would fail if the network cannot represent the function—no matter what method is adopted for learning. A simple example illustrating the basic representability problem is provided by the Boolean logic function *AND*. The function takes two inputs and produces one output which is *true* or 1 if both the inputs are *true* or 1; otherwise the output is zero or *false*. Can a neural representation be given to the Boolean *AND* function? Let the two inputs be x_1 and x_2; then the *AND* function, x_1. *AND.* $x_2 = y$ (say), will produce the truth table (Table 8.1).

Can we construct a perceptron with two input and one output neurons that represents x_1. *AND.* x_2 correctly? The answer is in the affirmative as can be demonstrated by considering the following perceptron (Figure 8.4c). x_1 and x_2 in Figure 8.4c are input signals (0 or 1) from neurons 1 and 2. The synaptic weights between the input and the output neurons are ω_1 and, ω_2 respectively. The net input to the output neuron is $net = \omega_1 x_1 + \omega_2 x_2$. Let $y = f(net)$ be represented by a step function with a threshold θ. Suppose we have chosen the weights $\omega_1 = 0.45$, $\omega_2 = 0.45$ and the threshold value chosen is 0.5. So $y = f(net) = 0$ if $net \leq 0.5$, and is equal to 1 if $net < 0.5$. With network parameters chosen, let us see if the truth table for x_1 *AND* x_2 is correctly reproduced by the perceptron:

It is clear from the entries in Table 8.2 and 8.1 that the perceptron represents the Boolean *AND* function correctly. Is it possible to represent every other Boolean function by a perceptron? Minsky and Papert [7] pointed out that such a representation is impossible in the case of the *XOR XOR* function. To understand the problem let us consider the truth table (Table 8.3) for the Boolean function.

TABLE 8.1

Truth Table for Boolean *AND* Function

x_1	x_2	y	Class	
0 (*f*)	0 (*f*)	0 (*f*)	P_1	
0 (*f*)	1 (*t*)	0 (*f*)	P_2	C_1
1 (*f*)	0 (*f*)	0 (*f*)	P_3	
1 (*t*)	1 (*t*)	1 (*t*)	Q_1	C_2

TABLE 8.2

Perceptron for Boolean *AND* Function

x_1	x_2	ω_1	ω_2	net	y
0	0	0.45	0.45	0	0
0	1	0.45	0.45	0.45	0
1	0	0.45	0.45	0.45	0
1	1	0.45	0.45	0.90	1

TABLE 8.3

Truth Table for Boolean *XOR*
Function

x_1	x_2	y	Class
0	0	0	C_1
0	1	1	C_2
1	0	1	C_2
1	1	0	C_1

A perceptron model for *XOR* function can be sought to be constructed using two input and one output neurons as displayed in Figure 8.4d. With net input defined as $\omega_1 x_1 + \omega_2 x_2$, the output $y = f(z)$ can be calculated once the form of the function $f(z)$ is defined. Supposing that $f(z)$ is a step function with a threshold θ, we can set out to find weights ω_1 and ω_2 and the threshold parameter θ so that y represents the same output values displayed in Table 8.3. However, no matter how much we try, the neural perceptron representation of *XOR* displayed in Figure 8.4d (two input + one output neurons) would fail to discover appropriate weights and thresholds to satisfy the truth table. Why is it so? To understand the root of the problem let us consider the distribution of the values of y in the space of the input variables (x_1, x_2). From Table 8.3 it is clear that y assumes only two values, 0 and 1, and there are two points P_1 and P_2 on the $x_1 x_2$ plane with $y = 0$ and two points Q_1 and Q_2 on the same plane with $y = 1$ (see Figure 8.5a).

The single-layer perceptron seeks to classify the four points into two classes C_1 and C_2, with C_1 containing the points (P_1, P_2) and C_2 containing (Q_1, Q_2) by constructing the straight line represented by $\omega_1 x_1 + \omega_2 x_2 = 0$. By changing ω_1, ω_2, and θ, the perceptron can alter the orientation (slope) and position of the straight line, but no matter what values are selected for ω_1, ω_2, and θ, no straight line can be drawn so that we have (P_1, P_2) on one side of the line and (Q_1, Q_2) on the other side. One side corresponds to the region where, $\omega_1 C_1 + \omega_2 C_2 < \theta$ while the opposite side corresponds to points satisfying the inequality $\omega_1 C_1 + \omega_2 C_2 > \theta$. The single-layer perceptron is therefore inadequate to represent the *XOR* function. The fact is that the *XOR* function is not linearly separable and therefore the perceptron fails to find a straight line partitioning $x_1 x_2$ plane as required for representing the *XOR* function. Why does the single-layer perceptron succeed in representing the Boolean *AND* function? A little reflection will assert that the *AND* function is a linearly separable function and therefore the perceptron is able to find a straight line that subdivides the $x_1 x_2$ plane into two parts C_1 and C_2 with C_1 containing three points (P_1, P_2, P_3) with $y = 0$ and C_2 containing the lone point (Q_1) with $y = 1$ as displayed in Figure 8.5b (also see Table 8.1 where the truth table for the *AND* function is shown). The straight line clearly separates the $x_1 x_2$ plane into two regions, one containing the points P_1, P_2, P_3 for which $y = 0$ and the other containing the point Q_1 for which $y = 1$ fulfilling the operational requirement of the *AND* function. The two examples discussed in this section could lead us to generally surmise that a single-layer perceptron can successfully represent only a linearly separable function meaning that it will be able to construct a hyperplane to partition an n-dimensional space into two regions as required by the function. How the weights and thresholds are obtained is a matter of detail and constitutes an issue entirely different from the problem of representability of a function by a single-layer perceptron. If the function is linearly separable, the perceptron will ultimately be able to find the solution. A look at Figure 8.5a immediately suggests that a curved surface such as $L_1 L_1'$ or $L_2 L_2'$ will subdivide the plane into two regions, one containing P_1, P_2 and the other containing the Q_1, Q_2. The implication

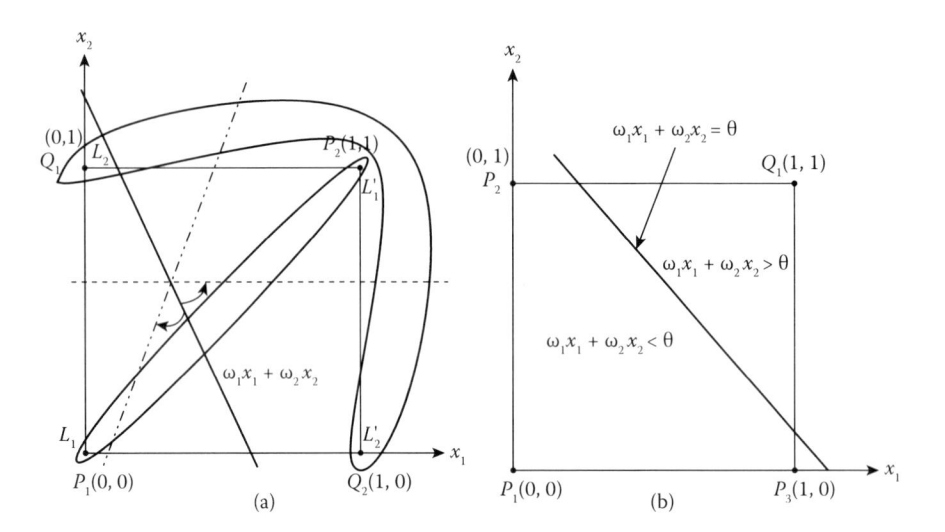

FIGURE 8.5

(a) *XOR* function: a curved surface such as L_1L_1' or L_2L_2' will subdivide the plane into the two regions, one containing P_1, P_2 and the other containing the Q_1, Q_2, (b) *AND* function: a straight line separates the x_1, x_2 plane into two regions, one containing the points P_1, P_2, P_3 and the other containing the point Q_1.

is clear—the function *XOR* is nonlinearly separable and the required classification can be achieved only by introducing an additional layer in the architecture of the perceptron—a single-layer perceptron will just not work for nonlinearly separable problems.

For the sake of completeness let us now consider a more general formulation of the perceptron training problem. Let there be m number of patterns represented by input vectors x_j ($j = 1, 2, ..., m$) x_j ($j = 1, 2, ..., m$) with n components (each representing a feature of the pattern) so that x_{ji} represents the value of the ith feature of the jth input pattern in the training set. Let the targeted output for x_j be t_j. The net input to the output neuron o_j at the instant t is

$$net_j(t) = \omega_0(t)x_{j0} + \omega_1(t)x_{j1} + \cdots + \omega_m(t)x_{jm}, \tag{8.21}$$

where $\omega_i(t)$ represents the ith component of the weight vector at the instant t for the jth input pattern in the training set. The output is $y_j(t)$ where

$$y_j(t) = f(net_j(t)) \tag{8.22}$$

and $x_{j0} = 1$ (the biasing element), and $f(z)$ is the appropriate activation function chosen for the problem. The initial weights at $t = 0$ are randomly assigned and have small magnitudes. The perceptron weight-updating formula or the learning rule then updates the weights as follows:

$$\omega_i(t+1) = \omega_i(t) + \alpha. \left(t_j - y_j(t)\right)x_{ji}; \; (i = 1, 2, ..., n; \, j = 1, 2, ..., m). \tag{8.23}$$

The learning rate α can be put equal to 1 without any loss of generality as any $\alpha \neq 1$ will simply rescale the weights. If the training set is linearly separable, the above updating formula is general and sure to converge and correctly classify all the vectors of the training sets.

Novikoff [8] proved the perceptron convergence theorem elegantly [9]. The understanding of the source of limitations of the single-layer perceptron led to rapid further developments in the ANN methods. The idea of introducing a hidden layer of neurons between the input and the output layers and optimizing the elements of weight vectors and thresholds (biases) by minimizing the output error (relative to a set of known solutions provided in the training set) quickly revolutionized the construction and the application of ANNs resulting in what became known as backpropagation networks (BPN); until now almost 90% of ANN applications are based on BPN.

8.5 Backpropagation Networks

8.5.1 The Basic Architecture

The architecture of an NN defines the pattern of connections among different neurons. The basic architecture of a BPN is a simple layered structure. There is an input layer of n_I neurons and an output layer of n_0 neurons. Interposed between these two layers are one or more hidden layers of n_H neurons (hidden because they receive signals only from the input layer neurons and not have any interface with the external world). A BPN with single hidden layer of neurons is depicted in Figure 8.6a and b.

The reader may note that each input neuron is connected to all the hidden layer neurons just as each hidden layer neuron is linked to all the output neurons. Each edge in the network carries a weight (ω or ω', ω'' etc.). The weight matrix) (W) connecting the input neurons with the hidden layer neurons has $n_1 \; x \; n_H$ number of elements (ω_{ij} with i ranging from 1 to n_I and j ranging from 1 to n_H). Similarly, the weight matrix (W') connecting the hidden layer neurons with the neurons of the output layer has $n_H \times n_0$ number of elements (ω'_{jl} with j ranging from 1 to n_H and l ranging from 1 to n_0). All these elements of weight matrices W and W' are basic unknowns of the problem. BPN learns by adjusting the elements of W and W' through supervised learning based on a set of input patterns for which the "supervisor" knows the correct output. One important characteristic of the BPN is that the flow of information within the network takes place only in the forward direction, i.e., from the input, to the hidden, to the output layer. Accordingly, the BPN is categorized as a feedforward or nonrecurrent network. So the basic characteristics of the backpropagation can be summarized as follows:

1. BPN is a multilayered feedforward (nonrecurrent) network. Information flows only in one direction.

2. In addition to the input and output layer of neurons, there are one or more hidden layers of neurons.

3. The input layer neurons are each connected with all the neurons in the hidden layer while each neuron in the hidden layer is connected to all the output layer neurons. If there are more than one hidden layer of neurons, each neuron of hidden layer 1 is connected with all the neurons of hidden layer 2, and so on.

4. The network learns by adjusting the weights in a supervised manner.

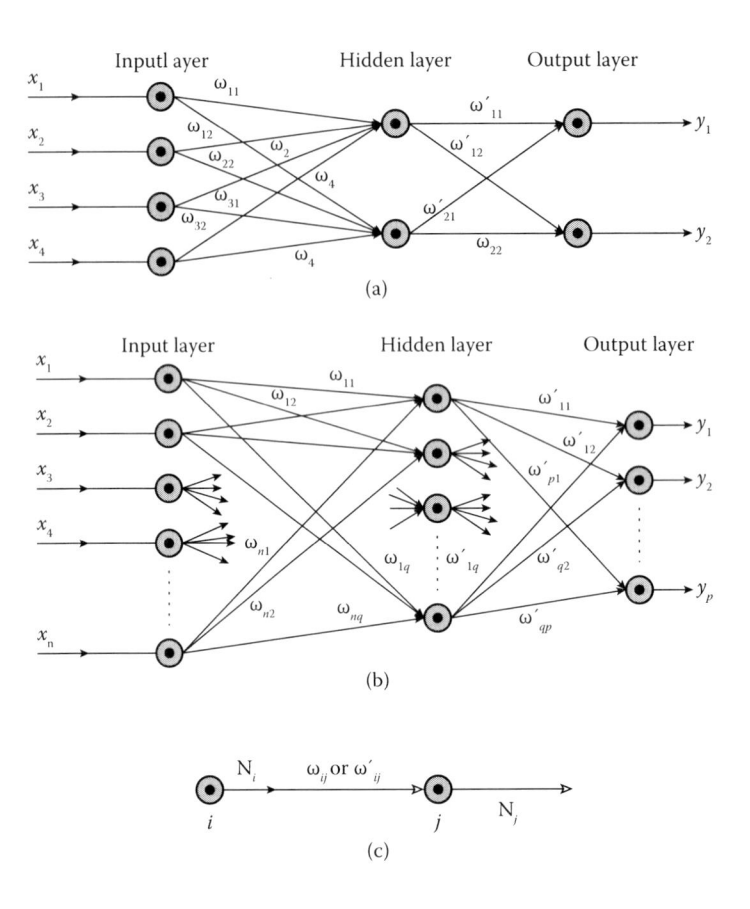

FIGURE 8.6

(a) A backpropagation network with single hidden layer of neurons with four input layer neurons, two hidden layer neurons, and two output layer neurons, (b) a general backpropagation network with single hidden layer of neurons, and (c) a pair of neurons, *i*, *j* belonging to two different interconnected layers of neurons of a BPN.

8.5.2 The Network Weights, Thresholds, and Activation Functions

Referring to Figure 8.6a we have a BPN network with four input layer neurons, two hidden layer neurons, and two output layer neurons. The input layer neurons (n_1, n_2, n_3, n_4) (n_1, n_2, n_3, n_4) do not do any calculation, they simply transmit the input signals (x_1, x_2, x_3, x_4) to the hidden layer neurons (h_1 and h_2) weighted by the weight matrix element (ω_{ij}) assigned to the edge connecting a particular input neuron (say *i*) to a particular hidden layer (say *j*). Similarly, the edge connecting a hidden layer neuron (say *j*) with an output layer neuron (say *i*) is assigned a weight ω'_{jl}. The input layer neurons are often represented by the input signals (x_1, x_2, ..., x_n) they are supposed to transmit to the next layer. Let us now ask, with reference to Figure 8.6a, what is the net input by the hidden layer neurons h_1 and h_2? Considering weights associated with edges connecting input and hidden layer neurons, we have

$$net_1 = \omega_{11}x_1 + \omega_{21}x_2 + \omega_{31}x_3$$

$$= \sum_{i=1}^{3} x_i \omega_{i1} \tag{8.24}$$

$$net_2 = \omega_{12}x_1 + \omega_{22}x_2 + \omega_{32}x_3$$

$$= \sum_{i=1}^{3} x_i \omega_{i2}. \tag{8.25}$$

More generally, $net_j = \sum_{i=1}^{n_I} x_i \omega_{ij}$, $j = 1, 2, ..., n_H$ for the jth hidden layer neuron connected to n_I number of input neurons. The outputs coming from the hidden layer neuron would be determined by the thresholding or activation function $f(net)$ that the network is using. Assuming it to be sigmoid with threshold value θ, we have

$$h_1 = f(net_1) = \frac{1}{1 + e^{-(net_1 + \theta_1)}}$$

$$h_2 = f(net_2) = \frac{1}{1 + e^{-(net_2 + \theta_2)}}. \tag{8.26}$$

More generally, for a network having n_I number of input layer neurons connected to n_H number of hidden layer neurons, the output from the jth hidden layer neuron is

$$h_j = f(net_j) = f\left(\sum_{i=1}^{n_I} x_i \omega_{ij}\right) = \frac{1}{1 + e^{-(net_j + \theta_j)}} = \frac{1}{1 + e^{-\left(\sum_{i=1}^{n_I} x_i \omega_{ij} + \theta_j\right)}}. \tag{8.27}$$

We note that the outputs coming from the hidden layer neurons serve as incoming signals or activation for the output layer neurons. As usual, these signals are weighted by the weight matrix elements (ω'_{jl}) associated with various edges interconnecting hidden-layer neurons with output neurons. The net input received by the lth output layer neuron (net'_l) is thus

$$net'_l = \sum_{j=1}^{n_H} h_j \omega'_{jl} \tag{8.28}$$

while the output from the lth output layer neuron is

$$y_l = f(net'_l) = \frac{1}{1 + e^{-(net'_l + \theta'_l)}} = \frac{1}{1 + e^{-\left(\sum_{i=1}^{n_H} h_j \omega'_{jl} + \theta'_l\right)}} \tag{8.29}$$

where we have again assumed a sigmoid form of activation function for the output layer neurons, with a threshold θ'. In BPN the weights associated with different edges ($\{\omega_{ij}\}$ with $i = 1, 2, ..., n_I; j = 1, 2, ..., n_H$ and $\{\omega'_{jl}\}$ with $j = 1, 2, ..., n_H; l = 1, 2, ..., n_O$) along with the activation threshold ($\{\theta_i\}$, $i = 1, 2, ..., n_I$ and $\{\theta'_l\}$, $l = 1, 2, ..., n_O$) are parameters of the network that are to be determined via supervised learning. This requires a fairly large number of input–output mapping that can be used to train the network.

8.5.3 The Learning Scheme of a BPN

Once the architecture has been defined for a particular BPN with precisely defined input, hidden and the output layers, and the interconnections among them, the question to address is: how does the network learn? We have already mentioned, the network learns by adjusting weights associated with activation functions used by different neurons. The learning can be viewed as a supervised training exercise in which the network is fed with pairs of correct input–output patterns (these pairs form the training set) and the network is allowed to adjust weights and thresholds iteratively so that it can predict the correct output for each of the inputs in the training set after the conclusion of training. Thus, with n_i input neurons (i_1, i_2, ..., i_{n_i}), an input pattern (P) is presented as an n-component vector X with components (x_1, x_2, ..., x_{n_i}). The elements of the weight matrices W and W' are initially assigned random values from preset range ($-b$ to $+b$, $b = 0.25$ or 0.50 or any such quantity) the output vector (Y) with elements (y_1, y_2, ..., y_{n_O}) is computed by the network and compared with the correct output vector—the so-called target vector t — with component (t_1, t_2, ..., t_{n_O}). The network computes the error vector ε with elements ε_1, ε_2, ..., ε_{n_o} where $\varepsilon = |t - y|$. The network then iteratively adjusts the weight matrix elements following some well-defined techniques so that the norm of error vector gradually reduces to a small value (ideally zero) for all the input patterns (P_1, P_2, P_m) and the corresponding target patterns. Since the output pattern for each input pattern is known to begin with, the training of the BPN is called supervised learning. It may be noted that in the BPN, the flow of input is only from the input $\rightarrow$ hidden $\rightarrow$ output layers of neurons. There is no reverse flow of information in apparent contradiction with the term backpropagation. The network estimates the error in the output produced by comparing it with the defined target and then goes back to readjust the weights and thresholds so that the output better matches with the target given (we will see later that the threshold can be treated as additional weights and optimized)—that is why the term used is backpropagation.

8.6 The Mathematical Formulation of the Backpropagation Method

Let us suppose that we have a training set of m input patterns (P_1, P_2, ..., P_m) for which the target output patterns are (T_1, T_2, ..., T_m). We suppose further that the input patterns are each represented by vectors $X^{(i)}$ with n real components—(x_1^i, x_2^i, ..., x_n^i) ($i = 1, 2, ..., m$) and the targeted outputs are also represented by vectors $T(i)$ with p real components—(t_1^i, t_2^i, ..., t_p^i). Our problem at the training phase is to find the network weights (and thresholds) so that every time one of the patterns is presented to the network, it produces the correct or the designated output pattern. When the training is completed the network should be able to classify a pattern of the same type (even if it is not present in the training set) correctly demonstrating that it has learned to do the classification task.

Let us first construct the BPN for handling the task. The network (Figure 8.6b) designed has m number of input neurons I_1, I_2, ..., I_m each receiving and transmitting one component of the input pattern vector $X^{(i)}$ (i.e., I_1 receives and transmits x_1^i, I_2 receives and transmits x_2^i, and so on). The m input neurons are each connected to q number of hidden layer neurons (h_1, h_2, ..., h_q) which in turn are each connected to the q number of output layer neurons (o_1, o_2, ..., o_q). The output produced from o_i is designated y_i (Figure 8.6b). The weight matrix W has elements ω_{ij} representing the synaptic strengths of the connections between the ith input neuron (I_i) and the jth hidden layer neuron. There are thus ($n \times q$) matrix elements of W to be determined. Similarly, the elements ω'_{jk} of the matrix W' represent the strengths of synaptic connections between the jth hidden layer neuron and the kth output layer neuron, there being $q \times p$ number of elements of the weight matrix W'

to be determined. In addition, there are thresholds $\{\theta_i\}$ associated with activation functions used by the hidden and output layer neurons that are also to be optimally determined. The net input received by h_j (jth hidden layer neuron) is obtained by summing all the incoming inputs $(x_1, x_2, \ldots, x_n)$ from the n input neuron multiplied by the synaptic weights of the relevant input-hidden layer neuron connections $(\omega_{ij}, i = 1, \ldots, n)$. Thus, we have,

$$net_j = \sum_{i=1}^{n} x_i \omega_{ij}, \; j = 1, 2, \ldots, q \tag{8.30}$$

we may represent the net inputs received by the hidden layer neuron as a q component vector $Net = (net_1, net_2, \ldots, net_q)$. We will assume that the hidden layer neurons are each equipped with sigmoid activation functions with a threshold $\{\theta_j\} \; i = 1, 2, \ldots, q$. These activation functions will shape the output $\{h_j\}$ emanating from the hidden layer neurons by the mapping

$$h_j = \frac{1}{1 + e^{-(net_j + \theta_j)}}. \tag{8.31}$$

These h_js will serve as the incoming activation for the neurons $(o_1, o_2, \ldots, o_p)$ of the output layer. The net input (net'_k) to the kth output layer neuron (o_k) is obtained by summing overall the activation $(h_1, h_2, \ldots, h_q)$ coming from the hidden layer neurons weighted by the corresponding element of the weight matrix W'. Thus, we have

$$net'_k = \sum_{j=1}^{q} h_j \omega'_{jk}, \; k = 1, 2, \ldots p. \tag{8.32}$$

The net input received by different output layer neurons can be represented as a vector with p-real components $Net' = (net'_1, net'_2, \ldots, net'_k, \ldots, net'_p)$. The output from the kth output neuron is shaped by the activation function used by it. If we continue to assume that the output neurons also make use of sigmoid activation functions with threshold $\{\theta'_k\}$, we have,

$$y_k = \frac{1}{1 + e^{-(net'_k + \theta'_k)}}. \tag{8.33}$$

The output vector Y^i for a given pattern (P_i) can be represented as a p-component vector

$$Y^i = y_1^i, y_2^i, \ldots, y_k^i \ldots, y_p^i. \tag{8.34}$$

The initial weights and thresholds are assigned values randomly from a small range of values ($-a$ to $+a$, $a \approx 0.25$, for example). With these randomly assigned values the network takes the first input pattern (P_1), calculates Y^i and computes the error by computing the square of the Euclidean distance $(t^i - Y^i)$ between the target and the actual (calculated) output vector, where t^i is the known correct output vector for the ith input pattern P_i. It is advantageous to write the error as $\varepsilon^i = \frac{1}{2} \sum_k (t_k^i - y_k^i)^2$, the factor $\frac{1}{2}$ is used for numerical simplicity while computing derivatives of ε with respect to the factor $\frac{1}{2}$ is cancelled.

If the error $\varepsilon^i < \delta$, where δ is a preset small number (user-defined), the network has been able to reproduce the output correctly for the pattern P_i in the first iteration. This is unlikely to happen. If $\varepsilon^i > \delta$, it is necessary to readjust the weights (we assume for now that all the threshold values are set equal to zero and only $\{\omega_{ij}\}$ or $\{\omega'_{jk}\}$ are available for adjustment). The process of adjusting the weights so that the error ε^i or ε (if we drop the superscript) is reduced constitutes a mode of learning that the backpropagation network uses [10]. The adjustment is usually made by a deterministic procedure—the well-known method of steepest descent. It requires us to compute the gradient of the error surface (ε) with respect to elements of the weight matrix. Let the steepest descent correction to the weight matrix element be represented as $\Delta\omega_{ij}(t)$ where

$$\Delta\omega_{ij}(t) = -\lambda\left(\frac{\partial\varepsilon}{\partial\omega_{ij}(t)}\right) \tag{8.35}$$

and where λ is a scalar representing the step length to be taken in the direction of the local gradient. The steepest descent corrected or adjusted weight vector at the next iteration becomes

$$\omega_{ij}(t+1) = \omega_{ij}(t) + \Delta\omega_{ij}(t) = \omega_{ij}(t) - \lambda\left(\frac{\partial\varepsilon}{\partial\omega_{ij}(t)}\right) \tag{8.36}$$

for $i = 1, 2, \ldots, n; j = 1, 2, \ldots, q$. These represent the iterative correction formulae for the strength of input-hidden layer corrections or edges. Similarly, we can set up the corresponding correction formula for the strength of the hidden-output layer connection $\omega'_{jk}(t)$. Thus

$$\omega'_{jk}(t+1) = \omega'_{jk}(t) + \Delta\omega'_{jk}(t) = \omega'_{jk}(t) - \lambda\left(\frac{\partial\varepsilon}{\partial\omega'_{jk}(t)}\right) \tag{8.37}$$

for $j = 1, 2, \ldots, q; j = 1, 2, \ldots, p$. Once we have at our disposal the analytical formulae for computing the various partial derivatives that appear in Equations 8.36 and 8.37 we have a deterministic means of improving all the weight matrix elements so that the error ε in producing the correct output vector for any one of the m-numbers of training patterns (P_i, $i = 1, 2, \ldots, m$) becomes smaller than a threshold (δ). However, to achieve that a value for λ must be fixed. Usually, λ is assigned a value equal to 0.75–0.85. For better and more stable convergence characteristics, the weight modification schemes represented by Equation 8.36 and 8.37 obtained by straightforward applications of the steepest descent procedure are usually modified (see Section 8.6.1).

8.6.1 Explicit Formula for $\Delta\omega_{ij}(t)$, $\Delta\omega'_{jk}(t)$

The calculation of the $\Delta\omega_{ij}(t)$ or $\Delta\omega'_{jk}(t)$ requires calculations of partial derivatives of the error ε with respect to the synaptic weight matrix elements. These derivatives are easily and analytically computed by successive applications of the chain rule of differentiation. In what follows, we provide elaborate derivation of all the necessary formulae, to be used in Equations 8.36 and 8.37, for a single hidden layer backpropagation neural network. The formulae will undergo obvious modifications for BPNs with more than one hidden layer. We note here that the use of BPNs with a single hidden layer of neuron is the most common practice, although, on occasions, one comes across the use of BPNs with more than one (usually two or three) hidden layers. The relevant formulae are derived assuming that sigmoid activation functions with thresholds are used. If some other form of activation is used, appropriate changes need to be made.

Let us consider a pair of neurons, i, j belonging two different interconnected layers of neurons of a BPN (Figure 8.6c) where the output (N_i) from neuron i serves as the incoming activation (input) for the neuron j, the weight associated with the interconnection ($i \to j$) or ($j \to k$) being ω_{ij} or ω'_{jk} depending on whether N_i is an input layer or a hidden layer neuron. The BPN architecture for a single hidden layer network ensures that if i belongs to the input layer, j must be on the hidden layer, and if i is on the hidden layer j must be on the output layer (the feedforward nature ensures there is no feedback coupling). Referring back to Figure 8.6b, N_i is just x_i when i is on the input layer and N_j is just h_j, whereas for i being on the hidden layer, N_i is nothing but h_i and N_j is just y_j—the output from the jth neuron on the output layer.

Let ε be the error vector in the output for a particular input pattern (P) at tth stage of iteration. We have, according to our definition of error,

$$\varepsilon = \frac{1}{2} \sum_{j=1}^{p} (t_j - y_j)^2 \tag{8.38}$$

t_j being the jth component of the target vector, and y_j as usual the output from the jth output layer neuron and the sum is over all the output layer neurons. The jth output layer neuron receives a net incoming activation from all the neurons (N_i) it is connected with. The net activation received by it is net_j where

$$net_j = \sum_{i=1} N_i \omega_{ij}.$$

The output from the jth neuron is y_j which is determined by the activation function $f(net)$ used by the jth neuron so that, we have

$$y_j = f\left(\sum_{i=1} N_i \omega_{ij}\right) = f_j(net_j).$$

Our purpose here is to determine the analytical form of the partial derivatives $\dfrac{\partial \varepsilon}{\partial \omega_{ij}}$ for various choices of i and j permitted by the network architecture. Let us concentrate on the general term $\dfrac{\partial \varepsilon}{\partial \omega_{ij}}$, which by application of the chain rule, can be written as (dropping the iteration index t)

$$\frac{\partial \varepsilon}{\partial \omega_{ij}} = \frac{\partial \varepsilon}{\partial net_j} \frac{\partial net_j}{\partial \omega_{ij}}$$

$$= \frac{\partial \varepsilon}{\partial net_j} \frac{\partial \sum_{i=1} N_i \omega_{ij}}{\partial \omega_{ij}} \tag{8.39}$$

$$= \frac{\partial \varepsilon}{\partial net_j} (N_i) = D_j N_i$$

$\Delta\omega_{ij}(t)$, the steepest descent correction to $\omega_{ij}(t)$ now becomes (viz. Equations 8.36 and 8.39)

$$\Delta\omega_{ij}(t) = -\lambda \frac{\partial \varepsilon(t)}{\partial \omega_{ij}(t)} \tag{8.40}$$

$$= -\lambda D_j N_i$$

where $D_j = \dfrac{\partial \varepsilon(t)}{\partial \omega_{ij}(t)}$. We have made some progress, but are yet to find an explicit formula for D_j. We make use of the chain rule of differentiation once again to make further progress, and write, in general

$$D_j = \frac{\partial \varepsilon}{\partial net_j} = \frac{\partial \varepsilon}{\partial N_j}\frac{\partial N_j}{\partial net_j}. \tag{8.41}$$

Note that D_j is a product of two partial derivatives. The physical meaning of the above expression is clear: the rate of change of the error (ε) with respect to the change in the net incoming activation to the jth neuron is a product of the rate of change of the error as a function of the output N_j from the neuron j, and the rate of change in the output from the jth neuron with changes in the incoming activation net_j to the jth neuron. At this point, the architecture of the BPN (feedforward, single hidden-layer structure) must be recognized and used for further progress. The structure suggests that we consider two possible scenarios:

1. The neuron j belongs to the output layer—that demands i to be on the hidden layer and N_j is nothing but y_j (the output from the jth output neuron).
2. The neuron j does not belong to the output layer—then it must be on the hidden layer and i must be on the input layer. N_j then can be identified as h_j (output from the hidden layer neuron j).

These two cases are to be processed separately. Considering the first scenario (j is on the output layer) we have:

1. **Scenario 1**

$$\frac{\partial \varepsilon}{\partial N_j} = \frac{\partial \varepsilon}{\partial y_j} = \frac{1}{2}\frac{\partial}{\partial y_j}\sum_{i=1}^{p}(t_i - y_i)^2 = -(t_j - y_j) \tag{8.42}$$

$$\frac{\partial N_j}{\partial net_j} = \frac{\partial f_j(net_j)}{\partial net_j} \tag{8.43}$$

$f_j(net_j)$ is the activation function for the jth neuron on the output layer. Let

$$f_j(net_j) = \frac{1}{1 + e^{-(net_j + \theta_j)}} \tag{8.44}$$

Then

$$\frac{\partial f_j(net_j)}{\partial net_j} = f_j(net_j)\left(1 - f_j(net_j)\right) \tag{8.45}$$

$$= y_j(1 - y_j)$$

$$\text{That is, } D_j = -\left(t_j - y_j\right)y_j\left(1 - y_j\right). \tag{8.46}$$

So, in the situation where i is a hidden layer neuron and j is an output layer neuron, we have the updating formula for the weight matrix elements ω'_{ij} connecting hidden-layer neuron i to output neuron j as

$$\Delta\omega'_{ij}(t) = -\lambda\delta_j N_i$$

$$= +\lambda(t_j - y_j)y_j\left(1 - y_j\right)h_i$$

$N_i \equiv$ output from hidden layer neuron ($i = h_i$) and

$$\Delta\omega'_{ij}(t+1) = \Delta\omega'_{ij}(t) + \lambda(t_j - y_j)y_j\left(1 - y_j\right)h_i. \tag{8.47}$$

2. Scenario 2

Let us now take up the case where j is in the hidden layer, so i must on the input layer. In this case $\dfrac{\partial\varepsilon}{\partial N_j}$ in Equation 8.41 cannot be evaluated as in the previous case. Here, we may exploit the chain rule of differentiation in the following way:

$$\frac{\partial\varepsilon}{\partial N_j} = \sum_l \frac{\partial\varepsilon}{\partial net_l}\left(\frac{\partial net_l}{\partial N_j}\right) \tag{8.48}$$

where the index l runs over all the neurons in the layer that follows the layer in which j is situated (in our case, l runs over all the neurons in the output layer). In the present case, net_l represents the net incoming activation to the lth neuron in the output layer from all the hidden layer neurons so that we have

$$net_l = \sum_{l'}\omega'_{l'l}N_{l'} \tag{8.49}$$

and

$$\frac{\partial net_l}{\partial N_j} = \frac{\partial}{\partial N_j}\sum_{l'}\omega'_{l'l}N_{l'},$$

$$\frac{\partial\varepsilon}{\partial N_j} = \sum_l \frac{\partial\varepsilon}{\partial net_l}\left(\frac{\partial}{\partial N_j}\sum_{l'}\omega'_{l'l}N_{l'}\right)$$

$$= \sum_l\left(\frac{\partial\varepsilon}{\partial net_l}\right)\omega'_{jl}N_j \tag{8.50}$$

$$= \sum_l D_l\omega'_{jl}N_j.$$

Thus with i in the input layer and j in the hidden layer, the correction formula for network weights for input-hidden layer connection becomes

$$\omega_{ij}(t+1) = \omega_{ij}(t) + \Delta\omega_{ij}(t) = \omega_{ij}(t) - \lambda D_j N_i$$

$$= \omega_{ij}(t) - \lambda \left(\frac{\partial \varepsilon}{\partial N_j}\right)\left(\frac{\partial N_j}{\partial net_j}\right) N_i \tag{8.51}$$

$$= \omega_{ij}(t) - \lambda \left(\sum_l D_l \omega'_{jl} N_j\right) N_j (1 - N_j) N_i.$$

Noting that with j in the hidden layer and i in the input layer, $N_i = x_i$ (input signal to the ith neuron) and N_j is just the output h_j from the jth hidden layer neuron, we can write

$$\omega_{ij}(t+1) = \omega_{ij}(t) - \lambda h_j(1 - h_j)x_i\left(\sum_l D_l \omega'_{jl} h_j\right) \tag{8.52}$$

where $D_l = \dfrac{\partial \varepsilon}{\partial net_l}$, l is a neuron in the output layer.

If there is more than one hidden layer (scenario 2) the equations have to be reworked—now there are three sets of weight matrices W (input–hidden1), W' (hidden1–hidden2) and W'' (hidden2–output), the elements of which represent input layer neuron to hidden layer1 neuron connections, hidden layer1 neurons to neurons in the hidden layer2 connections, and neurons in the hidden layer2 to neurons in the output layer connection. It would be useful for the reader to work out the equations for $\omega_{ij}(t+1)$, $\omega'_{ij}(t+1)$, $\omega''_{ij}(t+1)$ explicitly following the derivation presented here. So far we have considered the problem of optimizing the network weights only assuming that the threshold $\{\theta_j\}$ or $\{\theta'_j\}$ in the sigmoid activation function are all equal to zero. The optimization of activation thresholds can be easily done in the following manner, treating them as additional weights. The procedure becomes transparent if we recall that

$$f_j(net_j) = \frac{1}{1 + e^{-(net_j + \theta_j)}} \tag{8.53}$$

and $net_j = \sum_i \omega_{ij} x_i$ (x_i is the incoming activation to j)
which allows us to write

$$net_j + \theta_j = \sum_i^n \omega_{ij} x_i + \theta_j \omega_{i+1,j} \tag{8.54}$$

with $\omega_{i+1,j} = 1$, we have $net_j + \theta_j = \sum_i^n \omega_{ij} x_i + \theta_j$. So we can imagine that θ_j is the connection weight of an imaginary neuron, $n+1$ (say), the incoming activation from this imaginary neuron being always unity. Hence, we can adjust the threshold values $\{\theta_j\}$ as additional weights.

A number of questions that may have topmost priority in the reader's mind at this stage need to be addressed. One of the questions concerns whether any other optimization method (other than the steepest descent method) could be exploited by the network to learn how to adjust the weights. The answer is in the affirmative. Stochastic methods such as the method of simulated annealing, random mutation hill climbing, genetic algorithms, and swarm intelligence methods are all capable of guiding the BPN to adjust or optimize the network weights correctly. The SAM and RMHC are especially suitable as they are not population based, but work with a single solution string. SAM could be time-consuming as the $T \to 0$ limit must be reached slowly to ensure that the global minimum is reached. In both the cases (namely SAM and RMHC), the objective function is just the error $\varepsilon = \frac{1}{2} \sum_{j=1}^{p} (t_j - y_j)^2$ as already defined. In the RMHC, it can be converted into either a scaled or an unscaled fitness function $f = e^{-\gamma \varepsilon}$ (scales between 0 and 1), or perhaps $f_1 = \frac{1}{\varepsilon + \delta}$ ($<\infty$, δ = a small value).

The swarm intelligence- or GA-driven search for the optimal network connection is population based, but ensures good exploration of the search space of weights and thresholds and exploitation of the available information. Because of their population-based approach they are time-consuming and must be implemented in parallel. In fact, it would be a good idea to make use of hybrid methods in which the initial explorations are done by a GA- or swarm intelligence-based recipe. After the initial exploration phase when the f or f_1 values have increased significantly, one can switch over to a deterministic method such as the steepest descent or to a nondeterministic single string-based recipe such as the SAM or CARMHC for fine-tuning the weights and threshold—such hybrid methods are increasingly being tested to enhance the network learning process. The advantages of the stochastic methods lies in their simplicity. No information about the derivatives of the error with respect to weights and thresholds is necessary and their global search capabilities give us confidence that we are getting the best network weights and thresholds.

If we are using the deterministic steepest descent methods for finding the optimal weights and thresholds, we have to be careful in choosing the learning rate (λ in Equation 8.47 or 8.51) so that the minimum being approached is not overshot. There is a need for an initial search for a good λ. Even with a good choice of λ the steepest descent (SD) iterations may show some instability. The iterative SD search can be stabilized by mixing information of the previous $[(t - 1)$th] step with the information of the current (tth) step. Thus, in the SD procedure for adjusting the network weights (ω_{ij}s or ω'_{ij}s) we have, with λ as the learning rate ($\lambda > 0$)

$$\omega_{ij}(t+1) = \omega_{ij}(t) + \lambda \Delta \omega_{ij}(t) \tag{8.55}$$

$\Delta \omega_{ij}(t)$s are the current values for various weight correction terms. Let $\Delta \omega_{ij}(t-1)$ be the value of the corresponding correction terms in the preceding step. Then a modified scheme of learning can be constructed by mixing $\Delta \omega_{ij}(t)$ with $\Delta \omega_{ij}(t-1)$, each weighted by appropriate scalars. The modified weight-updating scheme then reads

$$\omega_{ij}(t+1) = \omega_{ij}(t) + \lambda \Delta \omega_{ij}(t) + \beta \Delta \omega_{ij}(t-1) \tag{8.56}$$

where λ is the learning rate and β (>0) is a constant called the momentum rate. The user must carry out a few exploratory calculations with several choices of (λ, β) parameters to find out the best combinations for stabilized and smooth passage to the minimum being sought. In fact, a good policy could be to dynamically adjust the learning and the momentum rates for a better network performance. In the literature, λ values in the range of 0.45–0.65 and β values in the range of 0.8–0.9 have been commonly used with success.

We have so far assumed that there is one pattern P which the BPN learns by adopting an appropriate learning protocol, (such as pure SD or a hybrid of pure SD with one of the many stochastic or soft optimizers) in the training phase. However, there would be a group of patterns $P_1, P_2, ..., P_n$ (such as the English alphabet, chemical structures, spectra of molecules) which the network must be trained with so it can later recognize and correctly classify when similar but not the same pattern is presented to the network.

Let $(t_1^i, t_2^i, ..., t_k^i, ..., t_p^i)$ be the p complementary target output vector (T^i) for the ith input pattern P_i. If the network output vector $Y(q)$ at the qth stage of iterative learning has components $(y_1^i(q), y_2^i(q), ..., y_p^i(q))$, the error in predicting the correct output for the ith pattern is

$$\varepsilon^i(q) = \frac{1}{2} \sum_j \left(t_j^i - y_j^i(q) \right)^2 . \tag{8.57}$$

In general, the iterations are assumed to have converged if $\varepsilon^i(q) < \delta_1$ at the qth stage of iterations (δ_1 a small preassigned number). Alternatively, we may demand that $|t_j^i - y_j^i(q)| < \delta_2$ for $j = 1, 2, ..., p$ ($\delta_2 > \delta_1$). When one or the other convergence condition has been satisfied, or a criterion based on a maximum number of iterations has been passed, the pattern i has been learned by the network. One starts with $i = 1$ (i.e., pattern P_1), randomly assigns weights, and starts the iteration (supervised learning), checks if the convergence condition has been achieved, and the weight matrix elements and thresholds at convergence hold the clue to the learned pattern ($i = 1$). The network then takes the second pattern as input ($i = 2$, P_2) and using the weights and thresholds obtained while the pattern P_1 was learned as the initial weights, the learning steps are repeated for pattern P_2, until appropriate convergence conditions are satisfied. The procedure is repeated for all the m-patterns $P_1, P_2, ..., P_m$ sequentially, using the same convergence criterion. Once all the patterns have been used as input, one epoch is complete. One then starts the second epoch—the difference from the first epoch is that the weights and the thresholds obtained for the last (mth) pattern (P_m) at convergence are used as the initial or the guessed weights for the first pattern in the second epoch. In this way, the training or supervised learning procedure is repeated over many epochs until the network is able to produce the correct output patterns for all the input patterns in the training set in just one step, signaling that learning is complete within the limits of the convergence criterion used. Once this stage is reached, the network will be able to recognize a similar (but not the same) pattern even if it was not included in the training set. For the BPN it does not matter whether we are dealing with a linear or non-linear problem—the only requirement for us is to be able to mold the problem as a pattern recognition/classification problem. That means we must be able to present the problem as an input to output pattern mapping problem. The question that arises now is: how do you present the input as a pattern? In what follows we discuss how to do it reasonably.

8.6.2 Feeding Input Patterns to the Network

Suppose that we wish to train the NN to recognize the electronic (absorption) spectrum of simple molecules like water (H_2O) or carbon dioxide (CO_2) or perhaps nitrogen dioxide (NO_2). The spectrum in each case represents a continuous plot of the intensity of light (I) absorbed against a range of frequencies (Figure 8.7). The spectrum is a pattern of relation between intensity and frequency and can be used as a signature of the particular molecule producing the spectrum.

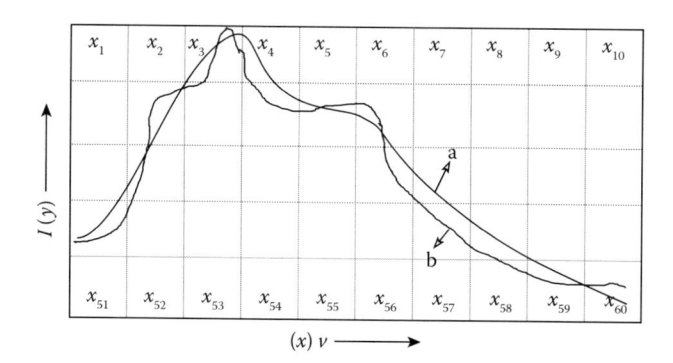

FIGURE 8.7

The spectrum in (a) represents the target pattern while the spectrum in (b) stands for an input pattern.

Now our objective is to feed the pattern in Figure 8.7 to the NN and make it learn the detailed features of the pattern so that when a similar spectrum—which may be somewhat noisy or distorted—is presented to the network, it can recognize the spectrum correctly and identify the molecule responsible for the spectrum. The spectrum in Figure 8.7a represents the target pattern while the spectrum in Figure 8.7b stands for an input pattern in the training set. To feed the input pattern (the noisy spectrum 8.7b), let us convert the two-dimensional ($I - v$ or $y - x$) plot into a one-dimensional array. We can do this $2D - 1D$ mapping by converting the plane of the paper displaying the spectra into $n \times m$ square boxes. Each such box may contain a small segment of the graph (the spectrum or pattern) or may be empty. In our case, $n = 6$, $m = 10$, so there are 60 boxes. Each box represents a value of the abscissa which has been divided into 60 intervals (x_1 to x_{60}). So x_i stands for the frequency v_i and the portion of the graph inside the ith box represents the value of I—the intensity of absorption at that frequency (v_i). To simplify, let us assign only two values to I_is—if the ith box is occupied by a segment of the curve, $I_i = 1$, otherwise it is equal to zero. Thus, the input spectrum (Figure 8.7b) has been broken down into 60 values of intensity, one each for 60 frequencies ($v_1 - v_{60}$). We will therefore require a single layer of 60 input neurons ($i_1, i_2, \ldots, i_k, i_{60}$) each being fed with the corresponding value of intensity (x_k for I_k). The BPN as we have stressed already uses a supervised learning model. That means, each input spectrum must be accompanied by the corresponding target or the correct spectrum, which in our example is represented by graph **a** in Figure 8.7. The target spectrum is also broken down into 60 pieces just as for the input spectrum—the value of intensity in each box (x_{ii}), representing the desired value (t_i) that the output neuron $\{O_i\}$ must produce when training is complete. Since in our notation output from neuron O_i is y_i, the network trains itself by reducing the error $\varepsilon = \frac{1}{2} \sum_{i=1}^{60} (t_i - y_i)^2$ by adjusting the weights associated with different edges. Suppose that we introduce a hidden layer of 20 neurons, together with the 60 input and 60 output neurons, the network has ($60 \times 20 + 20 \times 60$) = 2400 edges, so we have to optimize 2400 weight matrix elements along with activation thresholds, if any. The training set may be containing 20 such spectra each with its target, so that it is a huge computing problem and naturally time-consuming. We will consider these details in a subsequent section. We may point out, that not only spectra, but any other plot, handwritten characters, or structures can be represented similarly and fed to the network for training and classification.

8.6.3 Programming Hints

We are now in a position to outline how the BPN is trained in a stepwise fashion. Let us note that the maximum number of iterations allowed to the network to learn a particular pattern is called "max cycle" and is given a preset value. When all the patterns in the training set have been learned by the network (either converged, or have been iterated the max cycle number of times) an epoch of the training phase is over. At the end of the first epoch, the network will only partially remember the patterns already learned during the first epoch. After many epochs have elapsed, it is expected that all the patterns have been completely and permanently learned—which means that given any one of the input patterns present in the training set, the BPN will be able to correctly predict the target pattern corresponding to that specific input pattern without any iteration. The "learning" is encoded in the finally converged weights and threshold values.

We consider a BPN network which makes use of a sigmoid activation function. The network has a single hidden layer of q-neurons, n-input neurons, and q-number of output neurons. The $(n + 1)$th neuron in the input layer is an imaginary neuron (x_{n+1} with value 1). The weight associated with the edges between this imaginary input neuron with the q number of neurons in the hidden layer represents the value of the thresholds ($\theta_1, \theta_2, \ldots, \theta_q$) of the sigmoid activation function used by the hidden layer neurons. Similarly, the $(q + 1)$th neuron ($h_{q+1} = 1$) in the hidden layer is an imaginary neuron, the weight associated with the edges between this imaginary hidden layer neuron (h_{q+1}) and the p number of output layer neurons represents the thresholds ($\theta'_1, \theta'_2, \ldots, \theta'_p$) of the sigmoid activation function being used by the output layer neurons. Let us consider the network including the imaginary neurons—an extension of Figure 8.6c in which the thresholds have been represented as weights. That enables us to represent the supervised learning process by adjustment of various weights alone—without any reference to the thresholds.

Step 1. Initialize all the weights and thresholds by assigning small and uniformly distributed random values ($-a$ to $+a$) to them ($a = 0.25$ or 0.5 may be convenient choices).

Step 2. Begin the outer loop of iteration with a pattern from the training set (vector X) and the corresponding target vector (t):

- Set the outer loop counter.
- Do until the BPN can correctly reproduce the target vectors for each input pattern $x_1, x_2, \ldots, x_n$.
- Begin the inner iteration loop.

    For each input pattern executes steps i, ii, and iii until the norm of the error vector $|t - y| \le \varepsilon$ ($\varepsilon \sim$ a small value like 10^{-3}) or the max cycle number of iterations have been carried out.

    i.   Read in one input pattern X.

    ii. Use the weights to calculate the output vector y for the input X.

    iii. If $\|t - y\| \le \varepsilon$, return to step 2. If not, adjust the weights (backpropagate) so that the new y is closer to t; go back to the beginning of the inner loop.

The simple structure of the program described above is easy to implement. Here, the BPN learns one pattern at a time. The user can most conveniently use arrays as the data structures for neurons and weights. Instead of using one pattern at a time for learning, the program can be easily tweaked to handle all the patterns at the same time. In that case weights are adjusted to minimize the norm of the error vectors averaged over all the patterns: $\varepsilon_{av} = \dfrac{1}{n}\sum_i \|t_i - y_i\|$, n being the number of patterns in the training set. The n number of input patterns should be subdivided into n_t number of patterns constituting the training set and n_v number of patterns forming the validation set ($n = n_t + n_v$). Following the completion of training with n_t patterns, the BPN is tested with the n_v number of validation patterns. If the network predicts the target vector for the validation set correctly, the BPN is all set to be used for unknown patterns for classification. Although we have described a learning procedure that uses the SD method to adjust the weights, it is equally possible, as already mentioned, to use a soft-computing technique to find the optimal network weights. Simulated annealing, genetic algorithms, and swarm intelligence techniques have been exploited by various users leading to different realizations of hybrid networks. However, the most popular mode of learning adopted in BPN has been the SD technique. The training is generally time-consuming and considerable attention has been focused on finding means for accelerating convergence. One such technique involves scaling of the weight correction term $\Delta\omega_{ij}(t)$.

Let $\varepsilon(t)$ and $\varepsilon(t+1)$ represent the errors at the tth and $(t+1)$th iterations the corresponding gradient vectors, being $\vec{\nabla}\varepsilon(t)$ and $\vec{\nabla}\varepsilon(t+1)$. Let φ be the angle between the two gradient vectors and γ be a scalar (<1). $\Delta\omega_{ij}(t)$ values calculated by the SD procedure can be directly scaled up or down by setting

$$\Delta\omega_{ij}(t)(scaled) = e^{\gamma\cos\varphi}\Delta\omega_{ij}(t). \tag{8.58}$$

If $\varphi = 0$, we know that the error was decreasing in the two successive iterations $\big((t-1)\text{th, and }t\text{th}\big)$ along the same direction so that a higher (scaled-up) value of the weight correlation term $(\Delta\omega_{ij}(t))$ could be used to accelerate learning:

$$\Delta\omega_{ij}(t+1) = \omega_{ij}(t) + e^{\gamma}\Delta\omega_{ij}(t). \tag{8.59}$$

If $\varphi > 0$, but less than $\dfrac{\pi}{2}$, $\cos\varphi < 1$, we have

$$\Delta\omega_{ij}(t+1) = \omega_{ij}(t) + e^{\gamma'}\Delta\omega_{ij}(t) \tag{8.60}$$

$\gamma' = \gamma\cos\varphi < \gamma$, the scaling effect becomes smaller.

If $\varphi = \dfrac{\pi}{2}$, $\cos\varphi = 0$, and

$$\Delta\omega_{ij}(t)(scaled) = \Delta\omega_{ij}(t). \tag{8.61}$$

So that there is no effect of scaling.

If $\varphi > \dfrac{\pi}{2}$, $\cos\varphi < 0$, leading to a decrease in $\Delta\omega_{ij}(t)$ values due to scaling. Thus, scaling can ensure a proper choice of the size and sign of the weight correction terms depending on the history of error reduction in two consecutive learning steps. A judicious use of learning (λ) and momentum (β) rates together with scaling (ρ) parameters can achieve better convergence of the learning process. However, some experimenting with different choices of λ, β, and ρ values is needed to select the parameters properly for a particular problem.

Alternatively, these parameters can be dynamically or adaptively adjusted during the entire learning process. A pertinent question in this context concerns the order in which the "patterns" in the learning set should be presented to the BPN for learning. In general, the best practice would be to present the patterns in a randomized order. A rule of thumb tells us that "harder" patterns are better presented before the patterns perceived to be "easier" to learn. The choice of the order of presentation of patterns does have an effect on the rate at which the learning process converges for the entire training set. Along with the choice of order of presentation of the patterns, the initial choice of weights ($\omega_{ij}(0)$, $\omega'_{jl}(0)$) also has a bearing on the convergence rate and therefore computational labor. Although a perfectly random choice of $\omega_{ij}(0)$ s and $\omega'_{jl}(0)$ s from a given range ($-a$ to $+a$) generally works, the value of a itself should be chosen with care after analyzing the physical nature of the patterns to be learned. If neural network representations of similar patterns are already available, the corresponding weights can form the basis of initial choice of $\omega_{ij}(0)$ or $\omega'_{jl}(0)$. A realistic and educated choice of initial weights can beneficially affect the rate of convergence of the learning process.

We note here that the learning task involves finding the optimal weights of various interconnections among the neurons, so that the error (ε) is reduced to a minimum value in the global sense. In a nonlinear mapping, there may be multiple solutions—that is, there could be a number of local minima along with the global one. In that case, how does one guarantee that the global minimum has been located? The SD procedure generally converges to the nearest local minimum. A rule of thumb is to check the error magnitude after the iterations have converged. If the error (ε) is high even at convergence, one has reason to suspect that the iterations have converged to a local minimum. The converged weights are then perturbed randomly around the converged values and the iterations are continued until convergence has been achieved and ε has reduced further. This procedure could be repeated a number of times, until the error value at convergence becomes sufficiently small. It is in this context that the simulated annealing or the random mutation hill climbing methods can be exploited to move into the global minimum after the SD-based BPN has converged to a local minimum (suspected). The process of ensuring global convergence in this manner is time-consuming and has to be carried out very carefully so that the global minimum is not overshot.

We will end this section with a note on the BPN architecture to be chosen for representing the solutions of a particular problem. The number of input layer neurons is, of course, fixed by the problem itself. Similarly, the number of neurons in the output layer too, is fixed by the problem. The flexibility that is available to the network comes in the form of choosing the number of hidden layers and the number of neurons in each of the hidden layers. A majority of the BPNs make use of only one hidden layer—some use two, while the use of more than two hidden layers is rare. The designer must keep in mind that too few hidden layer neurons would compromise the flexibility of the network to represent solutions while the use of too many hidden layer neurons would inevitably increase the computational labor without adding much to the quality of the solutions being represented by the network. A good balance between the required flexibility and reasonable computational labor involved must be struck so that the network can represent the solutions fairly accurately at a reasonable computational cost. Once again a rule of thumb suggests that the number of hidden layer neurons should be half the number of neurons in the input layers. The user may choose to distribute these neurons in one or more hidden layers.

8.6.4 Why Neural Networks?

The use of ANNs in physicochemical or other problem marks a shift in the traditional computing paradigm. The ANNs, we may recall, are modeled after the brains of intelligent species (such as humans) and categorized as artificial intelligence (AI)-based methods. However, they are philosophically very different from the traditional symbolic AI and in a way complement symbolic AI. The scope of their applications to physical/chemical problems is rather wide and unique in many ways. An attractive feature of ANNs lies in that they can be easily parallelized (like many other soft-computing techniques). Thus, a problem or a model for which parallel algorithms are difficult to construct, can find parallel realization on ANNs in an indirect way. This is possible in BPN because the incoming activation for each neuron in a specific layer can be summed up, weighted by appropriate connection weights independently of the others in the same layer. The ease of parallelization and the ability to learn by simply adjusting the weights contribute to the advantages of using BPNs and their popularity. The BPNs (or other ANNs) have the capability to handle new patterns outside the list of learned patterns if the new ones are similar (but not the same). This ability to achieve a certain extent of generalization in pattern handling and recognition adds to the advantages of ANNs. The ANNs are tolerant to a certain level of noise in the input patterns and keep functioning even if a part of the network has been nonfunctional or has been damaged. The greatest plus point of ANNs is their ability to handle nonlinear problems for which traditional computing paradigms fail or mathematical solutions are not feasible. The input–output mapping realized in BPNs, for example, can be done regardless the type and severity of nonlinearity—the only requirement of the network is that the problem is presentable as a pattern. However, there are limitations or even disadvantages of the ANNs as problem-solvers. One commonly cited disadvantage of using ANNs is that the training can be excessively time-consuming and may ultimately fail to converge. On the other hand, once a pattern is learned after a long session of training, the converged weights may be copied and used on other machines. The greatest disadvantage of ANNs is that scaling it up is impossible—the whole work has to be redone on the new scaled-up network. The human brain has evolved so as to enable it to process input data (signals) from sensory organs. It performs much better when it comes to solving problems requiring perception and recognition of patterns than problems involving logical steps and huge data manipulation. The BPN and the other ANNs also share this quality with the living brains. The ANNs are not so useful when faced with tasks that can be solved by writing sequential computer programs and running them on digital machines. These similarities notwithstanding, the ANNs employ processing elements and architectures that have practically little in common with real brains. In principle, the ANNs are endowed with the power of a universal computer and can realize an arbitrary mapping of one vector space into another. Since physicists and chemists often deal with such problems, the ANNs have been exploited to correlate parameters embedded in the Hamiltonians with numerical values of specific properties in the hope that given a statistical sample of data points or a set of examples, the network will manage to acquire an idea of what the global mapping could look like.

The early work of Darsey et al. [10] and Androsiuk et al. [11] toyed with the idea of teaching the neural network about solutions of time-independent Schrödinger equations. They sought to train the network about the underlying correlation between parameters in the Hamiltonian—say a two-dimensional harmonic oscillator Hamiltonian, and the energy eigenvalues supported by it. The 2-*D* harmonic oscillator (mass m) Hamiltonian $H(x, y)$ has a potential $V(x, y)$ that is quadratic in the coordinates (x, y) and also in oscillator frequencies (ω_x, ω_y). The energy eigenvalues are linear in the frequency parameters (ω_x, ω_y). Thus we have

$$V\left(x, y, \omega_x, \omega_y\right) = \frac{1}{2} m\left(\omega_x^2 x^2 + \omega_y^2 y^2\right)$$

$$E\left(\omega_x, \omega_y\right) = A\omega_x + B\omega_y \tag{8.62}$$

where,

$$A = \left(n_x + \frac{1}{2}\right)\hbar, \ n_x = 0, 1, 2, \dots$$

$$B = \left(n_y + \frac{1}{2}\right)\hbar, \ n_y = 0, 1, 2, \dots \tag{8.63}$$

The training data set was prepared by specifying the values of the potential $V(x, y, \omega_x, \omega_y)$ on a rectangular (x, y) mesh for different values of ω_x, ω_y. These values are the input neuron values. The output neuron values (the targets) were the energy eigenvalues corresponding to each pair of ω_x, ω_y chosen. The backpropagation scheme was used to adjust the weights and the output values were checked against the targeted (known) energy eigenvalues. The iterations were continued until the error was sufficiently reduced. At this stage the authors felt that they had found an NN representation of the 2-*D* harmonic oscillator Hamiltonian and the network would predict correct energy eigenvalues for any combination of (ω_x, ω_y), even if such combinations are not explicitly included in the training set. This application of BPN to guess the correlation among Hamiltonian parameters and the energy eigenvalues may appear to be trivial; but it demonstrates the possibility of finding important correlations through a judicious choice of training set of parameters (of the Hamiltonian) and the eigenvalues that the "H" supports, and later exploiting the correlations so established by the network for predicting eigenvalues for new sets of the relevant parameters.

Conceptually very similar but computationally more ambitious exploitation of the BPN could be found in a series of papers published by Sumpter et al. [13–18]. These authors investigated the patterns of dynamics of internal energy flow in molecules by using data from molecular dynamics (MD) simulations. The ANN was trained to learn the relationship between phase-space points along a classical trajectory and energies of the different modes of vibrations in a four-atom molecule, e.g., stretches, bends, and torsions. These authors used a network that had 84 neurons (nodes) of which 24 were in the input layer, 4 were in the output layer and 56 nodes were in two hidden layers. The input layer neurons carried the input information in the form of values of coordinates and momentum of each of the four atoms ($6 \times 4 = 24$). The input vector was thus 24-dimensional. The output layer neurons represented the energies of the four modes. The network had 1648 interconnections and the training set comprised 2000 sets of data points. Sumpter et al. noted that the accuracy of the predicted energies varied in the 5–20% range. In recent years theoretical chemists have experimented with possible applications of ANN in constructing accurate representations of the PES of many-atom systems (clusters, for example), and used the ANN representation of the PES to compute the forces acting on different atoms. Once energy values and forces are determined, an MD simulation with ANN-predicted forces can be carried out at different temperatures and pressures, and simulation carried out to predict properties. However, the generation of a sufficient number of energy values at widely distributed positions on the PES, i.e., for a large number nuclear configurations of the many-atom species, by a suitably chosen quantum chemical method is time-consuming. The training of the network is also a time-consuming activity. Once the training is over, the computation of forces becomes almost instantaneous. We will review such applications later in this chapter. Presently, we will focus on some other types of NNs and their learning mechanisms. The first and the foremost in this category is the Hopfield networks.

8.6.5 Hopfield Networks and Associative Memory Formation

Hopfield [19,20] was able to establish a formal analogy between a network of bipolar neurons (Hopfield network) and a system of interacting two-valued Ising spins, and demonstrated how methods developed for handling a spin system could also be exploited for the purpose of understanding the learning process of a neural network and autoassociative memory formation in them. Let us first examine the architecture of a Hopfield network.

The Hopfield neural network (HNN) has a single-layered interconnected architecture (Figure 8.8) in which all the bipolar neurons (N_1, N_2, ..., N_n) are connected with each other (note the difference with BPN). The network works both in feedforward and feedbackward modes and belongs to the category of recurrent networks. In other words, each network neuron receives input signals from both external neurons and the other network neurons it is interconnected with. The interconnections are bidirectional (Figure 8.8).

Each input (external) x_1, x_2, ..., x_n assumes a discrete value (+1 or –1, say). The output y_1, y_2, ..., y_n, too assumes discrete binary values (+1 or –1). The number of neurons to be deployed in the HNN is determined by the patterns it is designed to handle. If the pattern can be represented by binary values stored in a two-dimensional $m_1 \times m_2$ array, the number of neurons in the HNN will have to be $n = m_1 \times m_2$. The interconnection weights (ω_{ij}s) are determined by m number of example patterns (input) represented as n-dimensional vectors $X^{(r)}$, $r = 1, 2, ..., m$, (the network (HNN) is set to learn the patterns) which are held fixed. The network weights do not change; what changes are the states of neurons so as to minimize the energy of the network (not defined yet). How are the synaptic weights determined? The only constraints on ω_{ij} are that $\omega_{ii} = 0$ (lack of self-coupling) and $\omega_{ij} = \omega_{ji}$ (symmetric).

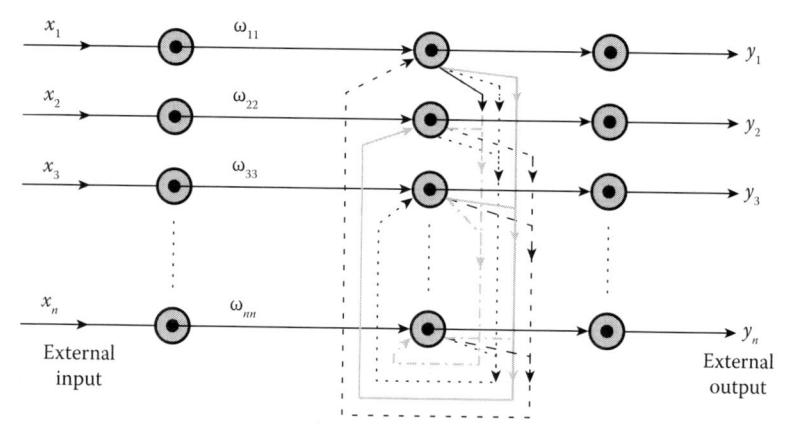

FIGURE 8.8
A Hopfield neural network of n binary neurons.

Let us consider the input vectors $\{X^{(p)}\}$, $p = 1, 2, \ldots, m$, the n components of which may assume values ± 1 only. The weights ω_{ij} s are determined as follows:

$$\omega_{ij} = \begin{cases} 0 \text{ if } i = j \\ \sum_{p=1}^{m} x_i^{(p)} x_j^{(p)} \text{ if } i \neq j \end{cases} \tag{8.64}$$

The symmetric nature of the W matrix is evident from construction; $\omega_{ii} = 0$ (for $i = 1, 2, \ldots, n$) signifies that the neurons are not self-connected. The HNN learns not by adjusting the weights, which remain fixed, but by flipping the neuronal values ($+1 \rightarrow -1$ or $-1 \rightarrow +1$) so that the network finds itself eventually in an energy minimum. The network energy E, also called the Lyapunov function at a time t is defined as follows:

$$\varepsilon(t) = -\frac{1}{2} \sum_{i=1}^{n} \sum_{j=1}^{n} \omega_{ij} x_i(t) x_j(t) + \sum_{i=1}^{n} \theta_i x_i(t) \tag{8.65}$$

$x_i(t)$s represents the neuronal state values at the time t (the time t just defines the stage of iteration as the network evolves to an energy minimum). θ_is represent the threshold values for neurons i which are often set to be zero so the energy assumes a slightly simpler expression:

$$\varepsilon(t) = -\frac{1}{2} \sum_{i=1}^{n} \sum_{j=1}^{n} \omega_{ij} x_i(t) x_j(t). \tag{8.66}$$

The formal connection with a network of 2-valued spins is apparent. Replace ω_{ij} by the spin–spin coupling constants J_{ij} (with $J_{ii} = 0$), the neurons x_i by spins $\vec{S}_i$, so that

$$\varepsilon_{spin}(t) = -\frac{1}{2} \sum_{i=1}^{n} \sum_{j=1}^{n} J_{ij} \vec{S}_i(t) \vec{S}_j(t). \tag{8.67}$$

How does the network change its state? Suppose that the network has been presented with m exemplar patterns $X^{(1)}, X^{(2)}, \ldots, X^{(m)}$ (each a vector with n components) from which the weight matrix elements have been determined following the recipe laid down in Equation 8.64. The stored weight matrix (W) has imprinted in it the exemplar patterns the network has already been presented with and is the source of the associative memory of the network. Imagine now that a new pattern vector X with n-elements $x_i(0)$, $i = 1, 2, \ldots, n$ has been fed as an input. Can the network associate the new input vector with one of the m pattern vectors already presented to it and learn by it through the process of generating appropriate synaptic weight matrix elements $\{\omega_{ij}\}$ which determine the strength of the interconnections among the network neurons. The network now tries to establish the association between the new input pattern and the ones already in memory by changing or flipping the states of individual neurons so as to minimize the network energy $\{\varepsilon\}$. The values of the individual input neurons are changed under the following schedule: let the current value of the ith neuron be $x_i(t)$ (this may be +1 or –1 in the bipolar representation we have chosen). The neuron i is flipped based on the value of net_i at the instant t that is $net_i(t)$, where $net_i(t)$ is defined as follows:

$$net_i(t) = \omega_{ij}x_j(t) \tag{8.68}$$

$x_j(t)$, $j = 1, 2, \dots, n$ being the neuronal values at the instant t. The updating of the ith neuron is dictated by the following logic (assuming $\theta_i = 0$) based on the activation of the neuron i:

$$x_i(t) \to x_i(t+1) = \begin{cases} 1, \text{ if } net_i(t) > 0 \\ x_i(t), \text{ if } net_i(t) = 0. \\ -1, \text{ if } net_i(t) < 0 \end{cases} \tag{8.69}$$

The neurons are updated one at a time taking care to ensure that all the neurons in the network are updated at the same average rate. Thus, for the next neuron, say neuron k, the already updated values of other neurons are used in computing $net_k(t)$ — this is therefore an asynchronous updating scheme. We could have computed all the $\{net_k(t)\}$, $k = 1, 2, \dots, n$ in one go and then updated all the neurons at the same time.

What happens when a neuron changes its state (that is, its value flips from +1 to –1 or vice versa)? We will show in what follows that the network energy $\varepsilon(t)$ decreases whenever a neuron is updated (i.e., flips). With the thresholds θ_i set to zero, the energy of the network is given by Equation 8.66. Suppose that at the current instant the qth neuron has been chosen for updating and based on the value of activation $net_q(t) = \sum_{j=1} \omega_{qj}x_j(t)$, this neuron has been flipped into the state denoted by x_q', i.e., $x_q(t+1) = x_q'$, $x_q(t) = x_q$. The energy of the network following the updating is $\varepsilon(t+1)$ while $\varepsilon(t)$ is the energy of the network before the updating. From the definition of ε, we have (dropping t on the right-hand side)

$$\varepsilon(t) = -\frac{1}{2}\sum_{i=1}^{n}\sum_{j=1}^{n}\omega_{ij}x_ix_j$$

$$= -\frac{1}{2}\sum_{\substack{i=1 \\ \neq q}}^{n}\sum_{\substack{j=1 \\ \neq q}}^{n}\omega_{ij}x_ix_j - \frac{1}{2}\sum_{j}\omega_{qj}x_qx_j - \frac{1}{2}\sum_{i}\omega_{iq}x_ix_q. \tag{8.70}$$

(Note: In the second step, we have split the sum into parts that contain the index q and a part that is free of the index q.)

Similarly, we can write for the energy following the updating

$$\varepsilon(t+1) = -\frac{1}{2}\sum_{\substack{i=1 \\ \neq q}}^{n}\sum_{\substack{j=1 \\ \neq q}}^{n}\omega_{ij}x_ix_j - \frac{1}{2}\sum_{j}\omega_{qj}x_q'x_j - \frac{1}{2}\sum_{i}\omega_{iq}x_ix_q'. \tag{8.71}$$

The change in the network energy is given by

$$\Delta\varepsilon = \varepsilon(t+1) - \varepsilon(t)$$

$$= \frac{1}{2}\sum_{j=1}^{n}\omega_{qj}x_{j}\left(x_{q}-x_{q}'\right) + \frac{1}{2}\sum_{i=1}^{n}\omega_{iq}x_{i}(x_{q}-x_{q}')$$

$$= \frac{1}{2}\sum_{j=1}^{n}\omega_{iq}x_{j}\left(x_{q}-x_{q}'\right) + \frac{1}{2}\sum_{i=1}^{n}\omega_{iq}x_{i}(x_{q}-x_{q}')^{*} \qquad (8.72)$$

$$= \sum_{i=1}^{n}\omega_{iq}x_{i}(x_{q}-x_{q}')^{**}$$

$$= net_{q}(t)\left(x_{q}-x_{q}'\right)$$

(*using symmetry of the weight matrix; **using the fact that j, i are dummy indices).
There are now two possibilities.

1. Suppose that the updating $x_q \to x_q'$ has changed the neuron value from +1 to –1. In that case $x_q - x_q' = 2$, i.e., >0. However, the flip can occur only if $net_q(t) < 0$ or negative, which means that $\Delta\varepsilon$ given by the product of $net_q(t)$ and $\left(x_q - x_q'\right)$ is < 0. So the flip is accompanied by a lowering of the energy of the network.

2. In the second case $x_q \to x_q'$ transition involves a change in neuronal values from –1 to +1. In that event $x_q - x_q' = -2$, i.e., < 0. However, this transition can be realized only if $net_q(t) > 0$ so that the product of $net_q(t)$ and $(x_q - x_q')$ is < 0 ensuring network energy reduction.

Thus, we have shown that the network energy is lowered whenever a neuron is updated. It turns out that after receiving a new pattern, the HNN tries to attain a state of the lowest energy possible by randomly but uniformly updating the network neurons. When such a state is reached and no further lowering of energy takes place, the HNN should have been able to associate the new input pattern to one exemplar pattern it had etched in its memory in the form of weight matrix elements constructed from the exemplar patterns. At this stage, the converged output neuronal values $\{x_i(t)\}$ can be sent to the output neurons $\{y_i\}$ (see Figure 8.8) for printing.

The implementation of a Hopfield network is rather simple and straightforward in principle. In practice, there are tricky points to be taken care of. One such point lies in the selection of the neuron for updating. The neurons must be selected randomly, uniformly, and independently of other neurons at the same average rate. An easily workable strategy can be constructed as follows: generate a random integer from 1 to n and select the neuron according to the integer so generated until all the neurons have been updated at least once. When this has been accomplished one epoch has elapsed. A new epoch of iteration then begins. To check if and when all the neurons have been updated at least once during the current epoch, it may be convenient to define an update vector U with n-components $U \equiv (u_1, u_1, \dots, u_n)$ with all the components set to zero at the beginning of an epoch. During the iterations, u_i is flipped from 0 to 1 if the ith neuron is updated. The square of the norm of the U vector is calculated ($norm^2 = u_1^2 + u_2^2 + \cdots + u_n^2$). As soon as the $norm^2$ attains the value n' it signals that the current epoch is over and a new epoch of iteration is started after ensuring that the update vector is reset to have zero values for all the components once again.

8.6.5.1 The Philosophy of the Hopfield Network

We have already seen how the network determines the weight matrix elements representing different edges of the network. Recall that for a set of m exemplar patterns $X^{(1)}, X^{(2)}, \ldots, X^{(m)}$ (each n-component vector with bipolar values: +1 or −1), the elements of the W matrix are computed as follows:

$$\omega_{ii} = 0$$

$$\omega_{ij} = \sum_{p=1}^{m} x_i^p x_j^p = \omega_{ji}, \ i = 1, 2, \ldots, n; j = 1, 2, \ldots, n. \tag{8.73}$$

If for a given pattern both $x_i = 1$ and $x_j = 1$, or $x_i = -1$, $x_j = -1$, $x_i x_j > 0$, so that the particular neuron pair (i, j) contributes a positive value to the corresponding weight matrix element (ω_{ij}). If for a majority of the m patterns $x_i x_j > 0$, ω_{ij} will have a fairly large positive magnitude, a majority of the training patterns strengthening the synaptic connection between the ith and jth artificial neurons. Such neuron pairs are called **excitatory** in nature. On the other hand, if $x_i = 1$, $x_j = -1$ or vice versa, the contribution of the neuron pair to ω_{ij} is < 0. If $x_i x_j < 0$ for some of the patterns and $x_i x_j > 0$ for others, there will be cancellation of positive and negative contribution to ω_{ij} coming from different patterns. As a consequence ω_{ij} may have a small value. The neuron pair is then called **inhibitory** in nature, and the corresponding synaptic connection strength as measured by the value of ω_{ij}, is weak. The learning in HNN thus follows the Hebbian dictum—"Neurons that fire together wire together" and "Neurons that do not fire together do not wire together." Thus, the values of the weight matrix elements encode, in a way, information about the m patterns presented to the network as the training set. The basic functional philosophy of the HNN is to find a way the network can associate a new input pattern (whole or partial) with or without error or noise to one of the m patterns $(X^{(1)}, X^{(2)}, \ldots, X^{(m)})$ it has already learned and stored in "memory" in the form of values of weight matrix elements (in the living brain, the storage would have been in the form of strong/weak synaptic connections between pairs of specific neurons). We note that the input pattern is presented to the network as values of the n components of a vector X rather than the addresses and the problem is to associate these "values" with information already stored in addresses $\{\omega_{ij}\}$. As we have already seen, the HNN achieves this association by randomly and uniformly changing the states of the network neurons following well-specified rules, so as to minimize the network energy to the extent possible. When the minimum energy configuration of the network neurons has been reached for the given input pattern, the network has been able to find the closest possible match for the input pattern with one of the patterns in the memory. Since the patterns are contents rather than the addresses themselves, and are not modified by the network during the entire process, the network is said to display associative or content addressable memory—an important functional attribute of the living brain.

The human brain displays tremendous ability to recall from memory by associating various types of inputs such as audio and visual. It is a common experience that we can remember the face of a friend we have not seen for many years by associating the voice or hearing the voice with the face, or vice versa. Even if the voice sample is noisy, or even if the voice has changed a bit, the process of association works. HNN is an example of achieving this association by well-defined mathematical steps and rules. Associative memory formation is a special attribute of the Hopfield network—it is absent in ordinary computers. We have seen (cf. Equations 8.68 and 8.69) how the association is made operationally, although there is no certainty that the iterations will always actually and surely converge to a stored or learned pattern. Note that we are making use of the term "learned pattern." We may qualify the term further by noting in addition that the learning (Hebbian learning) here is in fact an unsupervised learning through the network finding the closest match for the input. The philosophy that works behind the process of matching pattern identification by association may be given the following interpretation on the assumption the number (m) of exemplar patterns presented to the network is much smaller than the number of neurons that constitute the network, and that the exemplar patterns are completely uncorrelated. Recalling that for the kth neuron, the net input strength is

$$net_k = \sum_j \omega_{kj} x_j ; \quad (\omega_{kk} = 0). \tag{8.74}$$

Resolving the network weight matrix elements in terms of the m-input patterns (cf. Equation 8.73) we have

$$net_k = \sum_{\substack{j=1 \\ \neq k}}^{n} \left(\sum_p x_k^{(p)} x_j^{(p)} x_j \right)$$

$$= \sum_p x_k^{(p)} \left(\sum_{\substack{j=1 \\ \neq k}}^{n} x_j^{(p)} x_j \right). \tag{8.75}$$

Consider now the case where the new input pattern X is indeed very close to one of the m-exemplar patterns, say for $p = p_0'$ while it has no correlation with the rest of the exemplar patterns ($p \neq p_0'$). In that case, the sum $\sum_{\substack{j=1 \\ \neq k}}^{n} x_j^p x_j$ in Equation 8.75 will have a fairly large

value if $p = p'$ as $x_j^{p'}$ and x_j are expected to match for a majority of the n-values of j, and all these terms will add up to $(n - 1)$ in the best case scenario, i.e., when $X^{p'}$ and X match exactly. This large sum when multiplied by $x_j^{p'}$ (cf. Equation 8.75) will produce a large contribution to net_k value while the remaining terms coming from patterns with $p \neq p_0'$ may be expected to make a rather small contribution to net_k as there will be many cancellations of terms when X does not match X^p. Thus, during updating $X(t + 1)$ will move closer to $X^{p'}$ as iterations proceed. This is the philosophy of associated memory formation in the Hopfield network. The convergence of iterations is, however, incumbent on several facts such as the ratio of the number of exemplar patterns (m) to the number of neurons in the network (n) being $\ll 1$ (optimum ≈ 0.15), the technique adopted for implementing updating, values of $X_i(t = 0)$, or the complete absence of correlation among the exemplar patterns.

It is possible to view the updating of neurons in the Hopfield network as the evolution of network toward a stable fixed point of the network considered as a dynamical system. There may be multiple stable fixed points—the network evolves into one of them. Thus, there is always a possibility of discovering a local optimum, instead of the global one.

The Hopfield network can be used for solving any problem for which an energy function or Lyapunov function $\varepsilon(x)$ can be defined, x being an n-dimensional vector the component of which can assume only one of the two values (-1 and $+1$, or 0 and 1) at a particular instant of time (t) and the goal of the problem-solving exercise becomes equivalent to finding a stable fixed point of the network. The application domains include signal and image processing, graph coloring, data deconvolution, pattern matching, job scheduling, and resource allocation. The Hopfield network has been successfully adapted to solve NP-hard combinatorial problems like the traveling salesman problem by designing an appropriate network energy function incorporating necessary constraints that take care of such physical conditions, such as every city is visited only once during a valid tour or at a given time the traveler can be in one city only. We will now introduce an NN model that is capable of unsupervised learning and displays competitive learning and self-organization. The specific NN we have in mind is the Kohonen network [21]. We may note here that unsupervised learning is a very important component of learning by human beings. A child learns a lot of things by watching what is happening or how individuals around him behave or react to a situation. Obviously the child learns all these without being told what is right or wrong by a supervisor or an external agent. This mode of learning, called unsupervised learning, stands in complete contrast with supervised learning—a teacher or supervisor continuously monitors and guides the learner to the correct answer each time the student comes up with an answer. There is another mode of supervised learning in which the answers are graded occasionally according to their correctness and the student receives a score for his overall performance. In fact this is what happens in a typical classroom learning scenario where students are asked to sit for examination, quizzes, etc., once in a while. It is called a graded or reinforcement learning. A neural network is said to follow an unsupervised learning model if it can learn on its own without any external information or interference. If the neural network can organize itself to form usable and useful information without external intervention, it is said to display what is known as self-organization. Such networks can be exploited to share some of the features of the unsupervised human learning process. A type of unsupervised neural network can learn from competition; here, neurons compete with each other. The neurons that win strengthen their synaptic weights while the losing neurons have their connection weights weakened or left unchanged after competition. Through the process of competitive learning, these NNs can self-organize and therefore display a kind of unsupervised learning. The Kohonen neural network to be described below is endowed with a self-organizing ability through competitive learning and can be used for clustering and categorizing input patterns, or compressing and quantizing massive data. They have also found use in analytical chemistry for classifying metabolic and photochemical reactions.

8.7 Kohonen Networks

8.7.1 Architecture

The Kohonen neural network (KN) has a multilayered architecture as shown for example in Figure 8.9a where the network has two layers. The layer on the left constitutes the input layer of neurons while the one on the right is the layer of output neurons. The output neurons are called Kohonen neurons. The Kohonen neurons in Figure 8.9a are arranged in one dimension; they can also be arranged in higher dimensions.

8.7.2 Connection Pattern

The neurons are fully interconnected. That means, any neuron in the input layer is connected to any or every other neuron in the Kohonen layer with variable (but optimizable) weights assigned to different edges. The network is of nonrecurrent (feedforward)

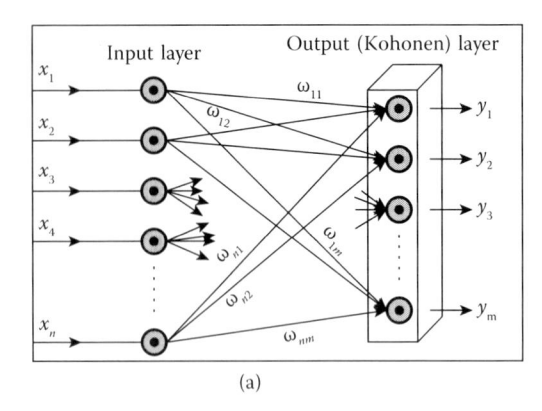 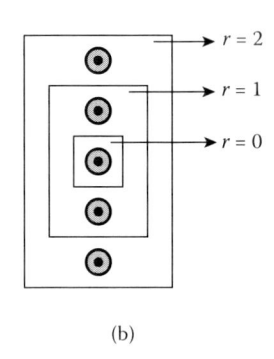

FIGURE 8.9

(a) A Kohonen neural network with one-dimensional Kohonen layer, (b) Kohonen layer neurons in one dimension with neighborhoods of different sizes defined by radius r.

type—i.e., the flow of information is only from the input to the Kohonen (output) layer. There is no feedback coupling in the network.

The input patterns are presented to the input layer neurons vectors X with n-real components (floating point numbers). The connection weights (ω_{ij}s) are also represented as real floating point members that are ultimately determined through competitive learning.

Since each neuron has to compete with other neurons for learning, it is necessary to define a neighborhood parameter (r) for each neuron in the output layer, i.e., the Kohonen layer. The parameter acts as a radius that encloses all the neighborhood neurons. Thus, in Figure 8.9b, the neighborhood parameter $r = 0$ means that the neuron marked " has no neighbors, while $r = 1$ implies it has two neighbors, $r = 2$ means there are four neighbors. The important point to be noted here is that the neighbors are not required to wrap around cyclically from one end to the other. That essentially means that the missing end-neurons are neglected.

The extension of the concept of neighborhood to two dimensions is straightforward and is represented by squares of different sizes enclosing the example neuron and its neighbors. Needless to mention is that the square may be replaced by other shapes, like the hexagon or octagon. In three dimensions, toroidal enclosures are often used to define the neighborhood. There are other versions of KNs which are architecturally and computationally different from what we are going to describe in this section. The computational procedure may be broken down into several steps as follows:

1. Initialization: Initial values of weight vector components are set. Also set are the learning rate, radius defining neighborhood, etc.

2. Present input pattern vectors to the network in a random sequence.

3. For the given input vector, select a winning neuron.

4. Modify weights for all neurons within the current radius or neighborhood of the winning neuron at the current learning rate (α). Go back to step 2 to enter another pattern.

5. Reduce the learning rate ($\alpha_{new} = \lambda \alpha_{old}$) and go back to step 2.

6. Reduce the radius (shrink the neighborhood) and go back to step 2.

7. Check for convergence—print winning neuron y_k for each input vector.

The initialization step involves assigning initial values of the weight parameters ω_{ij} ($i = 1, 2, ..., n; j = 1, 2, ..., m$), ω_{ij} representing the connection weight between the ith input layer neuron to the ith Kohonen neuron. The initial values of ω_{ij} are chosen randomly from a predefined range of small real values. The other two parameters defined at this step are the learning rate (α) and a neighborhood parameter (r). Typically, one may start with $\alpha = 0.809$ and $r = 3$ or 4.

Steps 2 to 7 are then executed for the n-component input vectors $X^{(i)}$ ($i = 1, 2, ..., N$), assuming that N input patterns are to be tried for clustering. At step 2 the n components of an input vector is entered or read in as ($x_1, x_2, ..., x_n$). Steps 3 and 4 are then executed to simulate the process of competitive learning by the network. Selection of a winning Kohonen neuron and weight modifications are to be carried out under a competitive learning (steps 3 and 4) scheme.

8.7.2.1 Selection of Winning Kohonen Neuron

Suppose that the current pattern vector (X^k) has been presented to the network through the input neurons. All the Kohonen neurons ($j = 1, 2, ..., m$) now compute their distance functions R_j ($j = 1, 2, ..., m$) from the n-component input vector where

$$R_j^k = D(\omega_j, X^k)$$

$$= \sum_{i=1}^{n} \left(\omega_{ij} - X_i^k\right)^2 \tag{8.76}$$

ω_j being the weight vector associated with the jth Kohonen neuron. Of course, this is not the only way that R_j^k can be computed. The method suggested is perhaps simplest. Once all the R_j^ks have been computed, the winning neuron is the one that has the smallest computed distance. Let the winning neuron be the jth neuron.

8.7.2.2 Weight Modification

At this step, only the weights associated with the winning neuron (suppose, j) and the neurons within the defined current neighborhood of j are modified. Suppose that the neighborhood distance is $r = 2$. Then the neighbors are $j - 1, j - 2, j + 1$, and $j + 2$th neurons (if $j - 2$ or $j - 1 < 1$, neglect the neighbors; also if $j + 1$ or $j + 2 > m$, the neighbors can be neglected). Having chosen the winning Kohonen neuron and its neighbors, the next step is to modify the relevant weights: the weight modification of the jth Kohonen neuron and its neighbors is done in the following manner:

$$\omega_j^{t+1} = \omega_j^t + \alpha\left(X^t - \omega_j(t)\right) \tag{8.77}$$

where ω_j^t is the n-component weight vector of the jth Kohonen neuron at the tth stage and X^t is the input vector at the same stage. The weight vectors of all other neurons (the losing neurons) are not updated or modified. The effect of the weight modification is to move the weight vectors associated with the winning neuron (j) and its neighbors ($j - 1, j - 2, j + 1, j + 2$) closer to the currently available input pattern vector. The extent of movement, however, depends on the current learning rate α.

Once the procedures laid down in steps 2–4 have been completed for all the input patterns (N), one epoch is complete. The learning rate (α) is now reset to a lower value and the whole process is repeated. After a preset number of iterations has elapsed, the neighborhood radius is reduced to $r-1$, and the procedures are continued until the weight vectors converge. Once convergence has been achieved the neural network should have been able to cluster the input vectors into several classes. The network achieves this without any external intervention, thereby displaying self-organization and self-learning ability. It is pertinent at this point to try to understand what the Kohonen network has been able to do physically. The network clearly stores the input vectors or patterns by modifying the elements of the weight vectors through competitive learning. As more and more input vectors are presented to the network, the weight vectors become more densely distributed in the regions where the network has more frequently received input patterns. This amounts to a clear clustering of the input patterns into several disjoint groups or classes automatically. The clustering becomes possible because the density distribution of the weight vectors at convergence closely approximate the density of distribution of the input (pattern) vectors. A consequence of this self-learning-based clustering is that similar input vectors will be grouped into the same cluster and the same winning neuron (y_j) will be fired each time a similar vector is presented.

The Kohonen network has been successfully applied to classify and monitor chemical reactions by linking databases for chemical reactions to NMR data. We may refer for example to the explorative work [22] on reaction classification based on ^{1}H NMR data of reactants and products. The authors considered the problem of classification of photochemical and metabolic reactions by self-organizing Kohonen networks. Differences between the ^{1}H NMR spectra of the reactants and products were exploited as descriptors of the reactions taking place and were presented to the network as input vectors. The ^{1}H spectra were simulated from the molecular structures by SPINUS software. The data set comprised a training set of 147 photochemical cycloaddition reactions and a test set of 42 reactions manually classified into seven classes. The network had a toroidal topology. A linearly decreasing neighborhood (triangular) scaling function was used; the learning rate varied between 0.0 to 0.1 while 50 to 100 epochs of learning were carried out. The winning neuron selection criterion was based on the Euclidean distance between the input vectors and the appropriate weight vectors. After the predictive models of classification were established on the basis of simulated NMR data, their applicability to reaction data gathered from mixed sources, both experimental and simulated, was carefully evaluated. The study clearly demonstrates the power and utility of the self-organizing Kohonen networks.

8.8 Neural Networks in Optimization

8.8.1 The Hopfield–Tank Model

So far, we have examined several neural network models in the context of pattern classification, associative memory formation, and self-organization. NNs have also been successfully used to solve hard optimization problems belonging to the NP-hard or even NP-complete categories. The first such application of ANN was reported as early as 1986 when Hopfield and Tank [23,24] proposed an ANN that could find good solutions for the traveling salesman problem, if not the best one. The successful application stirred a great deal of interest in problem-solving with NNs and led to the development of other types of networks for solving difficult optimization problems (e.g., Boltzmann machine).

Let us consider a one-dimensional problem for the purpose of introducing how the Hopfield–Tank model of neural computing works as it is easier to comprehend. The 1-*D* network consists of n-neurons (Figure 8.10) all interconnected with symmetric weights ($\omega_{ij} = \omega_{ji}$). There is no self-coupling ($\omega_{ii} = 0$, $i = 1, 2, ...$). The 1-*D* network is characterized by a network energy function E that reads

$$E = -\frac{1}{2}\sum_{i=1}^{n}\sum_{j=1}^{n}\omega_{ij}V_iV_j - \sum_{i=1}^{n}V_iI_i \tag{8.78}$$

where V_i is the output from the ith neuron and I_i is the external input to the ith neuron.

The weights and external inputs are given small constant values consistent with the problem type. The task is to computationally determine the outputs (V_is) so that the network energy function is minimized globally. The energy function looks very similar to the energy expression of a one-dimensional Ising chain with V_i, V_j replacing the Ising spins S_i, S_j and ω_{ij} by coupling constant J_{ij}. The V_is are not given discrete values such as (–1, 1) or (0, 1) directly but the algorithm nudges the output from each neuron to a discrete value such as 0 or 1 or –1 through the use of a suitable nonlinear fitter activation function. In the Hopfield–Tank model, the optimal solution, i.e., the output neuron values (V_is) minimizing the network energy are obtained by solving a set of n equations of motion describing the evolution of a set of auxiliary quantity u_i ($i = 1, 2, ..., O$), i.e., the net input to neuron i which, in turn, determines V_i through an appropriate activation function $f(u_i)$. Let us take for example

$$V_i = f(u_i) = \frac{1}{2}\left\{1 + \tanh\left(\frac{u_i}{u_0}\right)\right\} \tag{8.79}$$

where u_0 is a user-defined constant and $i = 1, 2, ..., n$. The choice of activation function restricts the output neuron values to the range (0, 1). If we wish to have output in the range (–1, 1), a different form must be chosen for the activation function $f(u)$. The equation of motion for $\{u_i\}$, $i = 1, 2, ..., n$ is set up as follows:

$$\frac{du_i}{dt} = -u_i + \sum_{\substack{k=1 \\ \neq i}}^{n}\omega_{ik}V_k + I_i. \tag{8.80}$$

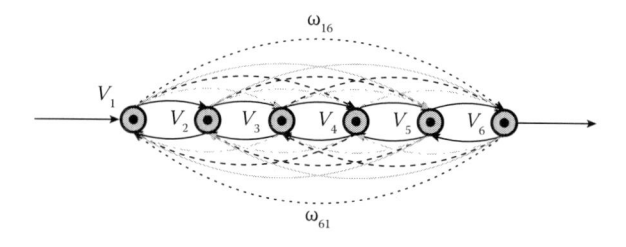

FIGURE 8.10
A one-dimensional Hopfield–Tank network.

I_i is the external input to the neuron I which may be likened to a magnetic field acting on the *i*th spin. Suppose that at the *t*th time-step the values of $\{u_i\} = u_i^t$, $i = 1, 2, \ldots, n$ with u_i^0s chosen randomly from a small range of values. For example, we may set $u_i^0 = u_0 \pm \Delta u$ where $\Delta u = a.u_0$ with $a = 0.1$ (say). Let Δt denote the increment of time in going from the current step to the next step. We have

$$u_i^{(t+1)} = u_i^t + \frac{du_i^t}{dt}.\Delta t \tag{8.81}$$

with $\dfrac{du_i^t}{dt}$ being estimated from the equation of motion, i.e., Equation 8.80. The $u_i^{(t+1)}$ is next used to estimate the output $V_i^{(t+1)}$ from Equation 8.79 as

$$V_i^{(t+1)} = \frac{1}{2}\left\{1 + \tanh\left(\frac{u_i^{(t+1)}}{u_i^0}\right)\right\}. \tag{8.82}$$

New $V_i^{(t+1)}$s then provide a new estimate of energy E^{t+1}. The iterations are continued until the network energy stops evolving any further. The search should have, by then, located the global minimum energy configuration of the neural network. The Hopfield–Tank algorithm can be implemented in the following stepwise manner:

8.8.1.1 Initialization

Step 1:

    a. Set the values of u_0, ω_{ij}s, and I_is (the external inputs) at $t = 0$ along with E^0 (a large positive number).

    b. Assign u_i^t values ($i = 1, 2, \ldots, n$) randomly from the range $u_0 \pm \Delta$, ($\Delta = 0.1u_0$).

Step 2: Compute V_i^t s as $V_i^t = \dfrac{1}{2}\left\{1 + \tanh\left(\dfrac{u_i^t}{u_0}\right)\right\}$ and estimate E^t from Equation 8.78.

Step 3: Check if $E^t \geq E^{t-1}$.

    a. If yes, terminate iterations. The current values of V_i^ts are to be printed as the optimal values of the search variables.

    b. If not, compute $\dfrac{du_i^t}{dt}$ from Equation 8.80 and update u_i^ts to u_i^{t+1}s by using Equation 8.81 and return to step 2. Iterate until termination condition is satisfied.

As an example, we may take up the problem of determining the optimal (lowest energy) spin-configuration on a two-dimensional square Ising lattice. The lattice is occupied by Ising spins with randomly chosen initial values (± 1). That means, $S(i, j) = \pm 1$ (randomly chosen) for $i = 1, 2, \ldots, n$; $j = 1, 2, \ldots, n$. The energy of the lattice is given by

$$E = -\frac{1}{2}\sum_i\sum_j\sum_k\sum_l J_{ij,kl}S(i,j)\cdot S(k,l) \tag{8.83}$$

where $J_{ij,kl}$ indicates the coupling or interaction strength between the spins at the lattice site (i, j) and that at the site (k, l). If a magnetic field is present, the spins will interact with the field and the energy expression will have an additive term. The modified energy equation reads

$$E = -\frac{1}{2}\sum_{i=1}^{n}\sum_{j=1}^{n}\sum_{k}\sum_{l}J_{ij,kl}S(i,j)\cdot S(k,l) - \sum_{i}\sum_{j}J_{ij}I_i \qquad (8.84)$$

where J_{ij} is the magnetic field being experienced by the spin at the site (i, l). It is now quite straightforward to map the problem to the Hopfield–Tank neural network-based optimization problem. The spins at lattice sites (i, l) or (k, l) are replaced with neurons V_{ij} or V_{kj}. The coupling strengths $J_{ij,kl}$ are replaced by the network weight factors $\omega_{ij,kl}$ (held fixed). Thus the network energy expression can be written as

$$E_{network} = -\frac{1}{2}\sum_{i}\sum_{j}\sum_{k}\sum_{l}\omega_{ij,kl}V_{ij}.V_{kl} - \sum_{i}\sum_{j}I_{ij}V_{ij}. \qquad (8.85)$$

There is no self-interaction or coupling among the neurons so that $\omega_{ij,ij} = 0$ $(i = 1,2,\ldots,n; j = 1,2,\ldots,n)$.

If only the nearest neighbor spins or neurons interact, further restrictions can be imposed on the elements of the weight matrix or equivalently on the spin-coupling constant matrix (J). The determination of spin orientation so as to minimize the energy globally can now be mapped onto the problem of finding the optimal values of the neurons at different sites (ij, kl) by solving the Hopfield–Tank equation of motion for the problem as follows $(J_{ij,kl} = \omega_{ij,kl})$

$$\frac{du_{ij}^t}{dt} = -u_{ij}^t + \sum_{kl}\omega_{ij,kl}V_{ij}.V_{kl} - \sum_{kl}I_{kl}V_{kl}$$

$$u_{ij}^{t+1} = u_{ij}^t + \frac{du_{ij}^t}{dt}.\Delta t \qquad (8.86)$$

$$V_{ij}^{t+1} = f\left(u_{ij}^{t+1}\right).$$

Note that $f\left(u_{ij}^{t+1}\right)$ here is an activation function that maps the net input to the neuron at the site ij onto the range $(-1, 1)$. A convenient choice of the activation function for use in the present problem is the sigmoidal bipolar function—that means, we can set

$$f\left(u_{ij}\right) = \tanh\left(\lambda u_{ij}\right); \ \lambda > 0. \qquad (8.87)$$

Although the problem at hand is a discrete optimization problem, the Hopfield–Tank neural network with an appropriate choice of activation function can handle it as a continuous optimization problem rather conveniently and smoothly, and often locate the global minimum successfully. It is possible to extend the method to handle 3-D Ising spin systems and find out the ground state. The Hopfield–Tank method has been very successfully applied to solve the traveling salesman problem [25, 26] and the n-queen problem which belong to the NP-hard category. It would be interesting to study its efficiency in handling, for example, the Coulomb glass problem.

8.9 Neural Network Potentials in MD Simulation

In an MD simulation of a system consisting of many interacting atoms, it is not possible to guess beforehand which nuclear configurations will be visited or sampled during the simulation run. On the other hand, for the simulation to proceed unhindered, we must have a means of computing the potential energy of the system and forces acting on every individual atom of the system at every configuration sampled in the course of the simulations. In an ab initio MD simulation, one possibility could be to perform a vastly large number of quantum chemical calculations covering a very wide range of possible nuclear arrangements of the N-atom system (i.e., structures) and use the stored data in MD simulation for computing energy and forces. However, even for a medium-sized system, the number of molecular electronic structures and energies that can be computed and kept stored even in a huge database are too limited. That means, there will be many configurations visited during the actual simulation for which the required information would not be available. Accurate multidimensional interpolation being an yet-to-be solved problem of numerical analysis, cannot be an option. Ab initio MD simulations therefore invoke the other option—i.e., compute the energies and forces on the fly by using density functional theory if the system is large, or utilize a wave function-based technique if the system is not too large. On the other hand, the MD calculations could very easily be carried out if the energy computed for many different atomic arrangements could be given a compact analytic representation in terms of nuclear positions, charge carried by atoms, etc. However, the form of the function could turn out to be generally too complicated to be derived analytically and used in simulations. Several alternatives have been explored to tackle the problem of suitably representing many-atom potentials. In one such approach, the problem has been handled by making approximations to construct the potential energy surfaces on which simulations can be carried out. Thus, the solutions of the Schrödinger equation (Born–Oppenheimer energies) are not directly used. Instead, rather simplified energy expressions derived on the basis of reasonable approximations and physical considerations are used. Such potentials abound in chemical literature. They vary in complexity and often fail to provide the needed accuracy. Another much used (and probably abused) approach has been to make use of what has become known as "classical force fields" which hinge on the idea of decomposing the total molecular energy of the system into a sum of pair, triple, and quadrupole interactions that are assumed to represent covalent bonds or covalent interactions. Coulomb and Lennard–Jones potentials are also added to take care of electrostatic and van der Waals interactions, which would also be generally present in the system. We may note here, that these low-dimensional bonding terms take into account only the immediate chemical environment of the atom in question. The neighboring atoms may, however, strongly modulate the bonding properties of a given atom. Thus, in addition to just specifying the atom type, additional information about the atom in question must be provided, for example, by specifying the functional group to which the atom belongs. With all these additional modifications the total energy in classical force fields is ultimately expressed as a sum of approximately additive components. These force fields are inadequate to describe bond-breaking or bond-making events for which specially designed reactive force fields have been proposed and used. While these potentials have been useful and are still being used, the search for alternatives continued culminating in the development of a new class of atomistic potentials which are very flexible as far as their functional forms are concerned, but are generally devoid of any physical meaning or rationale. We may perhaps consider them as purely mathematical potentials or machine-learning potentials. The strategy here has been to gather enough reference data generated by electronic structure calculation codes (quantum chemistry packages) at a particular level of sophistication, depending on the size and nature of the problem and then fit the data to some complicated analytical form. Once that is done, we have at our disposal a means of mapping the input information about a specific structure into an output energy (or some other property) without performing additional electronic structure calculations at the particular geometry or nuclear configuration even if it was not included in the filtering procedure. Among the purely mathematical potentials, the permutation invariant polynomial potentials and the modified Shepherd's potential [27–31] are noteworthy. The most popular and versatile machine-learning potential that has come up over the last decade or so is the artificial neural network potential (ANNP). If constructed with enough care so as to preserve the physical shape of the PES in question, these ANNPs can be very accurate (as accurate as the data used to train the network) and quite accurately represent systems for which no approximate physical potential is yet available. The type of NNs used in such constructions belong to the category of multilayered feedforward NNs and the learning method adopted has uniformly involved adjustment of the network connection weights by the backpropagation method already described in this chapter. These neural network potentials exploit a single multilayered feedforward network for constructing a direct mapping—i.e., producing a functional relation between the atomic or nuclear configuration and the potential energy of the system. The input to the neural network (received by the input neurons) is a vector (X) of atomic coordinates while the output is the corresponding energy $E(X)$ that the single output neuron produces as output. Consider, for example a triatomic system ABC, the structure of which is completely defined by the three geometrical parameters r_1, r_2, θ. The input vector for the ith structure is

$$X^i = \begin{pmatrix} r_1^i \\ r_2^i \\ r_3^i \end{pmatrix}.$$

One can carry out accurate quantum chemical calculations for many such configurations of the system (ABC) producing total energies $E_1, E_1, ..., E_i, ..., E_M$ where $E_i = E(X^i)$, $i = 1, 2, 3, ..., M$. A simple ANN that can be exploited to construct the neural network potential accurately representing the structure energy data is displayed in Figure 8.11. The network has three input neurons in the input layer, three neurons in the single hidden layer and only one neuron in the output layer. The connections are: each

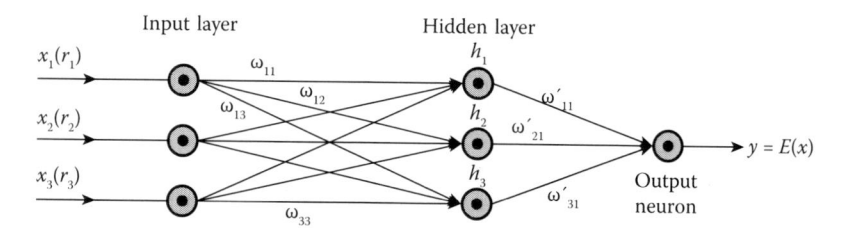

FIGURE 8.11
A simple ANN.

input neuron is connected to all the hidden layer neurons and each hidden layer neuron is connected to the single output neuron. The hidden layer neurons and the output neuron use suitable activation functions the threshold parameters of which are adjustable, as the network weights are.

More hidden layers can be added or their numbers in a particular layer can be increased for providing desired flexibility to the ANNP.

The weights $\{\omega_{ij}\}$ and $\{\omega_{kl}\}$s are adjusted so as to enable the network to accurately reproduce the energy $\{E(X^i)\}$ as the output $y(i)$ from the output neuron for the structures $\{X^i\}$ after the conclusion of the training. Once trained, the network is expected to be able to produce the correct energy $y = E(X^{new})$ for a fresh input configuration X^{new} (not present in the training set). The network behaves like a rather complex interpolating function—the so-called neural network potential function that is capable of yielding the correct energy of the new configurations presented to the network without any further electronic structure calculations. Formally, we may represent the ANNP as a nested function of the network weights and input (X) as follows:

$$y_i = \phi\left(\sum_k \omega'_{ik}\phi\left(\sum_l \omega_{kl}X_l + b_k^{(1)} \right) + b_k^{(2)} \right) \tag{8.88}$$

where the b_ks are the bias terms.

It is not difficult to realize that such a simple NNP is adequate for a low-dimensional problem—a problem for which the structure can be coded by a low-dimensional structure vector X. (Note that in the example given, X is a three-component or three-dimensional vector). A single feedforward ANN is, however, inadequate for representing the neural network potential when the structure vector is of much higher dimensionality. In such cases, the computational cost of training for one thing, rises exponentially with the increase in system size. The ANNP consisting of a single feedforward network remains applicable only to systems of the size that has been employed in constructing the potential, thereby making addition or removal of one or more atoms from the system impossible even when required (for example, during a reaction). There is a third, in a sense, more fundamental issue that concerns symmetry. In the example case of a water molecule, if the bond lengths r_1 and r_2 are interchanged, the chemical species (H_2O) does not change and therefore, the energy must remain unchanged. The connection weights of the neurons representing the atoms H_1 and H_2 with the hidden layer neurons may however be very different. If we change the order of the input neurons (X_1, X_2) mindlessly, the output may be quite different. In other words, the weight vector does not reflect the symmetry of the molecule that the ANNP seeks to represent automatically. A simple way out of this symmetry dilemma is to replace the internal coordinates by a new symmetrized set that leads to the complete loss of identity of all the chemically-equivalent atoms (H_1 and H_2 in the present example). As the size of the molecule becomes larger, the required symmetrizing transformation becomes more and more complex to construct and the process becomes unwieldy. Thus, a simple ANNP based on a single multilayered feedforward network with appropriately symmetrized, internal coordinates is viable only for small molecules such as H_2O, NH_3, or CH_4. For larger systems, the simple representation fails to be effective. For high-dimensional systems, a strategy has been developed by Behler and others [32–39] bypassing all the difficulties associated with single feedforward neural network-based representation of the high-dimensional potential. The basic philosophy of the method can be described

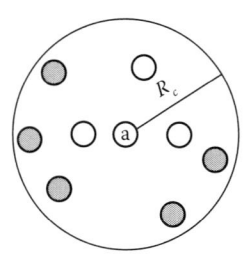

FIGURE 8.12
The chemical environment seen by the atom a and the cutoff radius.

as follows: the energy (E) of the system containing N atoms is represented as a sum of the energies of the constituent atoms (E_a)

$$E = \sum_{a=1}^{N} E_a. \tag{8.89}$$

The atomic energies E_a are, however, not the energies of the free atoms (E_a^0), but are shifted atomic energies, the shift being produced by the immediate chemical environment in which the atom a exists and with which it interacts. Thus, the environment takes into account the effects of interactions with all other atoms that exist within a cutoff radius (R_c) measured from the point where the atom a is located (Figure 8.12).

Usually, the extension of the chemical environment is a sphere of volume $\frac{4}{3}\pi R_c^3$ with $R_c = 10 - 20\Delta$ and the coordinates of all the other atoms present in it make up the structural "fingerprint" of the chemical environment. The energetic fingerprint of the environment manifests itself in the form of the energy shift $\Delta E_a = (E_a - E_a^0)$ due to interaction of atom a with the chemical environment that it sees around itself. The structural fingerprint of the environment is constructed from a set of many body functions called symmetry functions. These n_g-symmetry functions are collected into a vector with n_s components and fed as input to the input neurons of the atomic neural network for the atom a. For each atom of the system, there is one atomic neural network, the output from each being the corresponding atomic energy as modified by the chemical environment. The output energies from the atomic NNs are summed up by the output neuron of the molecule as a whole and calibrated against high-quality ab initio calculated energy of the system at the given configuration. The calibration is carried out by the backpropagation technique that adjusts the various connection weights so as to produce the "correct" shifted energies (E_a), the sum of which produces the ab initio energy accurately. Once the network has been trained with enough input configurations and their energies, the high-dimensional ANNP is ready for use in MD simulations for calculation of bulk properties with high accuracy. The high-dimensional ANN comprising N number of atomic NNs is schematically shown in Figure 8.13 below.

The high-dimensional ANNP has been profitably used in high-quality MD simulations. Behler and Parrinello [32] introduced for the first time the idea of constructing a generalized neural network representation of a high-dimensional potential energy surface, which could provide energies and forces as functions of atomic positions in systems of arbitrary sizes at a fraction of the computational cost involved in simulations in which energies are provided by DFT on the fly. The applications to bulk silicon provided exciting results.

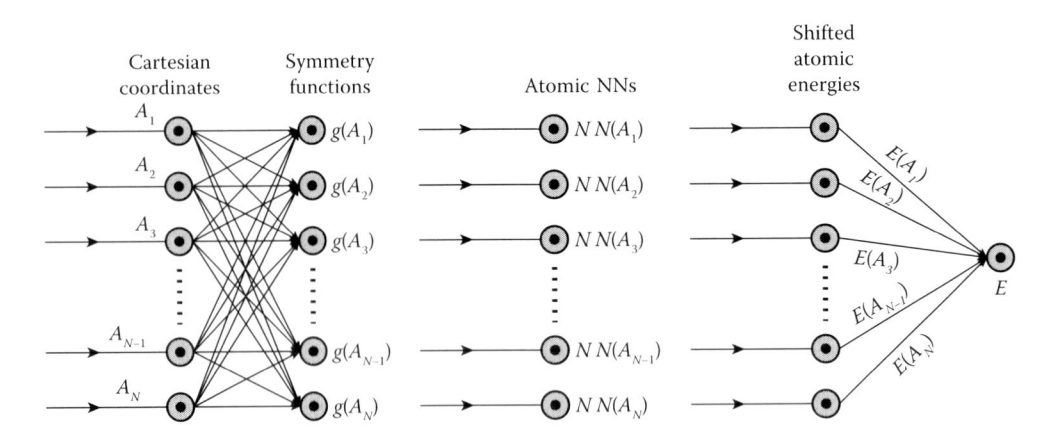

FIGURE 8.13
A high-dimensional neural network comprising N atomic NNs.

Eshet et al. [39,40] constructed an ab initio quality neural network potential for sodium by using the high-dimensional neural network representation of the ab initio interatomic potential energy surface. The resulting ANNP was found to be suitable for modeling high temperature, high pressure crystalline, and liquid phases of sodium. These authors demonstrated the unique ability of the ANNP to provide an ab initio quality description of a number of properties of liquid sodium, and bcc, fcc and c/16 crystalline phases over a range of fairly high pressure and temperature (pressure up to 120 GPa, and temperature up to 1200 K). The authors claim that the ANNP combines computational efficiency and accuracy in a unique manner leading to MD simulation of the physicochemical properties of sodium in the high pressure and temperature regimes with unprecedented accuracy. Khaliullin et al. [41,42] exploited the neural network mapping of ab initio PES to study the thermodynamics of graphite $\leftrightarrow$ diamond phase coexistence phenomenon. The ANNP deftly combines the accuracy of a first-principle description of both the phases with computational efficiency of empirical force fields and enables one to undertake MD simulations of ab initio quality for computing the thermodynamics of graphite $\leftrightarrow$ diamond coexistence. The agreement between the computed and experimental coexistence curve was found to be good when nuclear quantum effects were taken into account. The high-dimensional ANNP representation has now been perfected to an extent where systems containing thousands of atoms can be handled efficiently and accurately, and various problems of chemistry, physics, and materials science can be elegantly addressed.

Atomistic simulations can provide a wealth of information about processes taking place in condensed matter systems. A complete understanding of the underlying physical mechanisms of such processes demand quantitative analysis that can be carried out automatically on the data generated, identifying or detecting particular atomistic structures that appear or disappear at certain stages of the simulation. Once such structures are detected at the local level, it may be expedient to follow their motion in time to understand, for example, the dynamics of structural phase transition, crystallization of super cooled liquid, or how defect formation takes place in solids. Accurate identification and classification of local ordered and disordered structures thus constitute an important task in atomistic computer simulations. Geiger and Dellago [43] exploited the neural network-based approach recently developed by Behler and Parrinello [32] for constructing high-dimensional neural network potentials for computing energies and forces accurately at a low computational cost, and proposed a method capable of recognizing the local atomic arrangements from a set of predefined symmetry functions that characterize the chemical environment of a given atom in the system. Geiger and Dellago's algorithm for local structure detection in polymorphic systems is simple and flexible, and is independent of how the reference frame has been defined. The authors implemented the algorithm on the Lennard–Jones system, liquid water, and ice, and demonstrated how their neural network method very successfully detected amorphous and crystalline structures even in cases of complex atomic arrangements, for which the traditional structure detectors are not at all reliable. Critical to the structure detection is how carefully the symmetry functions have been defined for the problem being dealt with. These functions must be designed to carry the maximum possible information about the chemical environment seen by the atom i and remain invariant under rotations, translations, and permutations of identical atoms, yet contain sufficient details to enable detection and classification of local structures. The symmetry functions $G_k(r)$ $(k = 1, 2, \ldots, n_s)$ used by these authors proved to be sufficient for detecting local structures generated by the simulation quite accurately. $G_k(r)$s contain some tunable parameters which can be adjusted to make the functions adequately sensitive for the purpose of structure detection in the two cases addressed by them. They are functions of interparticle distance $\{r_{ij}\}$, the angles θ_{ijk} spanned by the three atoms i, j, and k (Figure 8.14) and R_c (the cutoff radius defining the chemical environment seen by the atom i.

Of the symmetry functions defined and used by Gelger and Dellago [43], $G_1(r) - G_5(r)$ have already been defined by Behler so we only consider $G_6(r) - G_8(r)$ defined for the specific problems they addressed. The three special symmetry functions are:

$$G_6^{(i)}(r) = \frac{1}{2^{\xi}} \sum_{j,k \neq i} \left(1 + \lambda \cos\theta_{ijk}\right)^{\xi} f_a\left(R_{ij}\right) f_a\left(R_{ik}\right)$$

$$G_7^{(i)}(r) = \frac{1}{2} \sum_{j,k \neq i} \sin\left[\eta\left(\theta_{ijk} - \alpha\right)\right] f_c\left(R_{ij}\right) f_c\left(R_{ik}\right) \qquad (8.90)$$

$$G_8^{(i)}(r) = \frac{1}{2} \sum_{j,k \neq i} \sin\left[\eta\left(\theta_{ijk} - \alpha\right)\right] f_b\left(R_{ij}\right) f_b\left(R_{ik}\right)$$

$R_c, \xi, \eta, \lambda, \alpha$ etc., are symmetry function parameters that are chosen carefully so that the symmetries characterizing local structures are correctly described. f_a, f_b, and f_c are

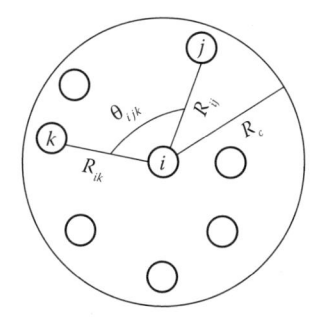

FIGURE 8.14
Geometry parameters used to define symmetry functions.

cutoff functions that are sharper than those used in energy calculations, and are defined as follows:

$$f_c(R) = \begin{cases} \left[1 + exp\{\alpha_c(R - R_c + \epsilon_c)\}\right]^{-1}, & \text{for } R < R_c \\ 0, & \text{otherwise} \end{cases}$$

$$f_a(R) = \begin{cases} \cos^2\left[\eta(R - \mu)\right], & \text{if } \mu - \dfrac{\pi}{2\eta} < R < \mu + \dfrac{\pi}{2\eta} \\ 0, & \text{otherwise} \end{cases} \tag{8.91}$$

$$f_a(R) = \begin{cases} \cos^2\left[\nu(R - a_l)\right], & \text{if } a_l - \dfrac{\pi}{2\nu} < R < a_l \\ 1, & \text{if } a_l < R < a_r \\ \cos^2\left[\gamma(R - a_r)\right], & \text{if } a_r < R < a_r + \dfrac{\pi}{2\nu} \\ 0, & \text{otherwise} \end{cases}$$

The use of a sharper cutoff function is necessary as the functions must be sensitive to variations in atomic positions within a small and well-defined region. The mean square error (χ^2) reduction in the backpropagation scheme was carried out by invoking the extended Kalman filter [44] which helped the search converge quickly with less tendency to get trapped in local minima. The work [43] brings into focus how crucial it is to construct appropriate symmetry functions for developing an ANN-based structure recognition system. Computation of symmetry functions consumes by far the largest amount of time in the ANN-based structure detection scheme. It is therefore expedient to keep the number of symmetry functions to the minimum possible value. An importance sampling can help in identifying the redundant symmetry functions, if any. A second important point to note is that the performance of the structure detection system can perhaps be enhanced by bringing in different structural fingerprints such as Steinhardt's bond order parameters as additional network inputs. There is a lot to be explored in this context and exciting developments may be anticipated. The cutoff distance R_c, which defines the region that is considered for defining the structural fingerprint in the form of symmetry function, is critically important in the ANN-based scheme of structure detection. The choice can delicately affect the accuracy as well as the spatial resolution of the structure detection or recognition systems based on ANN. Apart from the exciting developments in the fields of MD simulations of various properties using high-dimensional neural network potentials and ANN-based structure detection systems, ANNs have been traditionally exploited in theoretical and computational chemistry for prediction of ionization potentials, lipophilicity of a molecular species, physicochemical properties including mechanical properties of polymers by employing topological indices as neuronal inputs, relative permittivity, and oxygen diffusion of ceramic materials as function of ceramic composition [45–51], etc. The ANN has also been used to predict estimates of electronic correlation energies of diatomic molecules and all atoms up to radon [52]. All the calculations referred to made use of the BPN for calculation and the Kohonen network for visualization. Very recently, a machine-learning model trained on a database of ab initio calculated properties of thousands of organic molecules, was shown to be capable of predicting multiple electronic properties of the systems both in the ground and excited states [53]. The properties include atomization energy, polarizability, frontier orbital energies, ionization potential, electron affinity, and excitation energies. The machine-learning model employed by these authors is based on a deep multitask ANN that has the ability to exploit the underlying correlations among various molecular properties. The input to the network is simple and similar to what is fed as input to an electronic structure calculation code—that only the charges (z_A) carried by the nucleii (A) and the Cartesian coordinates of all the atoms are present in the molecule. For small organic molecules, the accuracy of values predicted by what the authors call "a quantum machine" is similar and sometimes even superior to the corresponding results of ab initio calculations. Considering that the prediction by the ANN requires little computational effort and input preparation is so simple, the ANN-based modeling of molecular properties can be extremely useful. Wang et al. [54] explored ANN-based prediction of heats of formation while Zheng et al. [55] experimented with the possibility of adopting an NN-based approach to the construction of a generalized exchange–correlation potential. Balabin et al. [56] exploited the idea of an ANN-based quantum chemical database that could quickly and accurately predict the DFT energies without having to undertake a detailed DFT calculation on the system of interest. Rupp et al. [57,58] investigated the possibility of accurately predicting molecular atomization energies with encouraging results. Machine-learning and DFT calculations were deftly combined by Hautier et al. [59] for tracing nature's missing ternary oxide compounds.

Neural network-based computing has now reached a stage where it is increasingly being exploited for predicting structure and functionality, or for predicting and designing structures that would support desired functionalities. It has thus grown into a potent tool for chemical research. The issue concerning the time consumed in training networks remains critical although rapid development in parallel or high-performance computing, especially the growth in GPU-based programming, may be expected to change the scenario drastically. One can therefore anticipate massive developments in ANN-based computing in electronic structure theory quantum dynamics for predicting spectra and state-to-state cross sections. Although the backpropagation network has been the most used network, we can anticipate increasing use of more sophisticated networks like the Boltzman machine, the Hopfield network, and the Elmann network for handling complex problems in chemistry [45,60–63].

References

1. MacGregor, R.J. 1987. *Neural and Brain Modeling. Neuroscience: A Series of Monographs and Texts.* London: Academic Press.
2. Haykin, S. 1998. *Neural Networks: A Comprehensive Foundation.* 2nd ed. Upper Saddle River, NJ: Prentice Hall PTR.
3. Konar, A. 2006. *Computational Intelligence: Principles, Techniques and Applications.* Berlin: Springer.
4. Munakata, T. 2007. *Fundamentals of the New Artificial Intelligence.* Munakata, T. (ed.). Texts in Computer Science. London: Springer.

5. Sarkar K. 2014. Soft-Computing algorithms for solving some problems of Quantum Chemistry. PhD thesis, Indian Association for the Cultivation of Science, Jadavpur University, Kolkata, India.

6. Rosenblatt, F. 1958. The Perceptron: A Probabilistic Model for Information Storage and Organization in the Brain. *Psychological Review* 65, no. 6: 386–408.

7. Minsky, M., and S. Papert. 1969. *Perceptrons*. Oxford, England: MIT Press.

8. Novikoff, A. 1962. On Convergence Proofs on Perceptrons. In *Proceedings of the Symposium on the Mathematical Theory of Automata*, April 24–26, vol. 12, pp. 615–622. New York: Polytechnic Institute of Brooklyn..

9. Block, H.D. 1962. The Perceptron: A Model for Brain Functioning. I. *Reviews of Modern Physics* 34, no. 1: 123–135.

10. Hebb, D.O. 1949. *The Organization of Behavior: A Neuropsychological Theory*. A Wiley Book in Clinical Psychology. New York: Wiley.

11. Darsey, J.A., D.W. Noid, and B.R. Upadhyaya. 1991. Application of Neural Network Computing to the Solution for the Ground-State Eigenenergy of Two-Dimensional Harmonic Oscillators. *Chemical Physics Letters* 177, no. 2: 189–194; 181, no. 4: 386.

12. Androsiuk, J., L. Kułak, and K. Sienicki. 1993. Neural Network Solution of the Schrödinger Equation for a Two-Dimensional Harmonic Oscillator. *Chemical Physics* 173, no. 3: 377–383.

13. Sumpter, B.G., C. Getino, and D.W. Noid. 1994. Theory and Applications of Neural Computing in Chemical Science. *Annual Review of Physical Chemistry* 45, no. 1: 439–481.

14. Sumpter, B.G., and D.W. Noid. 1996. On the Use of Computational Neural Networks for the Prediction of Polymer Properties. *Journal of Thermal Analysis* 46, no. 3–4: 833–851.

15. Sumpter, B.G., and D.W. Noid. 1996. On the Design, Analysis, and Characterization of Materials Using Computational Neural Networks. *Annual Review of Materials Science* 26, no. 1: 223–277.

16. Ulmer, C.W., II, D.A. Smith, B.G. Sumpter, and D.I. Noid. 1998. Computational Neural Networks and the Rational Design of Polymeric Materials: The next Generation Polycarbonates. *Computational and Theoretical Polymer Science* 8, no. 3–4: 311–321.

17. Otaigbe, J.U., Barnes, M.D., Fukui, K., Sumpter, B.G., and Noid, D.W. (2001). Generation, Characterization, and Modeling of Polymer Micro- and Nano-Particles. In *Polymer Physics and Engineering*, pp. 1–86. Berlin: Springer, https://link.springer.com/book/10.1007/3-540-44484-X.

18. Muckley, E.S., J. Lynch, R. Kumar, B. Sumpter, and I.N. Ivanov. 2016. PEDOT:PSS/QCM-Based Multimodal Humidity and Pressure Sensor. *Sensors and Actuators B: Chemical* 236: 91–98.

19. Hopfield, J.J. 1982. Neural Networks and Physical Systems with Emergent Collective Computational Abilities. Article. *Proceedings of the National Academy of Sciences* 79, no. 8: 2554–2558.

20. Gurney, K. 1997. *An Introduction to Neural Networks*. London: Taylor & Francis.

21. Kohonen, T. 1982. Self-Organized Formation of Topologically Correct Feature Maps. *Biological Cybernetics* 43, no. 1: 59–69.

22. Latino, D.A.R.S., and J. Aires-de-Sousa. 2007. Linking Databases of Chemical Reactions to NMR Data: An Exploration of 1 H NMR-Based Reaction Classification. *Analytical Chemistry* 79, no. 3: 854–862.

23. Hopfield, J.J., and D.W. Tank. 1986. Computing with Neural Circuits- A Model. *Science* 233, no. 4764: 625–633.

24. Hopfield, J.J., and D.W. Tank. 1985. "Neural" Computation of Decisions in Optimization Problems. *Biol. Cybern.* 52, no. 3: 141–152.

25. Wilson, G.V., and G.S. Pawley. 1988. On the Stability of the Travelling Salesman Problem Algorithm of Hopfield and Tank. *Biological Cybernetics* 58, no. 1: 63–70.

26. Feng, G., and C. Douligeris. 2000. Using Hopfield Networks to Solve Traveling Salesman Problems Based on Stable State Analysis Technique. In *Proceedings of the IEEE-INNS-ENNS International Joint Conference on Neural Networks. IJCNN 2000, July 27. Neural Computing: New Challenges and Perspectives for the New Millennium*, vol. 6, pp. 521–526. Como: IEEE.

27. Braams, B.J., and J.M. Bowman. 2009. Permutationally Invariant Potential Energy Surfaces in High Dimensionality. *International Reviews in Physical Chemistry* 28, no. 4: 577–606.

28. Jiang, B., and H. Guo. 2013. Permutation Invariant Polynomial Neural Network Approach to Fitting Potential Energy Surfaces. *The Journal of Chemical Physics* 139, no. 5: 54112.

29. Jiang, B., and H. Guo. 2014. Permutation Invariant Polynomial Neural Network Approach to Fitting Potential Energy Surfaces. III. Molecule-Surface Interactions. *The Journal of Chemical Physics* 141, no. 3: 34109.

30. Shepherd, A.J. 2012. *Second-Order Methods for Neural Networks: Fast and Reliable Training Methods for Multi-Layer Perceptrons. Perspectives in Neural Computing.* London: Springer.

31. Wu, W., G. Feng, Z. Li, and Y. Xu. 2005. Deterministic Convergence of an Online Gradient Method for BP Neural Networks. *IEEE Transactions on Neural Networks* 16, no. 3: 533–540.

32. Behler, J., R. Martoňák, D. Donadio, and M. Parrinello. 2008. Metadynamics Simulations of the High-Pressure Phases of Silicon Employing a High-Dimensional Neural Network Potential. *Physical Review Letters* 100, no. 18: 185501.

33. Behler, J. 2011. Neural Network Potential-Energy Surfaces in Chemistry: A Tool for Large-Scale Simulations. *Physical Chemistry Chemical Physics* 13, no. 40: 17930.

34. Jose, K.V.J., N. Artrith, and J. Behler. 2012. Construction of High-Dimensional Neural Network Potentials Using Environment-Dependent Atom Pairs. *The Journal of Chemical Physics* 136, no. 19: 194111.

35. Morawietz, T., V. Sharma, and J. Behler. 2012. A Neural Network Potential-Energy Surface for the Water Dimer Based on Environment-Dependent Atomic Energies and Charges. *The Journal of Chemical Physics* 136, no. 6: 64103.

36. Morawietz, T., and J. Behler. 2013. A Density-Functional Theory-Based Neural Network Potential for Water Clusters Including van Der Waals Corrections. *The Journal of Physical Chemistry A* 117, no. 32: 7356–7366.

37. Artrith, N., and J. Behler. 2012. High-Dimensional Neural Network Potentials for Metal Surfaces: A Prototype Study for Copper. *Physical Review B* 85, no. 4: 45439.

38. Behler, J. 2015. Constructing High-Dimensional Neural Network Potentials: A Tutorial Review. *International Journal of Quantum Chemistry* 115, no. 16: 1032–1050.

39. Eshet, H., R.Z. Khaliullin, T.D. Kühne, J. Behler, and M. Parrinello. 2012. Microscopic Origins of the Anomalous Melting Behavior of Sodium under High Pressure. *Physical Review Letters* 108, no. 11: 115701.

40. Eshet, H., R.Z. Khaliullin, T.D. Kühne, J. Behler, and M. Parrinello. 2010. Ab Initio Quality Neural-Network Potential for Sodium. *Physical Review B* 81, no. 18: 184107.

41. Khaliullin, R.Z., H. Eshet, T.D. Kühne, J. Behler, and M. Parrinello. 2010. Graphite-Diamond Phase Coexistence Study Employing a Neural-Network Mapping of the Ab Initio Potential Energy Surface. *Physical Review B* 81, no. 10: 100103.

42. Khaliullin, R.Z., H. Eshet, T.D. Kühne, J. Behler, and M. Parrinello. 2011. Nucleation Mechanism for the Direct Graphite-to-Diamond Phase Transition. *Nature Materials* 10, no. 9: 693–697.

43. Geiger, P., and C. Dellago. 2013. Neural Networks for Local Structure Detection in Polymorphic Systems. *The Journal of Chemical Physics* 139, no. 16: 164105.

44. Haykin, S. (ed.). 2001. *Kalman Filtering and Neural Networks.* New York: Wiley.

45. Goncalves, V., K. Maria, and A.B.F. da Silva. 2013. Applications of Artificial Neural Networks in Chemical Problems. In *Artificial Neural Networks—Architectures and Applications.* Suzuki K. (ed.), InTech. pp. 203–223. http://dx.doi.org/10.5772/51275

46. Sigman, M.E., and S.S. Rives. 1994. Prediction of Atomic Ionization Potentials I-III Using an Artificial Neural Network. *Journal of Chemical Information and Modeling* 34, no. 3 (May 1): 617–620.

47. Tetko, I. V., and V.Y. Tanchuk. 2002. Application of Associative Neural Networks for Prediction of Lipophilicity in ALOGPS 2.1 Program. *Journal of Chemical Information and Computer Sciences* 42, no. 5: 1136–1145.

48. Sumpter, B.G., and D.W. Noid. 1994. Neural Networks and Graph Theory as Computational Tools for Predicting Polymer Properties. *Macromolecular Theory and Simulations* 3, no. 2: 363–378.

49. Scott, D.J., P.V. Coveney, J.A. Kilner, J.C.H. Rossiny, and N.M.N. Alford. 2007. Prediction of the Functional Properties of Ceramic Materials from Composition Using Artificial Neural Networks. *Journal of the European Ceramic Society* 27, no. 16: 4425–4435.

50. Negarestani, A., S. Setayeshi, M. Ghannadi-Maragheh, and B. Akashe. 2003. Estimation of the Radon Concentration in Soil Related to the Environmental Parameters by a Modified Adaline Neural Network. *Applied Radiation and Isotopes* 58, no. 2: 269–273.

51. Ito, Y. 1991. Representation of Functions by Superpositions of a Step or Sigmoid Function and Their Applications to Neural Network Theory. *Neural Networks* 4, no. 3: 385–394.

52. e Silva, G.M., P.H. Acioli, and A.C. Pedroza. 1997. Estimating Correlation Energy of Diatomic Molecules and Atoms with Neural Networks. *Journal of Computational Chemistry* 18, no. 11: 1407–1414.

53. Montavon, G., M. Rupp, V. Gobre, A. Vazquez-Mayagoitia, K. Hansen, A. Tkatchenko, K.-R. Müller, and O. Anatole von Lilienfeld. 2013. Machine Learning of Molecular Electronic Properties in Chemical Compound Space. *New Journal of Physics* 15, no. 9: 95003.

54. Hu, J., X. Zhang, and Z. Wang. 2010. A Review on Progress in QSPR Studies for Surfactants. *International Journal of Molecular Sciences* 11, no. 3: 1020–1047.

55. Zheng, X., L. Hu, X. Wang, and G. Chen. 2004. A Generalized Exchange-Correlation Functional: The Neural-Networks Approach. *Chemical Physics Letters* 390, no. 1–3: 186–192.

56. Balabin, R.M., and E.I. Lomakina. 2009. Neural Network Approach to Quantum-Chemistry Data: Accurate Prediction of Density Functional Theory Energies. *The Journal of Chemical Physics* 131, no. 7: 74104.

57. Rupp, M., A. Tkatchenko, K.-R. Müller, and O.A. von Lilienfeld. 2012. Fast and Accurate Modeling of Molecular Atomization Energies with Machine Learning. *Physical Review Letters* 108, no. 5: 58301.

58. Hansen, K., G. Montavon, F. Biegler, S. Fazli, M. Rupp, M. Scheffler, O.A. von Lilienfeld, A. Tkatchenko, and K.-R. Müller. 2013. Assessment and Validation of Machine Learning Methods for Predicting Molecular Atomization Energies. *Journal of Chemical Theory and Computation* 9, no. 8: 3404–3419.

59. Hautier, G., C.C. Fischer, A. Jain, T. Mueller, and G. Ceder. 2010. Finding Nature's Missing Ternary Oxide Compounds Using Machine Learning and Density Functional Theory. *Chemistry of Materials* 22, no. 12: 3762–3767.

60. Raff, L.M., Komanduri, R., Hagan, M., and Bukkapatnam, S.T.S. 2012. *Neural Networks in Chemical Reaction Dynamics*. New York: Oxford University Press.

61. Baughman, D.R., and Y.A. Liu. 2014. *Neural Networks in Bioprocessing and Chemical Engineering*. San Diego, CA: Academic Press.

62. Perea, J.D., S. Langner, M. Salvador, J. Kontos, G. Jarvas, F. Winkler, F. Machui, et al. 2016. Combined Computational Approach Based on Density Functional Theory and Artificial Neural Networks for Predicting The Solubility Parameters of Fullerenes. *The Journal of Physical Chemistry B* 120, no. 19: 4431–4438.

63. Schneider, P., A.T. Müller, G. Gabernet, A.L. Button, G. Posselt, S. Wessler, J.A. Hiss, and G. Schneider. 2017. Hybrid Network Model for "Deep Learning" of Chemical Data: Application to Antimicrobial Peptides. *Molecular Informatics* 36, no. 1–2: 1600011.

9

Fuzzy Systems

9.1 Introduction

It may appear rather strange that we are addressing the subject of fuzzy systems in the context our discourse on shifting paradigms in scientific computing in physics and chemistry. Scientists are trained to accept something as true (truth value 1) or false (truth value 0) based on hard evidence. This crisp true false doctrine does not work everywhere, even in science. When faced for example with the problem of categorizing a given material or a molecule based on specific properties the answer may not always be amendable to such a crisp description. Thus, based on measured mechanical hardness, a material may be termed super hard, hard, not so hard, or soft. Take for another example the case of polarity of a molecule. The molecule, based on a well-defined criterion of polarity, may be categorized as highly polar, polar, not so polar, or nonpolar. Thus categorization demands, in both the examples, a more continuous transition in the truth value than a discrete one, like hard or soft, and polar or nonpolar. It is not just for categorizing a molecule or a material; even for describing the appropriate conditions for carrying out a chemical reaction we may find it convenient and appropriate to make use of a fuzzy description such as temperature should be very high, not so high, moderate, or low. Thus, the concept of fuzziness need not be forced into the domain of science—it naturally creeps into scientific discourse.

The central dogma of fuzziness is to admit a gradual and continuous transition rather than sticking to the traditional concept of abrupt or sharp change from one Boolean value to another. The concept of fuzziness has been exploited mathematically in developing fuzzy systems which include fuzzy sets, fuzzy logic, fuzzy algorithms, and fuzzy control.

In ordinary set theory, an element is either a member of the set or not. The membership is Boolean—either 0 (not a member) or 1 (a member). There is nothing in between—there is no gray area. Similarly, in ordinary or crisp logic, a proposition is either true or false. The idea that a proposition may be partially true or an element may be a member of a set to an extent or degree are inadmissible by ordinary logic or ordinary set theory. The idea of fuzziness allows us to generalize the traditional set theoretic or crisp logic-based description and analysis by incorporating in it the notion of partial truth in the context of logic or the idea of an element being a member of a set only to a certain degree or extent in the set theoretic paradigm. The traditional set theory propounded by George Cantor is rooted in Boolean or crisp logic. The fuzzy set theory advanced by L. Zadeh [1,2] is likewise rooted in fuzzy logic. The fuzzy set theory and the fuzzy logic can be viewed as extensions of their traditional counterparts which allow the science of continuum to enter what was traditionally believed to belong to discrete disciplines. These extensions have proved to be valuable and have found many applications. The horizon of applicability is still expanding.

Since "fuzziness" implies vagueness and uncertainty, an often-asked question concerns whether the fuzzy systems could have been handled by the traditional and well-established probability theory. The answer depends on what the source of the fuzziness or uncertainty is. Probability theory is adequate in handling uncertainties or fuzziness in systems or situations where it arises from underlying random processes or purely chance events. There are system or situations where the vagueness and uncertainty have a nonrandom origin—such as a lack of complete information about the system being investigated, the information that is available being not entirely reliable, or perhaps the language used to describe the system is inherently imprecise. Fuzzy set theory and fuzzy logic are systematizations that enable us to handle all such systems or cases of uncertainties, imprecision, and partial truth elegantly. In all such cases, a traditional mathematical model is usually hard to come by. Take for example, the case where the parameters defining system and the initial values of system variables of interest are uncertain, incomplete, imprecise, or vague. A traditional mathematical model would not be able to predict the future behavior of a system under such conditions. Systematization based on fuzzy set theory and logic provides avenues for predicting at least approximately the future behavior of the system even in the midst of all the uncertainties and impressions. Sometimes it may so happen that rules to be invoked for drawing inferences or making decisions about the system are fundamentally imprecise or incomplete. In such cases, traditional methods of handling the decision-making process would not work. Fuzzy logic could still enable us to arrive at a decision in such cases, by adopting fuzzy techniques for estimating system parameters that provide guidance in decision-making even though precise and complete information is unavailable. It is not just the lack of complete and precise information that creates problems. A system (almost all real-life systems are) may be too complex with strong nonlinearities. In handling such systems within the confines of the traditional paradigm, the only way to proceed is by linearization which very often obliterates specific features of the response of the system that are essentially due to the nonlinearity. Fuzzy techniques can handle such systems as it is without sacrificing the crucial nonlinear elements.

9.2 Fuzzy and Crisp Sets and Membership Function

In ordinary set theory, a set is a well-defined collection of elements such that an element of the universe of discourse (UOD) or "universe" is either a member of the set or not a member of the set. One does not fail to notice the inherent crispness in the definition itself. Let us consider symbolic representation of such a crisp set. Given a set A comprising elements $x_1, x_2, \ldots, x_n$, we symbolize the set as

$$A \equiv \{x_1, x_2, \ldots, x_k, \ldots, x_n\} \tag{9.1}$$

We may then say that $x_1, x_2, \ldots, x_k, \ldots, x_n$ are members of the set A. The membership of the set may be determined by some property $p(x)$ that the elements of the set A must possess. Symbolically this is represented by

$$A = \{x \mid p(x)\} \tag{9.2}$$

which translates into: A is a set of all x having the property encoded by $p(x)$. For example, A may be the set of all even numbers so its symbolic transcription is

$$A = \{x \mid x \text{ is an even number}\} \tag{9.3}$$

Out of the UOD, i.e., the collection of all numbers (integers), every number either belongs to the set A (if it is even) or does not belong to it (if it is odd). The UOD thus separates into two types, members and nonmembers of the set A. Based on the above definition there is no other possibility. A fuzzy set is a generalization of the concept of an ordinary or crisp set—a generalization that allows each member of the UOD to have a degree (m) of membership of set. The degree of membership or simply the degree (m) of an element is a real number [0,1]. Here $m = 0$ means that the element x is definitely not a member of the set while $m = 1$ implies that the element x completely belongs to the set. Note here that in the crisp set theory the degree of membership of an element could only assume either the value 0 or the value 1 while in the fuzzy set theory, the degree can assume any value between 0 and 1 (both inclusive) continuously. The degree of membership is determined by the value of the membership function $m(x)$ of the set in question. The membership function maps every element x of the UOD to a number m in [0,1]. The number m represents the degree of membership of x for the set. The idea of assigning every element in the set a degree of membership is central to the paradigm of fuzzy sets and fuzzy systems. It makes room for accommodating something that may not be black or white, but something in-between (gray, for example). From the very basic idea of the degree of membership, it becomes possible to define various operations and properties of the fuzzy sets and systems. Some of these have equivalents in the ordinary set theory and logic while some are exclusive—they are special features of fuzzy systems.

Let us consider for example the fuzzy set A of elements with a metallic character and the UOD being a collection of n number of elements ($n = 12$, say); the UOD is $U = \left(\text{Li, Na, K, Cs, Mg, Be, As, Sb, I, N, F, C} \right)$, while the fuzzy set A may be defined as

$$A \equiv \left\{ x \mid x \text{ is an element with metallic character} \right\}. \tag{9.4}$$

Based on some definite measure of metallic character we can assign to each element in U a membership $m(x)$ and describe the fuzzy set A as

$$A \equiv \{\text{Li/1.0, Na/1.0, ..., F/0.0}\}. \tag{9.5}$$

In the above representation, each entry (a symbol) is followed by a number m that measures the degree of membership of the element. The membership degrees for N and F are zero, which means that they are definitely not metallic and are not member of A. The degree for K, Na is 1, so that they entirely belong to the set A. As, Sb, and I, for example, have membership degrees that fall between 0 and 1 and are considered to have partial metallic character. The membership degree of x may be thought of as the value of a membership function $m(x)$ where x is a measure of some physical or chemical attribute of x. The fuzzy set A may also be represented as a collection of ordered pairs.

$$A = \{(x, m_A(x), x \in U)\}. \tag{9.6}$$

Each pair $(x, m_A(x))$ is called a singleton. In dealing with crisp sets the membership function becomes redundant and is omitted in the definition of the set. The reader will not fail to note that in our example set A, the elements in the UOD are discrete and finite in number, so the cardinality of U represented by $|U|$ is finite. Even with discrete members, the cardinality of U may be infinite as in the case where U is the collection of all natural numbers beginning with 0. Even if we define U as the collection of all real numbers between -1 and $+1$, the elements are nondiscrete; there are infinitely many of them in U, so the cardinality of U becomes infinite once again. Thus, discreteness or continuous nature of the elements of a set has nothing to do with its cardinality. It can be infinite both for sets with discrete or continuous members.

In our example set (A) we assigned discrete values (1, 0, 0.5, etc.) for the membership degree of each element without specifying any explicit functional form of $m_A(x)$. More often than not, the membership values (degrees) are better described by a continuous function $m(x)$. Consider for example a group of people (X) of different ages (x). We may wish to describe their youthfulness by assigning to each individual an age-dependent membership value [3]. It can be done by defining a continuous function of the age x of an element X as

$$m(x) = \begin{cases} 1.0 \text{ for } 0 < x \le 30 \\ \dfrac{1}{1 + \dfrac{(x-30)^2}{30}} \text{ for } x > 30. \end{cases} \tag{9.7}$$

Given the age x, the function $m(x)$ assigns a value $m(x)$ that expresses the degree of the youthfulness of an individual X. We can represent the values in a discrete tabular form as displayed in Table 9.1.

A second example, illustrating the concept of membership function $m(x)$ of a fuzzy set may be taken from chemistry which is a storehouse of such concepts. Take for example the group of elements listed in Table 9.2 and their electronegativity values, defined and

TABLE 9.1

Agewise Membership Degree in a Group

Age	Membership Degree
20	1.0
30	1.0
35	0.55
40	0.23
50	0.00

TABLE 9.2

Membership Function $m(x)$ of Elements with Different Electronegativity Values (x)

Element	Electronegativity	Value of the Membership Function $m(x)$
H	2.20	0.40
Si	1.98	0.51
Li	0.98	1.0
Fr	0.70	1.0
I	2.66	0.36
Na	0.93	1.0
Be	1.57	0.75
As	2.18	0.45
F	3.98	0.25
Se	2.55	0.39

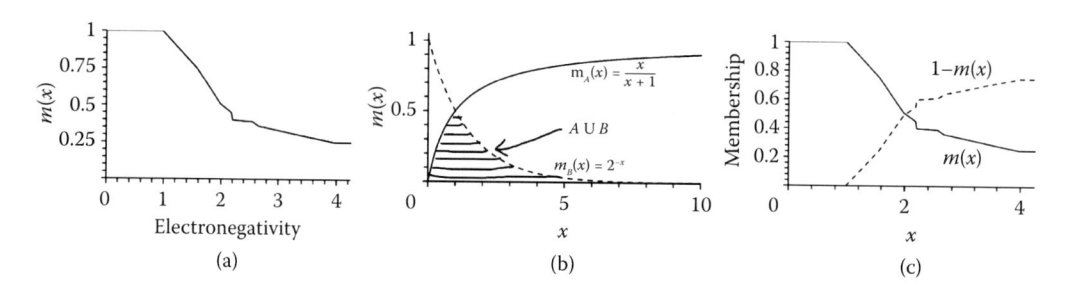

FIGURE 9.1
(a) Values of membership function defining the degree of an element for the set of metals based on electronegativity. (b) Displaying membership functions $m_A(x)$, $m_B(x)$ and identifying the fuzzy set $A \cup B$. (c) The complement of the set is defined by the membership function $(1 - m(x))$ which is obtained by reflecting the curve in (a).

measured on the Pauling's scale. Some of the elements have high values of electronegativity and some have lower values. Those with low electronegativity values are called metals and those with high electronegativity values are categorized as nonmetals.

There are elements for which the values of the electronegativity parameter may not be high or low—rather, they are in between. It is therefore not possible to categorize elements strictly into two classes, metals and nonmetals, but a more flexible—more fuzzy categorization would be more in keeping with the chemical reality. In such a situation, a membership function that assigns a degree of membership of an element to the set of elements with metallic properties can be more useful as well as meaningful. We define the membership function $m(x)$, where x is the electronegativity parameter of an element, as follows:

$$m(x) = \begin{cases} 1.0 \text{ if } 0 < x \le 1 \\ \dfrac{1}{1+(x-1)^2} \text{ if } x > 1 \end{cases}. \quad (9.8)$$

The degrees of membership calculated with the function defined in Equation 9.8 for 10 elements listed in Table 9.2 are reported under the third column of the same table. Li, Na, and Fr gave the degree equal to 1 and they may be rightly categorized as metals, while F and I have degrees equal to 0.25 and 0.36, respectively, so that they are to be treated as nonmetals. As and Se on the other hand have degrees equal to 0.45 and 0.39, respectively. They are thus expected to display some metallic as well as nonmetallic properties. Si behaves similarly. It is certainly possible to come up with even better membership function—but there is no denying that a crisp set classification will not reflect the chemical reality which is much better represented by a fuzzy classification scheme based on a membership function. The function $m(x)$ is displayed in Figure 9.1a as a continuous curve.

9.3 Relation in Fuzzy Sets

Basic fuzzy set relations are defined following the definitions of such relation for the ordinary or the crisp sets, using the membership function $m(x)$ for an element. They appear as generalizations of relation of ordinary sets and naturally subsume the crisp set counter parts.

Let us consider the equality relation of two fuzzy sets A and B. If $m_A(x) = m_B(x)$ for every x in $U(x \in U)$ we have $A = B$, $m_A(x)$ being the membership function for the fuzzy set A and $m_B(x)$ the membership function for the fuzzy set B. The equality relation for two ordinary sets A,B then becomes a special case of the equality relation for fuzzy sets A,B where either $m_A(x) = m_B(x) = 1$ or $m_A(x) = m_B(x) = 0$, depending on whether the element x is in the sets under consideration.

The concept of fuzzy subsets can be similarly defined using a fuzzy membership function $m(x)$. Thus, fuzzy set A is a subset of B (symbolically $A \subseteq B$) if and only if for every x in $U(x \in U)$, we have the condition $(m_A(x) \leq m_B(x))$ satisfied.

This definition of fuzzy subset also covers the subset definition in ordinary sets. Thus if x is a member of both A and B, $m_A(x) = m_B(x) = 1$, satisfying the condition $m_A(x) = m_B(x)$. On the other hand, for an element which is in B, but not in A, $m_B(x) = 1$, $m_A(x) = 0$ (satisfies the condition $m_A(x) < m_B(x)$). A corollary of the fuzzy subset relation is the condition for equality of two fuzzy sets: $A = B$, if and only if $A \subseteq B$ and $B \subseteq A$, mirroring a similar condition of equality in the case of crisp sets. Thus, these fuzzy set relations appear as generalizations of the corresponding crisp set relations with the qualifier that the membership functions in crisp sets have only two values, 0 and 1, while in fuzzy sets $m_A(x)$ can have any value between 0 and 1, both inclusive.

The idea of support of a fuzzy set A is important and illustrates its connection with crisp sets. Thus, the support of fuzzy set A is defined to be the crisp set comprised of every element x in the UOD (U) for which $m_A(x) > 0$, $m_A(x)$ being the membership function for A. If the fuzzy set A has only a single element of U as its support, the set is called a fuzzy singleton. Thus, if $U = (x_1, x_2, x_3, x_4, x_5)$ and x_5 is the support of the fuzzy set A with the membership function $m(x_i)$, we can formally write for the fuzzy singleton $A = \{x_5/m_A(x_5)\}$.

9.3.1 Basic Operations in Fuzzy Sets

Some of these basic operations have their counterparts in the domain of ordinary sets as well, while some are unique to the fuzzy set. We will first consider the fuzzy operations which have their parallels in ordinary or crisp set theory. They are union, intersection, and complement.

Let U be the UOD, x be an element of U and $m_A(x)$ be the degree of membership of x for the fuzzy set A. We can represent A as

$$A = \{x/m_A(x) | x \in U\}. \tag{9.9}$$

Similarly,

$$B = \{x/m_B(x) | x \in U\}. \tag{9.10}$$

The union of A and B is a new a fuzzy set represented symbolically as $A \cup B$ where

$$A \cup B = \left\{ x / \max\left(m_A(x), m_B(x)\right) \big| x \in U \right\}. \tag{9.11}$$

The intersection of A and B is again a new fuzzy set symbolized as $A \cap B$ where

$$A \cap B = \left\{ x / \min\left(m_A(x), m_B(x)\right) \big| x \in U \right\}. \tag{9.12}$$

The complement of A is a fuzzy set A^C where

$$A^C = \{x/(1 - m_A(x)) | x \in U\}. \tag{9.13}$$

A^C is thus a new fuzzy set with the membership function $(1 - m_A(x))$. Let us illustrate the three basic operations in discrete mode to bring out more clearly what they actually mean and how they operate. Consider two fuzzy sets A_1 and B_1 where

$$A_1 = \{x_1/0.4, x_2/0.6, x_3/0.0\}$$
$$B_1 = \{x_1/0.9, x_2/0.8, x_3/1.0\} \tag{9.14}$$

what is the fuzzy set A' produced by union of A_1 and B_1? The union $A_1 \cup B_1$ has three elements x_1, x_2, and x_3 for which the membership degree, by definition are

$$m_{A_1 \cup B_1}(x_1) \begin{array}{l} = \max(m_{A_1}(x_1), m_{B_1}(x_1)) \\ = \max(0.4, 0.9) = 0.9. \end{array} \tag{9.15}$$

Similarly,

$$m_{A_1 \cup B_1}(x_2) \begin{array}{l} = \max(m_{A_1}(x_2), m_{B_1}(x_2)) \\ = \max(0.6, 0.8) = 0.8 \end{array} \tag{9.16}$$

and

$$m_{A_1 \cup B_1}(x_3) = 1.0. \tag{9.17}$$

Thus

$$A' = A_1 \cup B_1 = \{x_1/0.9, x_2/0.8, x_3/1.0\}. \tag{9.18}$$

On the same pair of fuzzy sets, we can similarly work out what the fuzzy set is that represents the intersection of A_1 and B_1. Operationally, we proceed in the same manner to find the fuzzy set $A'' = A_1 \cap B_1$ comprising the three elements x_1, x_2, and x_3 by replacing the max operation by the min operation. It is easy to find that

$$A'' = A_1 \cap B_1 = \{x_1/0.4, x_2/0.6, x_3/0.0\}. \tag{9.19}$$

Finally let us try to find what the complements of A_1 and B_1 are. We make use of the membership degrees of the elements x_1, x_2, and x_3 in A_1 or B_1 to find out the complements A_1^C of A_1 and B_1^C of B_1. Recalling that the complement has membership degree $m^C(x) = 1 - m(x)$, we have

$$A_1^C = \{x_1/0.6, x_2/0.4, x_3/1.0\}$$
$$B_1^C = \{x_1/0.1, x_2/0.2, x_3/0.0\}. \tag{9.20}$$

It is also possible to make use of a continuous representation of the membership degree instead of the discrete representation used in the illustrations. Thus, we may consider for example fuzzy sets A and B defined on the interval $X = [0,10]$ of real number with membership functions $m_A(x)$ and $m_B(x)$ where

$$m_A(x) = \frac{x}{x+1}$$
$$m_B(x) = 2^{-x}. \tag{9.21}$$

The union of A and B is characterized by the degree membership defined by

$$m_{A \cup B}(x_1) = \max\left(m_A(x), m_B(x)\right)$$
$$= \max\left(\frac{x}{x+1}, 2^{-x}\right). \tag{9.22}$$

We can graphically represent the situation by displaying graphs of $m_A(x) = \frac{x}{x+1}$, $m_B(x) = 2^{-x}$ in the same figure and marking off the portions representing the set A, B and $A \cup B$ by noting that the membership degree of elements in $A \cup B$ is given by $\max\left(\frac{x}{x+1}, 2^{-x}\right)$ for any admissible value of x (see Figure 9.1b).

We can identify the union by taking the larger portions of the curves, i.e., portions identified by the solid lines. It is immediately apparent that $A \cap B$ (the interaction) is bounded by the dashed (----) portion in the same figure. The reader can plot the membership functions for the complements of A and B by noting that membership degree for A^C and B^C are following

$$m_{A^C}(x) = 1 - \frac{x}{x+1} = \frac{1}{x+1}$$
$$m_{B^C}(x) = 1 - 2^{-x} = \frac{2^x - 1}{2^x}. \tag{9.23}$$

We may close the discussion on the basic fuzzy set operations by displaying the membership function of the complement of the fuzzy set defined in Table 9.2 with the membership function displayed in Figure 9.1a. The complement of the set is defined by the membership function $(1 - m(x))$ which is obtained by reflecting the curve in Figure 9.1a about the value of $m(x) = \frac{1}{2}$ (see Figure 9.1c).

9.3.2 Properties of Fuzzy Sets

Some of the properties of fuzzy sets have their counterpart in the theory of crisp sets as expected. Many of these properties refer to the properties of the union, intersection, and the complements of fuzzy sets. We have listed these properties as follows:

$$\text{Commutability} \quad \begin{aligned} A \cup B &= B \cup A \\ A \cap B &= B \cap A \end{aligned} \tag{9.24}$$

$$\text{Associability} \quad \begin{aligned} A \cup (B \cup C) = (A \cup B) \cup C \\ A \cap (B \cap C) = (A \cap B) \cap C \end{aligned} \qquad (9.25)$$

$$\text{Distributivity} \quad \begin{aligned} A \cup (B \cap C) = (A \cup B) \cap (A \cup C) \\ A \cap (B \cup C) = (A \cap B) \cup (A \cap C) \end{aligned} \qquad (9.26)$$

$$\text{Idempotency} \quad \begin{aligned} A \cup A = A \\ A \cap A = A \end{aligned} \qquad (9.27)$$

$$\text{Double complement (Involution)} \left(A^C \right)^C = A \qquad (9.28)$$

$$\text{Transitivity if } A \subseteq B \subseteq C, \text{ then } A \subseteq C \qquad (9.29)$$

$$\text{De Morgan's laws} \quad \begin{aligned} (A \cup B)^C = A^C \cap B^C \\ (A \cap B)^C = A^C \cup B^C \end{aligned} \qquad (9.30)$$

Identity: To define this property, we recall first that an empty fuzzy set, symbolized as Φ, has none of the elements of the UOD (U) as its member, i.e., for all of them the degree is 0. Formally

$$\begin{aligned} \Phi = \left\{ x/0.0 | x \in U \right\} \\ U = \left\{ x/1.0 | x \in U \right\}. \end{aligned} \qquad (9.31)$$

With the definition of Φ and U in place we can define identity by the following properties of intersection and union of Φ or U with fuzzy set A. Thus, we have,

$$\begin{aligned} A \cup \Phi &= A \\ A \cap \Phi &= \Phi \\ A \cup U &= U \\ A \cap U &= A. \end{aligned} \qquad (9.32)$$

The proofs of all the properties reported in this section follow from definitions of union, intersection, and complement in a rather straightforward manner and readers are advised to convince themselves by working them out explicitly. The properties or operations of fuzzy sets described here have their counterparts in ordinary set theory. There are, however, some properties or operations special to the fuzzy sets, their crisp set counterparts being nonexistent. The following section is devoted to introducing such operation or properties.

9.3.3 Unique Fuzzy-Only Operations

We define here three operations of the fuzzy-only variety, namely concentration of a set A denoted by $CON(A)$, dilation of a set A represented as $DIL(A)$ and normalization of a fuzzy set A symbolized as $NORM(A)$. These operations are frequently used and deserve attention.

9.3.4 Concentration

Let A be a fuzzy set with elements x drawn from the UOD (U) and characterized by the membership function $m_A(x)$ where $0 \leq m_A(x) \leq 1$. Let $m_A^2(x)$ denote $m_A(x) \times m_A(x)$. Then the concentration of the fuzzy set A is defined as

$$\text{CON(A)} = \left\{ x / m_A^2(x) \mid x \in U \right\} \tag{9.33}$$

Note that $m_A^2(x) \leq m_A(x)$ since $0 \leq m_A(x) \leq 1$, the equality $m_A^2(x) = m_A(x)$ holding true only for $m_A(x) = 0$ or 1. Plots of $m_A(x)$ and $m_A^2(x)$, on the same figure (Figure 9.2a) then make the physical implication of $CON(A)$ clear.

$CON(A)$ has clearly a steeper and narrower graph than A. The concentration operation on A therefore produces a narrower graph which is also steeper as expected from the fact that the percentage of reduction of the membership degree on operation with CON is much higher for lower values of $m_A(x)$ compared to higher values of $m_A(x)$ as evident from the quantity $\dfrac{\left(m_A(x) - m_A^2(x) \right)}{m_A(x)} \times 100\%$. The CON operation becomes useful when we wish to focus on those elements x in A that satisfy the condition on the elements being "very tall," "very high," "very old" and so on. Referring back to the entries in Table 9.1, we can immediately realize what an application of CON to the set of elements with varying degrees of metallic or electropositive character will accomplish—it will generate the set of "very electropositive elements" or elements that are very metallic in character. Similarly, a set of polar molecules could be turned into another set of "very polar" molecules and so on. The operation dilation on the fuzzy set A or $DIL(A)$ does the opposite of what the concentration (CON) operation does on A. This becomes clear from the definition of $DIL(A)$ as given below:

$$\text{DIL(A)} = \left\{ x / \sqrt{m_A(x)} \mid x \in U \right\} \tag{9.34}$$

The operation produces a fuzzy set, the membership function of which is $\sqrt{m_A(x)}$ instead of $m_A(x)$ defined for the fuzzy set A. The square root operation increases the membership degree of an element (based on $m_A^{1/2}(x)$) very significantly if the original degree (based on $m_A(x)$) was low. Thus, if it was 0.04 based on $m_A(x)$, it becomes 0.2 based on $m_A^{1/2}(x)$—an

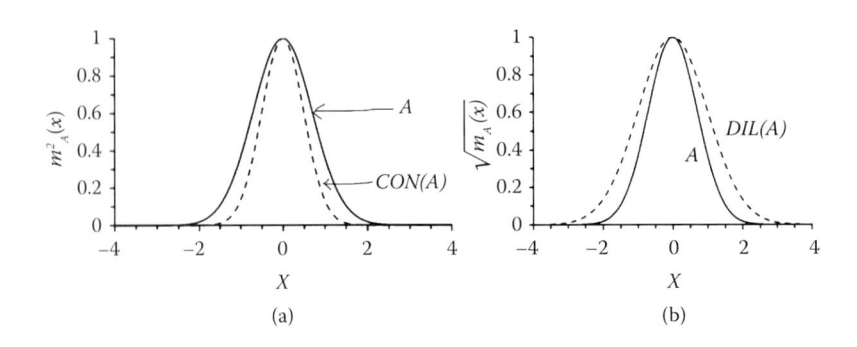

FIGURE 9.2
(a) Concentration of a set A: $CON(A)$, (b) dilation of a set A: $DIL(A)$.

increase of five times. The graph becomes flatter and wider following a dilation operation as depicted in Figure 9.2b.

The third operation unique to a fuzzy set is called normalization. Its operation on a fuzzy set A produces a fuzzy set termed $NORM(A)$ in which the membership function is normalized by the maximum value of the function $m_A(x)(max)$ defined for set A. Operationally, it amounts to dividing $m_A(x)$ by max. Formally the definition of $NORM(A)$ is

$$NORM(A) = \{x/(m_A(x)/max)|x \in U\} \tag{9.35}$$

with max = max $x \in U\{m_A(x)\}$.

As the reader may note, the operation of normalization is somewhat akin to the normalization of vectors and in a way reduces all fuzzy sets to the same base just as normalization of an arbitrary vector (x, y) will reduce it to a vector of unit length—the normalized vector being $\left(\dfrac{x}{\sqrt{x^2+y^2}}, \dfrac{y}{\sqrt{x^2+y^2}}\right)$.

Before leaving this section, we will introduce a different way of viewing fuzzy sets by using the ideas of union and singletons already defined. According to this view, a fuzzy set A can be regarded as the union of all its singleton members. Formally we can write

$$A = U_U \{x/m_A(x)\} \tag{9.36}$$

where U_U signifies unions over all the elements in the UOD (U). If the UOF is finite, let n be the cordiality of U ($n = |U|$). Then we may also write

$$A = \sum_{i=1}^{n} \{x_i/m_A(x_i)\} \tag{9.37}$$

where the summation sign (Σ) is to be interpreted as performing unions of the arguments over the domain $i = 1, n$. If the elements in U take on continuous values, Σ in Equation 9.37 may be replaced by the $\int$ sign as is done in ordinary calculus with the qualifier that $\int_U$ represents the act of taking the union of the argument $(x/m_A(x))$ over the domain of U. In that case we can write

$$A = \int_U \{x(m_A(x))\} \tag{9.38}$$

$\int_U$ is not to be confused with integration which in calculus is indicated by $\int dx$.

9.4 Fuzzy Relations

We have already introduced the idea of extending crisp or ordinary sets into the fuzzy domain by bringing in a membership function that generates degrees of membership, or simply degree for the set elements. One can expect therefore that ordinary set theoretic relations can also be extended to the fuzzy domain, producing fuzzy relations. It would be convenient and natural therefore to begin with a review of ordinary (nonfuzzy) relations and then look for their fuzzy counterparts.

9.4.1 Nonfuzzy Relations

An important idea in ordinary set theory is the Cartesian product and the relation defined on it. Let A and B be two crisp sets. The Cartesian product of A and B is denoted by symbol $A \times B$ where

$$A \times B = \{(a, b) | a \in A, b \in B\}. \tag{9.39}$$

Let the number of elements in A be n and that in B be m. By definition, $A \times B$ will contain nm elements. These elements are ordered pairs (a, b) in which every element of A and B have been picked up in the pair formation. Any subset of elements in $A \times B$ constitute a relation R from A to B. If A and B are nonempty sets and $A \neq B$, the Cartesian product is noncommutative, i.e., $A \times B \neq B \times A$.

Let us consider two sets $A = \{a, b\}$, $B = \{1, 2\}$ with cardinalities. $|A| = 2$, $|B| = 2$. Then

$$A \times B = \{(a, 1), (a, 2), (b, 1), (b, 2)\} \text{ with} |A \times B| = 4 = |A| \times |B|. \tag{9.40}$$

To make the idea clearer let us consider the set A of three people

$$A \{Ramesh, Bob, Angel\}, |A| = 3 \tag{9.41}$$

and another set, B, of four musical instruments,

$$B \{guitar, violin, piano, flute\}, |B| = 4. \tag{9.42}$$

The Cartesian product $A \times B$ can be represented as the collection or set of 12 ordered pairs of elements, as detailed below:

$$A \times B = \left\{ \begin{array}{l} (Ramesh, guitar), (Ramesh, violin), (Ramesh, piano), \\ (Ramesh, flute), (Bob, guitar), \ldots, (Angel, flute) \end{array} \right\}. \tag{9.43}$$

Instead of using the set form, we can also represent the same Cartesian product in a diagrammatic form or in a tabular form. We display the diagrammatic mode of representation first (Figure 9.3a).

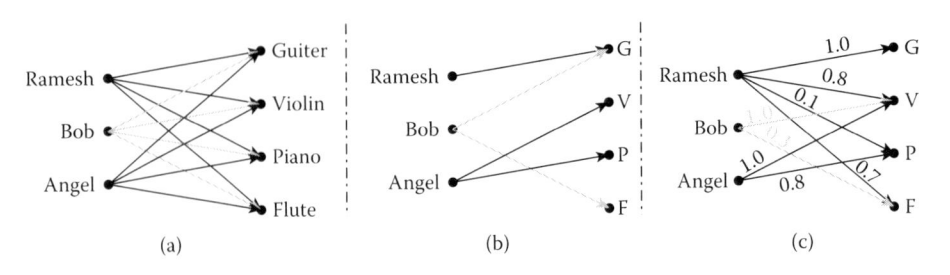

FIGURE 9.3
(a) The diagrammatic representation of $A \times B$ and (b) the diagrammatic mode of representation for the subset of elements in $A \times B$.

The diagrammatic representation of $A \times B$ is clear and easy to follow, and is often used to represent the cross product of two sets. The tabular form of representation is equally lucid as shown below:

$$
\begin{array}{c}
\overbrace{\qquad\qquad}^{B} \\
\begin{array}{c|cccc}
 & \text{Guitar} & \text{Violin} & \text{Piano} & \text{Flute} \\
\text{Ramesh} & \surd & \surd & \surd & \surd \\
\text{Bob} & \surd & & & \surd \quad A \times B \\
\text{Angel} & \surd & \surd & \surd & \surd \\
\end{array}
\end{array}
$$

As can be easily seen from the tabular representation, A × B has 12 elements obtained by combining the row entries with those in the column.

A relation R, as we have already stated, is a subset of the elements in the Cartesian product. In our example, we may define such a relation (R) by explicating the instruments played by each member, and then giving it a set theoretic representation; the relation R can be, for example

Ramesh	Guitar
Bob	Guitar, Flute
Angel	Piano, Violin

In set theoretic representation R has five elements from the cross product as shown below:

$$R \equiv \big\{ (\text{Ramesh, guitar}), (\text{Bob, guitar}), (\text{Bob, flute}), (\text{Angel, piano}), (\text{Angel, violin}) \big\}. \quad (9.44)$$

In tabular form the relation R is represented as follows (G for Guitar, V for violin, etc.). Rows are marked by persons and columns by the musical instruments.

$$
\begin{array}{c|cccc}
 & \text{Guitar} & \text{Violin} & \text{Piano} & \text{Flute} \\
\text{Ramesh} & \surd & & & \\
\text{Bob} & \surd & & & \surd \quad A \times B \\
\text{Angel} & & \surd & \surd & \\
\end{array}
$$

The diagrammatic mode of representation is straightforward and has five edges (Figure 9.3b).

9.4.2 Relation Matrix

For two sets A and B, relation R is as we have seen is a subset of elements in $A \times B$. If A has n elements $(a_1, a_2, \ldots, a_n)$ and B has m elements $(b_1, b_2, \ldots, b_m)$, then a relation $R(A, B)$ can be associated with a two-dimensional rectangular table or matrix R with n rows and m column. As an example, let us consider the set $A = \{1, 2, 3, 4\}$. $A \times A$ then has 16 elements. We may define a relation R as follows:

$$R = \left\{ \frac{(a,b)}{b} = a + 1 \,\middle|\, a, b \in A \right\}. \qquad (9.45)$$

Clearly then, R becomes a subset of three elements in $A \times A$ with

$$R = \{(1,2),(2,3),(3,4)\}. \tag{9.46}$$

The relation matrix in this case is a 4×4 matrix with all the elements except R_{12}, R_{23}, R_{34} set as zero, and $R_{13} = R_{23} = R_{34} = 1$.

$$R = \begin{array}{c} \\ 1 \\ 2 \\ 3 \\ 4 \end{array} \begin{array}{cccc} 1 & 2 & 3 & 4 \\ \left[\begin{array}{cccc} 0 & 1 & 0 & 0 \\ 0 & 0 & 1 & 0 \\ 0 & 0 & 0 & 1 \\ 0 & 0 & 0 & 0 \end{array} \right] \end{array}. \tag{9.47}$$

9.4.3 Operations on Nonfuzzy Relations

Let A and B be crisp sets. Given two relations R and S defined on $A \times B$ and represented by the appropriate relation matrices, the following operations are supported by R and S:

$$\text{Union } R \cup S \text{ where}$$
$$R \cup S(a,b) = \max\big(R(a,b), S(a,b)\big)$$
$$\text{Intersection } R \cap S \text{ where}$$
$$R \cap S(a,b) = \min\big(R(a,b), S(a,b)\big)$$
$$\text{Complement}: \bar{R} \text{ where}$$
$$\bar{R}(a,b) = 1 - R(a,b).$$

9.4.4 Composition of Relation

Let us consider ordinary or crisp sets A, B, and C, let R be a relation on $A \times B$ (or A to B), and S be a relation on $B \times C$ (or B to C). The composition of relations R and S is a new relation represented as $R \circ S$ with the definition that

$$R \circ S = \big\{(a,c) \mid (a,b) \in R, (b,c) \in S,\ a \in A,\ b \in B,\ c \in C\big\}. \tag{9.48}$$

The composition operation is not limited only to two relations, but can be extended to any number of relations, such as $S_1 \circ S_2 \circ S_3$ or $R_1 \circ R_2 \circ R_3$ and so on. The composition operation is associative which means $(S_1 \circ S_2) \circ S_3 = S_1 \circ (S_2 \circ S_3)$ so the use of parenthesis becomes redundant and we can straightaway write $S_1 \circ S_2 \circ S_3$.

A common form of composition relation is the max–min composition. It is convenient to make use of relation matrices to define the composition. Thus, with relations R and S already defined on $A \times B$ and $B \times C$, let R and S be the corresponding relation matrices. The max–min composition is then defined as $T = R \circ S$.

$$T(a,c) = \max_{b \in B}\big(\min\big(R(a,b),\ S(b,c)\big)\big) \tag{9.49}$$

To illustrate how it operates, let us consider R and S defined on the sets $\{1,3,5\} \times \{1,3,5\}$. We define R and S as

$$R : \{(a,b), b = a + 2\}, \; S : \{(a,b), a \leq b\}. \tag{9.50}$$

Then

$$R = \{(1,3),(3,5)\}$$
$$S = \{(1,3),(1,5),(3,5)\}.$$

The relation matrices for the two sets are

$$R = \; rows\begin{array}{c} 1 \\ 3 \\ 5 \end{array} \overbrace{\begin{array}{c} columns \\ \begin{array}{ccc} 1 & 3 & 5 \end{array} \end{array}}^{} \begin{bmatrix} 0 & 1 & 0 \\ 0 & 0 & 1 \\ 0 & 0 & 0 \end{bmatrix}; \; S = \; rows\begin{array}{c} 1 \\ 3 \\ 5 \end{array} \overbrace{\begin{array}{ccc} 1 & 3 & 5 \end{array}}^{columns} \begin{bmatrix} 0 & 1 & 1 \\ 0 & 0 & 1 \\ 0 & 0 & 0 \end{bmatrix}. \tag{9.51}$$

The max–min composition can now be used to find that

$$R \circ S = \begin{array}{c} 1 \\ 3 \\ 5 \end{array} \overbrace{\begin{array}{ccc} 1 & 3 & 5 \end{array}}^{} \begin{bmatrix} 0 & 0 & 1 \\ 0 & 0 & 0 \\ 0 & 0 & 0 \end{bmatrix} \tag{9.52}$$

The determination of $R \circ S$ involves determining the max–min product of R and S matrices wherein the row–column operation in matrix multiplication remains unchanged, but the product operation is replaced with min and addition by max. Thus

$$R \circ S(1,1) = \max\{\min(0,0) \; \min(1,0) \; \min(0,0)\}$$
$$= \max\{0,0,0\} = 0, \; \text{etc.} \tag{9.53}$$

Proceeding in this way, it is easy to find the other elements of the relation matrix

$$R \circ S - S \circ R = \begin{array}{c} 1 \\ 3 \\ 5 \end{array} \overbrace{\begin{array}{ccc} 1 & 3 & 5 \end{array}}^{} \begin{bmatrix} 0 & 0 & 1 \\ 0 & 0 & 0 \\ 0 & 0 & 0 \end{bmatrix}. \tag{9.54}$$

9.4.5 Fuzzy Relations from Ordinary Sets

We can now start extending the ordinary set relation to the fuzzy domain. As expected the membership function plays a crucial role in the process of extension. Let A and B be two crisp sets with the Cartesian product $A \times B$ (already defined) where

$$A \times B = \{(a,b)|\, a \in A, b \in B\} \tag{9.55}$$

A fuzzy relation (R, say) on $A \times B$ is a fuzzy subset of $A \times B$ as defined below:

$$R = \left\{ (a,b) / m_R(a,b) \middle| a \in A, b \in B \right\}. \tag{9.56}$$

Using the notion of singletons, we can write R also as a union of singletons over $A \times B$.

$$R = \cup_{A \times B} \left\{ (a,b) / m_R(a,b) \right\} \tag{9.57}$$

where $\cup_{A \times B}$ is the union over $A \times B$ and $m_R(a,b)$ is the membership function. Assuming that both A and B are finite with cardinalities $|A| = m$, $|B| = n$, we may adopt a different notation and write

$$R = \sum_{i=1}^{m} \sum_{j=1}^{n} \left\{ (a_i, b_j) / m_R(a_i, b_j) \right\} \tag{9.58}$$

Keeping in mind that Σ signifies taking set theoretic unions rather than summing numbers, we can, in the situation, represent R with an $m \times n$ matrix with $m_R(a_i, b_j)$ as elements. The rectangular relation matrix R then reads

$$R = \begin{bmatrix} m_R(a_1,b_1) & m_R(a_1,b_2) & \cdots & m_R(a_1,b_n) \\ m_R(a_2,b_1) & m_R(a_2,b_2) & \cdots & m_R(a_2,b_n) \\ \vdots & \vdots & \ddots & \vdots \\ m_R(a_m,b_1) & m_R(a_m,b_2) & \cdots & m_R(a_m,b_n) \end{bmatrix}. \tag{9.59}$$

If the sets A, B admit of continuous elements,

$$R = \int_A \int_B \left\{ (a,b) / m_R(a,b) \right\} \tag{9.60}$$

with $\int_A \int_B$ signifying taking unions rather than integration, as already emphasized.

Let us go back to the three persons and four musical instruments case and instead of associating a person with a particular instrument exclusively on a crisp "yes" or "no" basis, we may try to represent the association as a degree of membership to each person–instrument pair. With that we can define or represent the relation matrix R as

$$\text{Persons} \left\{ \begin{array}{c} \\ \\ \\ \end{array} \right. \begin{array}{c} \overbrace{\quad\quad\quad}^{\textit{Instruments}} \\ \begin{array}{cccc} G & V & P & F \end{array} \\ \begin{array}{l} \text{Ramesh} \\ \text{Bob} \\ \text{Angel} \end{array} \begin{bmatrix} 1 & 0.8 & 0.1 & 0.7 \\ 0 & 1 & 0 & 0.3 \\ 0 & 0.8 & 1.0 & 3 \end{bmatrix} \end{array}$$

As in the crisp set theoretic case, we can represent the person–musical instrument relation graphically as well, by introducing the notion of fuzzy graphs as displayed in Figure 9.3c. Fuzzy graphs, as depicted in Figure 9.3c, are directed graphs the vertices of which are represented by different elements of the two sets under consideration, while the graph edges correspond to different elements of the relation encoded by the relation matrix. Each edge is given a weight defining the membership degree of the specific pair connected by that edge. If the membership degree of a pair is 0, the corresponding edge does not appear in the fuzzy graph. The conversion of the relation matrix into a fuzzy graph or a fuzzy graph into a relation matrix is thus straightforward. We will illustrate the process with a simple example. Let A be a set where $A = \{a_1, a_2, a_3, a_4\}$ and R be a fuzzy relation on A where

$$R = \{(a_1, a_2) = 0.3, (a_1, a_3) = 0.3, (a_2, a_2) = 1.0, (a_2, a_4) = 0.6, (a_3, a_4) = 0.5, (a_4, a_2) = 0.2\}. \quad (9.61)$$

The relation matrix is easily formed as a 4×4 matrix

$$R = \begin{matrix} & \overbrace{\begin{matrix} a_1 & a_2 & a_3 & a_4 \end{matrix}} \\ \begin{matrix} a_1 \\ a_2 \\ a_3 \\ a_4 \end{matrix} & \begin{bmatrix} 0.0 & 0.3 & 0.1 & 0.0 \\ 0.0 & 1.0 & 0.0 & 0.6 \\ 0.0 & 0.0 & 0.0 & 0.5 \\ 0.0 & 0.2 & 0.0 & 0.0 \end{bmatrix} \end{matrix}.$$

The (directed) fuzzy graph (Figure 9.4) representing the relation R on A can also be easily drawn by using the R matrix.

9.4.6 Composition of Fuzzy Relations

Consider three ordinary sets A, B, and C. Assume that R is a fuzzy relation on $A \times B$, while S is a fuzzy relation on $B \times C$. The composition of R and S is then a fuzzy composite relation symbolically represented as $R \circ S$, where

$$R \circ S = \cup_{A \times C} \left\{ (a, c) \max_B \left\{ \min \left(m_R(a, b), m_S(b, c) \right) \right\} \right\}. \quad (9.62)$$

As in the case of crisp composition $\max_B \{y\}$ means the maximum value of y over all the elements of B. If the sets under consideration (A, B, C) are all finite, the definition of $R \circ S$ boils down to the formation of max–min product of the relation matrices R and S. The element of the max–min product of R and S are obtained by the normal rule of matrix multiplication except that the product operation of row elements of R and appropriate column elements of S is replaced by taking the min of $\left(R_{ik}, S_{kj} \right)_k$ and the subsequent

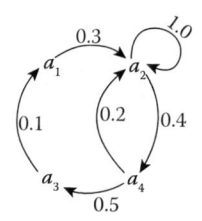

FIGURE 9.4
A fuzzy graph of the fuzzy relation R.

addition operation is substituted with taking the max of the elements generated by the min operation. The process can be elucidated with following example:

$$\text{Let } R = \begin{bmatrix} 0.2 & 0.6 \\ 0.4 & 0.8 \end{bmatrix}, \ S = \begin{bmatrix} 0.4 & 0.9 \\ 0.1 & 1.0 \end{bmatrix}. \tag{9.63}$$

Defined on $A \times B$, the element of R represents degrees of membership of (a_1, b_1), (a_1, b_2), (a_2, b_1) and (a_2, b_2). Similarly, element of S represents the degrees of (b_1, c_1), (b_1, c_2), (b_2, b_1), and (b_2, c_2).

The first element of $R \circ S$ represents the membership degree for the pair (a_1, c_1) which by the definition is obtained by first taking the min of $0.2(R_{11})$ and $0.4(S_{11})$, and the min of $0.6(R_{12})$ and $0.1(S_{21})$, and then taking the maximum of the two minimum generated. The two minima are 0.2 and 0.1 so the max(0.2,0.1) is 0.2. $R \circ S$ (1,1) is therefore 0.2. Similarly, $R \circ S$ (1,2) = 0.6, $R \circ S(2,1) = 0.4$, $R \circ S$ (2,2) = 0.8. In matrix form, we have the following:

$$R \circ S = \begin{bmatrix} 0.2 & 0.6 \\ 0.4 & 0.8 \end{bmatrix}. \tag{9.64}$$

Although the max–min definition of the fuzzy composition operation is popular and frequently used, it is neither unique nor universal. There are situations where the max–min operation would be better replaced with some other form of operation. Many alternatives have been proposed; the specific operation to be adopted largely depends on the problem at hand.

The fuzzy relations discussed so far were derived from crisp sets (A,B,C, etc.). We can also define and derive fuzzy relations from fuzzy sets as follows. Let A and B represent fuzzy sets of universe U_1 and U_2, respectively. The Cartesian product of fuzzy sets of A and B is defined as (fuzzy Cartesian product):

$$A \times B = \left\{ (a,b) / \min\left(m_A(a), m_B(b) \right) \mid a \in U_1, b \in U_2 \right\}. \tag{9.65}$$

If A, B were crisp sets, $m_A(a)$ or $m_B(b)$ would be 0 or 1, then $A \times B$ defined above would be reduced to the definition of cross product of ordinary sets A and B; but with fuzzy sets A and B, $A \times B$ defined above is a fuzzy relation from U_1 to U_2, and can be represented as a union over $U_1 \times U_2$:

$$A \times B = \cup_{U_1 \times U_2} \left\{ (a,b) / \min\left(m_A(a), m_B(b) \right) \right\}. \tag{9.66}$$

A fuzzy relation (R) from A to B can be defined in the following way:

$$R = \left\{ (a,b) / m(a,b) \mid m(a,b) \le m_A(a), \ m(a,b) \le m_B(b) \mid a \in U_1, b \in U_2 \right\}. \tag{9.67}$$

The Cartesian product defined here is binary. It can be generalized to a k-ary Cartesian product as well. Let $A_1, A_2, \ldots, A_k$ be fuzzy sets of universes $U_1, U_2, \ldots, U_k$ respectively. A generalized Cartesian product can then be defined as

$$A_1 \times A_2 \times \cdots \times A_k = \left\{ \begin{array}{l} (a_1, a_2, \ldots, a_k) / \min\left(m_{A_1}(a_1), m_{A_2}(a_2), \ldots, m_{A_k}(a_k) \right) \mid \\ a_1 \in U_1, a_2 \in U_2, \ldots, a_k \in U_k \end{array} \right\} \tag{9.68}$$

which can be expressed as a union over $A_1 \times A_2 \times \cdots \times A_k$.

A fuzzy binary relation can be similarly extended to generate fuzzy k-ary relation. An example of a fuzzy Cartesian product is provided in what follows. Let A be a fuzzy set with elements $\{a_1|0.2, a_2|6, a_3|0.35\}$ and B be another fuzzy set with $B = \{b_1|0.45, b_2|0.1\}$. The fuzzy Cartesian product $A \times B$ is then given by a 3×2 matrix R where

$$
R = A \times B = \begin{array}{c} \\ a_1 \\ a_2 \\ a_3 \end{array} \begin{array}{c} b_1 \quad b_2 \\ \begin{bmatrix} 0.2 & 0.2 \\ 0.4 & 0.1 \\ 0.35 & 0.1 \end{bmatrix} \end{array} \tag{9.69}
$$

The elements of the R matrix are computed as follows:

$$
\begin{aligned}
R(a_1, b_1) &= \min\left(m_A(a_1), m_B(b_1)\right) = \min(0.2, 0.45) = 0.2 \\
R(a_1, b_2) &= \min\left(m_A(a_1), m_B(b_2)\right) = \min(0.2, 0.7) = 0.2 \\
R(a_2, b_1) &= \min\left(m_A(a_2), m_B(b_1)\right) = \min(0.4, 0.45) = 0.4 \\
R(a_3, b_1) &= \min\left(m_A(a_3), m_B(b_1)\right) = \min(0.35, 0.45) = 0.35 \\
R(a_2, b_2) &= \min\left(m_A(a_2), m_B(b_2)\right) = \min(0.6, 0.1) = 0.1 \\
R(a_3, b_2) &= \min\left(m_A(a_3), m_B(b_2)\right) = \min(0.35, 0.1) = 0.1.
\end{aligned} \tag{9.70}
$$

9.4.7 Operations on Fuzzy Relations

Let R and S be fuzzy relations on $U \times V$. Then the following operations on R, S can be defined as

$$
\left. \begin{aligned}
&\textit{Union: } m_{R \cup S}(U, V) = \max\left(m_R(U, V), m_S(U, V)\right) \\
&\textit{Intersection: } m_{R \cap S}(U, V) = \min\left(m_R(U, V), m_S(U, V)\right) \\
&\textit{Complement: } m_{R^C}(U, V) = 1 - m_R(U, V)
\end{aligned} \right\}. \tag{9.71}
$$

9.5 Fuzzy Logic

9.5.1 Introduction

Logic embodies the science of reasoning. Mathematical or symbolic logic has evolved into a rather powerful computing paradigm which not only helps us describe events in the real world but also enables us to draw inference from a given set of facts. It turns out that logic, like the set theory, can be broadly classified into crisp and fuzzy logics. In fact, one can recognize several similarities between ordinary (crisp) set theory and ordinary (crisp) logic on one hand, and fuzzy sets and fuzzy logic on the other. Thus, just as in ordinary (nonfuzzy or crisp) set theory, an element either exists in a set or it does not; a proposition in ordinary logic either holds true or false. Similarly, the operations of union ($\cup$) and intersection ($\cap$) of ordinary sets have their equivalents in ordinary logic in the form of logical OR and logical AND operations. The complement operator of ordinary set theory has likewise a counterpart in ordinary logic in the form of the NOT operator; just as crisp sets are defined on a 2-state membership values (0 or 1), crisp logic is built on a 2-state truth value (true or false). We can then argue that fuzzy logic is an extension of ordinary logic, based on multiple truth values (not just true or false, but on, say, true, very true, false, partly true, or partly false), just as fuzzy sets are extensions of ordinary sets based on more than two membership values (not only 0 or 1 but any value from 0 to 1). The similarities or correspondences between fuzzy sets and fuzzy logic can be summarized as follows:

Fuzzy set	Fuzzy logic
Multiple membership values	Multiple truth values of a proposition
Union operation	OR operation
Intersection operation	AND operation
Complement operation	NOT operation

9.5.2 Fundamentals of Fuzzy Logic

In ordinary logic, truth values taken up by ordinary propositions or predicates are either true or false. Expressed numerically, the truth values acquired by propositions are either 0 or 1. In fuzzy logic, the truth values acquired can be more than two and expressed numerically, they are spread over the entire range 0–1. In fuzzy logic, we have to deal fuzzy propositions, that is, statements that can acquire fuzzy truth values.

Suppose, P is a fuzzy proposition; and $T(P)$ represents the truth value attached to P. We can then say that the fuzzy proposition P is associated with a fuzzy set (A) and the membership values associated with the elements of the fuzzy set A may be treated as the fuzzy truth values $T(P)$ of P. Symbolically, we can write

$$T(P) = m_A(a) \tag{9.72}$$

where

$$0 \leq m_A(a) \leq 1, \tag{9.73}$$

α being a member of the fuzzy set A.

Fuzzy logic, like ordinary logic, uses a number of connectives like negation ($\sim$), conjunction ($\wedge$) disjunction ($\vee$), and implication ($\Rightarrow$). We will focus on the connective called implication first. In ordinary logic, if A and B are propositional variables, implication is represented symbolically by $A \Rightarrow B$ which is interpreted as "if A then B" or "A implies B." A fuzzy implication on the other hand describes a relation between two fuzzy sets, rather than between two propositional variables. Thus although the fuzzy implication $A \Rightarrow B$ means "if A then B" or "A implies B," A and B here are fuzzy sets. The question is: how to mathematically define implication as a relation? It turns out there is no unique way of defining a fuzzy implication as a relation between fuzzy sets—the particular definition adopted is specific to the application type. In the context of fuzzy control in particular, the fuzzy implication $A \Rightarrow B$ is mathematically defined as the Cartesian product of the fuzzy sets A and B:

$$A \Rightarrow B = A \times B \tag{9.74}$$

If the universe of A is U_1, and that of B is U_2, the Cartesian product $A \times B$, as we have already introduced, can be written as

$$\begin{aligned} A \times B &= \left\{ (x,y) / \min\left(m_A(x), m_B(y) \right) \mid x \in U_1, y \in U_2 \right\} \\ &= \cup_{U_1 \times U_2} \left\{ (x,y) / \min\left(m_A(x), m_B(y) \right) \right\}. \end{aligned} \tag{9.75}$$

If we take a slightly more generalized case of a fuzzy implication such as "if A and B then C," the definition given for "if A then B" can be extended to the more general case as well. The extension leads to (A, B, C are fuzzy sets of universes U_1, U_2, U_3).

$$\begin{aligned} (A \text{ and } B) \Rightarrow C &= A \times B \times C \\ &= \left\{ (x,y,z) / \min\left(m_A(x), m_B(y), m_C(z) \right) \mid x \in U_1, y \in U_2, z \in U_3 \right\} \quad (9.76) \\ &= \cup_{U_1 \times U_2 \times U_3} \left\{ (x,y,z) / \min\left(m_A(x), m_B(y), m_C(z) \right) \right\}. \end{aligned}$$

These definitions have been found to be useful in fuzzy control. We emphasize again that the definition presented here is not unique or the only definition. Let us now turn to two other important concepts of fuzzy logic. One of the two is "fuzzy quantifiers," while the other is "fuzzy inference." In the ordinary or crisp logic, predicates are measured by quantifiers. Likewise, fuzzy logic propositions can also be quantified by fuzzy quantifiers. There are two classes of fuzzy quantifies: (a) absolute quantifiers and (b) relative quantifiers. In general, fuzzy quantifiers are expressions that allow us to express fuzzy qualities so as to give us an idea, though approximate, of the number of elements in a subset satisfying the given conditions. Absolute quantifiers tend to express fuzzy quantities over the total number of elements of a particular set, explaining whether this number is, for example, "very close to 50" or perhaps "much more then 20," "close to half," "about 200," "somewhere around 50."

Relative quantifiers express measurements over the total number of elements of a set that satisfy certain conditions depending on the total number of possible elements. The expressions like "about half," "the majority," or "the most" qualify as relative (fuzzy) quantifiers.

Fuzzy inference, also called approximate reasoning, on the other hand represents computing procedures adopted for evaluating fuzzy linguistic descriptions and drawing inferences. In other words, fuzzy inference is a process that allows us to formulate the mapping from a given input to an output making use of fuzzy logic. The mapping forms the basis of making decisions or recognizing a pattern. There are two important inference drawing procedures:

1. Generalized modus ponens (GMP)
2. Generalized modus tollens (GMT)

In a formal way, the GMP can be stated as follows:

If a is A then b is B
a is A'
$--------.$
b is B'

Note that A, A', B, B' are fuzzy quantities or terms. The linguistic statements above the horizontal line are analytically known. Below the line is what is not known analytically. On the other hand, GMT takes the form

If a is A then b is B
b is B'
$--------.$
a is A'

Here too, the linguistic statements above the horizontal line are analytically known, but the statement below the line is not. Note that A, B, A', B' are fuzzy terms. Before discussing fuzzy inference further, let us first introduce the compositional rule of inference which is important in this context. To illustrate, let us consider a fuzzy relation R on $U_1 \times U_2$. Let X be a fuzzy subset of U_1 and Y be a fuzzy subset of U_2. The compositional rule of inference then states that

$$Y = X \circ R \tag{9.77}$$

where $X \circ R$ represents the composition of X and R. X and Y can be regarded as vectors with components equal to the values of the corresponding membership functions. That means,

$$\begin{aligned} X &= \left\{ a/m_X(a) \,|\, a \in U_1 \right\} \\ Y &= \left\{ b/m_Y(b) \,|\, b \in U_2 \right\}. \end{aligned} \tag{9.78}$$

$X \circ R$ represent the max–min product of the vector X and the relation matrix R. In a more formal way, we can write

$$X \circ R = \left\{ b/\left[\max\left(\min\left(m_X(a), m_R(a,b) \right) \right) \Big| a \in U_1 \right] \Big| b \in U_2 \right\} \tag{9.79}$$

which can be elegantly condensed to

$$X \circ R = \cup_{U_2} \left\{ b/\max_{U_1}\left[\min\left(m_X(a), m_R(a,b) \right) \right] \right\}. \tag{9.80}$$

The procedure can be illustrated with an example where the vector

$$X = [0.4, 1, 0.2] \tag{9.81}$$

and the relation matrix

$$R = \begin{bmatrix} 0.6 & 0.8 & 0.2 \\ 0.9 & 1 & 0.3 \\ 0.4 & 0.7 & 1 \end{bmatrix}.$$

(9.82)

From own definition, we have

$$X \circ R = [0.4, 1, 0.2] \begin{bmatrix} 0.6 & 0.8 & 0.2 \\ 0.9 & 1 & 0.3 \\ 0.4 & 0.7 & 1 \end{bmatrix}$$

$$= [0.9 \ 1 \ 0.3]$$

$$= Y.$$

(9.83)

Thus, the composition $X \circ R$ defines Y which is said to have been induced by X and R. The compositional rule of inference discussed together with fuzzy implication constitutes a basis of fuzzy inference as can be demonstrated with the following rather simple example.

Suppose that A and X are fuzzy sets of a universe U_1, while B and Y are fuzzy sets of a universe U_2. We are given the premise "X is true" and the implication "if A then B" and wish to find Y as a conclusion. Following the procedure described already, we proceed by first computing the fuzzy implication $(A \Rightarrow B)$ as a fuzzy relation $R = A \times B$, then obtain Y as induced by X and with R

$$Y = X \circ R = X \circ (A \times B)$$

$$= \left\{ b / \left[\max \left(\min \left(m_X(a), m_A(a), m_B(b) \right) \right) \middle| a \in U_1 \right] \middle| b \in U_2 \right\}.$$

(9.84)

Alternatively, we could determine Y by directly applying the formula

$$Y = X \circ (A \times B)$$

$$= \cup_{U_2} \left\{ b / \max_{U_1} \left[\min \left(m_X(a), m_A(a), m_B(b) \right) \right] \right\}.$$

(9.85)

Returning to the GMP we note that we are required to compute the membership function of B'. To do that we can make use of the max–min composition of A' with relation matrix $R(a, b)$—the known implication relation (if–then). That means we can directly write

$$B' = A' \circ R(a, b)$$

(9.86)

Translated into an equivalent problem of computing membership function of B', we have

$$m_{B'}(y) = \max \left(\min m_{A'}(x), \ m_R(x, y) \right)$$

(9.87)

In the case of GMT, A' is computed similarly by forming the max–min composition of B' with $R(x, y)$ and writing [3,4].

$$A' = B' \circ R(x, y). \tag{9.88}$$

Expressed in terms of membership functions, we have

$$m_{A'}(x) = \max\left(\min\left(m_{B'}(y), m_R(x, y)\right)\right). \tag{9.89}$$

9.5.3 Fuzzy Rule-Based Systems

We have seen that fuzzy linguistic descriptions can be morphed into formal representations of systems made through fuzzy (if–then) rules. In general, they encode the knowledge about the system under consideration in the following form:

If (a set of conditions) holds true, then a set of consequents is inferred. Fuzzy "if–then" conditions are generally in the form

If (a_1 is A_1, a_2 is A_2, ..., a_n is A_n) then (b_1 is B_1, b_2 is B_2, ..., b_n is B_n)

where the linguistic variables a_i, b_j take on values from fuzzy sets A_i and B_j respectively. The collection of rules for a particular system is said to form a "fuzzy rule base." If the conclusion (C) drawn from the rule base (R) is the conjunction ($\wedge$) of all the individual consequents ($C_1, C_2, ..., C_i, ..., C_n$) of each rule, we can write

$$C = C_1 \cap C_2 \cap \cdots \cap C_i \cap \cdots \cap C_n \tag{9.90}$$

and the membership function takes the form

$$m_C(y) = \left\{\min\left(m_{C_1}(y), m_{C_2}(y), \ldots, m_{C_n}(y)\right) y \in Y\right\} \tag{9.91}$$

assuming that Y is the UOD.

It could well be that the conclusion (C) drawn from the rule base (R) is the disjunction of the individual consequent of each rule, if so then the following holds:

$$C = C_1 \cup C_2 \cup C_3 \cdots \cup C_n. \tag{9.92}$$

In terms of membership functions we have

$$m_C(y) = \left\{\max\left(m_{C_1}(y), m_{C_2}(y) \ldots m_{C_n}(y)\right) y \in Y\right\}. \tag{9.93}$$

9.5.4 Fuzzy Control

Let us consider a box into which inputs are being delivered and from which outputs are streaming out. We can also imagine several parameters resident in the box itself or coming in as a part of the input. The box serves as the controller, i.e., controls algorithmically the output for a given set of inputs and values of some relevant parameters. The basic problem of control (i.e., control engineering) is to develop an algorithm which is capable of mapping the input and the set of available parameters into an output—which should be the optimal output under the given situation. If we are considering a single input–single output system, a single controller (C) can be used. The input to the system may be viewed as a parameter which can be adjusted externally affecting the output. If we have a desired output in mind (set point in control terminology) and the actual output for the given input strays away from the desired output, the controller senses the deviation as an error signal and adjusts the input signal or input so that the desired output is obtained. The foregoing is an example of a feedback controller for a single input–single output system. If we are handling a multiple input–multiple output system, we will need more than one controller, but the basic principles remain the same. In fact, what we have described is an example of closed-loop control. Assuming that our system is a real-life water treatment plant, the controlled output may be the pH value of the treated effluent and the controller is a pH sensor cum controller. If the effluent pH is different from the targeted or the desired pH, it sends a signal to the plant to neutralize the effluent to the correct extent so that the pH of the effluent returns to normal, that is, the set point value. In the same way, there may be temperature controllers, pressure controllers, etc. We have so far assumed that the inputs and outputs are crisp quantities so the controller is a crisp controller. In real life, we encounter inputs or outputs, or rules that have fuzziness and require application of fuzzy logic for achieving appropriate control action. The resulting technique is called fuzzy control and the system (the box) executing the control is called a fuzzy controller. The control box has provisions for fuzzifications, facility of drawing fuzzy inferences making use of fuzzy if–then rules, and finally the provision for executing a well-defined defuzzification process. Most importantly, the controller here is capable of taking actions based on human-like descriptive judgment through the use of fuzzy rules such as "if A is high and B is low then the output (C) is high." The fuzzy controller is thus more human-like in action. We may now sum up the discussion by identifying the basic elements of fuzzy control systems:

1. Fuzzy input variable (x, y, say)
2. A set of if–then rules for achieving fuzzy control
3. Fuzzy output variable (z)
4. Defuzzification procedure

The basic task of fuzzy control is to apply the control rules to x, y to determine an appropriate value of z. The fuzzy variables that can represent the values of both input and output variables are rather limited in number. A commonly used set of fuzzy variables is listed in Table 9.3.

TABLE 9.3

Commonly Used Fuzzy Variables

Fuzzy Variable	Linguistic Meaning
PL	POSITIVE LARGE
PS	POSITIVE SMALL
PM	POSITIVE MEDIUM
ZE	ZERO
NS	NEGATIVE SMALL
NM	NEGATIVE MEDIUM
NL	NEGATIVE LARGE

In a specific application, we may not use all the fuzzy variables listed in Table 9.3 and work with a smaller subset of them, depending on the problem at hand and the nature of the control desired. Now that we have defined a set of fuzzy variables we must construct suitable membership functions for them. The suitability of the membership function is decided after considering different aspects of the control task—such as how fine or coarse the control actions should be, how fast the control should be achieved, how frequently the control has to be applied. The membership function is thus essentially user-defined. We must note, however, that a more complex membership function would consume more computation time for computing the membership degree and therefore it is better to keep the membership function as simple as possible, consistent with the type and extent of control desired. The membership functions $m(x)$ can be of different types, but serve the same purpose—they fuzzify the problem variable (x). A common and useful membership function is the so-called continuous triangular membership function which may be mathematically represented as

$$m(x) = \max\left\{\frac{a - |x - b|}{a}, 0\right\} \tag{9.94}$$

where a and b are constants with $a > 0$. The variable x may be kept normalized or unnormalized depending on the requirements of the control landscape. The a, b values are to be chosen accordingly. Let us assume that $a = 2$, $b = -2$, $-4 \le x \le 4$ and check what values $m(x)$ takes in the interval from $x = -4$ to $+4$ with $a = 2, b = -2$ and $a = 2, b = 2$, respectively.

From Figure 9.5, it is quite clear that the membership function $m(x)$ with $a = 2, b = -2$ (set I) can represent the fuzzy variable negative, small (NS). Thus, for $x = -2$, the membership degree is 1, while for $x = -4$, the degree is zero. In fact, outside the triangle $m(x)$ is zero everywhere. Note that the base width of the triangle is $2a$ and the location of the apex is determined by the parameter b. Similarly, the membership function with $a = 2, b = 2$ (set II) can be used to represent the fuzzy variable PS—the membership degree being 1 at $x = 2$, but 0 at $x = 0$ or 4 (Table 9.4). Once again outside the triangle (II) $m(x) = 0$, everywhere. We can construct triangular membership functions for other fuzzy variables such as PL and NL in a

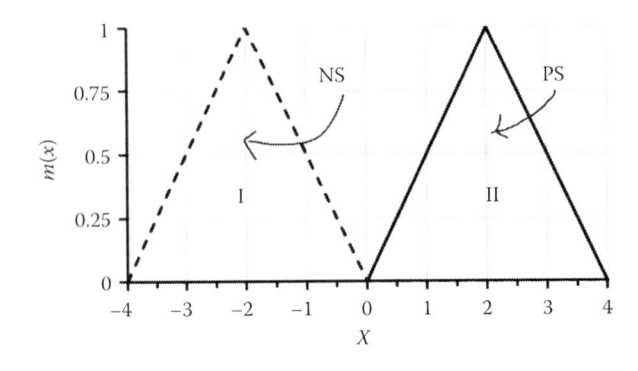

FIGURE 9.5
Triangular membership functions for sets (I and II).

TABLE 9.4

Membership Functions for Sets (I and II)

a, b (I)	x	m (x)	a, b (II)	x	m (x)
2–2	−4	0	2,2	−4	0
	−3	½		−3	0
	−2	1		−2	0
Set I	−1	½	Set II	−1	0
	0	0		0	0
	1	0		1	½
	2	0		2	1
	3	0		3	½
	4	0		4	0

Note: If we plot $m(x)$ against x for both the sets (I and II) in the same figure we get the following triangular graphs (I and II).

similar manner (the reader will do well to try). A more general form of triangular function with three parameters a, b, and c can also be used. Let

$$T(x;a,b,c) = \begin{cases} 0, x \leq a \\ \dfrac{x-a}{b-a}, a \leq x \leq b. \\ 0, c \leq x \end{cases}$$ (9.95)

Then, the corresponding triangular membership degree can be calculated using the max–min operation, as follows:

$$m(x) = \max\left(\min\left(\frac{x-a}{b-a}, \frac{c-x}{c-b} \right), 0 \right).$$ (9.96)

We list below several other forms of membership functions such as the trapezoid, Gaussian, and bell-shaped functions which also find use in fuzzy algorithms.

Trapezoid, membership function (MF):

$$TR(x;a,b,c,d) = \begin{cases} 0, \text{ if } x \leq a \\ \dfrac{x-a}{b-a}, \text{ if } a \leq x \leq b \\ \dfrac{d-x}{d-c}, \text{ if } b \leq x \leq c \\ 0, \text{ if } d \leq x \end{cases}.$$ (9.97)

Using the max–min operator, we have

$$m(x) = \max\left(\min\left(\frac{x-a}{b-a}, 1, \frac{d-x}{d-c} \right), 0 \right).$$ (9.98)

Gaussian MF:

$$G(x;c,\sigma) = exp\left(-\frac{1}{2}\left(\frac{x-c}{\sigma}\right)^2\right).$$

(9.99)

Bell-shaped MF:

$$T(x;a,b,c) = \frac{1}{\left[1+\left(\frac{x-c}{a}\right)^{2b}\right]}.$$

(9.100)

The choice of MF may be made after considering the nature of the problem at hand and the type of desired control action. Having discussed fuzzy variable MFs and fuzzy if–then rules, it is now time to consider means of using them to generate control action in a specific case.

Consider, for example, the case of efficient pH controller. Here we are dealing with two fuzzy input variables X (related to pH) and ΔX (rate of change of pH) and only one output variable Y (the amount of pH-controlling chemicals to be added). Let P_0 be the targeted pH value and P be the current pH. The input X to the controller measures the deviation $X = P - P_0$. The deviation varies from time to time and requires appropriate control action to restore P to its targeted value P_0. Let P_n and P_{n-1} be the measure pH values at the $(n-1)$ and the nth instants. Then we have, at the current (nth) instant the input pair:

$$X = P_n - P_0 \text{ (deviation from the targeted pH at the } n\text{th step)}$$
$$\Delta X = P_n - P_{n-1} \text{ (time rate of change of pH at the } n\text{th instant).}$$

(9.101)

The value of the output variable at the nth step is represented by Y_{n+1} and is expressed as a sum of Y_n (the current output) value and a correction term Z, i.e., we write

$$Y_{n+1} = Y_n + Z.$$

(9.102)

Z being the correction to the current output value (amount of chemicals) required to restore the pH to the target level. Theoretically, our basic task is to derive Z from the X and ΔX at the nth instant. To accomplish the task by fuzzy logic control, a fuzzy if–then table is used for mapping $(X, \Delta X)$ into n. Such a table has the following structure (Table 9.5). Rows indicate fuzzy values for X while columns describe fuzzy values for ΔX. The row–column intersection describes the fuzzy Z value derived by an appropriate fuzzy if–then rule for the X, ΔX combination. In Table 9.5, we have used only five fuzzy variables (additional variables may be included if finer control is desired), consistent with the type of control envisaged. Thus, for the input pair $(X, \Delta X) = (ZE, NL)$, the output (Z) is determined by the application of rule 1 (Table 9.5) which states if X is ZE and ΔX is NL then Z is PL. Thus, the entry at the $X - \Delta X$ intersection in this case is PL (Table 9.5).

The if–then rules used are summarized in Table 9.6

TABLE 9.5

Fuzzy If–Then Table Is Used for Mapping $(X, \Delta X)$ into Z

	ΔX				
	NL	**NS**	**ZE**	**PS**	**PL**
NL			PL		
NS			PS		
ZE	PL		ZE	NS	
PS			NL		
PL			NL		

TABLE 9.6

The Set of If–Then Rules Used for Constructing Table 9.5

Rule No (1)	Linguistic Content of the Rule
Rule 1	if X is ZE and ΔX is NL then Z is PL
Rule 2	if X is NS and ΔX is ZE then Z is PS
Rule 3	if X is NL and ΔX is ZE then Z is PL
Rule 4	if X is ZE and ΔX is ZE then Z is ZE
Rule 5	if X is PS and ΔX is ZE then Z is NL
Rule 6	if X is PL and ΔX is ZE then Z is NL
Rule 7	

Similarly, the combination $(X, \Delta S) = (NL, ZE)$ determines $Z = PL$ by invoking rule 3 which is appropriate for the particular combination (see Table 9.6). In this way, the other entries in Table 9.5 have been determined. Let us note here that Table 9.5, in principle, should have $5 \times 5 = 25$ entries whereas in practice one tends to work with a much smaller set of entries, typically 5–9, depending on the system and the types of fuzzy control variables. In our case, we have only seven entries. The missing entries, in fact, correspond to the situations that are unimportant or are unlikely to be encountered. Important also is the fact that the fuzzy if–then rules are actually used in conjunction with appropriate MFs for the fuzzy variables. Typically, these fuzzy variables do not represent a single value but values spread over a range, say a,b, with varying degrees of membership. It essentially means that each rule in Table 9.6 effectively covers a fairly wide range of values. Let us note further that the output Z is usually defined to represent the deviation from the current output rather than the output itself, and therefore has less critical effect on the system than the output itself. In consequence of all these factors, it is often possible to work with fewer if–then rules and assign Z to have zero value (crisp 0) at least for some of the missing entries, meaning that the current output value is fine and does not require any intervention or further control action. For other missing entries, we can appeal to common sense and fill in the table. For example, if X is PS and ΔX is NS, then Z is ZE (fuzzy zero), which appears to be entirely logical from a common-sense understanding of the practical situation. With Tables 9.5 and 9.6 in place, we are now in a position to elaborate how to go about computing Z from the information provided for X and ΔX, in a systematic stepwise manner. Three basic steps are to be executed:

1. **Fuzzification**

   This involves determining which fuzzy variables are to be assigned to X and ΔX based on their current values, such as $X = PS$ and $\Delta X = NS$. Once the appropriate fuzzification has been achieved, it is time for drawing the fuzzy inference.

2. **Drawing fuzzy inferences**

   This step involves four substeps:

   a. For the particular combination of fuzzy variables X and ΔX given as inputs to the control box, one must look up the fuzzy if–then rules table and identify the rule numbers, say i and j that would be applicable (say, $i = 3$, $j = 5$, meaning that rule 3 and rule 5 are applicable).

   b. In this substep, the weight (w_i) of each rule identified in substep 1 is calculated using MFs for the fuzzy variables f_1 and f_2 assigned to X and ΔX currently. The calculation involves evaluating w_i as

   $$w_i = \min\left(m_{f1}(X); m_{f2}(\Delta X)\right). \tag{9.103}$$

   c. Let f_3 be the fuzzy variable assigned to Z from the if–then table. The MF for Z, for each rule i is determined as

   $$m_i(Z) = w_i \wedge m_{f_3}(Z)$$
   $$m_j(Z) = w_j \wedge m_{f_3}(Z). \tag{9.104}$$

   d. The MF for Z is next computed as

   $$m(Z) = m_i(Z) \vee m_j(Z). \tag{9.105}$$

3. Once $m(Z)$ has been determined in substep (iv) of step b, we have at our disposal the output Z in fuzzy form meaning that we only know the range of values across which Z lies with a certain degree of membership. However, to be practically useful we must have a specific value of Z, say $\bar{Z}$, to effect control. One way to do this is to compute the center of gravity of $m(Z)$ as

$$\bar{Z} = \frac{\int m(Z)Z dZ}{\int m(Z) dZ}. \tag{9.106}$$

This process of mapping Z into $\bar{Z}$ is called defuzzification. Needless to mention, this is just one way of achieving defuzzification. In what follows we will focus on fuzzy control in the context of global geometry optimization of clusters.

9.6 Application to Cluster Geometry Optimization

The nature-inspired population-based metaheuristics such as GA- EP-, and PSO-driven optimization share many common features. They maintain populations of plausible solutions through random alterations and selection. They do, however, differ in the techniques used to represent the solutions, the types of alterations they adopt to produce new solutions and the methods they use for selecting the new solutions. One troubling feature of all their methods is the sensitivity of performance when the relevant parameters are changed, such as rate of cooling in SA, the crossover and mutation probabilities in GA, inertial weight and acceleration coefficients in PSO along with velocity limits (V_{max}, V_{min}). In GA, the candidate structures compete with each other for survival and in the process of breeding the fittest individuals transmit their genetic materials to the individuals of succeeding generations. Aided by a suitable selection scheme, and rare random mutations, a situation is created whereby the superindividual in the final population (at termination stage) is the best solution to the optimization problem being solved. Computational effort required in GA is quite substantial although only function evaluation is required. Take the SA method as an alternative: The problem of fixing a temperature and constructing an annealing schedule are cumbersome in SA. The presence of a high barrier at saddle points between minima may, at low temperatures, make the transition between minima quite difficult with kinetics rather than thermodynamics controlling the proceedings. The PSO, as already discussed (Chapter 5), is inspired by the metaphor of social interactions and communication among individuals of a population. The solutions evolve through cognitive and social learning of every individual in the swarm and the best solution eventually emerges through a delicate and dynamic balance between individual improvement (competition) and collective response (cooperation). The PSO occupies a special position among the population-based metaheuristics because of the ease of implementation, physical appeal, and excellent performance. Even then, with growing problem size and computational complexity, PSO suffers performance degradation. The traditional PSO may not therefore be the ideal choice for large-scale cluster geometry optimization. One can think of building hybrid algorithms of various kinds in which strong complementary features of different methods can be put together for better performance. Thus combining GA with the traditional Monte Carlo method, or GA with the steepest descent method [5–7], has led to hybrids with better performance in specific applications. GA and SA have complementary strengths so that a GA–SA hybrid could be expected to be more efficient than either of the two separately [8]. The success of hybrid methods fuels hope that development of more efficient hybrid algorithms would be possible and profitable in the context of global optimization problems. In the following section, we present a case study in which a version of completely adaptive random mutation hill climbing is hybridized with particle swarm optimization, in which inertia weight is either dynamically adjusted or determined by fuzzy logic-based adaptation.

9.6.1 Case Study I

The systems under investigation were LJ clusters (LJ_N) with specific values of N such as $N = 13, 19, 38, 98$, and 102 atoms. These clusters are known to possess global minima that are very hard to locate and therefore are good benchmark problems. In the simplest approximation, the potential energy of the cluster of N atoms is a $3N$-dimensional function $V(\vec{R})$ of the coordinates of all the particles where

$$V\left(\vec{R}\right) = \sum_{i=1}^{N-1} \sum_{j>i}^{N} v\left(\left|\vec{r}_i - \vec{r}_j\right|\right) \tag{9.107}$$

$v\left(\left|\vec{r}_i - \vec{r}_j\right|\right)$ is the interatomic pair potential of the atom pair i and j held motionless (assume $T = 0$). $V(\vec{R})$ represents the static $3N$-dimensional potential energy surface $\left|\vec{r}_i - \vec{r}_j\right|$ being the Euclidean distance between the ith and the jth atom $(V(\vec{R}) \equiv v(\vec{r}_1, \vec{r}_2, \dots, \vec{r}_N), \vec{r}_i = x_i, y_i, z_i)$. Although many types of pair potentials have been and can be used, it is necessary to impose some restrictions on $v(\vec{r})$ if $V(\vec{R})$ is to have satisfactory behavior over the entire configuration space. The following conditions are to be imposed:

$$v(\vec{r}) \to 0 \text{ as } r \to \infty$$
$$v(\vec{r}) \to \infty \text{ as } r \to 0$$
$$v'(\vec{r}_0) = 0 \text{ for a unique } \vec{r}_0 \text{ with } r_{\min} < \vec{r}_0 < \infty$$
$$v'(\vec{r}_0) < 0 \text{ and } v''(\vec{r}_0) > 0.$$

The pair potentials explored in the present context include the Lennard–Jones potential, the Buckingham potential, the Morse potential, the Gaussian potential, and the Born–Mayer potential. The LJ potential energy surface $V_{LJ}(\vec{R})$ is a superposition of the distinct LJ pair potential $v_{LJ}(\vec{r}_{ij})$:

$$V_{LJ}\left(\vec{R}\right) = \sum_{i=1}^{N-1} \sum_{j>i}^{N} v_{LJ}\left(\vec{r}_{ij}\right) \tag{9.108}$$

where

$$v_{LJ}\left(\vec{r}_{ij}\right) = 4 \in \left[\left(\frac{\sigma}{r_{ij}}\right)^{12} - \left(\frac{\sigma}{r_{ij}}\right)^{6}\right] \tag{9.109}$$

σ being the distance of separation between the atom pair at which the pair potential becomes zero, while $\in$ is the equilibrium well depth of the pair potential. The pair potential has two parts. The attractive part corresponds to induced dipole–induced dipole interaction proportional to $(r_{ij})^{-6}$ originating from van der Waals dispersion forces. This is a rather weak interaction that dominates the weak binding between closed-shell atoms such as Ar or Kr. The repulsive part corresponds to the London dispersion and quickly grows as $\vec{r}_{ij} \to 0$ preventing the collapse of the atom pair to a point under the action of the attractive forces. The repulsive part of the potential suggested purely on empirical grounds may have its origin in Pauli exclusion forces that are brought into play when two closed-shell atomic charge distributions begin to overlap.

The LJ potential energy surface looks simple but as an objective function in a global minimization problem it is nonconvex and highly combinatorial. The surface is known to support an exponentially increasing number of local minima (L_N) as the cluster size (N) increases. A rough estimate of the N-dependence of L_N is given by [9]

$$L_N = exp\left(-2.5176 + 0.3572N + 0.0286N^2\right). \tag{9.110}$$

Around $N = 100$, a simple calculation shows that the surface supports more than 4×10^{138} minima. No wonder that LJ clusters have played an extremely important role in the arena of global optimization by providing hard benchmark problems. The difficulties originate from the presence of both short- and long-range interactions in LJ clusters which makes the search for the global minimum (GM) difficult. LJ clusters of certain specific sizes have been known to pose severe challenges to a global optimizer. One such example is the LJ_{38} cluster. The GM of LJ_{38} is known to be a face centered cubic (FCC) truncated octahedron. A large barrier separates the FCC and icosahedral structures and the search space is dominated by liquid-like configurations with disordered polytetrahedral packing. In an unbiased search it is difficult to reach the GM of the LJ_{38} cluster which has an FCC structure with decahedral packing. The CARMHC algorithm introduced already [10–12] displays good global search capability when dealing with a fairly large coulomb cluster. Such clusters are, however, dominated by long-range attraction. As case studies, we report here the energy evolution profiles during the CARMHC-based search for the global minima in LJ_{27} (Figure 9.6a) and LJ_{38} (Figure 9.6b) [13] clusters along with snapshots of structures located in the intermediate stages. In each case the starting point was a randomly generated configuration of n-particles, each located within a small spherical volume with radius $R_0 = 0003$ nm. For LJ_{27}, the search converges to the GM smoothly in one shot. For LJ_{38}, the CARMHC failed to reach the GM in one run. It could locate the GM with 50 restarts and consumed 100 times the search time compared to LJ_{36} or LJ_{40} clusters in locating the global minimum.

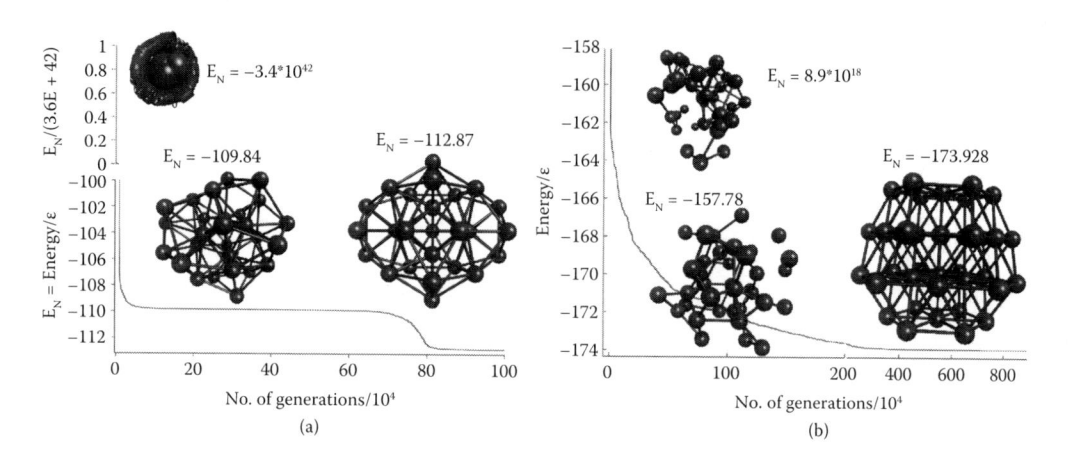

FIGURE 9.6

(a) The energy evolution profile for LJ_{27} cluster initialized with a random starting point. Evolution starts with a very compact structure and is then allowed to expand. All the particles are within the radius of 0.003 nm in the starting geometry. (b) The energy evolution profile (in the region close to the optimum structure) for LJ_{38} cluster. To locate the GM, several restarts from different geometry were required. (Courtesy of Sarkar, K., *Soft-Computing algorithms for solving some problems of Quantum Chemistry*, Indian Association for the Cultivation of Science, Jadavpur University Kolkata, India, 2014, 149.)

Clearly, many of the reconfiguring moves in LJ$_{38}$ were being wasted on the liquid-like structures. The difficulty experienced has been addressed by hybridizing CARMHC with a modified PSO method, the latter having enhanced ability to explore the search space while the former has excellent capability to exploit the information locally available.

Let us recall that in PSO, each particle in the swarm constantly accelerated toward the currently available global best position (g_{best}) and its personal best position (p_{best}) discovered so far. The velocity ($\vec{V}$) and position ($\vec{X}$) updating scheme read as follows:

$$V_i(t+1) = \omega V_i(t) + C_1\phi_1\left(p_i^{best} - X_i(t)\right) + C_2\phi_2\left(p_{gbest}(t) - X_i(t)\right)$$
$$X_i(t+1) = X_i(t) + \chi V_i(t+1). \tag{9.111}$$

χ is a damping parameter called the constriction coefficient [14] in SWARM literature. χ essentially has been introduced to exercise some control on the movement of particles through the search space. It is possible to set $\chi = 1$, and introduce a maximum velocity V_{max} parameter (if $V > V_{max}$ set $V = V_{max}$), as a cutoff. If V_{max} is large, it leads to better exploration of the search space (breadth search) while smaller V_{max} values facilitate better local exploration (depth search). The upper limit of velocity is imposed in all the dimensions. It prevents particles from moving too rapidly from one region in the search space to another, leading to swarm explosion. The introduction of inertia weight (ω) however does eliminate the need for clamping the velocity to V_{max} although numerical experiments [15] indicate that it is always good to limit V_{max} to X_{max}. A question then arises, whether ω should be fixed or dynamically adjusted as the search progresses. The general practice has been to reduce ω linearly from a starting value of 0.9 to a final value of 0.4 during the entire period of evolution. The dynamics of the change of inertia weight can be represented by

$$\omega(t) = \omega_s - \left[\omega_s - \omega_f\right]\frac{t}{t_{max}} \tag{9.112}$$

where ω_s is the starting value of ω and ω_f is the final value, t_{max} being the maximum number of generations (t). A rearranged form of the equation for weight controlling dynamics is easily seen to be

$$\omega(t) = \omega_f + \left[\omega_s - \omega_f\right]\left(1 - \frac{t}{t_{max}}\right). \tag{9.113}$$

The logic behind the introduction of inertial weight (ω) in the velocity updating equations is simple—to maintain the purported balance between the local and global exploration abilities. A larger value of inertial weight ($\omega(t)$) in the beginning of the search is expected to facilitate global exploration, while a smaller ω value at the terminal phases of the search would enhance local exploration. The acceleration coefficients C_1 and C_2 can also be dynamically adjusted as ω has been:

$$C_1(t) = C_{1,s} - \left(C_{1,s} - C_{1,f}\right) t/t_{max}$$
$$C_2(t) = C_{2,s} - \left(C_{2,s} - C_{2,f}\right) t/t_{max}. \tag{9.114}$$

The modified PSO-CARMHC algorithm with fuzzy adaptive control implemented in a case study can now be elaborated. Several modifications have been made in the standard PSO formulation which uses a population (n_p) of particles whose positions encode plausible potential solutions to the optimization problem of an N particle LJ cluster. The positional coordinates and velocity components of the kth particle of the swarm are strings S_k and V_k, respectively. The positional coordinates of particles are the $3N$ Cartesian coordinates of the N atoms in the cluster representing a cluster geometry (the kth geometry), while the velocity string V_k encodes the $3N$ Cartesian velocity components the kth particle is flying with. Thus

$$
\begin{aligned}
S_k = S\left(x_1^k, y_1^k, z_1^k; x_2^k, y_2^k, z_2^k; \ldots; x_N^k, y_N^k, z_N^k\right) \\
\equiv \left(\xi_1^k, \xi_2^k, \xi_3^k, \ldots, \xi_{3N-1}^k, \xi_{3N}^k\right) \\
\equiv 3N \text{ component position vector} \equiv R_N^k
\end{aligned}
\tag{9.115}
$$

$$
\begin{aligned}
V_k = S\left(v_{1x}^k, v_{1y}^k, v_{1z}^k; v_{2x}^k, v_{2y}^k, v_{2z}^k; \ldots; v_{Nx}^k, v_{Ny}^k, v_{Nz}^k\right) \\
\equiv \left(v_1^k, v_2^k, v_3^k, \ldots, v_{3N-1}^k, v_{3N}^k\right) \\
\equiv 3N \text{ component velocity vector} \equiv V_N^k.
\end{aligned}
\tag{9.116}
$$

9.6.1.1 Modification in Generation of Initial Coordinates

A seeding procedure has been followed. For an N-atom cluster we start with the coordinates of the $(N-1)$ atom cluster at its predetermined GM. The Nth atom is then randomly placed somewhere within the sphere of radius R_m where R_m is the maximum distance encountered in the $(N-1)$ atom cluster from its center. Only $\frac{n_p}{2}$ strings are generated in this manner. From each of these $\frac{n_p}{2}$ strings, two new solution strings are generated in the following manner: $\frac{N}{2}$ atoms are randomly selected and their position coordinates are copied onto the new string S_k'. The coordinates of the remaining $\frac{N}{2}$ atoms in S_k' are generated by reflections of the copied atoms on the XZ plane. The second string S_k'' is generated similarly with the remaining $\frac{N}{2}$ atoms (not copied onto S_k') of the original string S_k. In case N is odd, $\left(\frac{N}{2}+1\right)$ atoms are copied from S_k to S_k'.

9.6.1.2 K-Best Neighborhood Topology

The entire population of n_p strings (or particles) are then collected into n subgroups based on K-best neighborhood topology (every particle is connected to its K nearest neighbors in the topological space). With $K = 2$, we have a ring or circular neighborhood topology. On the other hand, if we choose $K = n_p - 1$, we encounter what is known as global best or g_{best} topology. The global version of PSO (i.e., PSO with the g_{best} topology) converges fast, but has the propensity to converge to a local minimum (the population loses diversity fast as every particle is influenced strongly by the g_{best} particle). The idea of using subgroup s is to retain the diversity of solutions as long as possible. From time to time, particles are exchanged between subgroups which enhance information sharing. For every particle,

g_{best} is chosen from its K neighbors. This version of PSO thus uses a dynamic neighborhood topology. In all the calculations $n_p = 40$ has been used whatever the cluster size may be.

9.6.1.3 Velocity Updating Scheme

Some modifications were made in the velocity updating equation:

$$V_i(t+1) = \omega V_i(t) + C_1\phi_1\left\{p_{ibest}(t) - \xi_i(t)\right\} + C_2\phi_2\left\{g_{best}(t) - \xi_i(t)\right\}. \tag{9.117}$$

The acceleration coefficients for the cognitive and social components of learning were assigned constant values of $C_1 = 2.05$ and $C_2 = 2.25$ after some experiments. ϕ_1 and ϕ_2 are uniformly distributed random numbers in the range $(0,1)$. The maximum cutoff velocity V_{max} was chosen as

$$V_{max} = \frac{1}{3}R_{max}^{N-1} \tag{9.118}$$

where R_{max}^{N-1} is the radius of the minimum volume sphere enclosing the $(N-1)$ atom cluster. It represents the maximum amount of change that a coordinate can undergo in one step. Not only the maximum velocity, the positional coordinates of the particles too were put under some constraint

$$-R_{max}^{N-1} \leq \xi_i^K \leq R_{max}^{N-1} \tag{9.119}$$

to avoid divergence and explosion; χ was set equal to 1 on the basis of initial experiments that seemed to indicate a propensity to converge to suboptimal regions of the search space with $\chi < 1$.

9.6.1.4 Inertial Weight Dynamics

Two basic schemes were devised with a view to striking a proper balance between the global and local exploration abilities of the algorithm.

1. The first was a combination of time varying (dynamic) weight factors which were randomly perturbed within a limit. We will call it the dynamic inertia weight method.
2. The second involved adaptive adjustment of the inertial weight factor based on a set of fuzzy if–then rules, i.e., based on a fuzzy inference system. We will call it the fuzzy adaptive inertia weight method.

9.6.1.4.1 Dynamic Inertia Weight Method

This method works in two phases. In the first phase the inertial weights $\left\{\omega_i^*\right\}$ for particles (i) in the beginning of the iterations are randomized around 0.5 by setting

$$\omega_i^* = 0.5 + \frac{rand}{2.0} \tag{9.120}$$

where *rand* is a uniformly distributed random number in the interval $(0, 1)$. That means the mean value of inertial weight (ω_i^*) would be 0.75. The random choice of inertia weight (ω_i^*)

tends to induce faster convergence during the early stages of search. In the second phase ω_i^* values are systematically reduced linearly as the number of iterations (t) increases following the recipe

$$\omega(t) = \omega^{end} + \left\{ \omega^{start} - \omega^{end} \right\} \frac{t_{max} - t}{t_{max} - t_1}$$ (9.121)

where ω^{start} *and* ω^{end} are the upper and lower bounds to ω. t_{max} is the maximum number of iterations allowed, and t_1 is the number of iterations performed in the first phase. The logic behind the use of linearly decreasing inertia weights in this phase is that the use of a larger value of inertia weights at the beginning of the second phase of search could allow the particles to explore a wider region of the search space before they start to converge. The overall convergence rate of the algorithm in this phase could be slower than in the earlier phases although it is expected to discover a better local minimum before getting trapped (in view of better explorations).

9.6.1.4.2 Fuzzy Adaptive Control of Inertia Weights

Fuzzy logic, as we have discussed already, is useful in representing expert knowledge and experience in the form of linguistic rules which can be easily understood by human beings. In this case study, a fuzzy inference technique is used to fix the inertia weight adaptively and nonlinearly. The method uses two fuzzy input variables, namely the current iteration stage t/t_{max} and the current relative velocity $[\Delta V_{av}(t)]$, and one fuzzy output variable (ω^{end} —the lower bound to the inertia weight). We make use of three triangular MFs corresponding to three fuzzy sets—low (L), medium (M), and high (H)—that the input variables could be members of. The fuzzy rules table displayed below (Table 9.7) indicates how to determine this fuzzy inertia weights.

Once a fuzzy assignment has been made for ω^{end} on the basis of Table 9.4, defuzzification is carried out to fix a numerical value for ω^{end} (ω_L^{end} or ω_M^{end} or ω_H^{end}). The actual value of ω_i^* is then determined as

$$\omega_i^* = \omega_{L|M|H}^{end} + \frac{rand}{3}.$$ (9.122)

Note that $\Delta V_{av}(t)$ has been computed as

$$\Delta V_{av}(t) = \left| V_{av}(t) - V_{av}(t-1) \right|.$$ (9.123)

TABLE 9.7

The Fuzzy Rule Table for Determining Fuzzy Inertia Weights

t/t_{max}	$\Delta V_{av}(t)$		
	L	**M**	**H**
L	H	H	M
M	H	M	L
H	M	L	L

where

$$V_{av}(t) = \frac{1}{3n_p N} \sum_{i=1}^{n_p} \sum_{j=1}^{N} V_{ij}(t).$$ (9.124)

9.6.2 Case Study I: Applications of Modified PSO-based Approaches to LJ Cluster

It is a matter of common experience that the simple (standard) PSO encounters difficulty in maintaining an appropriate balance between exploration and exploitation. The difficulty arises because the PSO appears to favor intensified search around the better solutions already discovered by the algorithm. PSO is therefore likely to miss the GM or if it does not do so, it takes a rather long time to reach it. The two modified PSOs described in SectionZ9.6.1 reveal some improvement in this respect when applied to discovering the GM in LJ clusters. It turns out that the modified PSO with dynamic inertia weight adjustment does not perform as well as the PSO with fuzzy logic-based adaptive adjustment of inertial weight when applied to larger LJ_N clusters with $N > 50$. Figure 9.7a through c

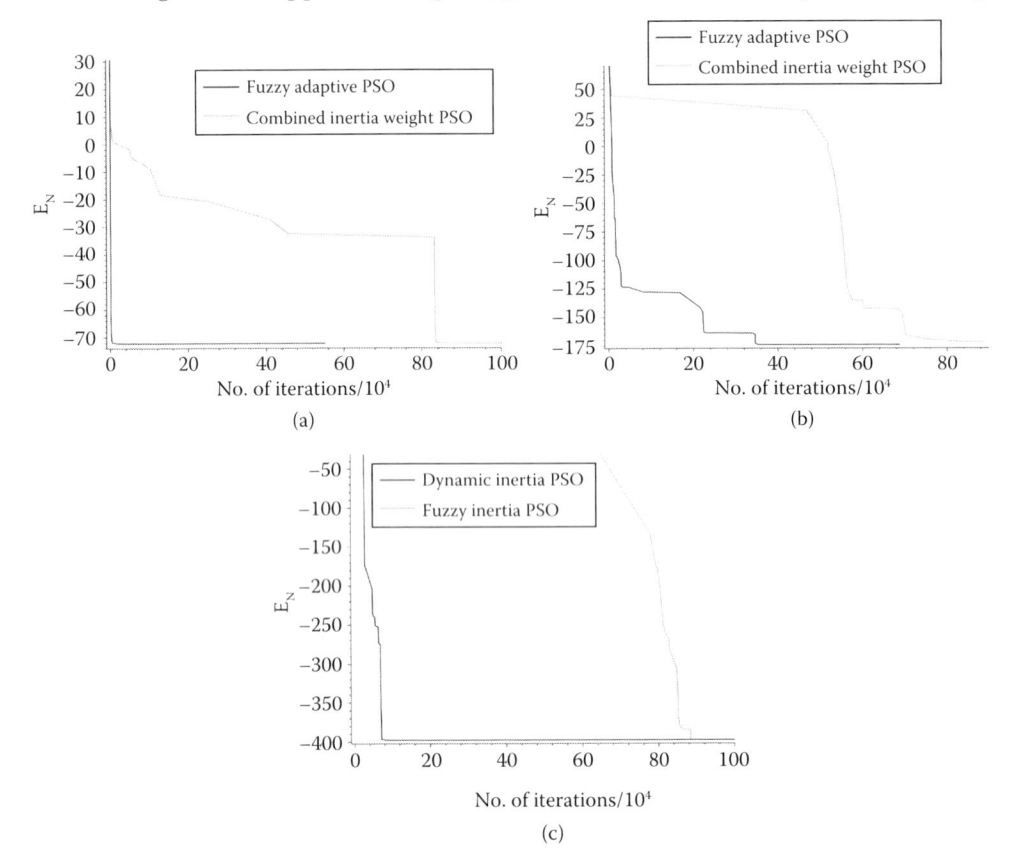

FIGURE 9.7
Energy evolution profiles of dynamic inertia weight-based PSO (gray) and fuzzy adaptive inertia weight-based PSO (black) for LJ clusters consisting of (a) 19 atoms, and (b) 38 atoms, and (c) 75 atoms. In the first two cases both the algorithms locate the GM in the long run. But in the third example the dynamic inertia weight-based PSO fails to move to the GM in a reasonable number of iterations, although in the long run it converges very rapidly. (Courtesy of Sarkar, K., *Soft-Computing algorithms for solving some problems of Quantum Chemistry*, Indian Association for the Cultivation of Science, Jadavpur University Kolkata, India, 2014, 156.)

displays the energy evolution profiles for the best strings in the population in the dynamically adjusted inertia weight-based PSO (in gray) as well as in the PSO with fuzzy adaptive adjustment of the same weight factor (in black). Three sets of calculations under each of the modified PSO schemes were carried out: (a) 19, (b) 38, and (c) 75-atom Lennard–Jones clusters. As the figures convey, both the algorithms succeed in locating global minima in the 19- or 38-atom clusters, but the fuzzy adaptive inertial weight-guided algorithm moves through the search space much more quickly toward the GM.

In the case of the LJ_{75} cluster, the fuzzy adaptive inertia weight method smoothly and quickly locates the GM, while the dynamic inertial (nonfuzzy) weight-based method takes a much longer time to move into the optimal region of the search space. Once in the optimal region, the algorithm homes in on the GM quickly. On the whole, therefore, the fuzzy logic-based adaptive adjustment of inertia weight enhances the efficiency of the PSO-based search for global minima in typically problematic LJ clusters. One would be curious to know how the inertia weight changes with generations in the dynamic (crisp) weight adjustment and the fuzzy adaptive weight-adjustment algorithms. Figure 9.8a and b display the variation in ω in typical runs involving (a) dynamic and (b) fuzzy logic-based inertial weight adjustment. In the dynamic ω-case, after an initial steep rise ω falls off linearly to a lower bound while in the fuzzy adaptive ω-case, there are random oscillations in ω around average values (depending upon the search stages) indicating that with fuzzy adaptation, the algorithm is able to adjust ω correctly in response to the requirements of the region of the search space it is currently exploring. This may enhance its ability to locate GM.

9.6.3 Case Study II: Applications of Hybrid PSOs to LJ Clusters

From case studies reported in the previous section it becomes evident that the fuzzy adaptive inertia weight-guided PSO could benefit from hybridization with a soft algorithm with proven efficacy of exploitation of information in the neighborhood of the optimal region. As already demonstrated in Chapter 4, the CARMHC method is a

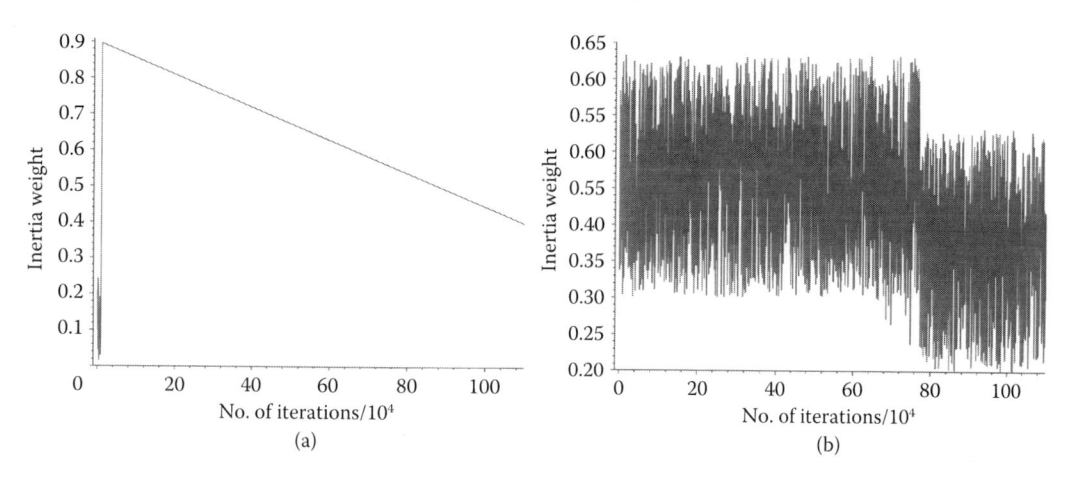

(a)

(b)

FIGURE 9.8
The way inertia weight changes during the course of evolution: (a) dynamic inertia weight-based PSO and (b) fuzzy adaptive inertia weight-based PSO for a system consisting of 38 particles. (Courtesy of Sarkar, K., *Soft-Computing algorithms for solving some problems of Quantum Chemistry*, Indian Association for the Cultivation of Science, Jadavpur University Kolkata, India, 2014, 156.)

single string (parent) evolutionary technique endowed with computational intelligence as well as efficiency—especially when the search has already moved into the optimal region and fine-tuning of the evolving solution is taking place. In the case study being reported here, the super individual emerging from the modified PSO (either with dynamic inertia weight or with fuzzy adaptive inertial weight selection) after an initial broad exploration of the search space is fed to the CARMHC algorithm for locating the GM. It turns out that the hybridization pays a good dividend. In addition, the deterministic steepest method has occasionally been used to explore the search space locally, adding further power to the hybrid algorithm. The flowchart of the algorithm is displayed in Figure 9.9.

Figure 9.10a and b depict representative results obtained for an LJ_{38} cluster when the search was carried out with (a) fuzzy inertia weight PSO + CARMHC and (b) dynamic inertia weight PSO+ CARMHC + SD-guided hybrid algorithm. The dynamic inertia weight-guided PSO finished the search at a relatively higher energy configuration compared to the GM (Figure 9.10a). The fuzzy inertia weight PSD significantly benefited from hybridization with CARMHC and SD and reached the GM in a single run. It is an established fact that large-scale optimization algorithms invariably exploit positive features of a number of algorithms to improve the global search efficiency, rather than depending on one algorithm. Our experience with PSO reflects the same fact—for large-scale optimization problems the PSO must be integrated with other nondeterministic or deterministic optimization methods along with the introduction of fuzzy adaptive control of the crucial PSO parameter called inertial weight.

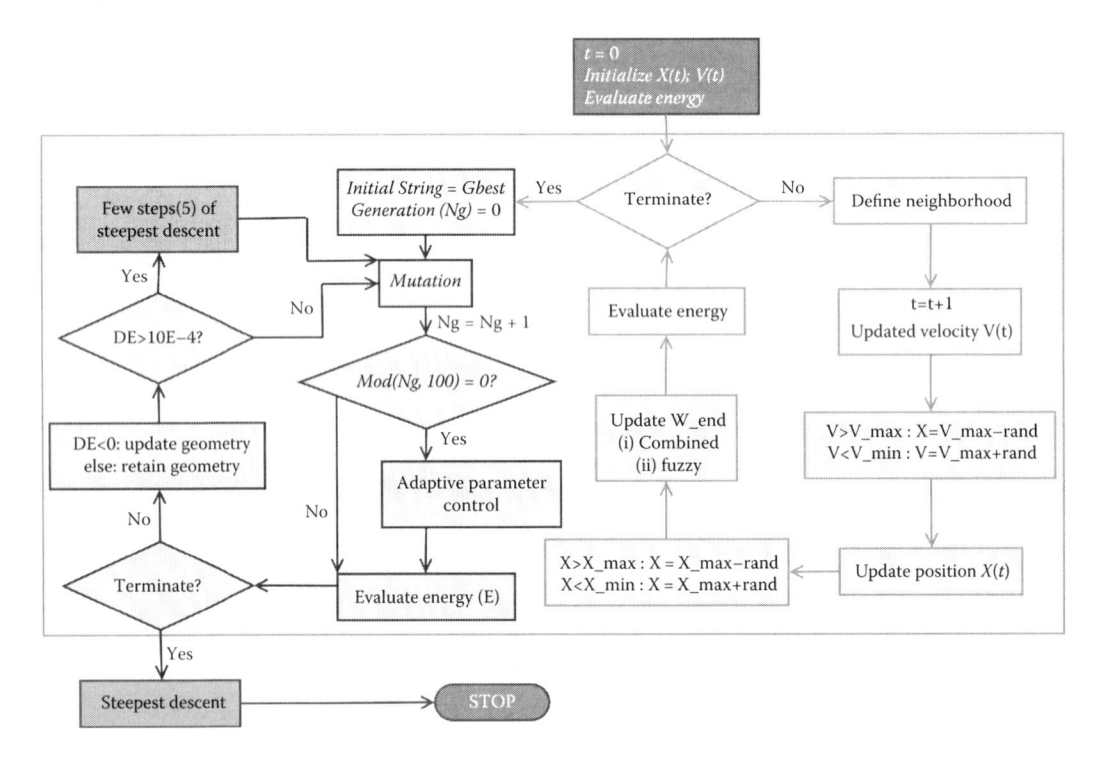

FIGURE 9.9
Flowchart for hybrid algorithm. (Courtesy of Sarkar, K., *Soft-Computing algorithms for solving some problems of Quantum Chemistry*, Indian Association for the Cultivation of Science, Jadavpur University Kolkata, India, 2014, 157.)

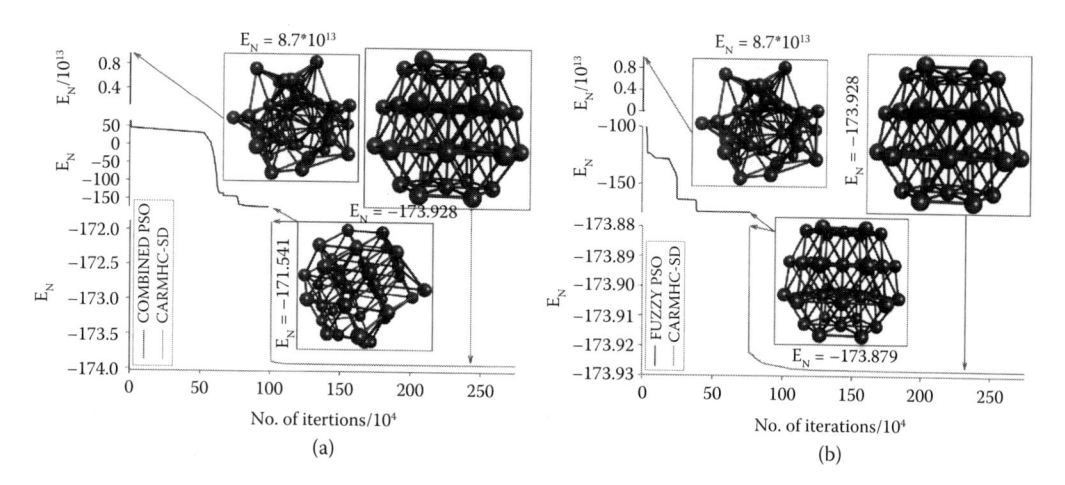

FIGURE 9.10

The energy evolution profile (in the region close to the optimum structure) for an LJ_{38} cluster with two different approaches: (a) The dynamic inertia weight PSO-based approach endowed with CARMHC and steepest descent method finished with relatively higher energy than (b) the fuzzy adaptive inertia weight PSO-based approach endowed with CARMHC and steepest descent method. In both approaches CARMHC along with occasional steepest descent steps helps the search to reach the GM. The dynamic inertia weight PSO-based approach however ends with relatively higher energy. (Courtesy of Sarkar, K., *Soft-Computing algorithms for solving some problems of Quantum Chemistry*, Indian Association for the Cultivation of Science, Jadavpur University Kolkata, India, 2014, 158.)

9.6.4 Case study III: Fuzzy Parameter Control in CARMHC Algorithm: Application to Metal Cluster

Understanding structural and electronic properties of nanometal clusters and alloys, and their correlation if any, enhances our ability to harness their full technological potential. Experiments and quantum chemical calculations, when combined carefully, can often lead to the emergence of the desired level of understanding. Thus, if the structure is known, spectral features can be calculated theoretically if such data are unavailable. Conversely, comparison between the predicted and experimentally observed spectrum (and possibly other properties) could lead to the identification of the correct structure. The problem of theoretically determining the ground state structure of metal clusters and alloys belongs to the NP-hard category. Soft computing provides an excellent avenue in such cases for conducting the search. Thus, computational studies on PES constructed with Gupta many-body potential invoking various soft-search heuristics have provided new insight into the geometrical features and segregation properties of several metal clusters and nanoalloys. For example, the experimentally observed tendency of Pd and Pt atoms to preferentially occupy the interior (core) and exterior (shell surface) sites, respectively, has been confirmed by computer simulations carried out on such systems. The different types of geometrical arrangement of metal atoms such as icosahedral, close packed cubic, decahedral, or even amorphous, assumed by the lowest energy isomers for Pt_n, Pd_n, or Pt_nPd_m clusters have been identified by computer simulations. Calculations by Rossi et al. [16] for example revealed that Pd_{34} and Pt_{34} clusters are incomplete dodecahedra, the same structural motif occurring also in the mixed Pd_nPt_m clusters for several choices of n and m. Calculations by Fernández et al. [17] also confirmed that the $Pt_{core}Pd_{shell}$ segregation predicted by simulations based on the Gupta potential are recovered only at higher levels of theory, although the ordering of the permutational isomers based on energy could differ depending on whether DFT or the Gupta potential [18] has been used for the simulations.

Compared to the monodispersed metallic clusters, bimetallic nanoalloys are of far greater complexity due to the presence of two different types of atoms which lead to the existence of many homotopes. The number of homotopes rises combinatorially as the cluster size and the ratio of the numbers of two types of atoms (in A_nB_m it is the n/m ratio) increase. The search for the global minima on the PES of bimetallic clusters is therefore an extremely challenging task that calls for exploring new search strategies. Before delving into the strategy itself, let us consider in brief detail the nature and parameterization of Gupta many-body potentials.

9.6.5 The Gupta Potential

A pair potential depends only on the distance r_{ij} separating the two interacting atoms. Although powerful and popular, the pair potential is inadequate to model systems with open shells where strong localized bonds may form (as in covalent crystals), or where there is a completely delocalized "electron sea" in which the ions "float" or "reside." In such systems, the description based on two body (pair) interactions alone fails miserably. The total energy of a crystal in the pair potential approximation is given by

$$\in_{pair} = \sum_{i=1}^{N-1}\sum_{j=i+1}^{N}\phi(r_{ij}).$$ (9.125)

If we switch over to a description based on many-body potentials, the total energy of the crystal reads

$$\in = \sum_{i=1}^{N}f(\rho_i) + \in_{pair}.$$ (9.126)

The first term in the equation for $\in$ is obtained by treating each atom (i) as an impurity embedded in a "host" represented by the rest of the atoms, which provides a medium of electron density ρ. The local density at the ith site is simply a superposition of the atomic electron densities contributed by all the surrounding atoms (j).

$$\rho_i = \sum_{j\neq i}\rho_j(r_{ij}).$$ (9.127)

$\rho_j(r_{ij})$ is nothing but the spherically averaged atomic electron density contributed by the jth atom at the site i at a distance r_{ij}. The additive pair potentials lead to a rather poor description of the interatomic interaction in metallic solids. Many-body interaction potentials have been devised to provide a consistently better description of the bulk and surface properties of metallic solids than is possible with the pair potentials. The tight-binding model is the simplest form of a many-body potential in which $\phi(r_{ij})$ represents the repulsion energy of the electron clouds of two neighborhood atoms (i,j), the atomic electron density being assumed to decay exponentially with distance r_{ij}. The cohesive properties of transition metals and their alloys originate from the rather large d-band density of states (DOS). The description of electronic DOS in terms of its moments is a natural tool to relate the electronic structure to the lattice topology, since moments are obtained by calculating the products of the matrix elements of the electronic Hamiltonian associated with closed paths of definite length. The kth moment $\{\mu_k\}$ can thus be interpreted as the contribution to the DOS coming from all possible closed paths involving k steps. The first moment μ_1 is easily related to the energy of the band center and fixes the energy scale; it can be set to zero in pure systems. The experimental binding energies of transition metals, in turn, appear to be roughly proportional to the average width of the DOS, as measured by the square root of the second moment $\{\mu_2\}$ of the density. The Gupta potential is based on the second moment approximation to the tight-binding model, the second moment being the sum of the squares of the hopping matrix elements or hopping integrals. Note that the hopping integrals are functions of radial distances between the atom pairs so the band energy can be written as

$$f(\rho_i) = F^B(i) = \left[\sum_{j \neq i} \xi^2(\alpha,\beta) exp\left\{ -2q(\alpha,\beta)\left(\frac{r_{ij}}{r_0(\alpha,\beta)} - 1 \right)^{-1} \right\} \right]. \tag{9.128}$$

α,β denote the element types of atom i and j, $r_0(\alpha,\beta)$ is the first neighbor distance, and ξ is the hopping integral depending only on the interacting atomic species. The parameter q also is determined by the interacting atomic species. In order to provide stability to the crystal, a repulsive interaction term must also be provided, usually in the form of a pair potential which is added to the bonding contribution. In the Gupta potential this interaction is described by a sum of Born–Mayer type of ion–ion repulsions:

$$\phi(i) = E^R(i) = \sum_{j \neq i} A(\alpha,\beta) exp\left[-p(\alpha,\beta)\left(\frac{r_{ij}}{r_0(\alpha,\beta)} - 1 \right) \right]. \tag{9.129}$$

The total cohesive energy in the Gupta potential is the sum of the repulsive and attractive many-body interactions. The A parameters are determined by fitting the experimental values of cohesive energy to those calculated. The potential parameters in the case studies being reported are summarized in Table 9.8.

In the case study that follows we have developed and implemented a new soft algorithm—a variant of the random mutation hill climbing method, called fuzzy adaptive random mutation hill climbing (FARMHC). The FARMHC algorithm is a single string-based mutation evolution-only technique of fitness maximization in which two controllable parameters namely the mutation probability (p_m) and mutation intensity (Δm) are both adaptively determined from past experiences of exploration and the current stage of evolution under a fuzzy inference system. In addition, a strongly elitist strategy is used in selection which means that a randomly mutated string is accepted and passed on as a new individual for the next generation only if its fitness value is higher. The FARMHC and CARMHC share some

TABLE 9.8

Gupta Potential Parameters Used in the Case Study Reported

Metal Type M+N	A(eV)	p	q	r_0 (Å)	ξ (eV)
Cu-Cu	0.0855	10.960	2.278	2.556	1.2240
Cu-Au	0.1539	11.050	3.0475	2.556	1.5605
Au-Au	0.2061	10.229	4.0360	2.884	1.790

common features, but the adaptive learning scheme used here is not the same scheme originally employed in CARMHC. Let us briefly sketch the steps involved.

1. Randomly generate one string (S) of floating point numbers, each representing one or the other of the $3N$ Cartesian coordinates of the N atoms present in the cluster and calculate its fitness ($f_{initial}$). p_m and Δm values are initialized.

2. The string (S) is mutated at a randomly chosen position with probability p_m and intensity Δm, both adaptively determined using a fuzzy inference system (call the mutated string S').

3. If the fitness of the mutated string $f_{new} > f_{initial}$ then replace S with S' and set $f_{initial} = f_{new}$.

4. Go back to step 2 until a suitably chosen termination criterion has been satisfied.

There is a provision for restart (i.e., going back to step 1) and generating a new string by randomly perturbing the current best string and resetting the p_m and Δm values. Steps 2 to 4 are then carried out in the usual manner with a view to discovering a better minimum. The fuzzy adaptive control scheme actually used may now be explicated.

9.6.6 Fuzzy Adaptive Control of Mutation Parameters

There are two fuzzy controllers (inputs) which have been leveraged to determine the fuzzy output parameters representing mutation probability and mutation intensity. The first controller represents the current stage of the evolution (t_g) with $t_g = t/t_{max}$, t being the current generation number and t_{max}, the maximum number of generations over which evolution will be carried out. The second fuzzy controller used here represents the fraction (call it ratio), of mutated strings (configurations) that has been accepted over the last 100 reconfiguring moves, i.e., generations. Thus $ratio = N_{ac}/100$. The input parameters t_g and ratio are given membership degrees of the three fuzzy sets—low (L), medium (M), and high (H)—to which they can belong by making use of three triangular MFs. The fuzzy rules table (Table 9.9) is then invoked to determine which fuzzy sets the output variable p_m and Δm belong to. A defuzzification is then carried out to assign numerical (crisp) values to p_m and Δm to be used at that stage of the evolutionary search.

Thus, when t_g is low (i.e., the search has just begun) large p_m and Δm values will lead to a better exploration of the search space while in the terminal phase, p_m and Δm values would be automatically reduced in order to assist the search to converge to the optima discovered so far. The algorithm described here has been applied to clusters involving up to 40 atoms of copper and gold. The energy evolution profile for Cu_{40} along the FARMHC search path is displayed in Figure 9.11 along with snapshots of structures discovered in the intermediate stages of the search. In a majority of the cases, FARMHC succeeded in discovering in a

TABLE 9.9

Fuzzy If–Then Rules Table for Determining p_m and Δm

$T_g = t/t_{max}$	Ratio		
	L	M	H
L	H	H	M
M	H	M	L
H	M	L	L

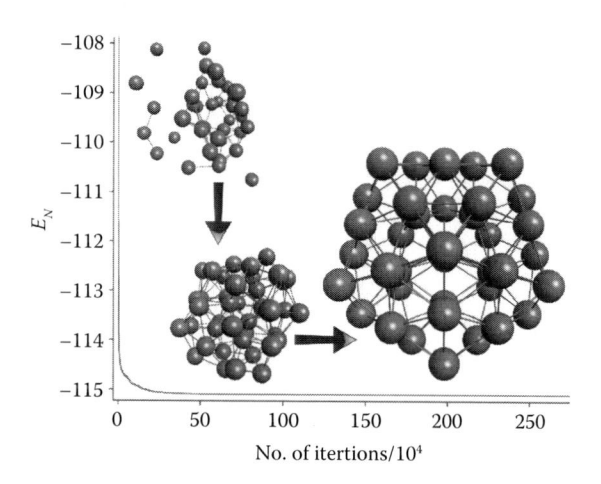

FIGURE 9.11

Energy evolution profile for 40 atom Cu cluster. (Courtesy of Sarkar, K., *Soft-Computing algorithms for solving some problems of Quantum Chemistry*, Indian Association for the Cultivation of Science, Jadavpur University Kolkata, India, 2014, 170.)

single run, the GM energy structure reported by Darby et al. [19], although, in some cases, the algorithm became stuck in local minima (see Table 9.10).

Pure copper clusters (Cu_m) have been, in general, found to have regular highly symmetric icosahedral geometries. At $m = 13$, a perfectly centered icosahedron defines the GM energy structure. All the clusters with $n = 7 - 12$ are found to be fragments of the 13 atom-centered-icosahedral geometry of Cu_{13}. Beyond $n = 13$, the added atoms start forming a second icosahedron and at $n = 19$, a double icosahedron appears as the GM energy structure. The icosahedral structural motif is not without exception. At $n = 40$, for example a decahedral minimum energy structure is discovered for Cu_n. In contrast to copper, the gold clusters are found not to adopt a single structural motif. In fact, they are mostly found to have rather low symmetry structure. Au_{12}, for example, has a flatter polytetrahedral structure while a majority of the larger gold clusters turn out to be amorphous. Au_{13} is an exception in having a centered icosahedral structure.

Compared to pure metal clusters, the bimetallic nanoalloys present a much harder optimization problem. This happens because of the increased complexity of the search space due to the presence of two different types of atoms, which opens up the possibility of formation of isomers based on the permutation of unlike atoms. In addition, isomers with different skeletal structures can occur. The relative lack of symmetry noticed among Au_n clusters vis-á-vis Cu_n clusters create an additional hurdle for the search for GM energy structures of Cu-Au nanoalloys. In fact, mixed clusters bear little or no structural resemblance to the corresponding pure clusters of the same size. In spite of the difficulties, the FARMHC algorithm was able to discover the GM energy structures of $(Cu\_Au)_n$ nanoalloys up to $n = 17$. The difficulties encountered at $n = 13, 14$ (see Table 9.10) apparently indicate that finer adaptive control on p_m and Δm along with occasional perturbation of the currently best geometry string could enhance the performance by increasing exploration potential. The flowchart of the FARMHC is displayed below (Figure 9.12).

The case studies presented in the preceding section are chosen to evoke interest in possible experimentation with fuzzy sets and fuzzy control not only in optimization methods, but also in other fields of research in physical and chemical sciences. Before leaving the present topic, we propose to present to the readers a brief account of a comparative

TABLE 9.10

The Minimum Energy Is Predicted by the FARMHC Algorithm for Cu_n, Au_n, and Cu_mAu_{n-m} Clusters

Cluster	Energy	Cluster	Energy	Cluster	Energy
Cu_{10}	24.300289	Au_{10}	32.160313	$Cu_5 Au_5$	28.849788
Cu_{11}	27.154276	Au_{11}	35.529095	--	--
Cu_{12}	30.277969	Au_{12}	38.922569	$Cu_6 Au_6$	35.402163
Cu_{13}	33.746311	Au_{13}	42.554931	--	--
Cu_{14}	36.264230	Au_{14}	45.958276	$Cu_7 Au_7$	42.190467
Cu_{15}	39.156871	Au_{15}	49.505451	--	--
Cu_{16}	42.008418	Au_{16}	53.057656	$Cu_8 Au_8$	48.735863
Cu_{17}	44.863249	Au_{17}	56.549399	--	--
Cu_{18}	47.851302	Au_{18}	60.010515	$Cu_9 Au_9$	55.161907
Cu_{19}	51.254009	Au_{19}	63.390827 (63.435109)	--	--
Cu_{20}	54.061470	Au_{20}	66.869621	$Cu_{10} Au_{10}$	62.125265
Cu_{21}	56.877228	Au_{21}	70.357450	--	--
Cu_{22}	59.829154	Au_{22}	73.920768	$Cu_{11} Au_{11}$	68.602768
Cu_{23}	63.138210	Au_{23}	77.478044	--	--
Cu_{24}	65.936807	Au_{24}	81.005936	$Cu_{12} Au_{12}$	75.451952
Cu_{25}	68.949164	Au_{25}	84.436362	--	--
Cu_{26}	72.065028	Au_{26}	87.933895	$Cu_{13} Au_{13}$	82.034540
Cu_{27}	74.981434	Au_{27}	91.501888	--	--
Cu_{28}	78.157439	Au_{28}	95.033548	$Cu_{14} Au_{14}$	88.796730
Cu_{29}	81.048961	Au_{29}	98.620399	--	--
Cu_{30}	83.917980 (84.012582)	Au_{30}	102.218298	$Cu_{15} Au_{15}$	95.448475
Cu_{31}	87.041581	Au_{31}	105.564459	--	--
Cu_{32}	90.273074	Au_{32}	109.142414	$Cu_{16} Au_{16}$	102.315599
Cu_{33}	93.297977	Au_{33}	112.662038	--	--
Cu_{34}	96.291701	Au_{34}	116.271123	$Cu_{17} Au_{17}$	109.152738
Cu_{35}	99.285280 (99.336863)	Au_{35}	119.773560	--	--
Cu_{36}	102.332958 (102.501833)	Au_{36}	123.307669	$Cu_{18} Au_{18}$	115.942364
Cu_{37}	105.597938	Au_{37}	126.964666	--	--
Cu_{38}	108.677139 (108.967591)	Au_{38}	130.540017	$Cu_{19} Au_{19}$	122.991484
Cu_{39}	112.057813 (112.088461)	Au_{39}	134.056743	--	--
Cu_{40}	115.095462	Au_{40}	137.572764	$Cu_{20} Au_{20}$	129.922694

Source: From Sarkar, K., *Soft-Computing algorithms for solving some problems of Quantum Chemistry,* Indian Association for the Cultivation of Science, Jadavpur University Kolkata, India, 2014, 171.

study of the hybrid fuzzy adaptive inertia weight PSO and CARMHC method with a hybrid of SA and GA methods. The method was developed at Indiana University (2013). The composite SA and GA (SAGA) [20] is a Fortran program implementing the simulated annealing and the genetic algorithm for global minimization of energy of Lennard–Jones clusters using a variety of packing motifs. Local minimization wherever required is done by either numerical or analytic gradient-based methods of minimization.

The annealing algorithm in SAGA is in the outer loop and reports the status of the search at the conclusion of each macroiteration. The first macroiteration of the search for the GM of a 23-atom LJ cluster performs 2300 (number of atom × 100) local geometry optimizations of the LJ_{23} cluster at an annealing temperature of 304.8 cm^{-1}. Out of the 2300 configurations sampled, 2025 were accepted by the Metropolis algorithm. The next macroiteration was

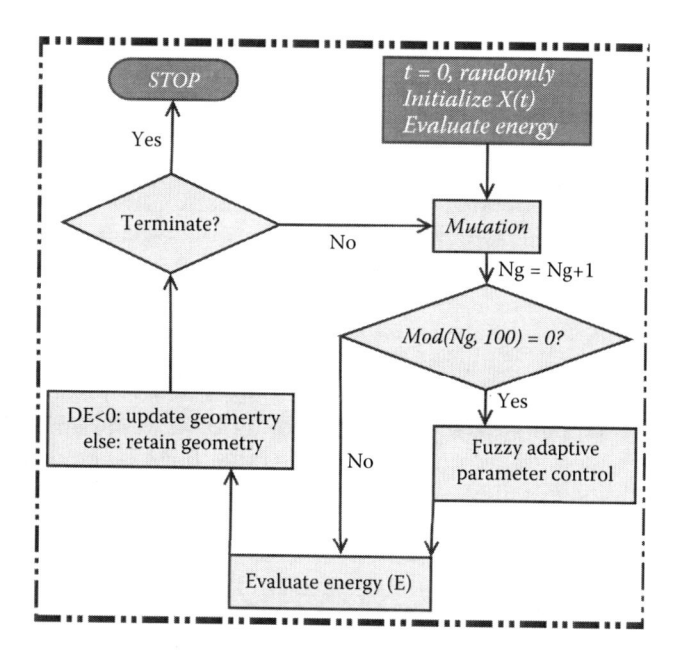

FIGURE 9.12
Flowchart for fuzzy adaptive random mutation hill climbing method. (Courtesy of Sarkar, K., *Soft-Computing algorithms for solving some problems of Quantum Chemistry*, Indian Association for the Cultivation of Science, Jadavpur University Kolkata, India, 2014, 169.)

carried out after scaling down the temperature by a factor of 0.75. The process was continued until the energy difference between the most and the least stable configuration was 0.0001 cm^{-1}. The annealing algorithm performed a total of 27,600 local optimizations before converging to the GM energy geometry ($E_{global} = -7483.3$ cm^{-1}). The trial configuration was generated by selecting one atom of the cluster and moving it to a different location on the surface of the cluster. A conjugate gradient minimization was carried out to identify the nearest local energy minimum on the LJ$_{23}$ potential energy surface. Once the local minimum has been identified, the selected atom is moved either to a random location on the cluster surface or onto a point on the principal rotation axes with equal probability. A move onto the principal rotation axis usually yields a trial configuration that is more spherical than the reference configuration and may have a lower energy. As we have mentioned already, a macroiteration consists of 23 × 100 macroiterations at a particular temperature with energy convergence being tested at the termination of a macroiteration. Unless convergence has been achieved, the temperature is reduced to $T_{new} = 0.75 \times T_{old}$ and the SA is allowed to proceed to the next macroiteration phase. The initial temperature is either user-specified or set by the algorithm itself by performing 10 × number of atoms reconfiguring moves and choosing the energy difference between the configurations of maximum and minimum energies as the initial (default option) temperature. It is possible that the SA macroiterations to converge to a configuration that has higher energy than a configuration sampled in one of the search steps. The SA is designed to always track the configuration with the lowest energy so far sampled. If it has converged to a configuration of higher energy, the SA search is restarted from the configuration of the lowest energy, ensuring that the SA always reports the lowest energy configuration ever visited during the search.

The GA implemented in SAGA is similar to the GA proposed by Deaven and Ho [21]. The GA in SAGA maintains a population consisting of a small (typically 20) number of cluster configurations or geometry strings (individuals in GA parlance). The GA begins by creating a diverse pool of configurations, the candidate structures being chosen from a selection of idealized icosahedral, cubic closed packed, hexagonal closed packed, and body-centered cubic configurations. A conjugate gradient optimization to the nearest local minimum is performed on every candidate structure. A carefully designed mating procedure is then employed to generate children configurations that inherit some traits of the two parent configurations. The two parents are randomly chosen from the population. The first parent is then made to undergo a random rigid rotation about its center of mass that coincides with the origin of the coordinate system. All atoms of the rotated parent 1 configuration lying below the xy plane are discarded. The second parent is also made to undergo a random rigid rotation and the atoms in the resulting configuration lying above the xy plane are discarded. A child configuration is subsequently assembled by continuing the upper half of the first parent with the lower half of the second parent with adequate care to ensure that the total numbers of atoms are conserved during the process. Furthermore, atoms lying at the interface of the two halves are moved to the surface of the cluster if close contacts beyond a limit are detected. The resulting child configuration is finally optimized to the nearest local minimum by invoking the conjugate gradient method. The resulting configuration serves as a child configuration and has the potential to enter the population by replacing one of the preexisting candidate structures. The algorithm targets the candidate of the highest energy for replacement if the child configuration has lower energy than the candidate having the highest energy in the population. The candidate is removed and the child assumes its position in the population. The child also replaces the candidate if it is a mutant. A child configuration of energy E_{child} is designated a mutant if a random real number drawn from the range (00–1.0) is less than the mutation rate (4% by default). Mutations help maintaining diversity in the population and allow the GA to sample a wider range of configurations. That increases the exploration ability of the search and facilitates descent into the GM. Had the mutants been consistently discarded, the algorithm would have developed an increased propensity to converge to a local minimum.

A comparison of the performance level of the SA or GA of SAGA with that of the fuzzy adaptive inertia weight PSO hybridized with CARMHC and steepest descent method described in the preceding section, when applied to the global energy minimization problem of LJ clusters of different sizes including the well-known difficult-to-locate clusters (LJ$_{38}$, LJ$_{75}$, or LJ$_{102}$, for example) is quite revealing. Figure 9.13a through c displays energy evolution profiles produced by different methods when applied to LJ$_{60}$, LJ$_{75}$, and LJ$_{102}$ clusters. All the three methods compared located the GM of LJ$_{60}$ (Figure 9.13a) albeit the hybrid of fuzzy adaptive inertia weight PSO with CARMHC + SD converged faster. The SA and GA of SAGA, however, failed to hit the GM of the harder benchmark problems posed by LJ$_{75}$ and LJ$_{102}$ clusters while the hybrid of fuzzy adaptive inertia weight PSO + CARMHC + SD smoothly converged to the GM in each case (Figure 9.13b and c). A few general observations on the performance of the methods compared would be useful to readers:

1. The annealing algorithm in SAGA generally required a huge number of iterations to discover the GM on the PES of LJ$_n$ clusters wherever it converged to the GM.

2. The GA in SAGA could locate the GM in all such cases, although the energy of each candidate structure in the population had to be locally minimized by the conjugate gradient method in every generation.

3. Both the SA and GA in SAGA failed to locate the GM on the PES of LJ$_n$ clusters with $n = 19, 38, 75,$ and 102. These clusters are notoriously well known to present a very difficult search landscape.

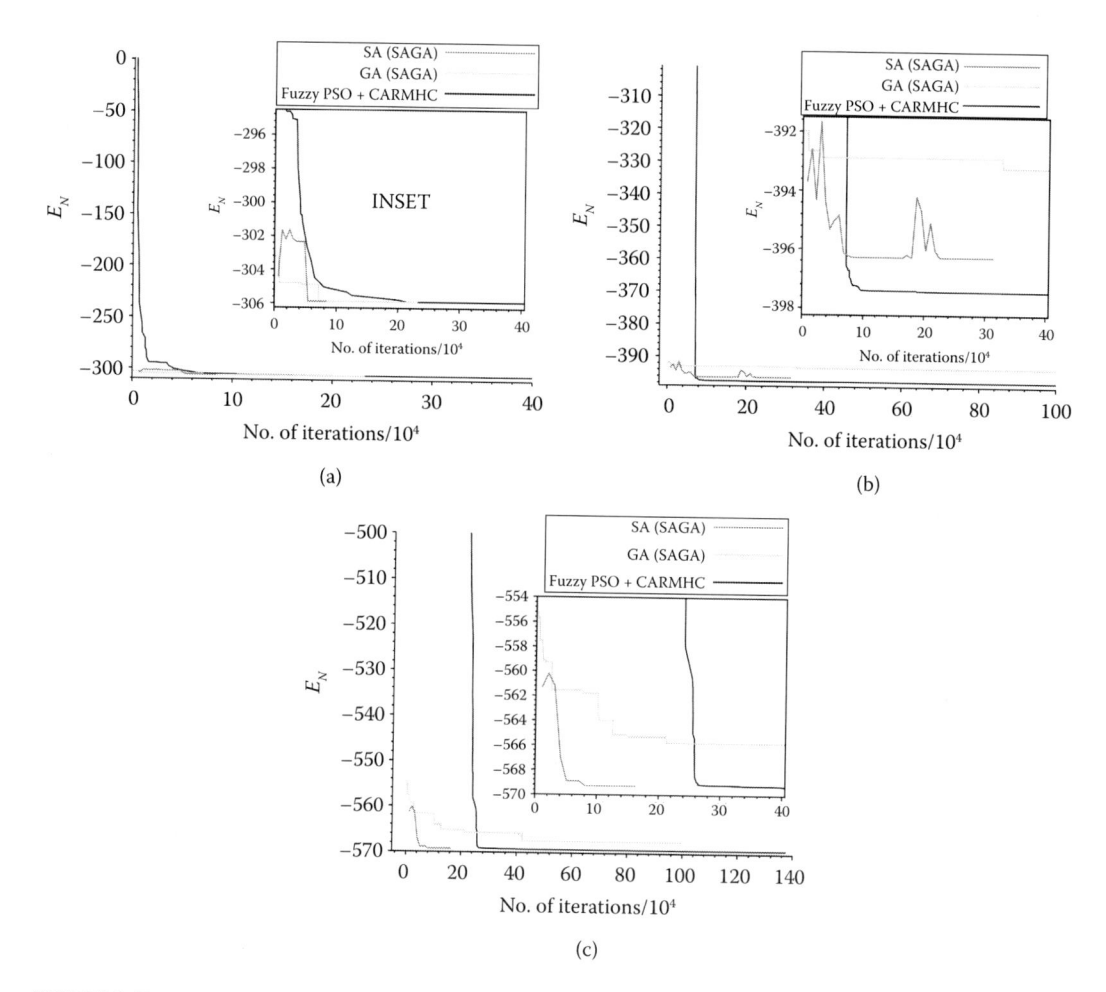

FIGURE 9.13

Comparison between the genetic algorithms in SAGA and the fuzzy adaptive inertia weight-based PSO (near optimal region). In inset, the optimal region has been shown in expanded scale. Genetic algorithms in SAGA start with a very good geometry whose energy is very close to the GM, whereas fuzzy adaptive inertia weight-based PSO starts with relatively higher energy and takes unnecessarily longer times at the end. That is why we thought to hybrid it with CARMHC. (a) LJ_{60}, (b) LJ_{75}, and (c) LJ_{102}. (Courtesy of Sarkar, K., *Soft-Computing algorithms for solving some problems of Quantum Chemistry*, Indian Association for the Cultivation of Science, Jadavpur University Kolkata, India, 2014, 160.)

4. The hybrid of fuzzy adaptive inertia weight PSO, CARMHC, and SD, however, succeeded in locating the GM of all the clusters rather smoothly including those that are known to present hard-to-search energy landscapes. It requires more generations (note, however, that it starts from rather high energy configurations) than the SA or GA of SAGA but consumes less CPU time. We must caution, however, that fuzzy adaptation as proposed in this chapter is not the panacea to solve all the difficulties encountered by soft optimization algorithms in hard (possibly NP-hard) problems. There is ample scope and need for further experimentation and fine-tuning of the algorithm for better adaptive response to the demands of complex search landscapes. It would also be useful to pay more attention to preparing the candidates for the starting generations.

9.6.7 Fuzzy Control in Mass Spectroscopy

Matrix-assisted laser desorption/ionization mass spectrometry (MALDI) has developed into a highly viable, sensitive, and versatile method for analysis of biomolecules or polymers. Conventional methods of ionization tend to cause extensive fragmentation of the target molecules like DNA, proteins, peptides, sugars, organic polymers, or macromolecules, and generates a lot of multiply-charged ions. MALDI like electron spray ionization is a method that allows soft ionization of the targeted large molecules in the gas phase allowing their subsequent analysis in the mass spectrometer. MALDI, let us briefly recall, operates in three steps:

1. The sample is mixed with a suitably chosen matrix material and the resulting mixture is applied or spread onto a metal plate.

2. The mixture on the plate is exposed to a pulsed laser which causes ablation and desorption of the sample as well as the matrix material.

3. The analyte molecules suffer protonation or deprotonation in the hot plumes of the ablated gases generating ions which are accelerated and directed into the mass spectrometer by control fields, for analysis/characterization. The MALDI has the potential for high throughput screening. The problem is that it requires time as well as human intervention. Thus, for obtaining mass spectrum of optimal quality, the laser fluence must be continuously adjusted when data acquisition is taking place to ensure that the system achieves close to the threshold level of ion production. That demands we must not only have a skilled operator, but also spend several minutes in data acquisition for each sample. Jensen et al. [22] demonstrated how fuzzy logic control of the laser fluence could lead to automated data acquisition in MALDI. Their control algorithm evaluates the signal intensity as well as mass resolution of the base peak and uses the information to maintain the ion signal intensity within the dynamic range of the data acquisition hardware, without sacrificing the high mass resolution. The authors clearly established that even for difficult samples such as the femto level peptide mixture, the data quality remains excellent when compared to manually obtained spectra. These authors used intensity and mass resolution as fuzzy input variables and employed "laser fluence change" as the output (fuzzy) variable. Below are the "fuzzy if–then rules" the authors make use of:

9.6.8 Intensity Rules

Rule 1. If intensity is very low, then fluence change is small positive.

Rule 2. If intensity is low, then fluence change is small positive.

Rule 3. If intensity is average, then fluence change is zero.

Rule 4. If intensity is high, then fluence change is small negative.

Rule 5. If intensity is very high, then fluence change is large negative.

The fuzzy rule for mass resolution used is "if low resolution, then fluence change is large negative."

Using these rules for controlling the input-to-output mapping the authors demonstrated the viability of fuzzy control in MALDI, with automated analysis of 78 chromatographic fractions with high mass accuracy. The stability, flexibility, and easy implementability of the fuzzy logic-based control in MALDI had been so successful that the authors were prompted to canvas for wider use of fuzzy control in mass spectrometry.

9.6.9 Fuzzy Data Clustering

The clustering process involves identifying natural grouping (clusters) within multi-dimensional data. The process requires use of some similarity or distance measures and is an important problem in the fields of pattern recognition and machine learning. A cluster is usually defined by a cluster center, also called cluster centroid. The main task is therefore to identify the cluster centroids (C_1, C_2, etc.). The process of data clustering is a hard-to-solve problem in the area of unsupervised pattern recognition. The difficulty is rooted in the fact that the clusters embedded in the data may have different shapes as well as sizes, of which no prior information is usually available to a clustering algorithm.

The crisp or hard clustering problem can be formally defined in the following way:

Suppose that a data set $P = (p_1, p_2, \ldots, p_N)$ is given where p_l is a pattern in the n_d dimensional feature space (i.e., p_l is a vector with n_d real components) and there are N such patterns in P. The clustering task is then to partition P into, let us say, K clusters $(C_1, C_2, \ldots, C_{k-1}, C_k)$ subject to the following conditions being satisfied:

1. $U_{k=1}^{K}(C_k) = P$ (ensures that each pattern has been assigned to one of the K clusters).

2. $C_k \neq 0$, for $k = 1, 2, \ldots, K$ (ensures that each cluster has at least one pattern assigned to it).

3. $C_k \cap C_{kk} = \phi$, if $k \neq kk$ (ensures that a pattern is assigned to one and only one cluster).

To accomplish the clustering, as we have mentioned already a similarity (distance) measure is necessary. The most popular and commonly used measure is the Euclidean distance $d(p_\alpha, p_\beta)$ between two patterns p_α and p_β in P defined as follows:

$$d\left(p_\alpha, p_\beta\right) = \left\{ \sum_{i=1}^{n_d} (p_{\alpha,i}, p_{\beta,i})^2 \right\}^{\frac{1}{2}}. \tag{9.130}$$

One of the popular methods of clustering is the so-called partitional clustering technique in which the algorithm tries to divide or partition the data set into a predefined number of clusters by minimizing a certain measure of error. It would therefore be apt to brand the clustering process by partitioning as an optimization problem. Such optimization problems commonly belong to the NP-hard category. The following steps are executed in clustering by partitioning:

Step 1. Randomly initialize the K cluster centroids $(C_1, C_2, \ldots, C_K)$.

Step 2. For each pattern p_i in the data set, calculate its membership $m(C_k | p_i)$ to each centroid C_k and also determine its weight $\omega(p_i)$.

Step 3. Recalculate the K-cluster centroids using the following formula:

$$C_k = \sum_{p_\alpha} \frac{m\left(C_k \mid p_\alpha\right) \omega\left(p_\alpha\right) p_\alpha}{\sum_{p_\alpha} m\left(C_k \mid p_\alpha\right) \omega\left(p_\alpha\right)}. \tag{9.131}$$

Note that $m(C_k|p_\alpha)$ is the MF that quantitatively measures the degree of membership of the pattern p_α to the cluster k with centroid C_k. The MF $u(C_k|p_\alpha)$ is required to obey following conditions:

1. $m(C_k|p_\alpha) \geq 0$, for $\alpha = 1, 2, \ldots, N; k = 1, 2, \ldots, K$.

2. $\sum_{k=1}^{K} m(C_k|p_\alpha) = 1$, $\alpha = 1, 2, \ldots, N$.

As we have already seen in this chapter, if the clustering is hard or crisp, $m(C_k|p_\alpha)$ can either be 1 or 0, emphasizing that the pattern p_α either belongs to the cluster C_k or does not. In fuzzy clustering the MF $m(C_k|p_\alpha)$ can have any value from 0 to 1. The weight function $\omega(p_\alpha) > 0$ reflects how important the pattern p_α is. The iterative K-means the algorithm is the most commonly used and popular method of crisp clustering of data based on the minimization of the following objective function O_k where

$$O_k = \sum_{k=1}^{K} \sum_{p_\alpha \in C_k} d^2(p_\alpha, C_k). \tag{9.132}$$

The K-means algorithm [23] begins with K centroids, the initial values of the centroids being chosen either randomly or based on specific information. Each pattern in the data set (p_α) is then assigned to the cluster centroid to which the pattern lies the closest (in Euclidean distance measurement). Once each cluster and its associated centroids have been identified, the centroids of K cluster are recalculated by using the standard formula (Equation 9.131). The process is repeated until one of the following convergence criteria has been satisfied:

1. The changes in centurial coordinates become smaller than a user-defined threshold.
2. A preset limit of maximum number of iterations has been reached.

The MF used in crisp K-means clustering can be defined as follows:

$$m(C_k|p_\alpha) = \begin{cases} 1, \text{ if } d^2(p_\alpha, C_k) = \arg\min_k \{d^2(p_\alpha, C_k)\} \\ 0, \text{ otherwise} \end{cases}. \tag{9.133}$$

The weight of p_α, i.e., $\omega(p_\alpha) = 1$, for $\alpha = 1, 2, \ldots, N$. As the reader can appreciate, the K-means clustering is rather easy to implement and has a time complexity $O(N)$, so it can handle large datasets. That the user must specify the number of clusters at the outset and that it can converge to a suboptional solution are some of the negative features that can limit its usefulness.

A much better approach is the fuzzy version of the K-means algorithm known as fuzzy C-means or fuzzy K-means algorithm [24]. The major difference between fuzzy C-means and the crisp K-means algorithms is that in the former each pattern p_α is assigned to each cluster (C_k) with a degree of membership. The idea is more realistic since in a real body of data, the clusters may have some overlapping regions. Along with fuzzy membership degree, the objective function that is to be minimized in fuzzy C-means clustering algorithm also differs from the K-means counterparts. The objective function of fuzzy C-means the algorithm is O_C:

$$O_C = \sum_{k=1}^{K}\sum_{\alpha=1}^{N} u_{k,p_\alpha}^r d^2\left(p_\alpha, C_k\right)$$

(9.134)

where r is the fuzzy exponent (≥ 1). The larger the value of r, the higher is the fuzziness of the algorithm. As usual m_{k,p_α} is the membership value (degree) of the pattern p_α to the kth cluster. $\{m_{k,p_\alpha}\}$ satisfy the following conditions:

1. $m_{k,p_\alpha} > 0$, $\alpha = 1, 2, \ldots, N; k = 1, 2, \ldots, K$.

2. $\sum_{k=1}^{K} m_{k,p_\alpha} = 1$, for $\alpha = 1, 2, \ldots, N$.

The MF in FCM is defined in the following way:

$$m_{k,p_\alpha} = \frac{\left\|p_\alpha - C_k\right\|^{-2/(r-1)}}{\sum_{k=1}^{K}\left\|p_\alpha - C_k\right\|^{-2/(r-1)}}.$$

(9.135)

The weight function $\omega\left(p_\alpha\right)$ is assigned the value of unity for $\alpha = 1, 2, \ldots, N$.

In practical applications, the fuzzy C-means algorithm almost always is found to have edge over the crisp K-means algorithm [25] although convergence to a local optimum cannot always be avoided. Needless to mention is the fact that like the crisp K-means algorithm FCM also requires the user to specify the number of clusters (K) in advance, and therefore shares the same disadvantage that K-means algorithm suffers from. However, FCM handles noise in the data better. A novel implementation of a data clustering algorithm within the framework of PSO has been very successful.

9.7 Fuzzy Sets in Computational Chemistry

Chemistry is replete with fuzzy concepts. In fact the very idea of molecular structure and hardness has been called fuzzy classical concepts. Anton Amann [26] examined if it was at all possible to understand the emergence of the fuzzy classical behavior at some appropriate limit where crisp set theory takes over and then calculate quantum deviation for finitely many degrees of freedom or finite nuclear masses within fuzzy set theory. Both local and global shape properties of molecules have been described using a fuzzy set formalism. Recently, fuzzy electron density modeling has been introduced in computational chemistry. The additive fuzzy density fragmentation (AFDF) scheme of Mezey [27] has been very useful. The simplest version of it is the Mulliken–Mezey AFDF method in the molecular electron density lego assembler (MEDLA) technique of Walker and Mezey [28] for generating ab initio quality electron density for macromolecules. An important application of electrostatic Hellmann–Feynman theorem within AFDF framework has led to a novel macromolecular geometry optimization technique [29].

References

1. Zadeh, L.A. 1965. Fuzzy Sets. *Information and Control* 8, no. 3: 338–353.
2. Zadeh, L.A. 1968. Fuzzy Algorithms. *Information and Control* 12, no. 2: 94–102.
3. Munakata, T. 2007. *Fundamentals of the New Artificial Intelligence*. Munakata, T. (ed.). Texts in Computer Science. London: Springer.
4. Rajasekaran, S., and G.A.V. Pai. 2003. Neural networks, fuzzy logic and genetic algorithm: Synthesis and applications (with CD). New Delhi, India: PHI Learning.
5. Chen, Y., Z. Cui, J. Yin, and Y. Tan. 2011. Global Minimum Structure Optimisation of Lennard-Jones Clusters by Hybrid PSO. *International Journal of Modelling, Identification and Control* 14, no. 4: 303.
6. Dugan, N., and S. Erkoc. 2009. Genetic Algorithm-Monte Carlo Hybrid Geometry Optimization Method for Atomic Clusters. *Computational Materials Science* 45, no. 1: 127–132.
7. Voglis, C., K.E. Parsopoulos, D.G. Papageorgiou, I.E. Lagaris, and M.N. Vrahatis. 2012. MEMPSODE: A Global Optimization Software Based on Hybridization of Population-Based Algorithms and Local Searches. *Computer Physics Communications* 183, no. 5: 1139–1154.
8. Pardalos, P.M., D. Shalloway, and G. Xue. 1994. Optimization Methods for Computing Global Minima of Nonconvex Potential Energy Functions. *Journal of Global Optimization* 4, no. 2: 117–133.
9. Hoare, M.R. 2007. Structure and Dynamics of Simple Microclusters. In *Advances in Chemical Physics*, Ilya Prigogine, Stuart A. Rice (ed.), vol. 40, pp. 49–135. New York: John Wiley & Sons.
10. Sarkar, K., and S.P. Bhattacharyya. 2013. Single String Based Global Optimizer for Geometry Optimization in Strongly Coupled Finite Clusters: An Adaptive Mutation-Driven Strategy. *The Journal of Chemical Physics* 139, no. 7: 74106.
11. Sarkar, K., and S.P. Bhattacharyya. 2013. Computationally Efficient Algorithm in Cluster Geometry Optimization. In *Solid State Physics: Proceedings of the 57th DAE Solid State Physics Symposium 2012*, vol. 1512, pp. 162–163, Bombay, India: AIP Publishing.
12. Sarkar, K., R. Sharma, and S.P. Bhattacharyya. 2010. Blending Determinism with Evolutionary Computing: Applications to the Calculation of the Molecular Electronic Structure of Polythiophene. *Journal of Chemical Theory and Computation* 6, no. 3: 718–726.
13. Sarkar K. 2014. Soft-Computing algorithms for solving some problems of Quantum Chemistry. PhD thesis, Indian Association for the Cultivation of Science, Jadavpur University, Kolkata, India.
14. Clerc, M., and J. Kennedy. 2002. The Particle Swarm—Explosion, Stability, and Convergence in a Multidimensional Complex Space. *IEEE Transactions on Evolutionary Computation* 6, no. 1: 58–73.
15. Eberhart, R.C., and Y. Shi. 2001. Particle Swarm Optimization: Developments, Applications and Resources. In *Proceedings of the 2001 Congress on Evolutionary Computation (IEEE Cat. No.01TH8546)*, 27–30 May 2001, Seoul, South Korea. vol. 1, pp. 81–86, South Korea: IEEE.
16. Rossi, G., R. Ferrando, A. Rapallo, A. Fortunelli, B.C. Curley, L.D. Lloyd, and R.L. Johnston. 2005. Global Optimization of Bimetallic Cluster Structures. II. Size-Matched Ag-Pd, Ag-Au, and Pd-Pt Systems. *The Journal of Chemical Physics* 122, no. 19: 194309.
17. Fernández, E.M., L.C. Balbás, L.A. Pérez, K. Michaelian, and I.L. Garzón. 2005. Structural Properties of Bimetallic Clusters from Density Functional Calculations. *International Journal of Modern Physics B* 19, no. 15–17: 2339–2344.
18. Gupta, R.P. 1981. Lattice Relaxation at a Metal Surface. *Physical Review B* 23, no. 12: 6265–6270.
19. Darby, S., T.V. Mortimer-Jones, R.L. Johnston, and C. Roberts. 2002. Theoretical Study of Cu–Au Nanoalloy Clusters Using a Genetic Algorithm. *The Journal of Chemical Physics* 116, no. 4: 1536–1550.
20. Glendening, E.D. 2012. *SAGA 1.0*. Terre Haute, IN: Department of Chemistry and Physics, Indiana State University.
21. Deaven, D.M., and K.M. Ho. 1995. Molecular Geometry Optimization with a Genetic Algorithm. *Physical Review Letters* 75, no. 2: 288–291.

22. Jensen, O.N., P. Mortensen, O. Vorm, and M. Mann. 1997. Automation of Matrix-Assisted Laser Desorption/Ionization Mass Spectrometry Using Fuzzy Logic Feedback Control. *Analytical Chemistry* 69, no. 9: 1706–1714.

23. Forgy, E.W. 1965. Cluster Analysis of Multivariate Data: Efficiency versus Interpretability of Classifications. *Biometrics* 21, no. 3: 768–769.

24. Bezdek, J.C. 1981. *Pattern Recognition with Fuzzy Objective Function Algorithms*. Boston, MA: Springer.

25. Hamerly, G., and C. Elkan. 2002. Alternatives to the K-Means Algorithm That Find Better Clusterings. In *Proceedings of the Eleventh International Conference on Information and Knowledge Management—CIKM '02*, November 4–9, McLean, VA. Nicholas, C., Grossman, D., Kalpakis, K., Qureshi, S., van Dissel, H., and Seligman, L. (eds.), vol. 4, pp. 600–607. New York: ACM Press.

26. Amann, A. 1997. Fuzzy Classical Structures in Genuine Quantum Systems. In *Fuzzy Logic in Chemistry*, Dennis H. Rouvray (ed.), pp. 91–138. San Diego, California: Elsevier.

27. Mezey, P.G. 1995. Methods of Molecular Shape Similarity and Topological Shape Design. In *Molecular Similarity in Drug Design*, P. M. Dean (ed.), pp. 241–268. Dordrecht, Netherlands: Springer .

28. Walker, P.D., and P.G. Mezey. 1993. Molecular Electron Density Lego Approach to Molecule Building. *Journal of the American Chemical Society* 115, no. 26: 12423–12430.

29. Mezey, P.G. 1999. Descriptors of Molecular Shape in 3D. In *From Chemical Topology to Three-Dimensional Geometry*, Balaban, A.T. (ed.), pp. 25–42. New York: Kluwer Academic Publishers.

10

Quantum and Soft Computing

10.1 Introduction

Soft computing as we have repeatedly stressed (Chapters 1–5) marks a remarkable shift in the computing paradigm. The deterministic philosophy of numerical analysis-based computing is being replaced by biology-inspired techniques of new artificial intelligence wherein a nondeterministic search for solutions to a problem plays a dominant role. A nondeterministic search philosophy does not, however, necessarily imply a shift from the classical theories of information processing or the ideas of classical physics based on which the electronic logic gates are engineered and computers that we see everywhere work. A second shift in the computing paradigm that involves a radically different scheme of representing and processing information has been slowly taking shape in the guise of quantum computing. Unlike soft computing, or traditional computing, quantum computing makes explicit use of specific quantum mechanical features in representing and processing information, and requires modifications in both software and hardware, for the new paradigm to work to its full potential. Such quantum features may perhaps be grafted into the framework of soft computing as well, yielding what may be called quantum versions of soft computing methods. The present chapter focuses on understanding the essential features of quantum computing and its possible interfaces with soft computing techniques. Before such interfacing of two new computing paradigms is considered, it is essential to understand what quantum computing stands for in terms of software, hardware, and computational complexities.

In a very broad way, we may take the stand that all computing are ultimately guided by quantum science. As Landauer [1] puts it, any information processed by a computer is physical, i.e., it concerns a physical system made up of atoms and molecules which are quantum mechanical entities. Even the hardware which provides the platform on which electronic gates and memory devices are implemented requires a material medium composed of atoms and molecules which are guided by the quantum rules of the microscopic world. In the ultimate analysis therefore, any form of computing and computers, big or small, can be linked to the quantum world. Even if such a broad view is accepted, it only points to an underlying implicit quantum connection.

Ordinary computers do not make any explicit use of the laws of quantum mechanics neither in the hardware design nor in the software, and are therefore considered classical in nature. Quantum computing or a quantum computer leverages certain special features of quantum mechanics to accomplish very complex tasks with incredible speeds unthinkable in the classical world of computation, thereby reducing the computational complexity of these problems to manageable limits. The special features of quantum mechanics that quantum computing exploits to achieve what is unachievable in classical computing are "quantum parallelism," "quantum interference," and the notion of "entangled states" or the idea of "quantum entanglement." In a way, they are linked to the principle of linear superposition of basis states and the principle of quantum measurement. We will take a close look at these features and principles in the following sections and subsections, and try to understand how quantum computing achieves the complexity advantage over classical computing and how quantum computers can do what classical computers cannot. The answers to these questions are still evolving and the final word has not yet been spoken. However, little that has been already been already demonstrated about the algorithmic superiority of quantum computing in solving problems considered intractable in classical computing is enough to excite the imagination and evoke interest in the new computing paradigm. In addition to the hint of awesome computing power that quantum computing could unleash, there is also a sense of inevitability about the emergence of quantum computing as the future tool for handling complex computing problems. The inevitability stems from the observed exponential rate at which the sizes of the devices on which the information is stored and processed fall every year. Gordon Moore's law [2] predicts that by 2020, the device size will reach atomic dimensions where quantum rules would automatically take over, forcing us to give up classical computing and take to quantum computing, whether we like it or not. Of course, Moore's law could fail due to unforeseen technological glitches impeding miniaturization beyond a limit. The experience so far does not indicate any such possibility whatsoever. Feynman [3] first noted that quantum phenomena may not be always amenable to simulation on classical computers because of exponential rise in the required resources. He went on to conjecture that quantum computers could be programmed to simulate any local quantum system—a claim that was supported by others [4]. The idea of qubits, quantum gates, and quantum computers soon took over.

10.2 Bits versus Qubits

Computing in any form involves storage and retrieval of information and processing it. The fundamental unit of storing information on a classical computer is called a "bit." A bit can exist in one of the two states at a time conventionally labeled as 0 and 1. The two states are unique and may physically correspond to the "off" and "on" positions of switches or gears, polarity, charge level, or any other attribute of a physical system that can be unambiguously assigned one of two values. Computation consists of processing the information by application of a well-defined sequence of operations called "gates" (electronic devices) which act on individual bits or a collection of bits producing new information. Of course the bits and gates are embedded in a material medium which can affect the "clock speed" and the size of the computer as well as its energy efficiency. For theoretical analysis, however, the fundamental computing power of the computer can be regarded as largely independent of the material medium on which bits and gates are implemented. It essentially means that two computers having the same storage capacity and supporting the same set of gate operations can be regarded as equivalent in terms of computing power. On a classical computer, one bit of information can be either in state 0 (the bit is off) or in state 1 (the bit is on). If two bits of information are there, 2^2, i.e., four classical states can be defined: "00," "01," "10," and "11." No other states are possible. Similarly, three classical bits can uniquely define 2^3 or eight states of the register "000," "001," "010," "100," "110," "011," "101," and "111." In general, the representation of an n-bit system in classical computing is just the concatenation of n one-bit systems. Thus there are only 2^n states of the classical n-bit register—no other states of the register exist. The situation is dramatically different in quantum computers. The fundamental unit of information in quantum computers is called a "qubit" (quantum bit) which can exist not only in the states $|0\rangle$ or $|1\rangle$, but also in infinitely more states formed by the superposition of the two fundamental computing basis states (labeled $|0\rangle$ and $|1\rangle$ following Dirac's bracket notation). Thus the one qubit quantum register can exist in any one of the states $|\psi\rangle$ where

$$|\psi\rangle = a_0|0\rangle + a_1|1\rangle \tag{10.1}$$

a_0, a_1 being complex numbers obeying the condition

$$|a_0|^2 + |a_1|^2 = 1. \tag{10.2}$$

If $a_0 = 1$, $a_1 = 0$, the qubit exists in the quantum state $|0\rangle$; if $a_0 = 0$, $a_1 = 1$, the qubit is in state $|1\rangle$. They are the quantum analogues of the classical states 0 and 1. If $a_0 \neq 0$, $a_1 \neq 0$, the qubit exists in states that have no classical analogues. In these superposed states, $|a_0|^2$ represents the probability that the qubit is found in the state $|0\rangle$ while $|a_1|^2 = 1 - |a_0|^2$ represents the probability of finding the qubit in the state $|1\rangle$. In the state of superposition of the qubit the quantities a_0, a_1 are evolving following the law of quantum mechanics (Schrödinger equation). A measurement on the qubit, however, collapses the superposition onto one of the two basis states or eigenkets $|0\rangle$ or $|1\rangle$ with a probability $|a_0|^2$ or $|a_1|^2$. The measurement outcome is thus known with the probability $|a_0|^2$ or $|a_1|^2$. If $a_0 = a_1 = \dfrac{1}{\sqrt{2}}$, for example, half the number of measurements on the qubit will find it in the state $|0\rangle$ and half the number of measurements will discover the qubit in the state $|1\rangle$. It is not possible to predict a priori which measurement will discover the qubit in which of the two basis states. The prediction about the state of superposition of the qubit is purely statistical in nature. This makes a complete departure from classical computing where a bit can be either in the state 0 or in the state 1 and a measurement will discover the bit in one of the two classical computing basis states with certainty.

Let us now consider the case of a 2-qubit quantum register. Two bits can exist in classical computers in one of the four states "00," "01," "10," and "11." A 2-qubit quantum register can exists in any one of the four combinations of fundamental basis states ($|0\rangle,|1\rangle$) i.e., $|00\rangle$, $|01\rangle$, $|10\rangle$, and $|11\rangle$ or in a superposition of all of them. Thus, in general, the states of the 2-qubit register can be most generally represented by the superposition

$$|\psi\rangle = a_0|00\rangle + a_1|01\rangle + a_2|10\rangle + a_3|11\rangle \tag{10.3}$$

where the superposition coefficients satisfy the relation (a_i s are complex numbers)

$$\sum_{i=0}^{3} |a_i|^2 = 1. \tag{10.4}$$

Once again, a measurement on the register in state $|\psi\rangle$ is expected to discover it in the state $|00\rangle$ with probability $|a_0|^2$, in the state $|01\rangle$ with probability $|a_1|^2$, in the state $|10\rangle$ with probability $|a_2|^2$, and in the state $|11\rangle$ with probability $|a_3|^2$. On measurement, the superposition collapses nondeterministically into one of the four combinations of the fundamental basis states with a probability defined by the square of the amplitude of the combination in the superposition. For an n-qubit register, the state is defined by the superposition of the 2^n combinations (products) of n fundamental basis states ($|0\rangle, |1\rangle$) and can be written as

$$|\psi\rangle = \sum_{k=0}^{2^n-1} a_k |i_1, i_2, ..., i_n\rangle_k \tag{10.5}$$

where $|i_1\rangle, |i_2\rangle, |i_n\rangle$ can each be one of the two fundamental basis states, $|0\rangle$ or $|1\rangle$. The representation of the states of an n-qubit quantum register therefore requires storage of 2^n superposition amplitudes $a_k (k = 0, 1, 2, ..., 2^{n-1})$. It is important to note that the state of n-qubit register is not just the concatenation of the n 1-qubit states $|i_1\rangle, |i_2\rangle, |i_n\rangle$ where $|i_1\rangle$ = either $|0\rangle$ or $|1\rangle$ (Boolean), but a superposition of all possible concatenations of the n 1-qubit states. This is in marked contrast with what is seen in classical computing where the register is definitely in one of the possible 2^n concatenations $(i_1, i_2, ..., i_n)$ of the n bits each of which assumes the value 0 or 1 uniquely—so the state of the classical register is defined as $i_1, i_2, ..., i_l, ..., i_n$ with $i_l = 0$ or 1, $l = 0, 1, 2, ..., n$.

The representation of the states of an n-qubit quantum register is therefore much more complex. The complexity is further increased when we recognize that all the n-qubits may not be independent (in classical computing they are indeed independent)—but may be entangled. The quantum entanglement has been exploited in a number of quantum algorithms with which we will make contact later in this chapter.

10.3 Gates versus Quantum Gates

In classical computers, gates are applied to the state of a classical register (a unique combination of bits, e.g., 1001) producing a new state in which the bit structure is altered (0110, say), producing new information. The classical *NOT* gate, for example, acts on a bit and flips it to 1 if it is 0 and to 0 if it is 1. Operationally, the gate is represented as follows (Figure 10.1a):

The gate as we can see acts on the bits. The quantum gates (q-gates) are different. The q-gates can be formally represented as matrices that act on the state ($|\psi\rangle$) of a quantum register (quantum computer) and alter the state to $|\psi'\rangle$, say. The action can be represented formally as multiplication of the q-gate matrix (M) with a column vector (A) containing the amplitudes (a_ks of Equation 10.5) of the basis states. The multiplication produces a new set of amplitudes (a'_K, say) defining the new state $|\psi'\rangle$ where

$$|\psi'\rangle = \sum_{k=0}^{2^n-1} a'_k |i_1, i_2, ..., i_n\rangle_k. \tag{10.6}$$

Since the squares of the absolute values of the new amplitude ($|a'_K|^2$) should add up to 1 ($\Sigma |a'_K|^2 = 1$) just as the squares of the absolute values of the old amplitudes ($|a_K|^2$) did ($\Sigma |a_K|^2 = 1$), and since a_ks are complex numbers, the q-gate matrices M are constrained to be unitary. That is

$$MM^\dagger = M^\dagger M = I. \tag{10.7}$$

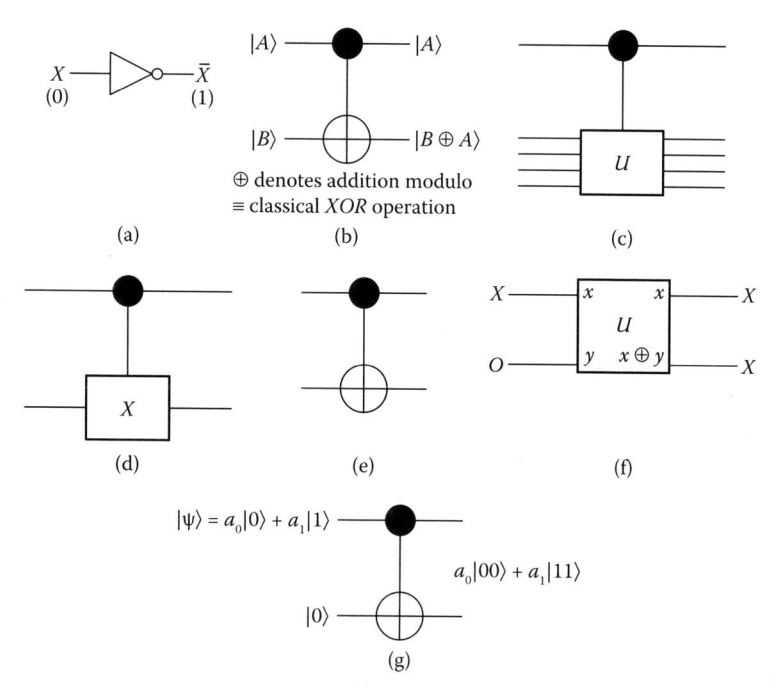

FIGURE 10.1
(a–g) Action of a classical 1-bit *NOT* gate.

$M^\dagger$ in Equation 10.7 is the complex conjugate transpose of M (the Hermitian adjoint). The action of q-gates therefore directly affects the amplitudes of the basis states without altering the basis states (qubit product states) themselves. As an example, let us consider the quantum version of the 1-bit *NOT* gate, called the *QNOT* gate. The single-qubit computing basis states are $|0\rangle$ and $|1\rangle$. The states of the qubit can be represented as column vectors v_1 and v_2 where

$$v_1 = \begin{pmatrix} 1 \\ 0 \end{pmatrix} \text{ and } v_2 = \begin{pmatrix} 0 \\ 1 \end{pmatrix}. \tag{10.8}$$

The components of the vectors represent the amplitudes of the computing basis states in the state of the register or the qubit. Thus the basis state $|0\rangle$ of the qubit is represented as

$$\left(|0\rangle \ |1\rangle \right) \begin{pmatrix} 1 \\ 0 \end{pmatrix} = |0\rangle \tag{10.9}$$

while the other basis state $|1\rangle$ is represented as

$$\left(|0\rangle |1\rangle \right) \begin{pmatrix} 0 \\ 1 \end{pmatrix} = |1\rangle. \tag{10.10}$$

The 2×2 *QNOT* matrix must therefore be able to act on the column vectors v_1 and v_2 and convert (flip) $v_1 \to v_2$ and $v_2 \to v_1$ which would be equivalent to flipping the qubit $|0\rangle$ to $|1\rangle$ and $|1\rangle$ to $|0\rangle$. A 2×2 unitary matrix M that is capable of effecting this transformation is easily found to be

$$X = \begin{pmatrix} 0 & 1 \\ 1 & 0 \end{pmatrix}. \tag{10.11}$$

It can be verified that $Xv_1 = v_2$ and $Xv_2 = v_1$. If the single qubit is in an arbitrary superposition of the two fundamental basis states $|\psi\rangle = a_0|0\rangle + a_1|1\rangle$, the column vector of amplitudes is

$$v = \begin{pmatrix} a_0 \\ a_1 \end{pmatrix}. \tag{10.12}$$

Action of X on v can be easily seen to flip the amplitudes $(a_0 \to a_1, a_1 \to a_0)$ as

$$Xv = X\begin{pmatrix} a_0 \\ a_1 \end{pmatrix} = \begin{pmatrix} a_1 \\ a_0 \end{pmatrix}. \tag{10.13}$$

Following the operation, the qubit is found in a new superposed state $|\psi'\rangle$ where

$$|\psi'\rangle = a_1|0\rangle + a_0|1\rangle. \tag{10.14}$$

This action demonstrates the nonclassical power of the *QNOT* gate.

Another single-qubit quantum gate is the so-called square root of *NOT* or *SRN* gate, the matrix representation of which can be easily constructed:

$$\underset{SRN}{M_1} = \frac{1}{\sqrt{2}}\begin{pmatrix} 1 & -1 \\ 1 & 1 \end{pmatrix}. \tag{10.15}$$

Suppose that the qubit is in the state $|0\rangle$ so that the amplitude vector v_0 is

$$v_0 = \begin{pmatrix} 1 \\ 0 \end{pmatrix}. \tag{10.16}$$

The application of M_1 on v_0 produces the vector v_1 as

$$M_1\begin{pmatrix} 1 \\ 0 \end{pmatrix} = v_1 = \begin{pmatrix} \dfrac{1}{\sqrt{2}} \\ \dfrac{1}{\sqrt{2}} \end{pmatrix} \tag{10.17}$$

which defines the qubit in a state $|\psi'\rangle$ which is an equal superposition of the basis state $|0\rangle$ and $|1\rangle$, i.e.,

$$|\psi_0\rangle \underset{\substack{M_1(SRN) \\ (\equiv|1\rangle)}}{\to} |\psi_1'\rangle = \frac{1}{\sqrt{2}}|0\rangle + \frac{1}{\sqrt{2}}|1\rangle. \tag{10.18}$$

Similarly,

$$M_1 \begin{pmatrix} 1 \\ 0 \end{pmatrix} = v_1 = \begin{pmatrix} \dfrac{1}{\sqrt{2}} \\ \dfrac{1}{\sqrt{2}} \end{pmatrix} \tag{10.19}$$

i.e.,

$$\left|\psi_0\right\rangle \underset{M_1(SRN)}{\xrightarrow{}} \left|\psi_1'\right\rangle = \frac{1}{\sqrt{2}}|0\rangle + \frac{1}{\sqrt{2}}|1\rangle. \tag{10.20}$$

The application of the *SRN* gate on one qubit therefore appears to randomize its value in the restricted sense that a measurement on the qubit following the action of the SRN will produce the value 0 or 1 with a probability of $\frac{1}{2}$ irrespective of its initial value.

What happens when the *SRN* (M_1) gate is applied twice on the vector v_0? It is easily verified that

$$M_1^2 v_0 = M_1 (M_1 v_0)$$

$$= \frac{1}{\sqrt{2}} \begin{pmatrix} 1 & -1 \\ 1 & 1 \end{pmatrix} \left\{ \frac{1}{\sqrt{2}} \begin{pmatrix} 1 & -1 \\ 1 & 1 \end{pmatrix} \begin{pmatrix} 1 \\ 0 \end{pmatrix} \right\} \tag{10.21}$$

$$= \begin{pmatrix} 0 \\ 1 \end{pmatrix} \equiv v_1 \equiv |1\rangle.$$

Similarly,

$$M_1^2 v_1 = \begin{pmatrix} -1 \\ 0 \end{pmatrix} \equiv -v_0 \equiv -|1\rangle. \tag{10.22}$$

Therefore, two successive applications of *SRN* gate on a qubit in state $|0\rangle$ will flip it to the state $|1\rangle$ while two successive applications of *SRN* on a qubit in state $|1\rangle$ will flip it to the state $|0\rangle$, i.e., the flipped qubit will acquire a phase which is supposedly immaterial from the perspective of measurement. Two successive applications of *SRN* gate on a qubit therefore produces a deterministic result—the qubit value is 0 if it was initially 1, and is 1 if its initial value was 0. That way it (SRN^2) simulates the effect of the *QNOT* gate explaining why M_1 is called the square root of *NOT* gate. The statement made earlier that the *SRN* gate acting on a qubit randomizes its value needs to be taken with circumspection—the precise result of action of the *SRN* gate may depend on the history of the qubit—a rather nonclassical result.

An important 1-qubit gate that is routinely used in quantum computing algorithms is known as the Hadamard gate (*H*). It is represented by the matrix

$$H = \frac{1}{\sqrt{2}} \begin{pmatrix} 1 & 1 \\ 1 & -1 \end{pmatrix}. \tag{10.23}$$

The action of H on a single qubit in state $|0\rangle$ or $|1\rangle$ produces the qubit in a state which is an equal superposition of the basic states $|0\rangle$ and $|1\rangle$. Thus, we have in terms of matrix vector representation

$$Hv_0 = \frac{1}{\sqrt{2}} \begin{pmatrix} 1 & 1 \\ 1 & -1 \end{pmatrix} \begin{pmatrix} 1 \\ 0 \end{pmatrix}$$

$$= \begin{pmatrix} \frac{1}{\sqrt{2}} \\ \frac{1}{\sqrt{2}} \end{pmatrix} \tag{10.24}$$

and

$$Hv_1 = \frac{1}{\sqrt{2}} \begin{pmatrix} 1 & 1 \\ 1 & -1 \end{pmatrix} \begin{pmatrix} 0 \\ 1 \end{pmatrix}$$

$$= \begin{pmatrix} \frac{1}{\sqrt{2}} \\ -\frac{1}{\sqrt{2}} \end{pmatrix}. \tag{10.25}$$

In terms of basis states the effect of applying the Hadamard gate (H) on a qubit in state $|0\rangle$ will produce the qubit in a state of superposition $\frac{1}{\sqrt{2}}\left(|0\rangle + |1\rangle\right)$, while acting on the qubit in state $|1\rangle$, it will produce the superposed state $\frac{1}{\sqrt{2}}\left(|0\rangle - |1\rangle\right)$. In either case, measurement on the qubit will produce the value 0 or 1 with equal probability $\left(p = \frac{1}{2}\right)$ revealing the nonclassical nature of the state. Two successive applications of the H-gate on the qubit in state $|0\rangle$ can be easily seen to force the qubit in the state $|0\rangle$ while two successive applications of H on the qubit in state $|1\rangle$ will recover the qubit in state $|1\rangle$. The H-gate is therefore operationally akin to what may be viewed as the "square root of identity." The result of the action of the H gate on a superposition of the single-qubit computing basis states, e.g., $a_0|0\rangle + a_1|1\rangle$ can be easily determined if we recall Equations 10.24 and 10.25. Thus we have

$$H\left\{a_0|0\rangle + a_1|1\rangle\right\} = a_0\left(\frac{|0\rangle + |1\rangle}{\sqrt{2}}\right) + a_1\left(\frac{|0\rangle - |1\rangle}{\sqrt{2}}\right). \tag{10.26}$$

Operationally, we can represent the action as

$$a_0|0\rangle + a_1|1\rangle \xrightarrow{H} a_0\left(\frac{|0\rangle + |1\rangle}{\sqrt{2}}\right) + a_1\left(\frac{|0\rangle - |1\rangle}{\sqrt{2}}\right).$$

A Hadamard gate is one of the most useful quantum gates and is routinely used as much for producing qubits in superposition of states (or in superposed state) as for decoding superpositions. Another single-qubit quantum gate that is useful is the z gate. A z gate can be seen as a 2×2 unitary matrix where

$$z = \begin{pmatrix} 1 & 0 \\ 0 & -1 \end{pmatrix}. \tag{10.27}$$

Clearly

$$zv_0 = z\begin{pmatrix} 1 \\ 0 \end{pmatrix} = \begin{pmatrix} 1 \\ 0 \end{pmatrix} = v_0$$

$$\text{and} \quad zv_1 = z\begin{pmatrix} 0 \\ 1 \end{pmatrix} = \begin{pmatrix} 0 \\ -1 \end{pmatrix} = -v_1. \tag{10.28}$$

Operationally, we can visualize the action of the z gate on a superposition $(a_0|0\rangle + a_1|1\rangle)$ as follows:

$$a_0|0\rangle + a_1|1\rangle \xrightarrow{z} a_0|0\rangle - a_1|1\rangle.$$

At this stage we must take a pause and wonder how many single-qubit quantum gates could be there? Since any 2×2 unitary matrix would qualify as a single-qubit quantum gate, the answer to the question is clear: infinitely many single-qubit gates can be designed. However, we only need a small subset of the infinitely many 2×2 unitary matrices (q-gates) in terms of which the action of an arbitrary single-qubit gate can be represented. The small subset of 2×2 unitary matrices defines the set of single-qubit quantum logic gates or simply the single-qubit quantum gates. An arbitrary 2×2 unitary matrix (U) can be represented as the following product of three 2×2 matrices together multiplied by a scalar:

$$U = e^{i\alpha/2} \begin{bmatrix} e^{-i\beta/2} & 0 \\ 0 & e^{i\beta/2} \end{bmatrix} \begin{bmatrix} \cos\frac{\gamma}{2} & -\sin\frac{\gamma}{2} \\ \sin\frac{\gamma}{2} & \cos\frac{\gamma}{2} \end{bmatrix} \begin{bmatrix} e^{-i\delta/2} & 0 \\ 0 & e^{i\delta/2} \end{bmatrix}. \tag{10.29}$$

On the right-hand side of Equation 10.29, $e^{i\alpha/2}$ represents a global phase shift. The second matrix represents a plane rotation while the first and the last matrices can be represented as rotations about the z axis. We will return to this aspect later. Before doing so, we will consider a number of 2-bit classical gates along with several 2-qubit quantum gates and their actions on information. There are only four states of a 2-bit classical registrar, namely 00, 01, 10, and 11. A 2-qubit quantum register can be in any one of the four fundamental states $|00\rangle$, $|01\rangle$, $|10\rangle$, $|11\rangle$, or in an arbitrary superposition of the four states

$$|\psi\rangle = a_{00}|00\rangle + a_{01}|01\rangle + a_{10}|10\rangle + a_{11}|11\rangle \tag{10.30}$$

with the stipulation that

$$|a_{00}|^2 + |a_{01}|^2 + |a_{10}|^2 + |a_{11}|^2 = 1. \tag{10.31}$$

The coefficients a_{ij} s are in general complex numbers called amplitudes. These amplitudes can be collected in a column vector v where

$$v = \begin{pmatrix} a_{00} \\ a_{01} \\ a_{10} \\ a_{11} \end{pmatrix}. \tag{10.32}$$

The arbitrary 2-qubit state $|\psi\rangle$ can then be expressed as a product of the row vector of the four basis states and the amplitude vector is

$$|\psi\rangle = \left(|00\rangle \ |01\rangle \ |10\rangle \ |11\rangle\right) \begin{pmatrix} a_{00} \\ a_{01} \\ a_{10} \\ a_{11} \end{pmatrix}. \tag{10.33}$$

The four basis states can thus be represented by the vectors v_0, v_1, v_2, and v_3 where

$$|00\rangle \equiv v_0 \equiv \begin{pmatrix} 1 \\ 0 \\ 0 \\ 0 \end{pmatrix}; \ |01\rangle \equiv v_1 \equiv \begin{pmatrix} 0 \\ 1 \\ 0 \\ 0 \end{pmatrix}; \ |10\rangle \equiv v_2 \equiv \begin{pmatrix} 0 \\ 0 \\ 1 \\ 0 \end{pmatrix}; \ |11\rangle \equiv v_3 \equiv \begin{pmatrix} 0 \\ 0 \\ 0 \\ 1 \end{pmatrix}. \tag{10.34}$$

The 2-qubit gates are designed to act on the column vectors to produce new amplitudes a'_{00}, a'_{01}, a'_{10}, and a'_{11} which define a new 2-qubit state $|\psi'\rangle$. The 2-qubit gates therefore can be represented as 4×4 unitary matrices (unitarity is essential to preserve the norm of the vector of amplitudes). The classical 2-bit gates are the AND, OR, XOR, NOR, and the NAND gates. Here we must refer to the very important result that the NAND gate is universal in the sense that any function on bits can be computed from compositions of NAND gates alone. This universality is missing in all other classical gates either alone or in combination.

Among the 2-qubit quantum gates, the controlled NOT gate—the CNOT gate—may be considered prototypical. The gate acts on 2-qubit states and conditionally changes the second qubit (the target qubit) if the control qubit is 1. Of the two qubits, the first qubit is called the control qubit. Operationally, therefore, the CNOT gate leaves the states $|00\rangle$ or $|01\rangle$ unaffected as the control qubit is zero; but it transforms the $|10\rangle$ state into $|11\rangle$ and the $|11\rangle$ state into $|10\rangle$ as the control qubit satisfies the condition. The gate matrix U_{CONT} (4×4 unitary) is therefore required to act on the vectors v_0, v_1, v_2, and v_3 with the following results

$$U_{CNOT} v_0 = v_0; \ U_{CNOT} v_1 = v_1$$
$$U_{CNOT} v_2 = v_3; \ U_{CNOT} v_3 = v_2. \tag{10.35}$$

These four results immediately fix the definition of U_{CONT} matrix which reads

$$U_{CNOT} = \begin{pmatrix} 1 & 0 & 0 & 0 \\ 0 & 1 & 0 & 0 \\ 0 & 0 & 0 & 1 \\ 0 & 0 & 1 & 0 \end{pmatrix}. \tag{10.36}$$

The so-called circuit representation of the CNOT gate is displayed in Figure 10.1b. The top line represents the control qubit, the bottom line specifying what happens to the target qubit. The action of CNOT gate in Figure 10.1b can be summarized mathematically as

$$|A, B\rangle \rightarrow |A, B \oplus A\rangle. \tag{10.37}$$

This indeed is a generalization of the classical XOR gate. That means the control qubit remains unchanged while the target qubit and the control qubit are "XORed" and the output is stored in the target qubit. We must point out here that the apparent similarity of the action of the CNOT gate with that of a generalized classical XOR gate must not be misconstrued to mean that XOR can be represented by a unitary quantum gate. The action of XOR is irreversible, i.e., from the output produced by the action of XOR gate on two bits, it is impossible to find out what the input bits were. The quantum gates being representable by unitary matrices are invertible (hence reversible) and therefore the irreversible XOR cannot be given a quantum unitary representation. The same argument applies to the classical (universal) NAND gate for which a corresponding unitary quantum gate cannot be designed. Is there a universal quantum gate then? The answer is yes, but in a slightly different sense. It has been shown [5] that any multiple qubit quantum logic gate can be designed using the 2-qubit CNOT gate and the other 1-qubit gates as components, providing the quantum parallel of the universality of NAND gate. Notwithstanding the important universality theorem just mentioned, special purpose multiple qubit quantum gates have been designed and utilized. One such gate is the 2-qubit SWAP gate. The gate matrix is

$$U_{SWAP} = \begin{pmatrix} 1 & 0 & 0 & 0 \\ 0 & 0 & 1 & 0 \\ 0 & 1 & 0 & 0 \\ 0 & 0 & 0 & 1 \end{pmatrix}. \tag{10.38}$$

The gate U_{SWAP} acts on the 2-qubit state $|01\rangle$ or $|10\rangle$ and swaps the states of the 2-qubits. It has no action on the other 2-qubit states $|00\rangle$ or $|11\rangle$.

$$U_{SWAP} v_0 = v_0; \; U_{SWAP} v_1 = v_2$$
$$U_{SWAP} v_2 = v_1; \; U_{SWAP} v_3 = v_3. \tag{10.39}$$

Yet another useful 2-qubit gate is the so-called controlled phase gate represented by the unitary matrix U_{cphase} acting on the space spanned by the $|00\rangle$, $|01\rangle$, $|10\rangle$, and $|11\rangle$ states. The U_{cphase} matrix has the following general structure

$$U_{cphase} = \begin{pmatrix} 1 & 0 & 0 & 0 \\ 0 & 1 & 0 & 0 \\ 0 & 0 & 1 & 0 \\ 0 & 0 & 0 & e^{i\alpha} \end{pmatrix}. \tag{10.40}$$

TABLE 10.1

Truth Table for Toffoli Gate

Input bits x y z	Output bits x' y' z'
0 0 0	0 0 0
0 0 1	0 0 1
0 1 0	0 1 0
1 0 0	1 0 0
1 1 0	1 1 1
1 1 1	1 1 0
0 1 1	0 1 1
1 0 1	1 0 1

Before leaving the brief discussion on the generation gates and qubits, it would be useful to introduce a 3-bit reversible classical gate called the Toffoli gate that works with three input bits, two of which are the control bits left unaffected by the gate action. The third bit is a target bit that is flipped if both the control bits are in the state 1. The truth table (Table 10.1) has the following structure:

The analogous quantum gate would be a unitary 8×8 matrix acting in the space spanned by the eight 3-qubit computing basis states $|000\rangle$, $|001\rangle$, $|010\rangle$, $|011\rangle$, $|100\rangle$, $|101\rangle$, $|110\rangle$, $|111\rangle$. The most general 3-qubit state $|\psi\rangle$ can be written as a superposition of the eight basic 3-qubit computing states: $|\psi'\rangle = a_0|000\rangle + a_1|001\rangle + a_2|010\rangle + a_3|011\rangle + a_4|100\rangle + a_5|101\rangle + a_6|110\rangle + a_7|111\rangle$, where as usual the amplitudes (complex numbers) a_i; $(i = 0, 1, 2, ..., 7)$ satisfy the normalization condition $\sum_{i=0}^{7} |a_i|^2 = 1$. Corresponding to the eight computing basis states, we can define eight unit vectors $v_0, v_1, v_2, ..., v_7$ where

$$v_0 = \begin{pmatrix} 1 \\ 0 \\ 0 \\ 0 \\ 0 \\ 0 \\ 0 \\ 0 \end{pmatrix}, \ v_1 = \begin{pmatrix} 0 \\ 1 \\ 0 \\ 0 \\ 0 \\ 0 \\ 0 \\ 0 \end{pmatrix}, \ v_2 = \begin{pmatrix} 0 \\ 0 \\ 1 \\ 0 \\ 0 \\ 0 \\ 0 \\ 0 \end{pmatrix}, \ v_3 = \begin{pmatrix} 0 \\ 0 \\ 0 \\ 1 \\ 0 \\ 0 \\ 0 \\ 0 \end{pmatrix}, \ ..., \ v_7 = \begin{pmatrix} 0 \\ 0 \\ 0 \\ 0 \\ 0 \\ 0 \\ 0 \\ 1 \end{pmatrix}. \tag{10.41}$$

The matrix for the quantum Toffoli gate must satisfy the following relations with respect to operation on the vectors $(v_i; i = 0, 1, 2, ..., 7)$

$$U_T v_0 = v_0; \ U_T v_1 = v_1; \ U_T v_2 = v_2; \ U_T v_3 = v_3;$$
$$U_T v_4 = v_4; \ U_T v_5 = v_5; \ U_T v_6 = v_7; \ U_T v_7 = v_6. \tag{10.42}$$

It is straightforward to recognize that the unitary 8×8 matrix U_T satisfying the conditions laid out is

$$U_T \equiv \begin{pmatrix} 1 & 0 & 0 & 0 & 0 & 0 & 0 & 0 \\ 0 & 1 & 0 & 0 & 0 & 0 & 0 & 0 \\ 0 & 0 & 1 & 0 & 0 & 0 & 0 & 0 \\ 0 & 0 & 0 & 1 & 0 & 0 & 0 & 0 \\ 0 & 0 & 0 & 0 & 1 & 0 & 0 & 0 \\ 0 & 0 & 0 & 0 & 0 & 1 & 0 & 0 \\ 0 & 0 & 0 & 0 & 0 & 0 & 0 & 1 \\ 0 & 0 & 0 & 0 & 0 & 0 & 1 & 0 \end{pmatrix}. \tag{10.43}$$

The action of the unitary Toffoli gate (U_T) effectively leaves all except the $|110\rangle$ and $|111\rangle$ states unaffected. The $|110\rangle$ state is converted to $|111\rangle$ state and $|111\rangle$ state is transformed into $|110\rangle$ state mimicking completely and reversibly the action of the classical Toffli gate on a 3-bit register. The quantum Toffoli gate is especially important. It has been shown that the gate can be used to realize the quantum versions of circuits with the irreversible (universal) NAND and FANOUT gates. It essentially means that all possible deterministic classical computations can be carried out also on quantum computers employing appropriately connected Toffoli gates [5]. The possibility immediately points to an advantage of quantum computers over classical ones with respect to energy requirements. The quantum gates are all reversible unlike many of their classical counterparts which are irreversible. The irreversibility of the classical gates leads to energy dissipation, while the reversibility of quantum gates ensures that no loss of energy takes place at all by way of dissipation. In terms of energy expenditure, therefore, quantum computers have a clear edge over the classical computers. While this assertion is true, we must add a word of caution. Quantum computers, reversible as they are, also too sensitive to noise to be stable (this would be true for many other reversible models of computation), therefore requiring some form of continuous measurement and error correction. The bits used for storage of the results of continuous measurement have to be erased from time to time to accommodate the results of new measurement. The erasure requires energy expenditure of $k_B T \ln 2$ per bit and must be accounted for in estimating the energy requirement for computing on quantum computers. The fact remains, however, that the total energy requirement of quantum computers would be much lower compared to the equivalent classical computers, in view of reversibility of the quantum gates.

10.4 Special Features of Quantum Gates

Quantum circuits comprise quantum gates joined by wires (not physical wire) following certain rules. For example, unlike classical circuits no feedback from one part of the quantum circuit to another is allowed (no loops). The wires are not allowed to be joined together into a single wire (classical FANIN) carrying the bitwise OR of the inputs as the operation is not reversible. The inverse of the operation (classical FANOUT) which results in copying of the qubits is also disallowed (see Section 10.4: no-cloning theorem). Let U be a unitary matrix acting on n-qubits. We can regard U as a quantum gate operating on those qubits. We can define a controlled U gate that has a single control qubit diagrammatically represented by a single line with a black dot (Figure 10.1c) while the n target qubits are represented by the boxed U. If the control qubit is 0, the target qubits remain unchanged. If it is 1, the U-gate acts on the target qubits and changes their states as desired. The prototypical example of the controlled U-gate is the already introduced CNOT gate which can, according to the convention just introduced, be represented diagrammatically as shown in Figure 10.1d. An alternative circuit representation of CNOT could be as in Figure 10.1e. Can there be a classical bit-copying circuit? The answer is yes, with the classical CNOT gate providing an easy means of copying the bit as shown in Figure 10.1f. The gate takes in the bit that is to be copied in an unknown state (say z) and a "scratchpad bit" initialized to 0. As Figure 10.1f conveys, the output is two bits, each in the same state ($x \oplus y$ is addition modulo 2). The question is: can there be an equivalent quantum circuit for qubit copying using the CNOT gate?

Suppose that we have qubit in an unknown (superposition) state $|\psi\rangle = a_0|0\rangle + a_1|1\rangle$ with $|a_0|^2 + |a_1|^2 = 1$. The input state of the two qubits is then

$$\left(a_0|0\rangle + a_1|1\rangle\right)|0\rangle \equiv a_0|00\rangle + a_1|10\rangle. \tag{10.44}$$

As we have already discussed, the action of the CNOT gate is to flip the second qubit only if the first (control) qubit is 1; therefore, the output from the operation of CNOT on ($a_0|00\rangle + a_1|10\rangle$) will be ($a_0|00\rangle + a_1|11\rangle$). Is the output a copy of $|\psi\rangle$? The answer is yes, if either $a_1 = 0$, $a_0 = 1$ or $a_1 = 1$, $a_0 = 0$. In the first case ($a_0 = 1$), the unknown state $|\psi\rangle$ was in the state $|0\rangle$ (classical equivalent of "0") and the output $|00\rangle$ carries an exact copy of the input qubit. In the second case, $a_1 = 1$, the input was in state $|1\rangle$ (classical equivalent of "1") and the output $|11\rangle$ carries an exact copy of the input. Thus, classical-type information ($|0\rangle$ or $|1\rangle$) can be copied by a quantum circuit employing a CNOT gate when the input is nonclassical information (i.e., both a_0 and a_1 are nonzero). The "output" is not an exact copy of the input in which case the output should have been $|\psi\rangle|\psi\rangle$, i.e., $a_0^2|00\rangle + a_1^2|11\rangle + a_0a_1|01\rangle + a_1a_0|10\rangle$ and not $a_0|00\rangle + a_1|11\rangle$, as shown in Figure 10.1g. Thus, the quantum version of a bit-copying circuit cannot copy exactly the unknown quantum state of a qubit. It can be proved that it is impossible to make a copy of an unknown quantum state and this assertion constitutes what has become known as the "no-cloning theorem." The proof proceeds along the following line of reasoning. Suppose that a quantum copying machine has two slots labeled "a" (the data slot) and "b" (the target slot). Let the slot "a" be in some unknown but pure quantum state $|\psi\rangle$ which is to be copied into the target slot "b" that exists in some standard pure state $|t\rangle$. The initial state of the quantum copying machine is $|\psi\rangle \otimes |t\rangle$. Suppose now that some unitary evolution U (a quantum circuit with a unitary gate) executes the copying process so that we have

$$U\left(|\psi\rangle \otimes |t\rangle\right) = |\psi\rangle \otimes |\psi\rangle. \tag{10.45}$$

Let us assume further that the same copying process also works for another arbitrary pure state $|\phi\rangle$ so that we must also have

$$U(|\phi\rangle \otimes |t\rangle) = |\phi\rangle \otimes |\phi\rangle. \tag{10.46}$$

Taking the inner product of the two equations and using $U^\dagger U = \mathbb{I}$, we get

$$\langle\psi|\phi\rangle = \left(\langle\psi|\phi\rangle\right)^2. \tag{10.47}$$

Equation 10.47 demands that either $\langle \psi | \phi \rangle = 1$ in which case $|\phi\rangle = |\psi\rangle$ (contrary to the assumption that $|\phi\rangle$ is arbitrary) or $\langle \psi | \phi \rangle = 0$ in which case $|\psi\rangle$ and $|\phi\rangle$ are orthogonal (again contrary to our assumption about $|\phi\rangle$) [5]. Thus, the quantum cloning device can only clone states that are mutually orthogonal which means that a general quantum cloning machine based on unitary evolution is impossible to construct. Thus, the state $|\psi_1\rangle = \dfrac{1}{\sqrt{2}}(|0\rangle + |1\rangle)$ and $|\psi_2\rangle = \dfrac{1}{\sqrt{2}}(|0\rangle - |1\rangle)$ can be copied as $\langle \psi_1 | \psi_2 \rangle = \langle \psi_2 | \psi_1 \rangle = 0$, i.e., the states are mutually orthogonal. The states $|\psi_1\rangle$ and $|\psi_2\rangle = |0\rangle$ cannot, however, be copied or cloned as $|\psi_1\rangle\,|\psi_2\rangle \neq 0$. A pertinent question is: can there be a nonunitary qubit copying circuit? It has been shown that such a circuit will copy an arbitrary state with finite loss in fidelity, thereby preventing exact cloning. The no-cloning theorem [6,7] has also been extended to states that are not pure states. So far, in this section, we have seen one very positive aspect of quantum computers working with unitary quantum gates (zero or very low energy expenditure) and also one negative aspect of it (inability to copy qubits in arbitrary quantum states), both the features being in stark contrast with what one experiences with a classical computer or in classical computation. We will now examine a third feature of quantum computing that is classically unimaginable, and endows a quantum computing machine with a spectacular ability to do things that are impossible to do on a classical machine. The origin of the special feature lies in what has been called "quantum parallelism"—an attribute intrinsic to qubit-based computing, quantum computing, or quantum algorithms.

10.5 Quantum Parallelism

Suppose that we are required to evaluate a function $g(x)$ for many (n, say) values of x. In classical computation we may distribute the task among n-processors (n circuits) and complete the evaluation of $g(x)$ simultaneously. One classical processor (circuit) cannot do the job of n-evaluations of $g(x)$ simultaneously. Quantum parallelism, to put it in an oversimplified manner, is the ability of quantum algorithms (quantum computers) to evaluate $g(x)$ simultaneously for many values of x on a single quantum processor (a single quantum circuit). Let us consider the possibility in detail. Suppose we are required to compute a function $g(x)$ that has a 1-bit domain and range. Let there be a 2-qubit quantum computer with the register initially in the state $|x, y\rangle$. The first is the data register while the second is the target register. It is possible to design a circuit containing a series of gates such that the input state $|x, y\rangle$ on being acted upon by the gates evolves into the state $|x, y \oplus f(x)\rangle$ where $\oplus$ stands for addition modulo 2. The evolution process can be represented by a unitary transformation (call it U_f) so that can write

$$|x,y\rangle \xrightarrow{U_f} |x,y \oplus f(x)\rangle, \quad U_f^\dagger U_f = U_f U_f^\dagger = 1. \tag{10.48}$$

Had we started with $y = 0$, the final state of the target qubit would be just the value $f(x)$. Thus we have the result that

$$|0,0\rangle \xrightarrow{U_f} |0, f(0)\rangle$$

$$|1,0\rangle \xrightarrow{U_f} |1, f(1)\rangle.$$

Suppose now that the data register is initially prepared not in the computational basis states $|0\rangle$ or $|1\rangle$, but in the superposed state $|x\rangle = \dfrac{1}{\sqrt{2}}\left(|0\rangle + |1\rangle\right)$ which, as we have already seen, can be created by applying the Hadamard gate to the computational basis state $|0\rangle$. The application of U_f to $|x\rangle$ will now produce the superposed state $(|\psi\rangle)$ where

$$|\psi\rangle = \frac{1}{\sqrt{2}}\left\{|0, f(0)\rangle + |1, f(1)\rangle\right\}. \tag{10.49}$$

The most remarkable feature of the superposed state $(|\psi\rangle$ is that it carries information about both $f(0)$ and $f(1)$ simultaneously, created by a single application of U_f to the state $|x, y\rangle$. It appears therefore that one can, in principle, evaluate the function $f(x)$ simultaneously for both the values of x (0 or 1) by applying a quantum circuit (U_f) only once on the superposition state of the data register, something that is classically impossible to do. This feature is a direct consequence of the quantum nature of the superposed state of the data qubit and has been called quantum parallelism. The quantum parallelism is special because a single circuit for measuring $f(x)$ evaluates the function simultaneously for different values of x, while classically we can do the same only by applying multiple $f(x)$-computing circuits. Note, however, that a direct measurement on the data qubit after application of U_f will yield either $f(0)$ or $f(1)$ with equal probability as the measured state collapses onto one of the two computational basis states. Thus, although the state $(|\psi\rangle$ carries information about both $f(0)$ and $f(1)$ simultaneously, a straightforward measurement yields only one of the two values with equal probability, and quantum parallelism, from a practical point of view, appears to be a mere theoretical construct devoid of any usefulness. Fortunately, it is not so. In order to achieve that we must modify the measurement technique. Before we go into the possibility of extracting information about both $f(0)$ and $f(1)$ through a modification of the measurement technique, it would be expedient to enquire what would be the situation if the data register contains multiple qubits, say n of them. It turns out that the process can be generalized to handle functions on any number of qubits by application of Walsh-Hadamard transform (or simply Hadamard transform), an operation involving the action of n Hadamard gates in parallel on the n qubits. For a 2-qubit data register plus 1-qubit output register with each qubit initially prepared in the state $|0\rangle$, two parallel applications of the Hadamard gate on the first two qubits, symbolized by the operator $H^{\oplus 2}$ lead to the output state

$$H^{\oplus 2}|\psi_0\rangle = H^{\oplus 2}\left(|0\rangle|0\rangle\right)|0\rangle = |\psi'\rangle \tag{10.50}$$

where,

$$|\psi'\rangle = \left\{\frac{|0\rangle + |1\rangle}{\sqrt{2}}\right\}\left\{\frac{|0\rangle + |1\rangle}{\sqrt{2}}\right\}|0\rangle$$

$$= \frac{1}{2}\left[|00\rangle + |01\rangle + |10\rangle + |11\rangle\right]|0\rangle. \tag{10.51}$$

A single application of U_f (the $f(x)$ computing circuit to $|\psi'\rangle$) will produce the result

$$U_f|\psi'\rangle = \frac{1}{2}\left[|00\rangle|f(00)\rangle + |01\rangle|f(01)\rangle + |10\rangle|f(10)\rangle + |11\rangle|f(11)\rangle\right]|0\rangle. \tag{10.52}$$

We can now quickly generalize the result to n-bit input (x') and 1-bit output $f(x')$ for performing parallel evaluation of the function $f(x)$ for all possible x. The steps involved are as follows:

1. Prepare the $n + 1$ qubit register in the state $|0\rangle^{\oplus n}|0\rangle$.

2. Apply Hadamard transform on the first n-qubits producing $\left[\left\{ \frac{(|0\rangle + |1\rangle)}{\sqrt{2}} \right\}, \ldots, n \text{ times} \right] |0\rangle$.

3. Apply the quantum circuit U_f once to the post-Hadamard transformed state producing the state

$$|\psi'\rangle = \frac{1}{\sqrt{2^n}} \sum_x |x\rangle |f(x)\rangle \tag{10.53}$$

where the summation is over all possible values of x.

Although values of $f(x)$ for all possible values of x have been generated by a single application of U_f, any measurement on the output state will force the equal superposition of 2^n possible states to collapse into only one of them with equal probability, and the corresponding value of $f(x)$ would be recovered—all other values would be lost. We must therefore modify the process of measurement on the superposed state created by the application of $H^{\oplus n}$. However, the fact that all the $f(x)$ values can be generated by a single application of U_f points to the operation of a quantum parallelism as opposed to classical parallel computation where 2^n parallel applications of the classical $f(x)$ computing circuit would have been necessary to generate all the values of $f(x)$. The quantum processor therefore would have a clear edge in the function computation provided we have a means to extract multiple values of $f(x)$ produced in step 3. Here, the phenomenon of interference can come to our rescue.

10.6 Quantum Parallelism and Interference

Suppose we have a 2-qubit register $|x, y\rangle$. The qubits of the register are initialized such that the input state is $|\psi_0\rangle = |01\rangle$. Applying the Hadamard gate to the first qubit we have it in the superposition state $\left\{ \frac{(|0\rangle + |1\rangle)}{\sqrt{2}} \right\}$. A second Hadamard gate is applied to the second qubit that is $|y\rangle \equiv |1\rangle$ into the superposition $\left\{ \frac{(|0\rangle - |1\rangle)}{\sqrt{2}} \right\}$. The net outcome of the application of two Hadamard gates is to create a state $|\psi_1\rangle$ where

$$|\psi_1\rangle = \left\{ \frac{(|0\rangle + |1\rangle)}{\sqrt{2}} \right\} \left\{ \frac{(|0\rangle - |1\rangle)}{\sqrt{2}} \right\}. \tag{10.54}$$

Suppose we are dealing with the state $|x\rangle \left\{ \dfrac{(|0\rangle - |1\rangle)}{\sqrt{2}} \right\}$ to which the function $f(x)$ comput-

ing circuit U_f is applied. This results in a state $(-1)^{f(x)}|x\rangle \left\{ \dfrac{(|0\rangle - |1\rangle)}{\sqrt{2}} \right\}$. Applying this result

to $|\psi_1\rangle$ we have

$$U_f|\psi_0\rangle = \pm \left\{ \frac{(|0\rangle + |1\rangle)}{\sqrt{2}} \right\} \left\{ \frac{(|0\rangle - |1\rangle)}{\sqrt{2}} \right\} = |\psi_2\rangle \ \text{if} \ f(0) = f(1) \tag{10.55}$$

and

$$U_f|\psi_1\rangle = \pm \left\{ \frac{(|0\rangle - |1\rangle)}{\sqrt{2}} \right\} \left\{ \frac{(|0\rangle - |1\rangle)}{\sqrt{2}} \right\} = |\psi_2\rangle \ \text{if} \ f(0) \neq f(1). \tag{10.56}$$

Application of a Hadamard gate to the first qubit in $|\psi_2\rangle$ then yields the state $|\psi_3\rangle$ where

$$|\psi_3\rangle = \pm |0\rangle \frac{(|0\rangle - |1\rangle)}{\sqrt{2}} \ \text{and} \ f(0) = f(1) \tag{10.57}$$

or the state $|\psi_3\rangle$ where

$$|\psi_3\rangle = \pm |1\rangle \frac{(|0\rangle - |1\rangle)}{\sqrt{2}} \ \text{and} \ f(0) \neq f(1). \tag{10.58}$$

The two cases can be combined into a single result by noting that

$$f(0) \oplus f(1) = 0 \ \text{if} \ f(0) = f(1)$$
$$f(0) \oplus f(1) = 1 \ \text{if} \ f(0) \neq f(1) \tag{10.59}$$

so that we can write

$$|\psi_3\rangle = \pm |f(0) \oplus f(1)\rangle \left(\frac{1}{\sqrt{2}} \ \{|0\rangle - |1\rangle\} \right). \tag{10.60}$$

The beauty of the result is that a single application of U_f followed by a measurement of the first qubit has enabled us to determine $f(0) \oplus f(1)$. That is, a single evaluation of $f(x)$ has yielded a property of $f(x)$ that depends on two possible values of x (namely 0 and 1). This would have been impossible to achieve on a classical computing circuit as the state "0" of a bit would have completely excluded the state "1" of the bit and vice versa. In the quantum case, as we have just described, the two possibilities were made to interfere with each other leading to the computation of $f(0) \oplus f(1)$, by applying the Hadamard gate appropriately. The result hints at the superior power of quantum computing. The algorithm we have described here is a variant of the original Deutsch's algorithm (see Nielsen and Chuang [5]). A more compelling evidence of the superior computing power of a quantum algorithm comes from consideration of the Deutsch–Jozsa algorithm (DJA). Before we describe the DJA, we will consider first a very special kind of 2-qubit quantum state called the Bell states and the very intriguing phenomenon of quantum entanglement.

We have already seen that the quantum 2-qubit register can exist in one or the other of the four computing basis states, viz. $|00\rangle$, $|01\rangle$, $|10\rangle$, $|11\rangle$ as well as in a superposition of them

$$|\psi\rangle = a_{00}|00\rangle + a_{01}|01\rangle + a_{10}|10\rangle + a_{11}|11\rangle$$

$$= c_0|00\rangle + c_1|01\rangle + c_2|10\rangle + c_3|11\rangle \tag{10.61}$$

with $\sum_{i=0}^{3}|c_i|^2 = 1$ (normalization condition).

Suppose that we have created a special superposed state $|\psi'\rangle$ in which $a_{00} = a_{11} = \dfrac{1}{\sqrt{2}}$. That is, $|\psi'\rangle = \dfrac{1}{\sqrt{2}}\{|00\rangle + |11\rangle\}$. Now, let us consider an act of measurement of the first qubit in the state $|\psi'\rangle$. If the result of measurement yields the value "0" (the state $|0\rangle$) for the first qubit we can infer without exception that the second too has the value "0." Similarly, if the measurement on the first qubit yields the value "1" (the state $|1\rangle$) the second qubit also must be in the state $|1\rangle$. This happens because on measurement, $|\psi'\rangle$ collapses to one of the two computing basis states of which $|\psi'\rangle$ is a superposition. In the first case it collapses to the state $|00\rangle$ while in the second instance the collapse takes it to the state $|11\rangle$. In each case, the probability is $\dfrac{1}{2}$. The point to note is that in either case, the result of measurement of the first qubit determines the state in which the second qubit finds itself. This happens no matter what the physical distance is separating the two qubits. The two qubits therefore appear to be correlated in some mysterious manner so that the result of measurement of the first qubit is immediately conveyed to the second one foreclosing its option to assume a value, or to be in a state independent of the first qubit. The superposition $|\psi'\rangle$ is said to be in an entangled state and the phenomenon is called quantum entanglement. The state $|\psi'\rangle$ of a 2-qubit system is one example of what has become known as a Bell state. The state, so simple in appearance or form, is responsible for many unusual features of quantum computation or quantum information processing. There are other examples of Bell states in 2- qubit systems. It is convenient to describe them by adopting the basis vector representation in place of representation in terms of product basis states. In Section 10.3, we have described the correspondence

$$v_0 = \begin{pmatrix} 1 \\ 0 \\ 0 \\ 0 \end{pmatrix} \begin{matrix} \equiv |00\rangle \\ \equiv \phi_0(1)\phi_0(2) \end{matrix}, \quad v_1 = \begin{pmatrix} 0 \\ 1 \\ 0 \\ 0 \end{pmatrix} \begin{matrix} \equiv |01\rangle \\ \equiv \phi_0(1)\phi_1(2) \end{matrix}, \quad v_2 = \begin{pmatrix} 0 \\ 0 \\ 1 \\ 0 \end{pmatrix} \begin{matrix} \equiv |10\rangle \\ \equiv \phi_1(1)\phi_0(2) \end{matrix}, \quad \text{and}$$

$$\tag{10.62}$$

$$v_3 = \begin{pmatrix} 0 \\ 0 \\ 0 \\ 1 \end{pmatrix} \begin{matrix} \equiv |11\rangle \\ \equiv \phi_1(1)\phi_1(2) \end{matrix}$$

v_0, v_1, v_2, v_3 constitute an orthonormal set of vectors. The most general form of a superposed state (normalized), can be represented as the vector

$$v = \frac{1}{\sqrt{c_{00}^2 + c_{01}^2 + c_{10}^2 + c_{11}^2}} \begin{pmatrix} c_{00} \\ c_{01} \\ c_{10} \\ c_{11} \end{pmatrix}. \tag{10.63}$$

We can construct four Bell states [8,9] for the 2-qubit register represented by vectors v_{B1}, v_{B2}, v_{B3}, v_{B4}, where

$$v_{B1} = \frac{1}{\sqrt{2}} \begin{pmatrix} 1 \\ 0 \\ 0 \\ 0 \end{pmatrix} \equiv \frac{1}{\sqrt{2}}\left(|00\rangle + |11\rangle\right) \equiv |\phi_{B0}\rangle$$

$$v_{B2} = \frac{1}{\sqrt{2}} \begin{pmatrix} 1 \\ 0 \\ 0 \\ -1 \end{pmatrix} \equiv \frac{1}{\sqrt{2}}\left(|00\rangle - |11\rangle\right) \equiv |\phi_{B1}\rangle$$

$$v_{B3} = \frac{1}{\sqrt{2}} \begin{pmatrix} 0 \\ 1 \\ 1 \\ 0 \end{pmatrix} \equiv \frac{1}{\sqrt{2}}\left(|01\rangle + |10\rangle\right) \equiv |\phi_{B2}\rangle \tag{10.64}$$

$$v_{B4} = \frac{1}{\sqrt{2}} \begin{pmatrix} 0 \\ 1 \\ -1 \\ 0 \end{pmatrix} \equiv \frac{1}{\sqrt{2}}\left(|01\rangle - |10\rangle\right) \equiv |\phi_{B3}\rangle.$$

The four Bell states provide particular examples of entangled states of 2-qubit registers. These states are manifestly orthonormal and can serve as basis states for expressing any arbitrary 2-qubit state as a linear combination of the Bell states. Suppose that the 2-qubits represent half integer spins. Then it is easy to show that the Bell states are eigenstates of the product spin operators $S_z(1)S_z(2)$ and $S_x(1)S_x(2)$. Successive applications of these product spin operators will separate any 2-qubit state into the four 2-qubit Bell channels. Since Bell states are eigenstates of product spin operators, the two qubits (spins) must be measured simultaneously. Measurement of a single qubit (spin) will destroy the entanglement. Since entanglement is leveraged to derive the special power of quantum computing it is necessary to maintain the entanglement all the while computation is being carried out. That means any measurement on a single qubit must be deferred until the quantum computing process is over. In the intermediate stages, only 2-qubit measurement can be undertaken as that will not destroy the entanglement.

10.7 Quantum Algorithms

Not too many quantum algorithms have been proposed or analyzed to date. We present here some of the algorithms that have attracted wide attention due to their superior abilities vis-á-vis the best classical algorithms for handling identical problems.

10.7.1 The Deutsch–Jozsa Algorithm

The algorithm does not have a direct technological implication but serves to demonstrate the unique features of quantum computing that cannot be duplicated on a classical computer. The problem that the Deutsch–Jozsa (DJ) algorithm [10] addresses can be posed as follows. Two persons, A and B, are participating in a game. They are located in two cities, a and b, separated by a very large distance. A selects a number x from the range $0 - (2^n - 1)$ and mails it to B who then uses the input (x) to calculate the value of a function $f(x)$. The function $f(x)$ has been chosen to be either a constant in which case $f(x)$ for any value of x is the same (0 or 1) or balanced, in which case $f(x)$ is equal to zero for half the input values of x and 1 for the remaining half of the values of x. After computing $f(x)$, B sends the results (0 or 1) to A each time he computes the value of $f(x)$. The question is how quickly can A confirm that the function chosen by B is balanced or constant (quickly here refers to the minimum number of queries made by A). Before trying to understand how the quantum DJ algorithm works, let us consider a classical algorithm for solving the problem. The minimum number of times A must know the values of $f(x)$ is $\left(\dfrac{2^n}{2} + 1\right)$ if the algorithm is classical. The argument is simple. Thus if A receives $\dfrac{2^n}{2}$ consecutive 0s (or 1s) followed by a 1 (or 0s) he can confirm that $f(x)$ is balanced. $\left(\dfrac{2^n}{2} + 1\right)$ consecutive 0s (or 1s) would have led A to conclude correctly that $f(x)$ is constant. The quantum algorithm on the other hand would have enabled A to confirm the nature of $f(x)$ correctly just after making a single query. The basic idea of the quantum DJ algorithm is the following. A and B would now be exchanging qubits instead of classical bits and B would be computing $f(x)$ by applying a unitary transform U_f exploiting of quantum parallelism in the process. Let us assume that A has a n-qubit register which he uses to store the query (x) and a single-qubit register which he simply uses to store the answer (value of $f(x)$). A first creates the query as well as the answer registers in superposed states. B then evaluates $f(x)$ using quantum parallelism and stores it in the answer register. A probes the states in the superposition with a Hadamard transform performed on the query register and completes the determination of the nature of $f(x)$ by making necessary measurements. The steps in the quantum DJ algorithm are summarized below:

1. Prepare all the qubits of the query register in the state 0 and the answer register in state 1 $\left(|\psi_0\rangle = |0\rangle^{\otimes n}|1\rangle\right)$.
2. Create superpositions of all the states of the query register and the answer register by applying Hadamard gates:

$$\left[|\psi_0\rangle \to |\psi_1\rangle = \left(\frac{1}{\sqrt{2^n}}\sum_{x=0}^{2^n-1}|x\rangle\right)\left(\frac{(|0\rangle - |1\rangle)}{\sqrt{2}}\right)\right]. \tag{10.65}$$

3. Calculate the function $f(x)$ using the so-called black box U_f (U_f executes the transformation $U_f\left(|x\rangle|y\rangle\right)=\left(|x\rangle|y\oplus f(x)\rangle\right)$ with $f(x)\in\{0,1\}$).

$$|\psi_1\rangle\rightarrow|\psi_2\rangle=\sum_x(-1)^{f(x)}|x\rangle\left[\frac{\left(|0\rangle-|1\rangle\right)}{\sqrt{2}}\right].$$
(10.66)

4. Perform Hadamard transform on $|\psi_2\rangle$ producing $|\psi_3\rangle$

$$|\psi_2\rangle\rightarrow|\psi_3\rangle=\sum_q\sum_x\frac{(-1)^{x.q+f(x)}|q\rangle}{\sqrt{2^n}}\left[\frac{\left(|0\rangle-|1\rangle\right)}{\sqrt{2}}\right].$$
(10.67)

5. Perform suitable measurement to obtain the final output q.

   Let us note further that:

   a. $x.q$ represents the bitwise inner product of x and q, modulo 2.

   b. The algorithm shows that the DJ problem can be solved by a single evaluation of $f(x)$ on a quantum computer.

   c. Quantum parallelism and interference play crucial roles in the emergence of the superior computing power of the quantum algorithm (see Nielsen and Chuang [5] for further details).

10.7.2 Grover's Algorithm (Quantum Search)

Grover's algorithm is a quantum search algorithm (QSA) for recognizing and extracting marked items from a large database with an efficiency that surpasses the efficiency of the best classical search algorithm for the same problem. The basic idea can be represented as follows.

Let us consider a database of N-elements of which M-elements are marked—i.e., they are the solutions of the search problem, with $1\leq M\leq N$. If we assume that $N=2_n$, (n an integer), we can also assume that every element in the search space can be assigned a unique index (a number) for every one of the individuals. Such an index will be an integer in the range 0 to $N-1$ and could be stored in an n-bit register. As a simple example, we can think of a search problem based on a function $f(x)$ where x is an integer in the range $0\leftrightarrow N-1$. The function has the property that $f(x)=1$ if x is a solution of the search problem while $f(x)=0$ if x is not a solution of the search problem. We will assume further that we have at our disposal a quantum oracle—a black box—that is endowed with the inbuilt capacity to identify the solutions of the particular search problem. The oracle is a unitary operator 0, the action of which on the computational basis states can be described as

$$O\left(|x\rangle|q\rangle\right)\rightarrow\left(|x\rangle|q\oplus f(x)\rangle\right)$$
(10.68)

where $|x\rangle$ denotes the index register, $|q\rangle$ is a single oracle qubit, $\oplus$ as usual represents addition modulo 2. If x is such that $f(x)=1$, the oracle qubit is flipped to 1 (the qubit is initialized to 0 in the beginning) but left unaltered if $f(x)\neq1$. To check if x is a solution to the search problem, one therefore has to prepare the state ($|x\rangle|0\rangle$), allow the oracle to act on the state and then check if the oracle qubit (0) has been flipped to 1. If it has been flipped, x is a solution to the problem at hand. We have so far not specified what happens inside the black box—the quantum oracle 0. It is unimportant at this stage as our main thrust is to ascertain the minimum number of calls to the oracle that would be needed to finish the search process. Before proceeding further we may consider the following simplification in the representation. The simplification consists of preparing the oracle qubit ($|q\rangle$) initially in the state $\frac{(|0\rangle - |1\rangle)}{\sqrt{2}}$ instead of the state $|0\rangle$ as described previously (a Hadamard gate applied to the qubit in the $|0\rangle$ state$|1\rangle$ would realize this change). The action of the oracle can then be specified as follows:

$$O\left(|x\rangle \frac{1}{\sqrt{2}}(0-1)\right) = (-1)^{f(x)}\left(|x\rangle \frac{1}{\sqrt{2}}(0-1)\right). \tag{10.69}$$

Noting that the oracle qubit remains unchanged in the process, we might as well drop the oracle qubit from the representation and describe the action of the oracle more simply as

$$O(|x\rangle) = (-1)^{f(x)}|x\rangle. \tag{10.70}$$

It now appears that the oracle identifies or recognizes a solution by shifting its phase. For an N-item data base with M solutions, it turns out (see references [11,12]) that we must call or apply the oracle $0\sqrt{N/M}$ times to obtain a solution on a quantum computer. If $M = 1$, then the number of oracle calls needed will grow as $\sim \sqrt{N}$ with an increase in the size (N) of the database. For a classical algorithm, the number of calls would grow linearly with N and thus there is a quantum speedup. The quantum speedup may not be spectacular in Grover's quantum search, but is at least substantial when compared to a classical search. We have so far left the details of steps involved in the action of the oracle unspecified. The details are briefly summarized in what follows.

Grover's algorithm works with one n-qubit register which is initially prepared in the "all 0 state" denoted by $|0\rangle^{\otimes n}$. The Hadamard transform next brings the register in the state of equal superposition ($|\psi\rangle$):

$$|\psi\rangle = \frac{1}{\sqrt{N}} \sum_{x=0}^{N-1} |x\rangle. \tag{10.71}$$

The QSA of Grover then repeatedly invokes Grover's operator or Grover's subroutine "G," the quantum circuit of which can be split into the following steps:

1. Implementation of the oracle
2. Application of Hadamard transform ($H^{\otimes n}$: number of operations $\sim \log N$)
3. Application of conditional phase shift whereby every computational basis except $|0\rangle$ gets a phase of (–1); operationally, we have

$$|x\rangle \rightarrow (-1)^{\delta x_0} |x\rangle. \tag{10.72}$$

One can show that the conditional phase shift operation can be brought in by a unitary operator $U = 2(|0\rangle\langle 0|) - 1$ which can be implemented by applying $O(n)$ gates.

4. Apply Hadamard transform: $H^{\otimes n}$.

5. The net outcome of executing steps (ii) to (iv) can be shown to be equivalent to the operator $(2|\psi\rangle\langle\psi| - I)$ where $|\psi\rangle$ is the equally weighted superposition. The Grover operator G (steps I to iv) can thus be compactly represented as

$$G = (2|\psi\rangle\langle\psi| - I)O.$$

O being the oracle. The application of G on a general state can be shown to lead to what has been called an "inversion about the mean" operation. The point to note is that the oracle has been called only once in the quantum search.

10.7.3 Shor's Algorithm (Quantum Factoring)

Factoring is an important and a difficult problem. Suppose that a large number (integer) is given and we are told that r is a product of two primes p and q ($r = p \times q$). How do we go about finding the prime factors of r? The obvious direct method implementable on a classical computer is to search through all integers from 2 to $\sqrt{r}$, successively try dividing r by the factors from 2 to $\sqrt{r}$ until we find the smaller of the prime factors (say p).

The other prime factor can then be obtained by dividing r by p. The number of trial divisions to be performed in this straightforward approach on a classical computer is clearly $\sim \sqrt{r}$. There are much better classical algorithms in which the computation labor involved in prime factoring has been brought down very significantly.

The best known classical algorithms for factoring consume time proportional to $2^{\sqrt[3]{n}\log(n)^{2/3}}$, n being the number of digits in the number to be factored into primes. In contrast, Shor's quantum factoring algorithm [13,14] has been shown to require time proportional to only $n^2 \log(n)\log\log(n)$. Clearly, the time complexity advantage that Shor's algorithm offers over classical algorithm is spectacular. The saving in computational labor is nearly exponential if we assume that the classical algorithms for factoring have already reached optimal performance level. To have a feel of the complexity advantage we may consider a large number containing, say, 5000 digits. Then somewhat crudely assuming that the classical and quantum time complexity functions described here are exact, and noting that the logarithms have to be taken to base 2, the best classical algorithm will require roughly 80 billion years for completing the task (factoring) on the assumption that the computer can process one instruction in one nanosecond. Under the same set of assumptions Shor's algorithm would be able to finish the same task in few seconds. Of course we have ignored the constants in the time complexity functions and assumed that the classical and quantum computers have the same clock rates (this may not be true). Such details will modify the numbers we have quoted without qualitatively affecting the exponential complexity advantage that Shor's algorithm executed on a quantum computer will provide. Having given a feel of the unharnessed power that quantum computing can offer in a factoring problem, we will now briefly describe Shor's algorithm (for details see references [13,14]).

The modern approaches to solve the factoring problem exploit the following property. Suppose a be coprime with the number N (to be factored). Then we may define a function $f_{aN}(j)$ where

$$f_{aN}(j) \equiv \left(a^j, \bmod N\right) \tag{10.73}$$

$f_{aN}(j)$ thus defined has the following two very important attributes:

1. $f_{aN}(j)$ is periodic with period P. Take for example $a = 2$, $N = 15$. Then the function f_{aN} takes on values 1, 2, 4, 8, 1, 2, ..., revealing that it has a period $P = 4$. Similarly if we take $a = 3$, $N = 14$, the function f_{aN} takes values 1, 3, 9, 13, 11, 5, 1, 3, ..., and the period P is 6. In the present context P is a very important quantity, as we will see later.

2. If P is even, the greatest common divisors of the pairs of numbers $(a^{\frac{P}{2}} + 1, N)$ and $(a^{\frac{P}{2}} - 1, N)$ are factors of N. For $a = 2$, $N = 15$, $P = 4$, the pairs of numbers are $(2^2 - 1, 15)$ and $(2^2 + 1, 15)$, i.e., (3, 15) and (5, 15) so that the greatest common divisors of the pair are 3 and 5, respectively. Indeed, the prime factors of 15 are 3 and 5.

Now the time complexity involved in calculating the period P on a classical computer is comparable with time complexity encountered in any other factoring algorithm. Peter Shor showed for the first time [13,14] that special features of quantum mechanics can be leveraged to do the job with surprisingly higher efficiency on a quantum computer. In what follows, we summarize the steps:

1. Consider a quantum register with n qubits—we may call it a quantum register of size n which is assumed to store information in binary form. Therefore, the n-qubit register can store numbers $j = 0, 1, 2, ..., 2^n - 1$. If $n = 3$, the register can obviously store the numbers 0, 1, 2, 3, ..., 7. Similarly, if $n = 2$, the numbers that the register can accommodate are 0, 1, 2, 3.

2. Let there be a control register (left) and a target register. The operation starts with preparing both the control and the target registers in the $j = 0$ state, i.e., all the qubits are put in the ϕ_0 state.

$$|\psi_1^{(n)}\rangle = |\phi_0^{(n)}\rangle|\phi_0^{(n)}\rangle. \tag{10.74}$$

3. Now the control register receives the integer series and the state becomes

$$|\psi_2^{(n)}\rangle = 2^{-\frac{n}{2}}\sum_{j=0}^{2^n-1}\left(|\phi_j^{(n)}\rangle|\phi_0^{(n)}\rangle\right). \tag{10.75}$$

4. A value for a is chosen (keeping in mind that a, N are coprimes) and for each value of j replace $|\phi_0^{(n)}\rangle$ in the target register in the expansion for $|\psi_2^{(n)}\rangle$ by $|\phi_{aN(j)}^{(n)}\rangle$ producing

$$|\psi_3^{(n)}\rangle = \sum_{j=0}^{2^n-1}\left(|\phi_j^{(0)}\rangle|\phi_{f_{aN}(j)}^{(n)}\rangle\right). \tag{10.76}$$

For $n = 4$, $N = 15$, and $a = 2$, $|\psi_3^{(n)}\rangle$, as defined in the previous step, assumes the form (the period is $P = 4$)

$$|\psi\rangle = \frac{1}{4}\left(|\phi_0^{(4)}\rangle + |\phi_4^{(4)}\rangle + |\phi_8^{(4)}\rangle + |\phi_{12}^{(4)}\rangle\right)|\phi_1^{(4)}\rangle$$

$$+ \frac{1}{4}\left(|\phi_1^{(4)}\rangle + |\phi_5^{(4)}\rangle + |\phi_9^{(4)}\rangle + |\phi_{13}^{(4)}\rangle\right)|\phi_2^{(4)}\rangle$$

$$+ \frac{1}{4}\left(|\phi_2^{(4)}\rangle + |\phi_6^{(4)}\rangle + |\phi_{10}^{(4)}\rangle + |\phi_{14}^{(4)}\rangle\right)|\phi_4^{(4)}\rangle \tag{10.77}$$

$$+ \frac{1}{4}\left(|\phi_3^{(4)}\rangle + |\phi_7^{(4)}\rangle + |\phi_{11}^{(4)}\rangle + |\phi_{15}^{(4)}\rangle\right)|\phi_8^{(4)}\rangle.$$

6. We are now in a position to measure the target register. The measurement produces one of the four values that $f_{aN}(j)$ can assume (1, 2, 4, 8) while information about others is completely lost. Let the value that survives measurement on the target register be $\lambda = f_{aN}(j)$. After measurement therefore we need retain only those terms (states) in the control register that are associated with $|\phi_\lambda^{(n)}\rangle$ in the superposition that $|\phi_3^{(n)}\rangle$ represents. The post-measurement state can therefore be represented as $|\phi_4^{(n)}\rangle$ where (assuming that $2^n/P$ is an integer)

$$|\phi_4^{(n)}\rangle = \sqrt{\frac{P}{2^n}}\left(\sum_{r=0}^{\frac{2^n}{P}-1} \phi_{j=rp+q}^{(n)}\right)\phi_\lambda^{(n)}. \tag{10.78}$$

7. If the period P has to be established from the information contained in $|\phi_4^{(n)}\rangle$, the residue q must be eliminated from the right-hand side of $|\phi_4^{(n)}\rangle$. This can be achieved by performing a Fourier transform on the control register (i.e., on the state $|\phi_4^{(n)}\rangle$). The transform produces the state $|\phi_5^{(n)}\rangle$

$$|\phi_5^{(n)}\rangle = \sqrt{\frac{P}{2^n}}\left(\sum_{L=0}^{2^n-1}\sum_{r=0}^{r<\frac{2^n}{P}} \exp\left[iL(rP+q)\,\pi/2^{(n-1)}\right]\phi_L^{(n)}\right)\phi_\lambda^{(n)}. \tag{10.79}$$

Since $\dfrac{2^n}{P}$ is assumed to be an integer, the factor $\displaystyle\sum_{r=0}^{r<\frac{2^n}{P}}\exp\left[iL(rP+q)\,\pi/2^{(n-1)}\right]$ vanishes unless $L=0$, or is an integral multiple of $\dfrac{2^n}{P}$. Using this result, we arrive at a compact expression for $|\psi_5^{(n)}\rangle$ where

$$|\psi_5^{(n)}\rangle = \left\{\sum_{r=0}^{r<\frac{2^n}{P}} c_{\lambda,rP}\;\phi_{rP}^{(n)}\right\}\phi_\lambda^{(n)}. \tag{10.80}$$

As an example, we may take the case where $n = 4$, $a = 2$, and $N = 15$. Fourier transform on $|\psi\rangle$ of Equation (10.1) then produces

$$\left|\psi'\right\rangle = \left\{ c_{\lambda,0}\left|\psi_0^{(4)}\right\rangle + c_{\lambda,4}\left|\psi_4^{(4)}\right\rangle + c_{\lambda,8}\left|\psi_8^{(4)}\right\rangle + c_{\lambda,12}\left|\psi_{12}^{(4)}\right\rangle \right\}. \tag{10.81}$$

The reader may notice that all the subindices in the expression for $\left|\psi'\right\rangle$ that represents the control register are multiples of the period P (0P, 1P, 2P, 3P).

8. A measurement on the control register could therefore lead to the identification of the period P which in turn would lead to the factorization of N. The operations may be repeated until it has been possible to establish the period. We may mention that the Fourier transform in step 6 can be very efficiently implemented by fully exploiting the quantum parallelism which leads to the quantum Fourier transform. The efficiency of quantum prime factoring largely stems from the highly efficient implementation of the quantum Fourier transformation. The point that must not be missed is that Shor's algorithm can be implemented only on a quantum computer which is still a far cry. Prime factoring which has important consequences in cryptography, has never been proved to be classically impossible to achieve in polynomial time. The best known classical algorithms for factoring to date are however characterized by nonpolynomial (exponential) growth in computing time as the number to be factored becomes larger and larger. The possibility of factoring being done in polynomial time on quantum machines severely challenges the security of standard protocols followed in secure cryptography.

10.8 Quantum Soft Computing

Soft computing, as we have repeatedly stressed (Chapter 1) marks an important shift in the paradigm of traditional analysis-based computing. A second shift in computing paradigm emerged when fuzzy set and logic-based computing was realized. Quantum computing marks a third remarkable shift in the computing paradigm which when fully implemented could unleash a kind of computing power never heard of in the domain of classical (fuzzy or crisp) computing. An important question that comes up at this stage concerns blending the superior features of two different paradigms to generate a new approach wherein the strong features of both the components of the new method constructively contribute to produce superior algorithms.

Let us consider classical genetic algorithms in which the potential solution strings are assemblies of 0s and 1s in the binary representation. If the solution strings (individuals in the population) are n-bit strings, there are 2^n possible strings out of which only N ($\ll 2^n$) individuals are chosen (randomly as well as based on educated guesswork) as individuals in the initial population. As we have discussed already (in Chapter 2), the GAs are endowed with an implicit parallelism which enables the search algorithm to track the fitness of a combinatorially explosive number of substrings (schemas) while actually and explicitly evaluating the fitness values of only N number of strings present in the evolving population. As the schema theorem and building-block hypothesis surmise, the exploration ability of GAs has roots in the implicit parallelism of the GA search. A possibility to blend the implicit parallelism of GA with the inherent quantum parallelism that quantum computing is endowed with is therefore an exciting one. Although several efforts have been made in this direction the full and practical realization of a quantum genetic algorithm and its application to realistic systems is still in the realm of the future.

In what follows we explore several possibilities, explaining in the process how and where features of quantum computing could be introduced or exploited in the GA search for the solution of a problem. Let us consider a toy problem, the potential solutions of which are represented as 3-bit binary strings in a classical GA (CGA). A randomly chosen individual (S_1, say) may have the following bit structure:

$$S_1 \equiv (0\ 1\ 0). \tag{10.82}$$

In a qubit representation on a quantum computer, each of qubits may be in a superposition of $|0\rangle$ and $|1\rangle$ with amplitude α_i and β_i, respectively (*i* being in the *i*th qubit, $i = 1, 2, 3$).

The α_i, β_is, as we have already clarified, obey the normalization condition $\alpha_i^2 + \beta_i^2 = 1$ and the string S_1 in a quantum GA could therefore be represented as

$$S_1^q = \left[\left(\alpha_1 |0\rangle + \beta_1 |1\rangle \right) \left(\alpha_2 |0\rangle + \beta_2 |1\rangle \right) \left(\alpha_3 |0\rangle + \beta_3 |1\rangle \right) \right]. \tag{10.83}$$

Clearly $S_1^q \equiv S_1$ only if the superposition amplitudes are chosen as $\alpha_1 = 1$, $\beta_2 = 1$, and $\alpha_3 = 1$, and the rest zero. Similarly, with $\beta_1 = 1$, $\beta_2 = 1$, $\alpha_3 = 1$, and the rest zero, it is possible to realize another classical binary GA string, chromosome, or individual, (say S_2) where

$$S_2^q = (1\ 1\ 0) = S_2. \tag{10.84}$$

In fact, with appropriate choices for α_1, α_2, α_3 (and therefore β_1, β_2, β_3) we can realize all the $2^3 = 8$ classical strings that can be constructed. In addition, infinitely many more superposed states can be realized. If the right-hand side of S_1^q in Equation 10.83 is expanded, we have

$$\begin{aligned}
S_1^q = {} & \alpha_1\alpha_2\alpha_3 |0\rangle|0\rangle|0\rangle + \beta_1\beta_2\beta_3 |1\rangle|1\rangle|1\rangle \\
& + \alpha_1\alpha_2\beta_3 |0\rangle|0\rangle|1\rangle + \alpha_1\beta_2\alpha_3 |0\rangle|1\rangle|0\rangle \\
& + \alpha_1\beta_2\beta_3 |0\rangle|1\rangle|1\rangle + \beta_1\alpha_2\alpha_3 |1\rangle|0\rangle|0\rangle \\
& + \beta_1\beta_2\alpha_3 |1\rangle|1\rangle|0\rangle + \beta_1\alpha_2\beta_3 |1\rangle|0\rangle|1\rangle
\end{aligned} \tag{10.85}$$

where $\alpha_i^2 + \beta_i^2 = 1$, for $i = 1, 2, 3$. It appears therefore that the single qubit string S_1^q is in a superposition of all the possible classical 3-bit states, and carry much more information than a typical 3-bit classical string. The huge information content of S_1^q is not of immediate use as any act of measurement on the string S_1^q will force the superposition into one of the eight classical states with a probability measured by the square of the amplitude of that state in the superposition. For example, the state S_1^q may be thrown into the state $|0\rangle|1\rangle|0\rangle$ (i.e., $|010\rangle$) with a probability $|\alpha_1\beta_2\alpha_3|^2$. Thus, even though the superposition contains the state of the highest fitness value, there is no guarantee that the act of measurement will force collapse onto the state of the highest fitness. A quantum genetic algorithm must take care of the purely quantum mechanical problem of measurement on a general superposition of states, if the much larger information content of a superposed qubit string is to be exploited. If $\alpha_1 = \alpha_2 = \alpha_3 = \beta_1 = \beta_2 = \beta_3 = \dfrac{1}{\sqrt{2}}$, S_1^q will represent an equal superposition of all the 2^3 classical states with amplitude equal to $\dfrac{1}{\sqrt{2^3}}$ for each state. A measurement would therefore force the superposition to collapse onto one of the eight classical 3-bit states with equal probability $p = \left(\dfrac{1}{\sqrt{2^3}} \right)^2 = \dfrac{1}{8}$. The information about the best string thus remains elusive although the information could be realized with a probability $\dfrac{1}{8}$ after a single measurement on S_1^q in the present example. If we are using N qubit registers which have been put in the state of equal superposition of all the 2^n classical n–bit states, the probability (p) of retrieving a particular classical state after a single measurement would have been $p = \left(\dfrac{1}{\sqrt{2^n}} \right)^2$. With $n = 100$, for example, p becomes vanishingly small and the act of such a measurement is therefore unusable. However, one thing becomes clear—we may represent the potential solutions as floating point strings of values of α_is and β_is in a pairwise fashion as shown below:

$$S^q = \begin{pmatrix} \alpha_1 & \alpha_2 & \alpha_3 & \cdots & \alpha_i & \cdots & \alpha_n \\ \beta_1 & \beta_2 & \beta_3 & \cdots & \beta_i & \cdots & \beta_n \end{pmatrix} \tag{10.86}$$

with $\alpha_i^2 + \beta_i^2 = 1$, for $i = 1, 2, 3, \dots, n$. Thus if all the α_is are 1 and β_is are 0, the string S^q becomes

$$S_1^q = (1\ 1\ 1 \dots 1 \dots 1) \tag{10.87}$$

while a choice of $\beta_i = 0$, for $i = 1, 2, 3, \dots, n$ will lead to the string (in a bit representation)

$$S_2^q = (0\ 0\ 0 \dots 0 \dots 0). \tag{10.88}$$

It is possible to store the amplitudes (α_i, β_i) in an one-dimensional array, the first half storing the n-values of α_is while the second half stores β_is. The reduction of the array of $2n$ values of amplitude to an n-bit classical string can be done by measuring the amplitude of the string as follows:

For $i = 1, 2, \dots, n$

generate a random number r_i in $[0, 1]$

if $r_i > |\alpha_i|^2$ then

set the ith bit to 1

else

set the ith bit to 0

endif

enddo

Let us note therefore that a population of n_p amplitude strings can be easily reduced to a population of n_p classical bit strings (randomly generated) by following the procedure just described. If we choose the initial string with all the amplitudes α_i equal to $\dfrac{1}{\sqrt{2}}$

as the reference, the resulting string describes a state with equal superposition of all the possible classical n-bit strings and the measurement procedure describe above will recover randomly such states with probability $p = \left(\dfrac{1}{\sqrt{2^n}}\right)^2$. If n_p measurements are made,

we will get n_p chromosomes or individuals with the n qubits either in the state "0" or "1." The superposition has collapsed, yielding different individuals randomly with equal probability as an act of quantum measurement would have. Since these chromosomes have classically well-defined bit structure, they can be evaluated by a classical fitness evaluation routine yielding the fitness values $f_1, f_2, \ldots, f_k, \ldots, f_{n_p}$. Let k best correspond to the index of the best string in the initial population. The next task is to induce evolution in the population. One way that has been experimented with some success [15,16] involves the application of quantum rotation gates to the reference string (S_p) from which the population of n_p bit strings were generated through quantum measurement. The gate $U(\Delta\theta)$ is a unitary matrix that acts on the amplitudes (α_i, β_i) of the ith qubit in the qubit string S_p and generates new amplitudes (α_i', β_i') for the superposed ith qubit. If only one qubit (i) in the string is acted on by the rotation gate $U(\Delta\theta_i)$, it can be likened to a mutation process. In that case the site "i" for mutation must be selected with a probability p_m where p_m denotes the inherent mutation probability of the quantum strings. The question now is: how to choose $\Delta\theta_i$? The choice of $\Delta\theta_i$ has two components—the choice of the magnitude $|\Delta\theta_i|$ and the sign or the direction of change in the rotation angle. Here, a marked departure is made. The quantum mutation, unlike the classical mutation operation which acts on a randomly chosen bit and flips it with a probability p_m (without any reference to the fitness of the other strings in the population or its own fitness), acts more globally by choosing $\Delta\theta_i$ so as to nudge the string S_q toward the current best string (S_{kbest}) in the population. Note that the action of the rotation gate $U(\Delta\theta_i)$ is expressed by the following relation between the old and the new amplitudes of the ith qubit in S_p (with $\Delta\theta_i = 0.01\pi - 0.05\pi$).

$$
\begin{pmatrix} \alpha_i' \\ \beta_i' \end{pmatrix} = U(\Delta\theta_i) \begin{pmatrix} \alpha_i \\ \beta_i \end{pmatrix}
$$

$$
= \begin{pmatrix} \cos(\Delta\theta_i) & -\sin(\Delta\theta_i) \\ \sin(\Delta\theta_i) & \cos(\Delta\theta_i) \end{pmatrix} \begin{pmatrix} \alpha_i \\ \beta_i \end{pmatrix}.
$$

$$(10.89)$$

The newly superposed qubit string S_q' is subjected to a measurement (as done previously) and the new classical string, i.e., chromosome x' that is produced is evaluated for its fitness (f'). If $f(x') > f_{kbest}$, then x' replaces the current best string x_{kbest} in the population. If $f(x') < f_{kbest}$, the direction of the rotation is adjusted so that the reference qubit string (S_q') evolves in a direction that is likely to lead to the emergence of a better individual. The adjustment strategy adopted by the authors involves consulting a look-up table (see the review by Lahoz-Beltra [15]) and then fixing the sign of $\Delta\theta_i$. So far, we have assumed that the qubit string S_q has single qubits as genes. In many problems it is not possible to represent the solutions as chromosomes having single-qubit genes. In those cases, multiple qubit genes must be used. Let there be m genes in the chromosomal representation of the potential solutions, and let us also assume that l qubits are allocated to express one gene. Then each chromosomal individual in the population has a total of $m.l$ qubits. The pth member of the population S_p at the tth generation is then given the following multiple qubit representation:

$$S_p(t) \equiv \begin{bmatrix} \alpha_{11}^t & \alpha_{12}^t & \cdots & \alpha_{1l}^t & \cdots & \alpha_{m1}^t & \alpha_{m2}^t & \cdots & \alpha_{ml}^t \\ \beta_{11}^t & \beta_{12}^t & \cdots & \beta_{1l}^t & \cdots & \beta_{m1}^t & \beta_{m2}^t & \cdots & \beta_{ml}^t \end{bmatrix} \qquad (10.90)$$

where for each of the loci the normalization condition prevails:

$$\left|\alpha_{ij}^t\right|^2 + \left|\beta_{ij}^t\right|^2 = 1, \text{ for } i = 1, 2, \ldots, m; j = 1, 2, \ldots, l. \qquad (10.91)$$

As before, we can set all the α_{ij} s and β_{ij} s equal to $1/\sqrt{2}$ so as to create an equal superposition of all the classical (2^N, $N = ml$) bit strings which is subjected to a quantum measurement process that leads to a collapse of the superposed qubit string randomly to one of the 2^N classical bit strings with equal probability. The rest of the steps or procedures remain the same. Malossini et al. [17] in 2008 proposed a quantum genetic optimization algorithm which exploits the power of quantum computing in fitness evaluation as well as in the selection procedure. The work demonstrates how one can take advantage of the quantum phenomena to achieve a speedup of the classical computational procedure for handling the same problem. In particular, these authors reveal how the quantum genetic optimization algorithm outperforms its classical counterpart when fitness function is varying between the genetic steps. The algorithm is based on the Durr–Hoyer quantum algorithm [18] for finding the minimum in an unsorted table of data. It relies on the observation that the Durr–Hoyer algorithm can be stopped and the partial results used for constructing a quantum selection procedure.

Recently Sun et al. [19] proposed and implemented a novel quantum genetic algorithm for optimization of a multidimensional function $f(x)$ with $x \equiv (x_1, x_2, \ldots, x_N)$. The authors named it the variable boundary coded quantum genetic algorithm (vb-QGA). They proposed a departure from the standard and traditional binary-coded strings for representing the chromosomes. Instead, they make use of qubit chromosomes that collapse onto vb-coded chromosomes. They also introduced a new adaptive schema for selecting the rotation angles for construing the quantum gates to be applied to the chromosomes for refining the solutions. In view of the importance of global multidimensional function minimization in physical and chemical sciences, we propose to describe Sun et al.'s work in some detail so that readers are encouraged to experiment on further modifications that could improve the performance even more. The basic idea is to define the state of a qubit ($|\psi\rangle$) as a superposition of two kets $|x'\rangle$ and $|x''\rangle$:

$$|\psi\rangle = \alpha|x'\rangle + \beta|x''\rangle \qquad (10.92)$$

where $|x'\rangle$ and $|x''\rangle$ represent the lower and upper limits of the variable x which appears in the function $f(x)$, α and β are most generally complex numbers satisfying the condition.

$$|\alpha|^2 + |\beta|^2 = 1. \qquad (10.93)$$

Since $f(x)$ is a function of N number of variables, $x_1, x_2, \ldots, x_i, \ldots, x_N$, one must define the upper and lower limits of each of the variables. That means, we must define the following boundary relations:

$$x_i^l \le x_i \le x_i^u. \tag{10.94}$$

Sun et al. [19] define the chromosome k in the generation t as a collection of N substrings, one for each of the variables x_i. The qubit for x_i in the kth string in generation t is defined as follows:

$$q_{k,i}^t = \begin{bmatrix} \alpha_{k,i1} & \alpha_{k,i2} \\ \beta_{k,i1} & \beta_{k,i2} \end{bmatrix}. \tag{10.95}$$

The full chromosome in the qubit representation then reads

$$q_i^t = \begin{bmatrix} \alpha_{k,11}^t & \alpha_{k,12}^t \\ \beta_{k,11}^t & \beta_{k,12}^t \end{bmatrix} \begin{bmatrix} \alpha_{k,21}^t & \alpha_{k,22}^t \\ \beta_{k,21}^t & \beta_{k,22}^t \end{bmatrix} \cdots \begin{bmatrix} \alpha_{k,N1}^t & \alpha_{k,N2}^t \\ \beta_{k,N1}^t & \beta_{k,N2}^t \end{bmatrix}. \tag{10.96}$$

The length l of the qubit chromosome in this representation is clearly l. Once the reference string (k) is generated (the α^t, β^t values are chosen randomly, subject to $|\alpha^t|^2 + |\beta^t|^2 = 1$), Sun et al. then defined their measurement procedure (quantum measurement) on the string q_k^t which led to its collapse onto a variable boundary-coded chromosome. For any specific qubit (say q_i^t), with

$$q_i^t = \left[\alpha_{k,ij}^t, \beta_{k,ij}^t \right]^t; j = 1, 2 \tag{10.97}$$

a random number $r_{k,ij}^t$ in [0,1] is generated. If $r_{k,ij}^t \le |\alpha_{k,ij}^t|^2$ the process of measurement reduces the qubit into the state $|x_i^l\rangle$; if not, the qubit is forced into the state $|x_i^u\rangle$. With the substring for the variable x_i in chromosome k defined in the manner already stated, the collapse following the measurement forces it into a substring of a variable boundary coded chromosome (they call it $vb_{k,ij}^t$) which may be one of the following four possibilities:

$$vb_{k,ij}^t \equiv \left\{ \left[x_i^l, x_i^l \right], \left[x_i^l, x_i^u \right], \left[x_i^u, x_i^l \right], \left[x_i^u, x_i^u \right] \right\}; j = 1, 2. \tag{10.98}$$

The decoding of the variable boundary chromosome is a bit tricky. Noting that $\Delta x_i = \left(x_i^0 - x_i^l \right)$, the four possibilities that could be realized post measurement, mean that the value of x_i could lie in one of the four regions, each having a width $\Delta x_i / 4$. Once the value of x_is has been drawn from the appropriate region, the string is completely defined so that its fitness $f(x_k)$ can be calculated easily by the classical process of fitness evaluation. The improvements of the strings are carried out by a quantum mutation process that involves the applications of unitary rotation gates on the qubits. The authors recommend the use of a constant rotation angle $\theta = 0.05\pi$, but adopt an adaptive strategy for fixing the sign of the rotation angle $\theta_{k,ij}^t (j = 1, 2)$ that exploits knowledge about the fitness value of the best-fit string in the previous generation. The look-up table is replaced by the following process:

For the rotation gate $U_{k,ij}^t(\theta_k^t)$, the following definition of the gate is used:

$$U_{k,ij}^t\left(\theta_k^t\right) = \begin{bmatrix} \cos(\theta_{i,ij}^t) & -\sin(\theta_{i,ij}^t) \\ \sin(\theta_{i,ij}^t) & \cos(\theta_{i,ij}^t) \end{bmatrix}. \tag{10.99}$$

The angle of rotation $\theta_{i,ij}^t$ is defined, including sign by the authors (Sun et al.) as

$$\theta_{i,ij}^t = Sgn\left\{\alpha_{k,ij}^t \beta_{k,ij}^t \left[f\left(b^{t-1}\right) - f(x_k^t)\right] \frac{b_i^{t-1} - x_{k,i}^t}{x_i^u - x_i^l}\right\} \times 0.05\pi \tag{10.100}$$

$f(x_k^t)$ is the fitness of the kth string in the tth generation (x_k^t). b^{t-1} is the best-fit string in the $(t-1)$th generation and $f(b^{t-1})$ is the corresponding fitness value of that string. $x_{k,i}^t$ is the value of the variable x_i in the kth solution string in generation t, while b_i^{t-1} is the ith element of the best-fit string in the previous generation. $Sgn\{\cdot\}$, defines the sign of the argument appearing within $\{\}$. The following rule applies in fixing the output of the $Sgn\{\cdot\}$ function:

$$Sgn\{\cdot\}\begin{cases} = 1, & \text{if } \alpha_{k,ij}^t \beta_{k,ij}^t \left[f\left(b^{t-1}\right) - f(x_k^t)\right] > 0 \\ = -1, & \text{if } \alpha_{k,ij}^t \beta_{k,ij}^t \left[f\left(b^{t-1}\right) - f\left(x_k^t\right)\right] < 0 \\ = \pm 1, & \text{if } \alpha_{k,ij}^t = 0 \text{ and } \left(b_i^{t-1} - x_{k,i}^t\right)\left[f\left(b^{t-1}\right) - f\left(x_k^t\right)\right] < 0 \\ = \pm 1, & \text{if } \beta_{k,ij}^t = 0 \text{ and } \left(b_i^{t-1} - x_{k,i}^t\right)\left[f\left(b^{t-1}\right) - f\left(x_k^t\right)\right] > 0 \\ = 0, & \text{otherwise} \end{cases} \tag{10.101}$$

The authors tested the algorithm on eight test functions and concluded that it is superior in performance to classical genetic algorithms. The use of floating point as opposed to binary qubit representation marks a departure from traditional practice and the use of boundary value coded qubit strings enhances the flexibility and power of the search algorithm. One can anticipate further improvement in QGA in the context of multidimensional function optimization, in several directions (see the review by Lahoz-Beltra [15]).

Rylander et al. [20] in 2001 examined the possibilities relating to further development of QGA, and the difficulties standing in the way of practical realization or implementation of QGA. These authors in particular suggested a means of exploiting the phenomenon of quantum entanglement among a set of quantum registers. We will examine their proposal closely and in some detail to explicate where they propose to differ from the other QGAs discussed so far and where they propose to exploit quantum parallelism and entanglement.

Let us start with a set of N quantum registers (n-qubits) marked $R_0, R_1, \ldots, R_{N-1}$, each representing one individual of the population of size N. A Hadamard transform then converts each register into an equal superposition of all the N-individuals. Each register is thus made to represent all possible individuals, albeit in a state of equal superposition. Rylander et al. then suggested that a fitness measuring function (a quantum black box) is applied to each of the N-quantum registers in superposed states and the fitness values obtained are stored in a second set of N quantum registers $RR_0, RR_1, ..., RR_i, ..., RR_{N-1}$. The quantum fitness measuring system with one application, can and does evaluate all the individuals in the superposition (see DJ algorithm [10]). The second set of registers (RR_is) will therefore contain all the possible fitness values. The fitness measurement produces entanglement between the registers $\{R_i\}$ and $\{RR_i\}$. A second measurement is now applied to each of the second registers (RR_is). The superposition now collapses to a single value of fitness with a probability determined by its amplitude in the superposition. A random fitness value (f_i) therefore appears in each of RR_is following the second measurement on RR_i. Now the entanglement between the registers $\{R_i\}$ and $\{RR_i\}$ comes into play due to which the register R_i does not collapse to a single individual, but to a superposition of a subset of all those individuals having the specific fitness value, say f_i, found in RR_i. Could we predeterminedly measure the maximum fitness value for all the values stored in the second set of quantum registers RR_i? Unfortunately, as Rylander stresses, that is not possible. The outcome of the measurement is a randomly selected fitness value with the value of the maximum fitness being as probable as any other value stored in RR_i. Rylander's proposal is to improve the individuals represented by the quantum register (R_i) by the application of crossover in the manner it is used even in classical GA. The only difference is that here the crossover will take place between the quantum register R_i (which is in a state of superposition all these individuals in the population that have the fitness value f_i) and R_j (which is similarly in a state of superposition of all the individuals having the fitness value f_j). The process lends to all the possible individuals that can be generated by qubit exchange at the designated crossover site chosen with a preset crossover probability. The entire procedure is repeated until a termination criterion is satisfied. However, this proposal, to date remains a theoretical claim only in as much as quantum computers of appropriate sizes are yet to be available to implement the algorithm. The challenge will be to maintain the entanglement between the two sets of registers for a long enough duration so that the computing processes can be carried out. The implementation of the crossover operation as suggested is beset with considerable difficulty.

Gu et al. [21] suggested a novel parallel quantum genetic algorithm. These authors blended some ideas based on parallel evolutionary computing and features of quantum science to create a parallel quantum evolutionary computing model. They make use of some demes (subpopulation and some universes or groups of population) which appear as structures in a super star-shaped topology. A new migration scheme based on the penetration theory is developed to control the rate and direction of population migration adaptively between demes. The authors also propose and implement a novel quantum crossover process among the universes. A quantum evolution is executed on every deme using suitable improvement operators like a new quantum rotation gate and a catastrophe operator. The authors have tested their algorithm on a job-shop scheduling problem. More general applications are necessary to establish the supremacy of their algorithm as a problem solver of the general kind. We must mention here that Narayanan and Moore [22] in their seminal work (1996) described a quantum-inspired genetic algorithm to solve the traveling salesman problem where they blended the ideas of evolutionary parallel computing and concepts of quantum mechanics for the first time. As the authors noted in their paper, their purpose was not so much to present a formal exposition of quantum-inspired GA as it was to demonstrate the feasibility of such methods and more importantly, the novel shift in the computing paradigm that it entails. The authors reflected on the complexity class that such algorithm could belong to. To quote the authors "It is possible that quantum-inspired genetic algorithms fall somewhere between PSPACE, NP, and BPP (complexity classes). The relationship between NP and BPP is not currently known, and quantum-inspired genetic algorithms may offer insight into the precise relationship between NP and BPP."

A severe drawback that plagues quantum computing is the problem of constructing and maintaining a stable superposition of even a small number of qubits. It is no less hard to apply unitary gates on the qubits. The difficulty has its roots in the problem of decoherence [23]. Several proposals have been advanced from time to time on the ways a quantum computer can be constructed. Theoretically it is sufficient to be able to construct the set of universal gates like the CNOT and one qubit rotation gate, for building up a quantum computer. The most significant advancement toward the building of a quantum computer came in the form of the idea of constructing fault-tolerant quantum gates [24].

A second approach to quantum computing emerged when Farhi [25] proposed to exploit the quantum adiabatic theorem [26] for the purpose of quantum computing leading to the new paradigm of adiabatic quantum computation (AQC). The new AQC scheme has been shown to be more robust against noise [27]. The idea of AQC is simple: represent the system by a Hamiltonian H_0, prepare the system in its known ground state (lowest energy eigen state of H_0), and deform the system's Hamiltonian sufficiently slowly $[H(t) = H_0 + \lambda(t)(H - H_0)$ with $\lambda(t)$ varying slowly] ensuring that the system is never excited out of the ground state during the course of the evolution process; at the conclusion of the evolution the system will be in the ground state of the new Hamiltonian (H, say) produced at the conclusion of the evolution. Thus, if the problem to be solved can be mapped into the problem of finding ground state ($|\psi\rangle$) of a suitably designed Hamiltonian (H), and if an initial Hamiltonian H_0 and its ground state $|\psi_0\rangle$ are known, we can invoke the quantum adiabatic theorem to evolve the initial state $|\psi_0\rangle$ into the desired solution $|\psi\rangle$. It has been suggested that AQC has a kinship to a local search [28]. It has also been proved that it is possible to simulate any AQC by a series of reversible (unitary) gates. In fact, equivalence between the AQC and the circuit model was established in 2008 [29]. The equivalence follows from the fact that for any algorithm based on the circuit model, there is a history vector that encodes the entire series of actions of the concatenated (quantum) unitary gates on the chosen initial vector. If the history vector could be written as the ground state $|\psi_0\rangle$ of a certain Hamiltonian (H_0) the circuit computation can be represented by an adiabatic model of performing quantum computation (for details, see Aharonov [29]). The equivalence proved by Aharonov led to several adiabatic quantum computing protocols to solve the Grover's search problem, the problem of graph isomorphism, the Deutsch–Jozsa problem, etc. There is, however, no direct route still available by which one can translate a circuit model solution to a solution produced by adiabatic computing. The proof of equivalence does not provide any clue to such a route. The difficulty of transcribing the solution from one model (circuit model) to another (adiabatic quantum computing) has roots in the fact that the exponential of a sum of noncommuting Hamiltonians is not at all equivalent to the sum of exponentials of those noncommuting Hamiltonians [30]. That means our ability to design a simple set of universal gates in the circuit model to do a computation does not guarantee that it will be equally easy to do it in adiabatic computing; in fact, it turns out to be much harder.

Let us now turn to a rather deep analytic relation that has been shown to exist between the adiabatic quantum process and the process of simulated annealing [31,32]. In what follows we briefly sketch the arguments leading to the purported equivalence between simulated annealing and adiabatic computation, sometimes called classical to quantum mapping. Suppose that we have a set of n Boolean variables $(\sigma_1, \sigma_2, ..., \sigma_n) \equiv \bar{\sigma}$ with

$\sigma_i = \pm 1$. Clearly, the configuration space has 2^n dimensions. In this configuration space we can define an energy function $E(\bar{\sigma})$ where

$$E(\bar{\sigma}) = \sum_{i,j} J_{ij} \sigma_i \sigma_j \tag{10.102}$$

J_{ij}s being the coupling strengths between pairs of the Boolean variables. Let $S(\beta)$ be a stochastic process such that the equilibrium distribution (call it $P(\beta, \bar{\sigma})$) is given by

$$P(\beta, \bar{\sigma}) = e^{-\beta E(\sigma)} / z(\beta) \tag{10.103}$$

where

$$z(\beta) = \sum_{\sigma} e^{-\beta E(\sigma)} \tag{10.104}$$

and β is $(kT)^{-1}$.

It is possible to associate a quantum pure state $|\bar{\sigma}\rangle$ for each classical state $\bar{\sigma}$, and define a stochastic matrix S^β that is consistent with the principle of detailed balance. The demands $S^\beta(\sigma)$ and $P(\beta, \sigma_i)$ must satisfy

$$S^\beta(\bar{\sigma}_i | \bar{\sigma}_j) P(\beta, \bar{\sigma}_j) = S^\beta(\bar{\sigma}_j | \bar{\sigma}_i) \pi(\beta, \bar{\sigma}_i). \tag{10.105}$$

Using the S^βs, we can set up a Hamiltonian matrix with elements given by

$$H_\beta(\bar{\sigma}_i | \bar{\sigma}_j) = \delta_{i,j} - \sqrt{S^\beta(\bar{\sigma}_i | \bar{\sigma}_j) S^\beta(\bar{\sigma}_j | \bar{\sigma}_i)}. \tag{10.106}$$

It turns out [30–33] that the unique state of H_β (lowest energy eigen state) is

$$|\psi_\beta\rangle = \frac{1}{\sqrt{z(\beta)}} \sum_{\sigma} e^{-\beta E(\sigma)/2} |\bar{\sigma}\rangle \tag{10.107}$$

which corresponds to the Gibbs state of energy E at the temperature defined by β. Of critical importance in this discourse is the transverse field component, for example $\sum_i \sigma_i^x$, for it naturally leads to a correspondence between the stochastic transition matrix and the Hamiltonian. Once the correspondence has been established, it is natural to enquire how slow should the cooling rate be to ensure that the ground state is reached by the quantum annealing process. Morita and Nishimori [32] showed that the optimal cooling rate for quantum annealing should not be faster than

$$T(t) \sim \frac{n}{\log t} \tag{10.108}$$

which is very close to the optimal cooling rate in classical simulated annealing that ensures realization of the ground state.

It would be reasonable to assume that the correspondence between the stochastic transition matrices and Hamiltonian subsumes the classical simulated annealing into the set of quantum annealing such that the choice of a special quantum annealing path would lead to the classical simulated annealing process [34].

10.9 Quantum Annealing

As we have already mentioned briefly in Chapter 4, quantum annealing was proposed as an improved alternative to the method of simulated annealing (classical) for locating the global minimum in nonergodic systems such as the spin glasses [35]. In such systems, the minima of energy or the appropriate cost function may be separated from each other by barriers with energy of the order n $[O(n)]$ so that at any finite temperature, thermal fluctuation would practically take infinite time before pushing up the system over the barriers to the true ground state. In what follows we describe the basic steps of the quantum annealing.

The first step in quantum annealing involves mapping the problem at hand onto an equivalent physical problem for which the energy (cost function) can be represented by some Ising model Hamiltonian H_0 where

$$H_0 = -\sum_{i<j} J_{ij} S_i^z S_j^z - \sum_i h_i S_i^z \tag{10.109}$$

h_i is the magnetic field experienced by the spin at the site indexed by i. The second very important step is to render the system dynamic. It is achieved by adding an appropriate noncommuting Hamiltonian H_1 to the system represented by $H_1(t)$ which can cause spin flip and varies with time. A convenient choice for the present case could be

$$H_1(t) = -\gamma(t) \sum_i \eta_i S_i^x \tag{10.110}$$

η_i is the tunneling strength at the ith site, $S_i^x (= \frac{\hbar}{2} \sigma_i^x)$ causes spin flip while $\gamma(t)$ is a switch that brings in time dependence to the system. The total Hamiltonian reads

$$H = H_0 + H_1(t) \tag{10.111}$$

and defines the evolution equation for the system as described by the corresponding time dependent Schrödinger equation:

$$i\hbar \frac{\partial}{\partial t} |\psi\rangle = (H_0 + H_1(t)) |\psi\rangle. \tag{10.112}$$

A little reflection suggests that the Schrödinger Equation 10.112 describes the tunneling dynamics of the system among the eigen states of H_0. If we start from the ground state of $H(t=0)$ and $\gamma(t)$ is allowed to change very slowly (adiabatically) and become zero as $t \to \infty$, the evolution does not allow the system to depart significantly from the instantaneous eigen state of the Hamiltonian $H(t)$ at every instant so that eventually the system moves into an eigen state of H_0 which may be the ground state of the system described by H_0. $H_1(t)$ is designed to bring quantum fluctuations in the system and the slow reduction of $\gamma(t)$ ($\gamma(t) \to 0$) from a high initial value is reminiscent of slow cooling in the classical annealing algorithm. The quantum fluctuation induces tunneling among states localized in regions separated by high but narrow (thin) barriers which become transparent to the quantum annealing process. The dynamics therefore visit every state, making the free energies or the cost landscape fully ergodic [35]. Altshuler et al. [36], however, felt that the quantum annealing could fail due to Anderson localization. In 2001, Rajak and Chakraborty [37] demonstrated the advantages of a quantum annealing search for the ground state(s) of Ising spin glass systems with tunable transverse and longitudinal fields. Martonak et al. [38] have shown that the path integral Monte Carlo quantum annealing method performs better in solving the TSP problem for 1002 cities. The error reduction in quantum annealing took place much faster than in classical simulated annealing-based search for the solution. Jorg et al. [39] examined the time complexity of the adiabatic algorithm in the context of following first-order phase transition. It turns out that the time complexity of the adiabatic method is exponential. However, the time complexity achievable in the adiabatic algorithm is polynomial when it comes to second-order phase transition. It has been suggested that by adding a tunneling term in the adiabatic algorithm, even the first-order phase transition can be handled in polynomial time. The advantages of introducing a quantum tunneling component in the Hamiltonian was demonstrated in 1999 [40] in the context of the search for the ground state of a disordered ferromagnet [41]. At this point it is important to turn our attention to the problem of constructing quantum hardware (quantum computers) on which the quantum algorithm can be fully implemented. A quantum algorithm running on classical machines is not a worthwhile proposition.

10.10 Quantum COMPUTERS: The Exact State of Affairs

The physical realization of quantum computers requires (a) a physical representation of qubits with coherence time that is much larger than the computing time, (b) a physical mechanism for realizing controllable unitary evolution of the qubits, i.e., controllable unitary gates, (c) a method of preparing the initial states of the qubits, and (d) a means of performing a projective measurement of the final qubit states. It has been so far possible to construct small-scale quantum computers based on various kinds of physical qubits. For example, there have been the single photon quantum computers [42], trapped ions [43], nuclear spins [44], quantum dots [45], neutral atoms in optical lattices [46], and qubits based on superconducting circuits [47]. These quantum computers demonstrate the workability of particular concepts, but hardly qualify for providing the basic hardware for large-scale quantum computation.

The first breakthrough, though controversial, came from a privately owned Canadian company—Wave System Inc., which announced the launch of the first commercially available quantum computer named D-Wave One (2011). The machine contained 128 qubits and was followed up in 2013 by D-Wave Two, which had 512 qubit boards. D-WaveE Two was procured by Google. These machines use flux qubits which are cooled down to 20 mk by a dilution refrigerator. The core is kept in high vacuum of 10^{-10} atmosphere and shielded from the terrestrial magnetic field down to 5×10^{-5} Tesla. The D-Wave computers claimed to be the first working model of a commercial quantum computer, exploiting quantum annealing or quantum adiabatic computing for doing the computation. We may say that the D-Wave machines can operate either as quantum annealers or as adiabatic computers with programmable transverse magnetic fields for tuning. The D-Wave machines have been the subject of sharp controversy, for it is hard to establish if the machines are indeed quantum computers, or if there is a detectable quantum speedup.

The difficulty lies in the fact that it is hard to ensure that the time evolution indeed conforms to the exact requirements of the quantum adiabatic theorem. It appears that the D-Wave machines under the presently available working conditions probably belong to a regime that lies somewhere between the thermic and the adiabatic regimes. It has been suggested that such an intermediate regime could in fact enhance the probability of arriving at the correct result even though excitation into higher states could take place [48]. That is possible as excitation is followed by relaxation into the lower state thereby increasing exploration. A big question concerns the lack of universality; for there is no natural set of universal gates capable of generating the entire spectrum of the adiabatic model. The lack of universality forces one to categorize the machine as something akin to the one-purpose analog machines i.e., as quantum analog computers. It has been suggested that the D-Wave computers working adiabatically can solve, in principle, any NP-hard problem. The equivalence between adiabatic and quantum circuit computations [45] would tend to suggest that the adiabatic model can be further generalized to handle any NP-hard problem that the circuit model is capable of solving. There is no easy way, however to translate a circuit model into an equivalent adiabatic model as the latter lacks the representability in terms of universal gates. Much more work in this area can be expected to provide new insight.

As we have mentioned already the coherence time of the qubits used in D-Wave machines must be longer than the time taken by the algorithm to perform the computation. Typical coherence time of the SQUID is about 10 ns while the annealing (quantum) requires at least 5–15 ns [46]. D-Wave machines therefore operate in a semithermodynamic, semiadiabatic regime. There has been significant progress in increasing the decoherence time of the flux qubits [49] used in D-Wave systems raising hope that soon such machines would start working in a completely adiabatic quantum computing mode.

A third issue concerning the D-Wave computers is its scalability. Essentially, the issue boils down to the complexity of building a coherent circuit for adiabatic computation vis-á-vis any possible gain from superior algorithmic complexity. A scalable architecture for adiabatic quantum computing of NP-hard problems has been proposed by Kaminsky and Lloyd (2004) [50]. The trick used by these authors was to translate NP-hard problems into the problem of maximum independent sets for which a suitable Hamiltonian was constructed. A lot more needs to be done in the area for gaining a better insight into the scalability problems relating to AQC.

The two other rather vexing issues concerning quantum computers are the "quantum speedup" and the so-called quantumness of the computers such as D-Wave One or D-Wave Two. Both the issues are problematic and no fool-proof answers have so far emerged. We refer readers to a number of recent publications that address these issues [51–53]. It is hard to come to a definite conclusion as to the quantumness of the D-Wave machines or their ability to achieve significant and definite quantum speedup. However, the concepts of quantum annealers and quantum adiabatic computers, we hope will continue to spawn further research in quantum computers and quantum computing and pave the way to a real quantum computer for executing quantum algorithms with spectacular speedups.

10.11 Automatic Quantum Computer Programming

A pertinent question at this point may be to enquire whether automatic evolution of hardware and software could solve the problems of designing optimal hardware for quantum computers, along with appropriate software which can be implemented on the hardware so designed. It appears that some progress has been made on this front by invoking a soft computing option—the concept of genetic programming to be specific as a tool. Bautu and Bautu [54] have recently explored the possibility of designing quantum circuits by adopting the method of genetic programming. In their work, the quantum circuit is presented as a sequence of quantum gates which are encoded as strings $S_n[p]$ with S representing the gate type, n encoding the target qubits $0 \le n \le 2^{N-1}$, N being the number of qubits. p provides a variable sized list of additional parameters. These authors set the qubits in the standard base 2 representation, with n marking the qubits that the gate is designed to act upon. An individual of the population is an ordered list of quantum gates and the effects of an application of a quantum circuit on a quantum register is obtained by applying gates in the specific order of their appearance in the individual.

Fitness-proportioned selection was used to provide a bias to the GP evolution. The best individual was always preserved. The mutation operator in the method proposed works on an individual by randomly selecting a gate and replacing it with a new gate (chosen from a pool of elementary quantum gates). The crossover operator operates by swapping the gates appearing after the selected crossover point in one individual with those appearing in another individual (selected randomly) after the same crossover point. The authors designed an optimization procedure that favors circuits containing a fewer number of elementary quantum gates. They also designed an editing operator that removed unused gates in an individual.

The quantum GP algorithm was tested on the problem of designing quantum circuits for production of 2, 3, 4, maximally entangled qubits in the form $\frac{1}{\sqrt{2}}\{|00...0\rangle + |11...1\rangle\}$. The input register was always prepared in the state $|00...0\rangle$ and the target was to produce the output register in the designated entangled state. The set of gates used to form the individuals were the CNOT, NOT SN (square root of NOT), H (Hadamard), simple rotation NR(θ), and the SWAP gates. A population size of 300 individuals was used with a mutation rate of 1%, crossover rate of 70%, and gene editing rate of 10%. It is interesting to note that the GP was able to find the perfectly matching circuit (the correct solution) in ≤ 15 generations when the gene editing rate $\approx 10\%$. If the editing rate was kept less than 1%, the circuit designed usually had redundant gates.

As the preceding work demonstrates, automatic quantum computing based on GP can be useful in quantum circuit design. Spector [55] in his excellent book on automatic quantum computer programming has carefully examined several limitations of traditional GP in the context of large-scale use of automatic quantum computer programming for solving difficult problems [56]. A particularly vexing limitation of the traditional genetic programming concerns the restriction that the evolved programs can manipulate only a single data type. A version of GP called "strongly typed GP" was developed by Montana [57,58] which allowed for the evolution of programs capable of manipulating multiple data types although the facility comes with a cost—it can strongly affect the evolutionary dynamics. The lack of facilities for the expression of subroutines or other modular code structures in traditional Lisp-based GP is another shortcoming when it comes to GP-driven automatic quantum computer programming. The most popular scheme for incorporating modules in GP has been the automatically defined functions (ADF) scheme proposed and experimented with by Koza [59–61] who later showed how one can add another layer to the GP enabling the ADF architecture to evolve during a run. Yet another shortcoming of the traditional GP relates to the evolution of programs or, for that matter, any other executable structures that cannot be easily mapped onto the traditional Lisp-based representation of programs. This deficiency affects the GP-based representation of recurrent neural networks, for example since the NNs (they are programs in a sense) do not have the tree structure that is used to encode individuals in GP. Researchers have tried to provide a remedy by extending the genetic programming to have some developmental features. The basic idea has been to retain the traditional program representation, but drop the requirement that the evolved program itself is the solution to the problem at hand. On the contrary, the evolved program when executed produces an individual or an executable object or a program that leads to the solution of the problem through the standard fitness evaluation route. The so-called development is achieved by various means—the most commonly adopted strategy being to initiate each developmental phase with a minimal embryo and then allow it to grow by adding function calls. Spector [55] has contributed very significantly to the enterprise by developing a new specifically designed programming language call PUSH, which can be used very conveniently in evolutionary and genetic computation systematics for evolving computer programs. The advantage of PUSH lies in its simple syntax which allows one to implement the mutation and recombination operator with great facility for manipulating and generating programs. In addition to the capability of processing multiple data types without syntax restrictions, it supports entirely new evolutionary computing paradigms such as "autoconstructive evolution" in which genetic operators are allowed to evolve. In fact, a PUSH GP genetic programming system that evolves PUSH programs has a lot of attractive features (see, for example L. Spector [55]), and is ideal for experimenting with automatic program evolution. PUSH GP can often come up with an unusual solution. In one example is provided by Spector, who evolved a program to solve the so-called ODD problem—that is, the problem of determining whether the input to the program is odd or not. The automatic route achieved the objective by using the input number to the program as an index and subsequently evaluating a property of the code at that index. It is unlikely that a human programmer would ever approach the solution in this way. PUSH GP works in a self-adaptive manner in the context of determining the number and architecture of modules and the data types that should be used in a solution. All these features are determined dynamically and automatically as an integral part of the evolutionary process. PUSH GP has been provided with a means of avoiding evolutionary stagnation which could cripple further improvement of a solution if a program produced only exact clones of itself. The stagnation avoidance resulted from the use of a no-cloning rule (exact clones are not allowed to be passed on to the next generation). The usefulness of an autoconstructive evolution system is not completely known as of now, but one can guess that such a system will probably outperform the traditional GP systems by greater and better adaptation of their reproductive mechanisms, and adaptation of their representations to the problem environments.

The automatic program evolution techniques discussed so far can also be used to evolve quantum computing programs. To do that one must first determine what forms the evolved quantum program could take (it could also be a developed program). If we knew, for example, that the program that will solve a problem is a single-qubit rotation gate, we could represent the evolving program making use of four real-valued parameters of a $U2$ gate described earlier in this chapter. The problem-solving gate can then be automatically evolved by using a chromosome consisting of four real-valued parameters. For more interesting problems, the solutions will be in the form of program that operates on many qubits. Mindlessly extending what is used to handle the 1-qubit problems to start with is not a good idea. It is more convenient and sensible here to represent the quantum program as a sequence of otherwise well-understood quantum gates that act on multiple qubits. An attractive way of doing it can be to use the QGAME representation of quantum programs developed by Spector.

Barnum et al. [62] adopted the technique to design quantum circuits for OR and AND of OR. We will briefly describe their work in what follows. Suppose that a physical process P acts on two orthogonal input states $|0\rangle$ and $|1\rangle$ producing output $|f(0)\rangle$ and $|f(1)\rangle$ where f is just a classical Boolean function form $\{0,1\}$ to $\{0,1\}$. If P is a quantum unitary gate we can start with a suitably prepared quantum state as an input and then check if quantum computation (a sequence of quantum gate applications) would allow us to learn more about the function than is classically possible. We can for example, find out the parity $f(0)XORf(1)$. The quantum circuit designed by Deutsch was the first ever concrete demonstration of the greater-than-classical power of quantum computing (see Section 10.5). Barnum et al. further confirmed the greater-than-classical power of quantum computing by designing the quantum circuits which compute $f(0)ORf(1)$ with a single call to the subroutine f (i.e., the physical process P). The novelty of the work lies in the authors' use of automatic evolution of the circuit by quantum genetic programming. Unlike the XOR circuit, the OR circuit does not produce the absolutely correct result but it always produces a result that is better than what is classically possible. It would be of great interest to see if automatic quantum GP can handle more difficult problems with equal facility, or if can be interfaced with quantum adiabatic computing or quantum annealing. We anticipate exciting developments in this area in the near future.

References

1. Landauer, R. 1999. Information Is Inevitably Physical. In *Feynman and Computation: Exploring the Limits of Computers*, ed. A.J.G. Hey, pp. 77–92. Cambridge, MA: Perseus Books.
2. Moore, G.E. 1965. Cramming More Components onto Integrated Circuits. *Electronics* 38, no. 8: 114–117.
3. Feynman, R.P. 1982. Simulating Physics with Computers. *International Journal of Theoretical Physics* 21, no. 6–7: 467–488.
4. Lloyd, S. 1998. Universal Quantum Simulators: Correction. *Science* 279, no. 5354: 1113h–1117.
5. Nielsen, M.A., and I.L. Chuang. 2010. *Quantum Computation and Quantum Information*. Cambridge: Cambridge University Press.
6. Wootters, W.K., and W.H. Zurek. 1982. A Single Quantum Cannot Be Cloned. *Nature* 299, no. 5886: 802–803.
7. Dieks, D. 1982. Communication by EPR Devices. *Physics Letters A* 92, no. 6: 271–272.
8. Bell, J.S. 2004. *Speakable and Unspeakable in Quantum Mechanics: Collected Papers on Quantum Philosophy. Collected Papers on Quantum Philosophy*. Cambridge: Cambridge University Press.
9. Bell, J.S. 1964. On the Einstein Podolsky Rosen Paradox. *Physics* 1: 195–200.
10. Deutsch, D., and R. Jozsa. 1992. Rapid Solution of Problems by Quantum Computation. *Proceedings of the Royal Society A: Mathematical, Physical and Engineering Sciences* 439, no. 1907: 553–558.
11. Grover, L.K. 1996. A Fast Quantum Mechanical Algorithm for Database Search. In *Proceedings of the Twenty-Eighth Annual ACM Symposium on Theory of Computing—STOC '96*, pp. 212–219. New York: ACM Press.
12. Grover, L.K. 1997. Quantum Mechanics Helps in Searching for a Needle in a Haystack. *Physical Review Letters* 79, no. 2: 325–328.
13. Shor, P.W. 1997. Polynomial-Time Algorithms for Prime Factorization and Discrete Logarithms on a Quantum Computer. *SIAM Journal on Computing* 26, no. 5: 1484–1509.

14. Shor, P.W. 1994. Algorithms for Quantum Computation: Discrete Logarithms and Factoring. In *Proceedings 35th Annual Symposium on Foundations of Computer Science*, pp. 124–134. Washington, DC: IEEE Computer Society.

15. Lahoz-Beltra, R. 2016. Quantum Genetic Algorithms for Computer Scientists. *Computers* 5, no. 4: 24.

16. Laboudi, Z., and S. Chikhi. 2012. Comparison of Genetic Algorithm and Quantum Genetic Algorithm. Article. *Ccis2K.Org* 9, no. 3: 243–249.

17. Malossini, A., E. Blanzieri, and T. Calarco. 2008. Quantum Genetic Optimization. *IEEE Transactions on Evolutionary Computation* 12, no. 2: 231–241.

18. Durr, C., and P. Hoyer. 1996. A Quantum Algorithm for Finding the Minimum. *arXivno*. July: 1–2.

19. Sun, Y., and H. Xiong. 2014. Function Optimization Based on Quantum Genetic Algorithm. *Research Journal of Applied Sciences, Engineering and Technology* 7, no. 1: 144–149.

20. Rylander, B., T. Soule, J. Foster, and J. Alves-Foss. 2001. Quantum Evolutionary Programming. In *Proceedings of the 3rd Annual Conference on Genetic and Evolutionary Computation*, pp. 1005–1011. GECCO'01. San Francisco, CA: Morgan Kaufmann.

21. Gu, J., X. Gu, and M. Gu. 2009. A Novel Parallel Quantum Genetic Algorithm for Stochastic Job Shop Scheduling. *Journal of Mathematical Analysis and Applications* 355, *20–22 May 1996*, no. 1: 63–81. Nagoya. DOI: 10.1109/ICEC.1996.542334.

22. Narayanan, A., and M. Moore. 1996. Quantum-Inspired Genetic Algorithms. In *Proceedings of IEEE International Conference on Evolutionary Computation*, May 20–22, pp. 61–66. Nagoya: IEEE.

23. Namiki, M., H. Nakazato, and S. Pascazio. 1998. *Decoherence and Quantum Measurements*. World Scientific.

24. Shor, P.W. 1996. Fault-Tolerant Quantum Computation. In *Proceedings of the 37th Annual Symposium on Foundations of Computer Science*, pp. 56–65. FOCS '96. Washington, DC: IEEE Computer Society.

25. Farhi, E., J. Goldstone, S. Gutmann, and M. Sipser. 2000. Quantum Computation by Adiabatic Evolution. arXiv preprint Quant-ph/0001106. https://arxiv.org/abs/quant-ph/0001106

26. Messiah, A. 1978. *Quantum Mechanics, Volume 2*. Advanced Quantum Mechanics.[Translated from the French by J. Potter], Amsterdam: North Holland Publication Company.

27. Childs, A.M., E. Farhi, and J. Preskill. 2001. Robustness of Adiabatic Quantum Computation. *Physical Review A* 65, no. 1: 12322.

28. van Dam, W., M. Mosca, and U. Vazirani. How Powerful Is Adiabatic Quantum Computation? In *Proceedings 2001 IEEE International Conference on Cluster Computing*, pp. 279–287. Las Vegas, Nevada: IEEE Computing Society.

29. Aharonov, D., W. van Dam, J. Kempe, Z. Landau, S. Lloyd, and O. Regev. 2008. Adiabatic Quantum Computation Is Equivalent to Standard Quantum Computation. *SIAM Review* 50, no. 4: 755–787.

30. Tamir, B., and E. Cohen. 2016. *Notes on Adiabatic Quantum Computers* (December 7). 1512.07617v4. https://arxiv.org/abs/1512.07617.

31. Boixo, S., A. Datta, M.J. Davis, S.T. Flammia, A. Shaji, and C.M. Caves. 2008. Quantum Metrology: Dynamics versus Entanglement. *Physical Review Letters* 101, no. 4: 40403.

32. Morita, S., and H. Nishimori. 2008. Mathematical Foundation of Quantum Annealing. *Journal of Mathematical Physics* 49, no. 12: 125210.

33. Boixo, S., G. Ortiz, and R. Somma. 2015. Fast Quantum Methods for Optimization. *The European Physical Journal Special Topics* 224, no. 1: 35–49.

34. Boixo, S., E. Knill, and R. Somma. 2009. Eigenpath Traversal by Phase Randomization. *Quantum Information and Computation* 9, no. 9–10: 833–855.

35. Ray, P., B.K. Chakrabarti, and A. Chakrabarti. 1989. Sherrington-Kirkpatrick Model in a Transverse Field: Absence of Replica Symmetry Breaking due to Quantum Fluctuations. *Physical Review B* 39, no. 16: 11828–11832.

36. Altshuler, B., H. Krovi, and J. Roland. 2010. Anderson Localization Makes Adiabatic Quantum Optimization Fail. *Proceedings of the National Academy of Sciences* 107, no. 28: 12446–12450.

37. Rajak, A., and B.K. Chakrabarti. 2014. Quantum Annealing Search of Ising Spin Glass Ground State(s) with Tunable Transverse and Longitudinal Fields. *Indian Journal of Physics* 88, no. 9: 951–955.

38. Martoňák, R., G.E. Santoro, and E. Tosatti. 2004. Quantum Annealing of the Traveling-Salesman Problem. *Physical Review E* 70, no. 5: 57701.

39. Jörg, T., F. Krzakala, G. Semerjian, and F. Zamponi. 2010. First-Order Transitions and the Performance of Quantum Algorithms in Random Optimization Problems. *Physical Review Letters* 104, no. 20: 207206.

40. Brooke, J. 1999. Quantum Annealing of a Disordered Magnet. *Science* 284, no. 5415: 779–781.

41. Seoane, B., and H. Nishimori. 2012. Many-Body Transverse Interactions in the Quantum Annealing of the P-Spin Ferromagnet. *Journal of Physics A: Mathematical and Theoretical* 45, no. 43: 435301.

42. Knill, E., R. Laflamme, and G.J. Milburn. 2001. A Scheme for Efficient Quantum Computation with Linear Optics. *Nature* 409, no. 6816: 46–52.

43. Cirac, J.I., and P. Zoller. 1995. Quantum Computations with Cold Trapped Ions. *Physical Review Letters* 74, no. 20: 4091–4094.

44. Vandersypen, L.M.K., M. Steffen, G. Breyta, C.S. Yannoni, M.H. Sherwood, and I.L. Chuang. 2001. Experimental Realization of Shor's Quantum Factoring Algorithm Using Nuclear Magnetic Resonance. *Nature* 414, no. 6866: 883–887.

45. Imamog¯lu, A., D.D. Awschalom, G. Burkard, D.P. DiVincenzo, D. Loss, M. Sherwin, and A. Small. 1999. Quantum Information Processing Using Quantum Dot Spins and Cavity QED. *Physical Review Letters* 83, no. 20: 4204–4207.

46. Brennen, G.K., C.M. Caves, P.S. Jessen, and I.H. Deutsch. 1999. Quantum Logic Gates in Optical Lattices. *Physical Review Letters* 82, no. 5: 1060–1063.

47. Mooij, J.E. 1999. Josephson Persistent-Current Qubit. *Science* 285, no. 5430: 1036–1039.

48. Dickson, N.G., M.W. Johnson, M.H. Amin, R. Harris, F. Altomare, A.J. Berkley, P. Bunyk, et al. 2013. Thermally Assisted Quantum Annealing of a 16-Qubit Problem. *Nature Communications* 4: 1903.

49. Stern, M., G. Catelani, Y. Kubo, C. Grezes, A. Bienfait, D. Vion, D. Esteve, and P. Bertet. 2014. Flux Qubits with Long Coherence Times for Hybrid Quantum Circuits. *Physical Review Letters* 113, no. 12: 123601.

50. Kaminsky, W.M., and S. Lloyd. 2004. Scalable Architecture for Adiabatic Quantum Computing of Np-Hard Problems. In *Quantum Computing and Quantum Bits in Mesoscopic Systems*, Leggett, A.J., Ruggiero, B., and Silvestrini P. (eds.), pp. 229–236. Boston, MA: Springer.

51. Boixo, S., T.F. Rønnow, S.V. Isakov, Z. Wang, D. Wecker, D.A. Lidar, J.M. Martinis, and M. Troyer. 2014. Evidence for Quantum Annealing with More than One Hundred Qubits. *Nature Physics* 10, no. 3: 218–224.

52. Weigel, M., H.G. Katzgraber, J. Machta, F. Hamze, and R.S. Andrist. 2015. Erratum: Glassy Chimeras Could Be Blind to Quantum Speedup: Designing Better Benchmarks for Quantum Annealing Machines [Phys. Rev. X 4, 021008 (2014)]. *Physical Review X* 5, no. 1: 19901.

53. Albash, T., W. Vinci, A. Mishra, P.A. Warburton, and D.A. Lidar. 2015. Consistency Tests of Classical and Quantum Models for a Quantum Annealer. *Physical Review A* 91, no. 1: 42314.

54. Bautu, A., and E. Bautu. 2007. Quantum Circuit Design by Means of Genetic Programming. *Romanian Journal of Physics* 52, no. 5–7: 697–704.

55. Spector, L. 2004. *Automatic Quantum Computer Programming*. Vol. 7. Genetic Programming. Boston, MA: Kluwer Academic.

56. O'Neill, M., and C. Ryan. 2003. *Grammatical Evolution: Evolutionary Automatic Programming in an Arbitrary Language*. Genetic Programming, London: Kluwer Academic Press.

57. Montana, D.J. 1995. Strongly Typed Genetic Programming. *Evolutionary Computation* 3, no. 2: 199–230.

58. Montana, D.J. 1994. Strongly Typed Genetic Programming. Technical Report BBN 7866 Cambridge, MA: Bolt Banek and Newman.

59. Koza, J.R. 1992. *Genetic Programming: On the Programming of Computers by Means of Natural Selection*. Cambridge, MA: MIT Press.

60. Koza, J.R. 1994. *Genetic Programming II: Automatic Discovery of Reusable Programs*. Cambridge, MA: MIT Press.

61. Koza, J.R., D. Andre, F.H. Bennett, and M.A. Keane. 1999. *Genetic Programming III: Darwinian Invention & Problem Solving*. 1st ed. San Francisco, CA: Morgan Kaufmann.

62. Barnum, H., H.J. Bernstein, and L. Spector. 2000. Quantum Circuits for OR and of ORs. *Journal of Physics A: Mathematical and General* 33, no. 45: 8047.

Index